Garden and Landscape History

"TURBULENT FORESTERS"

Garden and Landscape History
ISSN 1758-518X

General Editor
Tom Williamson

This exciting series offers a forum for the study of all aspects of the subject. It takes a deliberately inclusive approach, aiming to cover both the 'designed' landscape and the working, 'vernacular' countryside; topics embrace, but are not limited to, the history of gardens and related subjects, biographies of major designers, in-depth studies of key sites, and regional surveys.

Proposals or enquiries may be sent directly to the editor or the publisher at the addresses given below; all submissions will receive prompt and informed consideration.

Professor Tom Williamson, School of History, University of East Anglia, Norwich, Norfolk NR4 7TJ, UK.

Boydell & Brewer, PO Box 9, Woodbridge, Suffolk, England, IP12 3DF, UK.

Previous publications are listed at the back of this volume.

"TURBULENT FORESTERS"

A LANDSCAPE BIOGRAPHY OF ASHDOWN FOREST

BRIAN SHORT

THE BOYDELL PRESS

First published 2022
The Boydell Press, Woodbridge

ISBN 978-1-78327-707-0

The Boydell Press is an imprint of Boydell & Brewer Ltd
PO Box 9, Woodbridge, Suffolk IP12 3DF, UK
and of Boydell & Brewer Inc.
668 Mt Hope Avenue, Rochester, NY 14620–2731, USA
website: www.boydellandbrewer.com

A CIP catalogue record for this book is available
from the British Library

This publication is printed on acid-free paper

Printed and bound in Great Britain by
TJ Books Ltd, Padstow, Cornwall

MIX
Paper from
responsible sources
FSC® C013056

CONTENTS

ILLUSTRATIONS

Figures

Tables

Full credit details are provided in the captions to the images in the text. The author and publisher are grateful to all the institutions and individuals for permission to reproduce the materials in which they hold copyright. Every effort has been made to trace the copyright holders; apologies are offered for any omission, and the publisher will be pleased to add any necessary acknowledgement in subsequent editions.

ACKNOWLEDGEMENTS

AS HAS BEEN SAID so many times before, there is a huge number of people to whom I owe thanks, largely because I have known the Forest for so long. The late Peter Brandon introduced me to Ashdown, sending me as a new doctoral student to spend two weeks walking and learning the Forest through the soles of my boots and through his erudition.

It is also a great pleasure to thank several generations of Forest staff including those who have alerted me to some of the practicalities of Forest maintenance, especially Hew Prendergast, Chris Marrable, Gerry Lee, Rich Allum and Katherine Meade, and the many local history groups who have welcomed me and demonstrated the great wealth of knowledge that they possess. I am particularly indebted to past or present staff at the East Sussex and Brighton and Hove Record Office, especially Christopher Whittick, and staff at the National Archives, National Portrait Gallery and the British Library. I also thank my former colleagues at the University of Sussex, particularly Evelyn Dodds, Vic Gammon, Carl Griffin, Alan Lester, Julian Murton, Mick Reed, David Robinson and David Streeter, and two dearly missed colleagues, Alun Howkins and John Lowerson. Many of the maps included here are the work of Sue Rowland and I am grateful for her professionalism and support. From elsewhere I also thank Daz Beatson (National Trust), Richard Coates (University of West of England), David Crook, Naomi Sykes and Charles Watkins (University of Nottingham), Mark Gardiner (University of Lincoln), Andy Margetts (University College London), Nigel Walford (Kingston University), Angus Winchester (Lancaster University) and, more locally, Pat Arnold (the Friends of Ashdown Forest), Sue Berry, Vivienne Blandford, Colin and Judy Brent, Dave Brooker, Chris Butler, Martin Berry and members of the Ashdown Forest Research Group, Bob Foreman (Sussex Biodiversity Record Centre), Peter Freeland, Jeremy Hodgkinson, Jaime Kaminski, the late Peter Kirby, the late Michael Leppard, Peter Macleod, Natalie Mclean (East Sussex County Council), Amanda Reeves (Sussex Wildlife Trust), Kelly Sharp (Wealden District Council), Bob Toynbee and Heather Warne. The wealth of knowledge and resources held by the local history societies around the Forest is immense, and I particularly wish to thank Mike Parcell and Chris Sugg (Hartfield), David Ridley, Charlie Stevenson and Mollie Smith (Nutley and Maresfield), Gill Rolfe (Danehill) and Rosemary Mulady and Tracy Atchison (Fairwarp). Furthermore, many students have accompanied me to the Forest or enlightened me through their own researches: Jeff Baker, Jonathan Douch, John Irons, Malcolm Lill, Carol Lockwood, Geoffrey Mead, Linda Merricks, Lara Phelan and Sharon Hall Shipp. I am particularly grateful for the extremely close and helpful attention to earlier drafts of the text paid by Richard Hoyle (University of Reading), to Tom Williamson (University of East Anglia) and an anonymous reader, and to Caroline Palmer, Elizabeth McDonald and Christy Beale at Boydell & Brewer.

So many others could have been mentioned but, finally, I remain grateful as ever to my family, to James for his much-needed IT skills and especially Valerie, for her patience, support, and for keeping me grounded throughout the very many years that this book has been in preparation.

I also gratefully acknowledge the generous grants towards the costs of production from the Marc Fitch Fund and from the Friends of Ashdown Forest.

EDITORIAL CONVENTIONS AND ABBREVIATIONS

THE PUBLISHED AND UNPUBLISHED sources are many and varied, and difficulties can arise when attempting to synthesise historical, archaeological and geographical conventions. Times and measurements change: metrication rendered the 14,000 acres of Forest within the pale to under 7,000 hectares. But here the imperial and pre-metric currencies of acres, miles, pounds, shillings and pence will be used, as befitting an historical study. More difficult are the archaeological equivalences of BC or BP, and there, if unavoidable, I have felt it appropriate to stay with the original dating employed. Finally, I have included footnotes, mostly for references that are rendered fully at first instance but thereafter in shortened form. Those sources dealing substantially with Ashdown are listed in the bibliography, which contains a little more information on many of the sources, particularly unpublished material. A glossary is added to assist with possible unfamiliar words. Finally, please note that I have used capitals for the Forest (when specifically addressing Ashdown) and for Commoners and the Conservators of Ashdown to emphasis their distinctiveness.

BL	British Library
BPP	British Parliamentary Papers
ESBHRO	East Sussex Brighton and Hove Record Office (at the Keep, Falmer)
ESHER	East Sussex Historic Environment Record
LBSCR	London, Brighton & South Coast Railway Company
ODNB	*Oxford Dictionary of National Biography* (Oxford: Oxford University Press)
PRO	Public Record Office
SAC	*Sussex Archaeological Collections*
SCM	*Sussex County Magazine*
SER	South Eastern Railway
SNQ	*Sussex Notes and Queries*

SRS	Sussex Record Society
TNA	The National Archives
U of Sx Spec Colls, AFC I and II	University of Sussex Special Collections: Volumes prepared for the Ashdown Forest Case 1876–1882
VCHSx	Victoria History of the County of Sussex
WSRO	West Sussex Record Office

PREFACE

> Put very simply our texts are not mirrors which we hold up to the world, reflecting its shapes and structures immediately and without distortion. They are instead creatures of our own making, though their making is not entirely of our own choosing.[1]

THE PAST IS PASSIVE; it can be moulded, interpreted, reinterpreted. Natalie Zemon Davies introduced her book on Martin Guerre by stating 'What I offer you here is in part my invention, but held tightly in check by the voices of the past.'[2] And so this book is also in part my invention, my interpretation of the history of Ashdown Forest, culled from over 50 years of association in various capacities. But I have always been an outsider, though never made to feel so, and have never lived on or even near the Forest. This doesn't necessarily lead to any more or less profound thinking about Ashdown, it doesn't make it any more or less objective. But it has for so long been a presence in my own life that now is perhaps the right time to deal with a mountain of collected material, to assemble, collate, interpret and set out my latest thoughts on what to me is a fascinating landscape.

This book is concerned to trace and understand the conflicts and antagonisms taking place over centuries which have resulted in producing Ashdown Forest, an area of outstanding beauty, now greatly valued but once a working landscape overlooked or scorned by outsiders.

This particular landscape, the Ashdown Forest in Sussex, will not only be traced historically but also placed alongside other similar environments to consider more general issues affecting this and other such landscapes. As such it is a study of a particular type of landscape that happens to be in Sussex; but being in Sussex, in the Weald and in South East England certainly mattered, and still does, as the pressures exerted on Ashdown were magnified by its presence in this more crowded part of Britain and being relatively accessible from London.

1 D. Gregory and R. Walford, 'Introduction: making Geography', in D. Gregory and R. Walford (eds), *Horizons in Human Geography* (Basingstoke: Macmillan Education, 1989), 2.

2 Natalie Zemon Davis, *The Return of Martin Guerre* (Cambridge, MA: Harvard University Press, 1983), 5.

This is a biography of landscape, and as in any biography it is essential to trace antecedents. In this case we need to examine the forces of change that have produced the modern landscape. Such forces might be those of the natural elements, such as the destructive wind of 1987, or the many and varied human impacts. It is thus a study of the interaction of the natural forces and human cultural landscapes and processes, combining within the particular location of the Sussex Weald.[3]

This book is a synthesis, of others' work (both published and unpublished) and my own; of personal communications and digital material; of archival documents, maps, illustrations and field observation. I have drawn upon a large range of sources and put my own interpretation upon them, while respecting those authorities whose expertise in particular themes or periods is far greater than my own. In this sense it is *intertextual*: one text draws upon a predecessor which itself draws upon a corpus of other texts, published or unpublished, or upon the landscape itself, which may be 'read' as yet another text.

There are very many secondary sources, too many to cite individually here, but which are included in the bibliography. In no particular order of precedence, we can list important types of publication, all of will be drawn upon and fully referenced where they appear in the text, rather than here.

Chronologically, there are first the local histories of the constituent parishes within which Ashdown is situated. This has been a longstanding interest, exemplified in solid Victorian manner by the writings of the Rev. Edward Turner or a clerical successor, J.L. Ward Petley, both of Maresfield; or later Catherine Pullein on Rotherfield or Ursula Ridley on West Hoathly. Then there are analytical and thematic surveys, such as that by John Irons on the historical ecology of Ashdown. There are archaeological investigations, allied to fieldwork and documentary analysis, many carried out earlier in the twentieth century by men such as Ivan Margary, Ernest Straker and Fred Tebbutt but also in the twentieth and twenty-first centuries by Wealden iron enthusiasts and by Chris Butler and associates. There are surveys of particular periods, such as the early modern period by Linda Merricks or the nineteenth century by the present author. And there are those that touch amply on Ashdown but which are part of wider investigations, such as J.L.M. Gulley's thesis on the Wealden landscape, Richard Coates' study of Sussex place-names or Lara Phelan's comparative study of Ashdown and the New Forest. There are also the previous histories of Ashdown, such as those by Garth Christian, Barbara Willard or Roger Penn; or even the seemingly quirky but very informative *Natural World of Winnie-the-Pooh*. There also websites displaying the interests of different local history groups, among them those relating to Hartfield, Danehill and Maresfield. Finally, there is the more recent 'grey' literature, with limited and specific readerships, such as planning documents or the studies of Ashdown's flora.

---

3 For landscape biography see J. Kolen and H. Renes, 'Landscape biographies: key issues', in J. Kolen, H. Renes and R. Hermans (eds), *Landscape biographies: geographical, historical and archaeological perspectives on the production and transmission of landscapes* (Amsterdam: Amsterdam University Press, 2015), 21–47.

In terms of its structure this book adopts a simple logic. Its chapters, except for chapter 2, follow on more or less sequentially to build up a picture of Ashdown and its people through time. In this sense it owes much to Fernand Braudel's *La Longue Durée* in taking a long-term approach to the structures and constraints on human life on the Forest. Beneath the recurrent crises and cycles of life there were 'old attitudes of thought and action, resistant frameworks dying hard, at times against all logic'.[4]

But we should be aware that, simple as this narrative structure seems, there are pragmatic issues surrounding even this approach. To start with, how should we 'cut up' the historical evolution of Ashdown into chapters? Time, as far as we can discern, is a continuous phenomenon, and time-depth is essential in our study. But, here, a straightforward chronology has been adopted that uses well-known historical moments or periods to demarcate chapters: 1066; the Victorian period; the post-Second World War years. But, as the amount of information expands after the medieval period, so the chapters slow down in time. The early modern period is therefore covered by three chapters, each taking a different theme – society, economy, landscape – and, similarly, the hundred years or so following Victoria's accession merit four chapters.

These are, of course, matters of individual decision. Another author might choose rather different arrangements or indeed ask different questions of the evidence. Indeed, a quite different approach might begin with the present landscape of Ashdown and 'peel away' the layers of history to arrive at what one geographer referred to as the Wealden *urlandschaft* – the primitive original landscape.[5] So to pretend that there is just one 'correct' version of the evolution of Ashdown is a fantasy that goes against much of the literary, ethnographic and historical thought of the past 30 years.

And yet, in using such a framework, we must remember that the Forest itself did not march to a national timescale. What was significant in the national consciousness might not register within the heathy topography of the Forest. The 'Glorious Revolution' bringing William of Orange to the throne was a distant event, and far less significant than the decree and award of the early 1690s that established the outlines of Ashdown's enclosures and common land. However, there cannot be an absolute division between national and local events. Indeed, the former were often to be played out in some form or another on the Ashdown stage: the Civil War and Commonwealth years of the 1640s and 1650s is one example, and the impact of wartime military operations throughout the centuries another.

And as with any biography it has also proven difficult at times to 'pigeon-hole' Forest life into convenient blocks of time (and thence chapters). Landscape histories have a lifespan of their own, with changes sometimes rapid, but often only slowly evolving over centuries. It can be challenging knowing best how to attribute a particular landscape feature to a particular period – an issue faced in most landscape histories, for there

---

4 F. Braudel, 'Histoire et sciences sociales: La Longe Durée', *Annales. Histoire, Sciences Sociales*, 13.4 (1958), 725–53.

5 J.L.M. Gulley, 'The Wealden Landscape in the early Seventeenth Century and its Antecedents' (Univ. of London unpub. PhD, 1960), 488–500. The technique is difficult to handle successfully and has had few adherents.

are very few single-phase landscapes in England.[6] A landscape is in continuous flow through time, but here I have arbitrarily cut up that flow into blocks for our analytical purposes. There are survivals and residual landscape features continuing to play their role in subsequent periods, reminding us of the depth of history that is one of the best-known attributes of the English landscape. The word 'palimpsest' is often used to describe its multi-layered features, and 'sedimentation' to allude to the layering of cultural landscape elements over time.

It has also been the intention in this volume to describe the *form* of the landscape but at the same time to set out the many *processes* by which change was occurring. This synthesis of form and process runs through the volume, alongside that between *ecology* and *culture*.[7]

Context is also important. Again, biographies require contexts, and I have therefore sought to place Ashdown within its wider Wealden context, or within the history of South East England, or even on a much wider stage. And, as with the biography of a person, we need to show the material form – what did the subject look like? – but also evaluate the subject's personality. Of course, the subject here, Ashdown Forest, did not possess agency. That would stray into a fetishism implying that places themselves make decisions. It is the people, the agents, who make these decisions, whether acting individually or in concert with others and within the constraints of nature, economy and culture. And so this book will attempt to show how agency has impacted upon the face of Ashdown. Again, notice the reference to a human face as applied to a landscape and remember how Thomas Hardy used the same device for Egdon Heath – 'A face on which time makes but little impression' – in *The Return of the Native* (1878).

Finally, there are issues of limits, biographical and spatial. When we trace individuals and their relationship to Ashdown it is often difficult to draw back from following their, perhaps fascinating, lives and remember that we are primarily interested not, say, in the genealogy of the Sackville family, but in their interactions, and those of their fellow residents, with or through the Forest. And a different kind of limit faces us in that the history of Ashdown entails investigating the archival sources and these will inevitably lead us into the parishes and manors constituting the Forest, by which the archives are catalogued. But some of the parishes were (and remain) large and sweep away from the Forest itself and 'beyond the pale'. In this volume a judgement has been made not to pursue everything within all the constituent parishes but to try to keep the material as close to the Forest as possible.

---

6 G. Fairclough, 'Protecting time and space: understanding historic landscape for conservation in England', in P. Ucko and R. Layton (eds), *The Archaeology and Anthropology of Landscape* (London: Routledge, 1999), 119–34.

7 For approaches to landscape that see such studies diverging as either theoretical and process-orientated or particular, empirical and historical see M. Johnson, *Ideas of landscape* (Oxford: Blackwell, 2007).

CHAPTER 1

INTRODUCTION: A FOREST LANDSCAPE

THE AREA THAT BECAME Ashdown is located at the centre of the Weald of South East England. The Forest lies roughly halfway between London and the south coast, accessible from London but remote enough within its Wealden fastness to ensure some degree of privacy. It lies in northern Sussex, close to the borders of Kent and Surrey, and between the modern towns of Crawley, East Grinstead, Tunbridge Wells, Crowborough, Uckfield and Haywards Heath (Figure 1.1). As we shall see, this location has resulted in considerable pressure on this, the largest remaining open heathland in the south-east. In fact, although the area is much given over to commuterdom and has a thriving population, the central Weald has historically been somewhat marginal to administration and legal oversight owing to its distance from the county towns of all three counties: Chichester, Lewes and Maidstone. There was no major trading routeway through here – only the river headwaters and ancient roads that skirt the Forest. The original medieval Ashdown covered about 13,000 acres with a 23-mile circumference and was thus a substantial landscape feature.

WHY 'FOREST'?

Modern visitors to Ashdown are frequently perplexed by its relative lack of trees – and not just visitors, for some residents also object to the Conservators cutting back encroaching trees and scrub from the heathland. The problem is that there was historically absolutely no formal connection with our modern use of the word 'Forest' indicating an extensive area covered by trees. The Norman forests were there to serve the royal elite. They may well have been wooded, but it is more likely that the landscape was interspersed with open 'launds' or more scrubby vegetation, and ideal for hunting. Indeed, at one time the royal forests were also controversially extended across all manner of countrysides, primarily as a money-raising venture for the king. The word adopted for these areas was the Latin '*foris*': 'outside, from outside', meaning that they were beyond ordinary law and the normal world of country people.

Forest jurisdiction was by no means limited to unpopulated areas. Forests were superimposed onto older patterns of land management, and royal rights became increasingly unpopular with both the nobility and the powerless lower orders. Forest laws, at least in the earlier years of the new Norman rulers, were stringent and

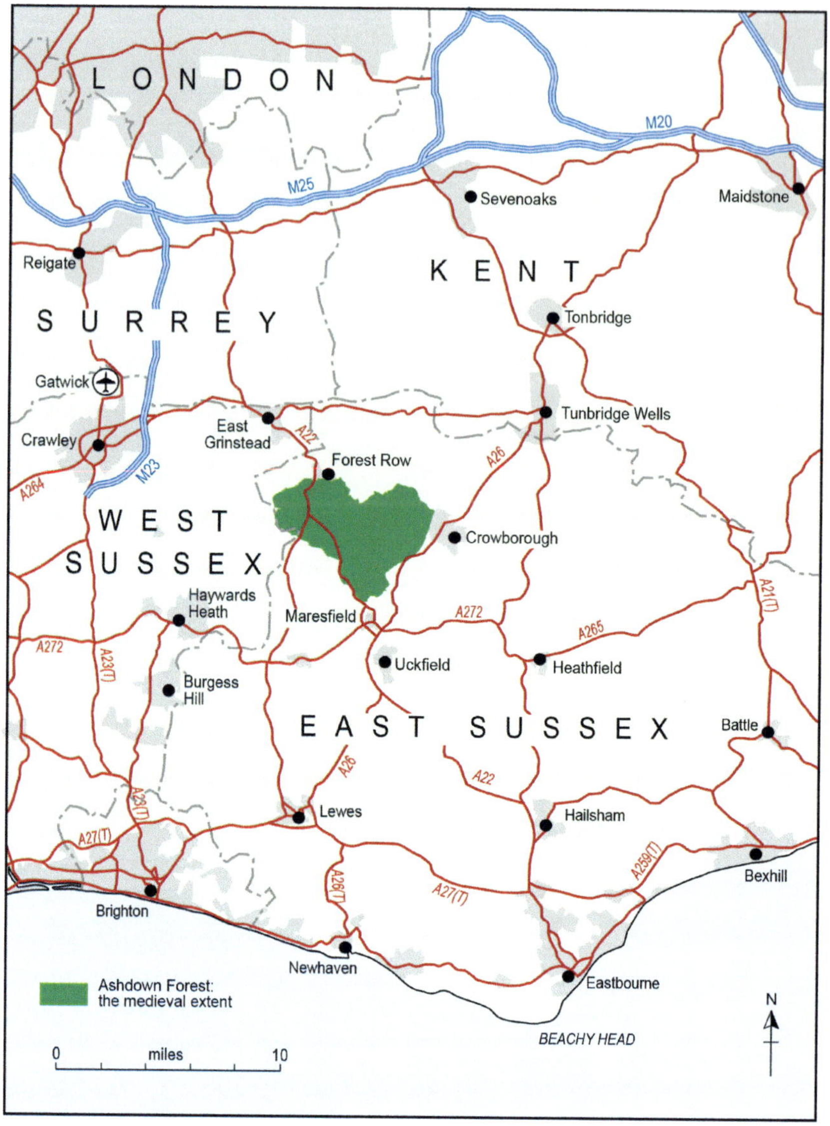

Figure 1.1. The location of Ashdown Forest (source: author).

enforced by a large body of officials, and were seen by later historians as symbolic of the Norman 'yoke', characterised by harsh attitudes towards the conquered subjects.

However, and most importantly, no evidence has been found to indicate that Ashdown was ever governed by these forest laws, even though, from time to time, it reverted to royal possession.

Instead, it was one of the many private forests, technically a chase, and indeed surrounded by an enclosure making it into a park within which deer could be reared and hunted. What remains a mystery is exactly why the word 'Forest' was ever attached to Ashdown. As a chase Ashdown Forest was one of the 'inferior franchises': alongside parks and warrens but below the importance of a royal forest. However, this distinction is sometimes hard to grasp, as there were also royal chases; and there were occasions when the king granted a forest to a noble, when it became technically a chase, only for the crown to regain possession at a later stage. There has been understandable confusion, therefore, about whether Ashdown was a royal forest, and such confusion can still be found, even in legal publications.[1] Grants were made to royal relatives, including queens for the duration of their lives, as well as to lay nobles or even the church. We should also remember that Charles II, following his accession, correctly or not, granted Ashdown to a loyal supporter 'with leave to disafforest', a procedure bitterly resented locally and to which we shall return at a later stage.

The concept of forest, chase and park was in fact complex:

> To discover the forests we must get behind 'royal forest' and other simplifying abstractions. In doing so we re-encounter the contradictions and inconsistencies that those legal figments were intended to resolve ... particular forests with different names, and different ones with the same name; parts of forests denoted as separate ones; franchises of forest, chase, park or warren in crown and subjects' hands; forests alternating as chases and warrens and boundaries that are hard to locate on today's maps.[2]

Because of changes to the extent of such landscapes through time, it is difficult to give a once-and-for-all figure for the number of and area covered by forests within England and Wales. An estimate in 2005 showed there to have been, at various times, as many as 199 forests and chases in England and 96 in Wales, covering vast countrysides.[3] We might also remember that Ashdown is also close to three other Sussex 'Forests': St Leonard's, Worth and Waterdown. And, equally, none was ever a forest in the strict sense of that term.[4]

Ashdown's variety of habitats is superficially similar to many other such areas across England. Perhaps better known as areas of lowland heath are the New Forest (the ancient forest covering 93,000 acres), the largest tract of lowland heath in Britain, together with the commons of east Devon, the Dorset heaths, those on the Tertiary

1 For example, N. Ubhi and B. Denyer-Green, *Law of Commons and of Town and Village Greens*, 2nd edn (Bristol: Jordans, 2006), 29. This issue will be fully discussed in chapter 4.

2 J. Langton, 'Seeing the Forests for the Trees', <http://info.sjc.ox.ac.uk/forests/tw.html> [accessed 1 February 2022].

3 J. Langton and G. Jones, 'Deconstructing and reconstructing the forests: some preliminary matters', in J. Langton and G. Jones (eds), *Forests and chases of medieval England and Wales c.1000–c.1500* (Oxford: St John's College Research Centre, 2010), 1–13; Figure 1 in J. Langton and G. Jones (eds), *Forests and Chases of England and Wales c.1500 to c.1850* (Oxford: St John's College Research Centre, 2005), viii.

4 See M. Weir-Wilson, 'The Making of St Leonard's Forest: a New Landscape History', *Southern History*, 38 (2016), 106–29.

deposits of the Hampshire and London Basins and the East Anglian heathlands, notably the Breckland. But Ashdown (6,150 acres currently administered by the Conservators) is the largest such area in South East England. All were formerly more extensive: the Suffolk Sandlings, for example, originally covered about 23,000 acres but this has shrunk to 5,400 acres today.[5]

The ecological and recreational significance of Ashdown will be dealt with in later chapters but historically the forest was also highly significant for the fact that the hunting ground was superimposed onto an older pattern of settlement and agrarian organisation revolving around the communal grazing and collecting activities of those who came to profess common rights. The intermixture of elite hunting with such rights was to provoke an ongoing conflict over many centuries, and this, and other conflicts, provide much of the substance of this volume. Again, such conflicts in myriad forms could be seen in the other similar environments, so this provides a case study of conflict in a marginal setting but a setting which came to acquire great significance.

Thus, although marginal for agriculture, the Ashdown landscape is and was significant for several reasons:

Historical – the setting for conflict between those with common rights and those without; resistance to attempted enclosures; elite interests versus working foresters and small cottagers.

Ecological – shifting vegetation patterns through time, now offering the largest area of open heathland in South East England, with its associated fauna, some very rare.

Recreational – an exercise in the preservation and management of open space in the pressurised South East, and associated conflicts.

THE NATURE OF THE COMMON

As well as being a 'forest', a chase or a park, Ashdown was also common land. The extent and nature of common rights will also be writ large throughout this book. These were rights to use land owned or tenanted by other individuals, and those possessing these rights were known as commoners. The common land was distinguished from that which was enclosed around Ashdown by hedging, and the enclosed areas were free of common user rights. Common land still covers about 3 per cent of England, but in 1500 it covered as much as 50 per cent. Commons might be primarily heathland, as at Ashdown, but they might also be partially wooded (again as at Ashdown) or they might be grassland, or fen or coastal pastures that might be subject to seasonal flooding. We should point out here that other types of common land – common arable, often set out in strips, and common meadow – were not present here.

Ashdown came to be regarded as the common 'waste' of Duddleswell manor, but it was actually intercommoned, since the manor of Maresfield, as well as some 'foreign' manors, also enjoyed rights over it. Manors, the institution most commonly

---

5 N. Webb, *Heathlands* (London: Collins, 1986), 53–77. This remains a basic text for the natural history of lowland heaths. But for an overview of British heathland types see also C. Chatters, *Heathland* (London: Bloomsbury, 2021).

associated with medieval feudalism, appointed officials through the manorial courts, which actually outlived feudalism and continued in some cases into the twentieth century. They were controlled by a manorial lord and his officers through a court baron nominally composed of larger tenants (the homage) to deal with business such as land use and tenurial issues. A series of court rolls or books recorded the proceedings.[6] There were also subsidiary courts such as the woodmote court, or court leet, which administered local justice, although these tended to decay and lose their functions to the court baron or to the public Quarter Sessions after the end of the medieval period. The extent to which Ashdown was governed by byelaws made outside the manorial court is unclear, as we know little about any subsidiary governance associated with the Forest-edge villages such as Maresfield.

We should also remember that we are dealing for the most part with hamlets and scattered settlements, as in much of the Weald. A manor could encompass several such settlements but, conversely, a settlement might have divided lordships and its inhabitants owe obligations to more than one manor. The manors here administered scattered properties on and around the Forest. And all was made even more complex by the fact that by the seventeenth century the poor law functions had been assigned to the parish rather than manor, but there was no neat correspondence between the extent of a manor and a parish. Whatever the case, the poorer inhabitants, the cottagers, landless and women, remained relatively powerless. That is not to say, though, that they might not find subtle means of protest and insubordination in this area of divided manorial control. The complexity of local attitudes to access to commons' resources can be addressed thus:

> Forest dwellers do not come in standardised homogenised forms: rather they comprise complex hierarchically structured groups and communities which may simultaneously display aspects of group solidarity, unity and shared objectives with respect to perceived threats from the rest of the world, but which nevertheless manifest internal stresses and frictions arising from inequalities in power, status and economic strength.[7]

This was written about actual (wooded) forest in parts of the developing world, but it represents an entirely apt characterisation of the Ashdown community which is to be explored in this volume.

At various times the Ashdown Commoners were, by custom, entitled to:

Pasture: cattle, stinted or otherwise controlled in number, but not sheep (originally).

Pannage: for specified numbers of pigs and at certain times. Oak and beech were therefore prized as supplying acorns and beech mast, to be eaten off the ground.

Estovers: gathering of underwood and small branches, birch, willow, hazel or alder, for fuel or repairing fences or buildings.

6 ESBHRO, AMS 7127; ADA/ACC 3628.

7 M. Doornbos, A. Saith and B. White, 'Forest lives and struggles: an introduction', in M. Doornbos, A. Saith and B. White (eds), *Forests: Nature, People, Power* (Oxford: Blackwell, 2000), 7.

Litter: bracken, heather and gorse to thatch and litter down cattle yards.

Turbary: peat and turf for domestic use.

Common in the soil: marl for improving poor soils; sand and gravel, and stone from specified quarries.

The historical rights of the manorial owners – the Sackvilles, De la Warrs or Gages – included the taking of minerals, sporting rights and the planting and cutting of trees.

But not all cottagers and small farmers enjoyed common rights. These adhered to the holding, not the individual – indeed, there were instances of Forest inhabitants wishing to relinquish their common rights because they were no longer used and entailed the payment of a Forest rate – but such requests were refused by the Ashdown Conservators in the nineteenth century as being contrary to local custom, although they were later allowed in the twentieth century. So, common rights were not for all, but were reserved for manorial copyholders and leaseholders, and not for cottagers or squatters who might otherwise have claimed access through living on the Forest.[8] Exactly how common rights were assigned in the first place remains something of a mystery, but they would certainly have been in operation before legal memory (before 3 September 1189). In later years such rights were granted if use over a long period could be demonstrated but the suspicion must be that initial rights were allocated over time to resident properties that did intercommon on the Forest. Custom and memory then played a role in developing and defending such rights. The importance of custom will be emphasised at many points in the following account as disputes arose and all sides sought to demonstrate precedent to suit their argument. Custom was not, as many wanted to believe, handed down unchanged from time immemorial but was fluid, open to interpretation, mutable across time and across space.

THE TRAGEDY OF THE COMMONS?

The question of common rights and the use of Ashdown's common resources raises an issue that will resurface at various points in this book. More than 50 years ago the concept of 'the commons' was the subject of work by the ecologist Garret Hardin, who developed the theme to apply it to many communal entities, tangible or intangible, and whether or not they have any degree of regulation, such as global climate, oceans or other environmental 'commons'. Hardin's much-cited work of 1968 on 'the Tragedy of the Commons' explored the thesis that individuals' decisions under a laissez-faire system of governance may not result in the best outcome for an entire population. Anxieties felt by many in the late 1960s over, for example, global population growth outstripping the world's resources would not necessarily translate into individuals therefore restricting their own families. There existed a cognitive dissonance between what people believed was right for wider mankind and the decisions they made in their own lives.

---

8 For more details on such issues see E.P. Thompson, *Customs in Common: Studies in Traditional Popular Culture* (New York: The New Press, 1993), 130.

What is particularly interesting for our purposes about Hardin's work is that he used the example of an area of common land. He explored the idea that each herdsman using a common would maximise their own herd sizes, but this could lead overall to overgrazing and depletion of the common's resources – hence 'tragedy'. Each man, he wrote, is locked into a system that compels him to increase his herd without limit – in a world that is limited. 'Ruin is the destination toward which all men rush, each pursuing his own best interests in a society that believes in the freedom of the commons.' He argued that freedom in a commons setting brings ruin to all. Any appeal to the conscience of those with user rights might find some responding but others might then multiply their herds all the more. Hardin's answer was the abandonment of free commons, thus infringing upon personal liberty, which might therefore result in commons becoming private property.[9] This, of course, was one logic of the enclosure movement, whose history is too long and complex to be detailed here. But we might add in passing that enclosure did not necessarily mean the exercise of power over others by excluding them from common land. It could also mean squatting and small-scale encroaching onto the commons by the poorer members of society – either with or without the acquiescence of the commons' owner. Each such small enclosure would spell the loss of resources for the community of Commoners.

Hardin's concerns are far from esoteric when the common uses of Ashdown Forest are examined empirically. At various times we shall find that an individual's attitudes to the common could indeed include taking from the landscape what could be found and appropriating it for private use. And such actions, by all sectors of society, whether rich or poor, when multiplied many times over and through many generations, might indeed lead to the impoverishment of the resources and 'tragedy' for Ashdown. This is a thesis to which we shall return in the final chapter, when we can consider more fully the history of this landscape.

LANDSCAPE AND CULTURE

Ideas about landscape will surface continuously throughout this book, so it is worth considering the concept of landscape and what is meant by the term. It is at once a word many will find obvious, but whose ambiguity and imprecision have left others struggling for a more precise definition. 'Landscape' has multiple layers of meaning. The word covers both conceptual and physical matters. There is an environmental and material presence, the Forest, that can be described and explained – although we should always remember that our descriptions and explanations are themselves a product of our own times and culture. A modern description of Ashdown would not perhaps have impressed William Cobbett in the 1820s or an 'improver' of the 1650s. But the landscape is also a social product, a human construct, arising from the transformation of a former, more 'natural' environment, and it is this meaning, also requiring understanding, that will be examined throughout this book. The Forest landscape is an environment itself constantly in flux, as were the people creating it. Different

9 G. Hardin, 'The Tragedy of the Commons', *Science*, 162 (1968), 1243–8.

groups have left their authorship upon Ashdown, in part constraining the activities of later generations. The relict landscape of one generation influences succeeding generations in their attitudes and activities. But, equally, many older imprints have been lost or damaged in just the way that Second World War Canadian tanks training on the Forest obliterated residual landscape features of the prehistoric and earlier centuries. Upon the Ashdown landscape, cultures have left expressions of their ideas that we might 'read', whether prehistoric barrows, hillforts, road systems, enclosures, the damming of streams or hilltop plantations. Some elements are functional, others symbolic, but their instigation or removal reminds us that Ashdown's landscape – as all other landscapes – is constantly changing.

The relationship between Ashdown's material landscape pattern and the culture and society of its inhabitants will be emphasised in this book. Ashdown will be seen as a product not just of its physical setting or of the quotidian economics of making a living in such a marginal space but of social practice, whether that be medieval hunting pursuits, early modern attempts at 'improvement', fashions in landscape appreciation or custom and memory in the defence of common user rights by those who had, or claimed, common rights. All such elements were manipulations, shaping the patterns on the ground, interacting with the topography and soils and moulding the ecology. So, as with Nicola Whyte's work on early modern west Norfolk, this book aims to '[understand] the landscape as a lived environment imbued with multiple and diverse meanings and associations particular to the time and place'.[10]

Political and ideological aims must also be taken into account, since Ashdown's landscape was both a manifestation of such aims and a target for those who disagreed with them. It provided the medium through which conflicts took place. Events on the national stage also played out around Ashdown – medieval dynastic affairs, the Reformation, the civil and religious conflicts of the seventeenth century, even the threat and the reality of international warfare all affected Ashdown's people and their landscape. The growing power of centralised state authority was also manifest, as were the changes wrought by changing industrial technologies, improved communications and growing populations.

An important distinction is sometimes made between the observer and the participant. The former may talk of the aesthetics and beauty of a landscape, perhaps as a painter or photographer, often with the actual people almost edited out, but the insider and participant may have little time to appreciate the beauty of a view, being more concerned with gaining a mundane livelihood – 'the practical as opposed to the contemplative'.[11] There is little mention of the beauty of Ashdown before the age of the leisured Victorian visitor and even then there were those who, like William Cobbett, were more concerned to point to the unproductive nature of the environment than to its landscape riches.

---

10 N. Whyte, *Inhabiting the Landscape: Place, Custom and Memory, 1500–1800* (Oxford: Windgather, 2009), 3.

11 D. Cosgrove, *Social Formation and Symbolic Landscape* (London: Croom Helm, 1984), 34–5.

In the following chapters Ashdown's landscape biography will be revealed, through time and with regard to its changing spatial patterns. It remains a prime example of a type of landscape that was once marginal, avoided, even despised, but which is now valued and heavily protected. Quite why and how this happened will be the concern of the rest of this volume.

CHAPTER 2

THE NATURAL CAPITAL OF ASHDOWN

> The ground itself, where we can read the information it affords, is ... the fullest and most certain of documents ... for without a clear and vivid realisation of the physical structure ... the incidents of the life which men have lived in it can have no interest or meaning.[1]

WHATEVER IMPACT PEOPLE AND animals may have on a landscape, the basic outlines are set by its physical features, its location, geology, topography, soils and drainage – its natural capital. This natural capital is drawn upon by humans over time: soils and vegetation, for example, can be seen as the outcomes of intervention over either long or short timescales.

TOPOGRAPHY AND GEOLOGY

Ashdown lies on the High Weald, a ridge arcing from around Horsham in the west eastwards to coastal Hastings, a region of valleys and watersheds (Figure 2.1). Altitudes can reach above 600 feet, as at Crowborough Beacon (792 feet), Gills Lap (670 feet), King's Standing (716 feet) and Wych Cross (658 feet). Much is between 300 and 450 feet, and a central watershed, marked locally by the Stonehill Road, sees the north side falling to the Medway Valley, thence northwards to the north Kent coast, with quite steep slopes dissected by numerous small valleys with springs. The south side has gentler slopes and drains to the River Ouse, flowing south through Lewes to the Channel. The Millbrook stream, rising in a boggy area at Wych Cross, originally formed part of the Ouse headwaters, but has been captured by the Medway headwaters, leaving a sharp elbow of capture as the stream now turns northwards, passing through the Isle of Thorns, Pippingford, Newbridge and under Poohsticks Bridge before joining the Medway at Withyham.

The Forest is centred on two main massifs, one around Wych Cross and the other from Camp Hill to Gills Lap, the massifs being separated by the lower ground of the Millbrook valley. Much of the land below 450 feet comprises valley slopes of varying

[1] J.R. Green, *The Making of England* (New York: Harper, 1882), vii–ix.

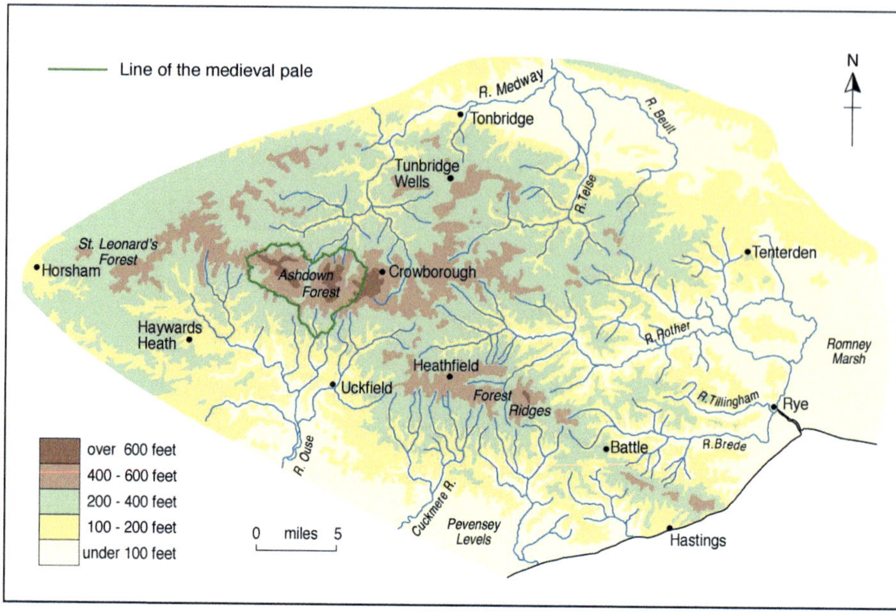

Figure 2.1. Ashdown: the setting (source: author).

degrees of steepness. Although this High Weald ridge stands above the surrounding Low Weald clays to north and south, the bedrock is by no means resistant to erosion. Indeed, small headwater streams have cut back through headward erosion to produce quite steep-sided valleys around the edges of the Forest. These are the ghylls, often with their own microclimates and vegetation communities, and with fast-flowing streams during much of the year, which have featured, as we shall see, as important elements in Ashdown's economic history.

This has historically been a somewhat isolated area, with challenging topography, and to understand this we need to examine the geology. The High Weald in general has three main sedimentary formations collectively known as the Hastings Sandstone, or Hastings Group, laid down by rivers in cycles of flood-plain and lake-lagoon deposits in a Wealden basin, a vast depocentre, during the early–mid Cretaceous Period, between 145 and 125 million years ago. The area lay approximately10° further south than at present but the climatic conditions fluctuated greatly, although for much of the period there was a humid sub-tropical climate and swampy environment. The sediments originated from three main sources, now referred to as Cornubia (in the west), Armorica (to the south) and a London-Ardennes Massif (to the north). These sediments are the arenaceous (sand or sandstone) fine-grained Tunbridge Wells Sand, the argillaceous (containing clay) grey mudstones and siltstones of the Wadhurst Clay and, the oldest of the group, the thick sequence of the arenaceous Ashdown Sandstone (Table 2.1).

Table 2.1. The solid geology of the Ashdown area.

Hastings Group	**Description**	**Strata thickness (feet)**
Upper Tunbridge Wells Sandstone	Sands, sandstones, silts and impersistant clay beds	220
Upper Grinstead Clay	Red and grey clays with thin siltstones	0–30
Cuckfield Stone	Sandstones, calcareous sandstones and siltstones	0–30
Lower Grinstead Clay	Grey clay with thin limestones, thin calcareous sandstones, ironstone nodules	0–30
Ardingly Sandstone	Massive and thickly bedded sandstone and sandrock	0–70
Lower Tunbridge Wells Sand	Silts and sandstone; locally thin clay beds	50–150
Wadhurst Clay	Soft grey mudstone with thin siltstones and silty mudstones, limestones and ironstone seams	110–235
Ashdown Sandstone	Silts and sands with thin clay seams becoming more important towards the base	650–750

Source: C. Bristow and R. Bazley, *Geology of the country around Royal Tunbridge Wells* [London: HMSO, 1972], 2.

Almost all the Ashdown Forest itself lies on the Ashdown Sandstone, hence the outcrop name. The so-called 'Top' Ashdown Sandstone and Ardingly Sandstone were particularly sought after for building purposes. A subsidiary band of clay within the Ashdown Beds appears at Marlpits Farm.[2] A small outcrop of Wadhurst Clay occurs in the far south of the Forest, between Dodds Bank and Fairwarp, with mudstones and shales, and with nodules of ironstone at its base. Another small area lies in the far north of the Forest at the revealingly named Claypits Gate, and the lower part of the clay formation can be seen near West Hoathly at Philpots Quarry. The iron is in the form of siderite, often seen in suspension as a red colouring in the small Forest streams. The Wadhurst Clay was a source for marl, and the many small pits arising can still be seen. In fact the clay contains little calcium carbonate to neutralise the acidic soils of Ashdown, but it was nevertheless extensively used.

Finally, a small band of Lower Tunbridge Wells Sand is also found in the south of the Forest between Dodds Bank and Nutley, yielding silts and silty sandstone overlain by massive sandstones, the Ardingly Sandstone, the highest-quality local building stone, capable of being worked as ashlar. This resulted in many small quarries. At

2 For the difficulties of farming such clays see p. 377.

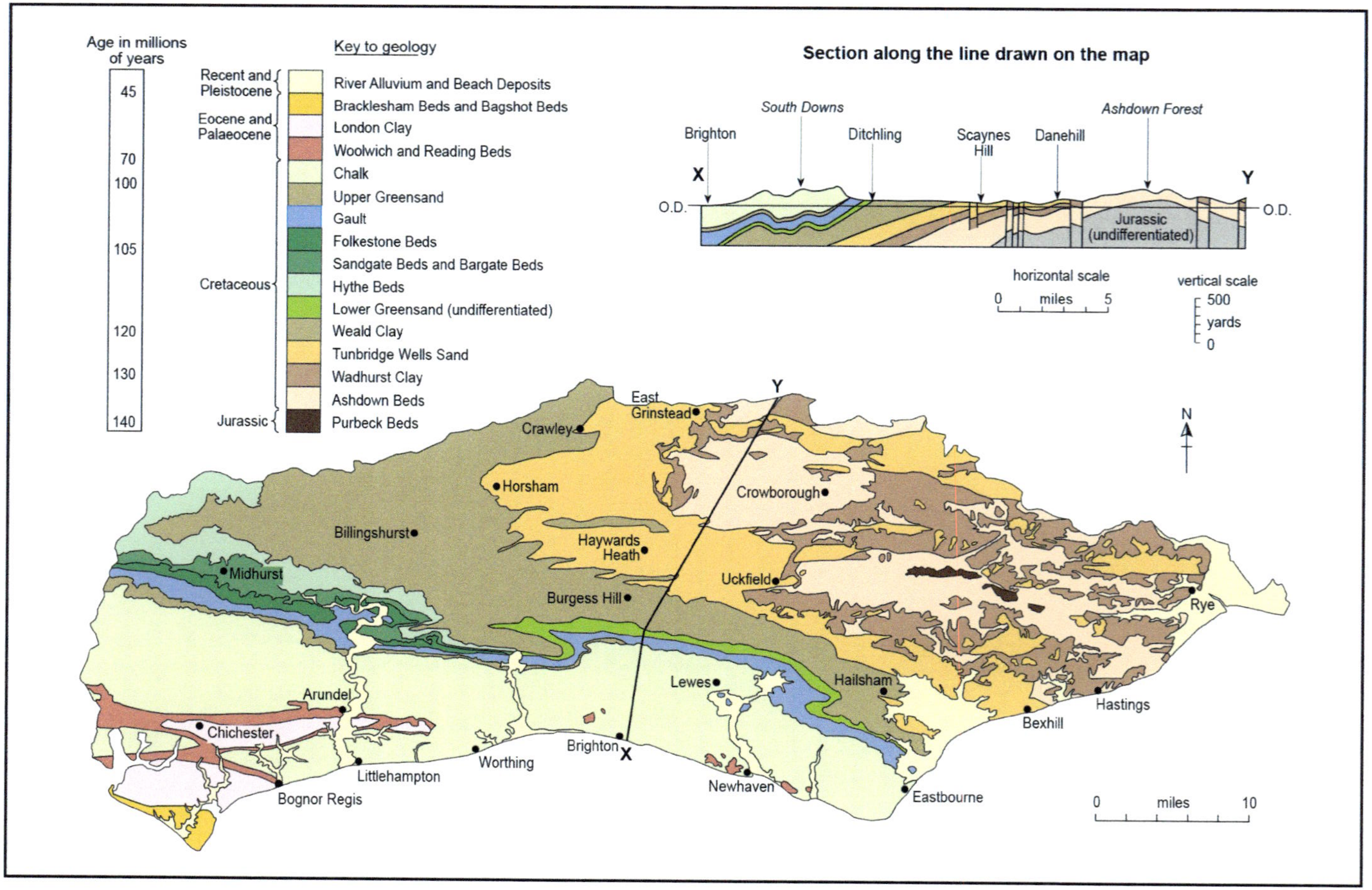

Figure 2.2. The geology of Sussex (source: based on R. Williams, 'Geology' in K. Leslie and B. Short (eds), *An Historical Atlas of Sussex* (Chichester: Phillimore 1999), 3).

Philpots Quarry, West Hoathly, it is worked beneath an overburden of Grinstead Clay. Such commercial exploitation produced several varieties of building stone, the most common being a massive brown stone used for stone wall infill between Ardingly Sandstone quoins and dressings. Hartfield church uses an attractive sandstone with brown to red bands caused by iron-oxide staining. This stone has been used for many pre-twentieth-century buildings around Ashdown, including in the late sixteenth-century Marshalls Manor, near Maresfield, in the walls and mullioned windows of West Hoathly manor house, built by the Infield family of Gravetye in the seventeenth century, and in the internal pillars and external quoins of West Hoathly church. Sackville College in East Grinstead, erected from stone brought from Old Buckhurst on the eastern side of Ashdown, uses Lower Tunbridge Wells sandstone, with Ardingly sandstone for the dressings. Well-known large Elizabethan houses such as Wakehurst Place, Borde Hill, Gravetye Manor and Nymans are also constructed to varying degrees from Ardingly sandstone.[3]

We should also notice that the Priest House museum (West Hoathly) is roofed with the heavy dark grey Horsham stone-slate, from a bed of sandstone at the base of the Weald Clay near Horsham itself. The Weald Clay otherwise surrounds the Hastings Group at the centre of the Weald. However, Horsham stone's heaviness largely precluded its more general use in eastern Sussex, although some Lewes buildings did employ it as a roofing stone.

Philpots and Hook Farm Quarries, on the western edge of Ashdown, offer evidence of Wealden rocks being deposited in a Mediterranean type of climate, and:

> exciting glimpses of a hot sun 130 million years ago, heavy but spasmodic rainfall over the London uplands and a fringing sandplain, not too far from the sea, occupied by shifting streams and populations of now extinct plants and animals subject to many vicissitudes of changing geography, climate and biological interaction.[4]

The original horizontal or gently sloping deposits were subsequently folded and faulted to produce a more complex geology. Ashdown lies on the Wealden or Crowborough Anticline, running east–west through the Forest and part of the greater Weald-Artois Anticline resulting from the compression from the Alpine Orogeny, beginning about 65 million years ago with the collision of the Eurasian and African plates. To the north of the anticline were the Withyham and Homestall faults and to the south was a complex of faults at Sheffield Forest, Danehill, Fletching, Hendall Wood, Burnt Oak and Greenhurst. All these now lend complexity to the modern landscape, and both the Ouse and Medway, in their highest reaches, have tributaries following the faulted structures.

---

3 R. Birch and R. Cordiner, *Building Stones of West Sussex* (privately published, 2014), 28–41. Philpots, worked since the nineteenth century, remains active today. And see R. Cordiner and A. Brook, *Building Stone Atlas of Sussex* (privately published, 2017), 14–19.

4 J. Radley and P. Allen, 'The Wealden (non-marine Lower Cretaceous) of the Weald Sub-basin, Southern England', *Proceedings of the Geologists' Association*, 123 (2012), 245–318, quotation from p. 295.

Such is the basic geological outline of the Ashdown area, but it should also be remembered that although no Quaternary ice sheet came anywhere close, the impact of periglacial landforms was considerable. Indeed, the South East as a whole has been termed 'a relict periglacial landscape' as it experienced recurrent periglacial conditions throughout the Quaternary.[5] Small, semi-permanent snowfields may have capped Crowborough Beacon, surrounded by frozen soils. Frost and wind erosion have shaped the outlines of the Ardingly Sandstone crags surrounding the Forest, together with landscape features created by the deposition from meltwater, valley bulging and cambering of massive beds of sandstone resulting from freeze–thaw action, and the deposition of Head. The latter, mostly from local material, is widespread, and its ongoing deposition mantles the solid geology on the ghyll sides and floors. Silty and sandy Head from the Upper Tunbridge Wells Sand thus masks the Wadhurst Clay outcrops in the upper reaches of the Ouse between Horsted Keynes and Lindfield.[6]

Dating probably from between 30,000 and 15,000 years BC, a thin mantle of loessic (or loessic brickearth), wind-blown silty material must also be envisaged over much of Ashdown. Originally blown from the north-east, from the North Sea lobe of the Late Devensian ice sheet, these deposits were reworked and dispersed through the processes of solifluction and colluvial and alluvial action. Most would have been incorporated into valley deposits, transported by the rivers and now seen along the valleys of the upper Medway in the north of the Forest, and to a lesser extent along the upper Ouse to the south-west.[7]

SOILS

Deriving from the different circumstances of deposition, complicated by faulting, folding and post-glacial actions, Ashdown's soils are extraordinarily varied. Because of the predominance of silt in the Hastings Beds impeded drainage is as common as on the clays, producing stagnogleys across much of the area. Many soils are degraded, often as a result of human intervention, and where woodland has been destroyed humus is lost and soil instability increases. To this extent Ashdown's soils are socially determined: they result from choices made over many generations, and which have favoured some parts at the expense of others as inputs of manure or marl have maintained a less acidic soil chemistry than elsewhere.

Ironically such 'beneficial' treatment is not wanted on today's open Forest, where the acid-loving heathland is so prized.[8] The result of human intervention here 'has been the accentuation of differences between soils developed on different parent

5 I am grateful to Julian Murton for advice on loessic deposits. The quotation is from his lecture to the South East section of the Geologists' Association (26 November 2016).

6 R. Gallois and B. Worssam, *Geology of the Country around Horsham* (London: HMSO, 1993), 95.

7 R. Scaife and P. Burrin, 'Floodplain Development in and the Vegetational History of the Sussex High Weald and some Archaeological Implications', *SAC*, 121 (1983), 1–10. P. Burrin, 'Loess in the Weald', *Proceedings of the Geologists' Association*, 92 (1981), 87–92.

8 For 'socially constructed soils' see C. Reboul, 'Determinants Sociaux de la Fertilité des Sols', *Actes de la Recherche en Sciences Sociales*, 17 (1977), 85–112.

materials and the creation of differences within soils developed on the same parent material but subjected to different levels of exploitation and husbandry'.[9] The view from Friends Clump northwards towards Old Lodge exemplifies this feature, where the improved grassland near the lodge contrasts remarkably with the same parent soils nearer Friends Clump, which are covered in open Forest vegetation.

However, because the Ashdown Beds can also be free draining, podzolisation occurs in many higher areas, whereby humic material is leached away from plant roots to accumulate as a hardpan lower in the soil horizon. These degraded soils are highly acidic and are characterised by loose white fine sands on the higher interfluves, but becoming sticky and clay-like when wet. The Hastings Beds outcrop generally is deficient in calcium and potassium as well as many metallic elements.

There are also extensive seepage zones and shifting springs where alternate seams of clay and shale occur, as in the junctions with the Wadhurst Clay, leading to poor drainage and small oligotrophic mires with few available plant nutrients. Where overland flows occur this can lead to shallow gulley formation, especially where soils have been compacted by trampling along pathways. Indeed, extensive and complex gullying occurs on a Second World War temporary airstrip on the western side of the Forest.

In terms of soil classification, the Poundgate Association soils are the most widely distributed around Ashdown, especially on the gentler slopes of the watershed. These are gley-podzols with fine sandy or silty upper horizons, but with slowly permeable gleyed silty horizons below showing characteristic mottling. They are very acidic and now largely under open heath.[10] On the lower slopes are Cranbrook series soils, deep, silty and waterlogged gleys, and around the numerous springs and flushes, and beneath wetter heathland, are soils of the Ashdown series, waterlogged gleys with an organic surface horizon and greater loam content.

There are also sandy brown earths, the Curtisden soils, with slowly permeable subsoils, often on drift deposits, and, although experiencing some seasonal waterlogging, brown earths are well-drained soils developing under deciduous woodland, with a leaf layer. They almost certainly constituted the major soil type on Ashdown prior to human modification and may still be found beneath relict stands of oak with an understorey of holly, remnants of an earlier more widespread vegetation pattern. But most of these once-extensive brown forest soils have deteriorated, depleted of nutrients as woodland has been felled. Finally, there are also peaty gleys, the Wickham I series, occurring at Colemans Hatch and Fairwarp and confined to valley slopes and waterlogged valley bottoms.[11]

---

9 D. Robinson and R. Williams, 'The Soils and Vegetation History of Sussex', in Geography Editorial Committee, *Sussex: Environment, Landscape and Society* (Gloucester: Alan Sutton Publishing, 1983), 124.

10 M. Jarvis, R. Allen, S. Fordham, J. Hazelden, A. Moffat and R. Sturdy, *Soils and their use in South East England*, Soil Survey of England and Wales Bulletin 15 (Harpenden, 1984), 236–9.

11 S.G. McCrae and C.P. Burnham, 'The soils of the Weald', *Proceedings of the Geologists' Association*, 86 (1975), 593–610; Cranfield University, *The Soils Guide* <https://www.cranfield.ac.uk/themes/environment-and-agrifood/landis/soils-guide> [accessed 11 September 2021].

For the agriculturalist this is a dismal story: soils with low nutrient values and general infertility, with waterlogging in winter and hardness and cracking during hot summers. The longstanding practice of heath burning has also reduced the supply of nutrients to the soil, and initial invasion by grasses and low shrubs offered little more nutrient. The soil structure is weakened and water retention capacity lowered. Animals attracted to the succeeding young shoots trample the soils, impeding rainfall infiltration but also allowing wind erosion of open soils in dry weather. The high soil acidity (falling to as low as pH 3) precludes activity by earthworms and microscopic bacteria, meaning that the incorporation of organic matter takes much longer and a layer of leaf litter frequently rests on the surface.

This unpromising picture is reinforced by climatic conditions. Ashdown experiences a temperate climate, mild and damp, much like the rest of the Weald. But it receives rather more rainfall than elsewhere in Sussex, normally over 35 inches, with heavier rainfall on the higher slopes around Crowborough. There is a more extended period of spring and early summer frosts and the growing period is perhaps ten days shorter than in more low-lying areas. Microclimates complicate these generalisations, with north- and south-facing slopes having varying conditions. Severe weather events would have been felt keenly on the more exposed higher parts, and the flooding from the River Uck in 2000 affected Uckfield particularly badly, as did the July 2007 floods at Newbridge. The 1987 storm wreaked havoc. And, of course, there have been more general climatic variations affecting the Forest to lesser or greater extents, such as the 'Little Ice Age' of 1500–1750 or the crisis years of 1315–22. And the extent of modern climate change and atmospheric pollution is a question for the present day.

VEGETATION OF THE OPEN HEATH

Lowland heathland of north-west Europe can be found ranging from the north coast of Spain via western France to northern Germany, Jutland and southern Scandinavia, although the main affinity of the Wealden heathlands, based on their flora, is with the heaths of western France and Spain.[12] In the United Kingdom there remain something over 180,000 acres of lowland heath, of which over half is in England, with the eastern counties providing optimum habitats for the growth of heather (*Calluna*). In all, 23 counties in England and Wales have some heathland and a 1993 estimate put the UK contribution to Europe's heathland at 20 per cent.

It was once held that areas of heath were natural deserts, devoid of useful resources for mankind, unlike the woodlands and river valleys. But this has clearly never been true: for at least 5,000 years the resources have been used. Grazing has long been a feature of these areas, with woodland clearance facilitating greater animal numbers. However, much has changed: grazing pressures have now ceased as farming has become more specialised and far less reliant on the grazing of heathland, especially

[12] D. Streeter, 'Biogeography, Ecology and Conservation in Sussex', in Geography Editorial Committee, *Sussex: Environment, Landscape and Society* (Gloucester: Alan Sutton Publishing, 1983), 136–7.

as visitor numbers have increased along with family dogs, which worry the sheep and cattle. And so plant succession has seen woodland again taking hold. Controlled burning, once a staple method of ensuring new shoots of heather, has declined and fragmentation of heathland has occurred as a result of development.

There was always a balance to be held between access and use of the heathland resources on the one hand and over-use and impoverishment on the other. Some control had to be exercised over the cutting and collecting of turf, peat and vegetation, numbers of grazing animals and the quarrying of stone or other mineral resources. We know little about systems of heathland management before the medieval period in north-west Europe, although we assume that originally there was a freedom to use resources that were common to all. But as populations grew these common resources had to be more carefully managed for present and future generations. The more turf was removed, for example, the less there was for grazing.

By the 1980s the United Kingdom's heathland was reduced to just 40 per cent of its 1945 extent. Between 1800 and 2000 no less than 80 per cent of lowland heath was lost, initially to farming, but then in the twentieth century to housing, roads, out-of-town supermarkets, industrial estates or the trees of the Forestry Commission; it was even churned up by mountain bikers and four-wheel drive enthusiasts.[13] And on the remaining heaths 65 per cent of plant and animal species have been lost.

The picture is not a happy one: in Sussex the past hundred years have seen a reduction in heathland of 70 per cent, accompanied by a decrease in average patch size, leaving by far the greatest remaining concentration on Ashdown (Figures 2.3 and 2.4). There are now more heathland habitat patches (increased from 250 to 755) and the average patch size has reduced from about 39 acres to under four acres over the past hundred years.[14] Fragmentation is now particularly acute on the Dorset heaths, where the once-vast heathland has been reduced to about 15,000 acres in over 100 fragments, and 30 per cent adjoins the Poole/Bournemouth conurbation, rendering it susceptible to disturbance, fire and eutrophication.

But Britain nevertheless still has a good proportion of the world's lowland heath, so it is vitally important to understand its historical complexity in order more fully to appreciate its present and future maintenance.[15] But to come anywhere near this objective it will be necessary to learn more about these relationships and how they have changed through time. This is one aim of this present book.

Left untouched by humans, the climax vegetation over most of Ashdown, as with most similar environments in southern England, would be a mixed woodland of beech and oak. With climatic warming from about 11,000 BC following the last glaciation, vegetation began to develop, covering much of the Weald in woodland – at first birch, pine and hazel, followed by oak and elm by about 7000 BC – forming what Rackham

13 J. Parry, *Heathland* (London: National Trust, 2003), 57.

14 Sussex Wildlife Trust, *A Vision for Nature and Wellbeing in Sussex* (2017), 11–12, <https://assets.sussexwildlifetrust.org.uk/Files/swtvssummary-reportfinalweb.pdf> [accessed 11 September 2021].

15 I. Newton, *Farming and Birds* (London: William Collins, 2017), 428–32.

Figure 2.3. Ashdown heathland with the North Downs in the distance. Scots pine with *Calluna vulgaris* (common heath or ling) and *Erica cinerea* (bell heather) (source: courtesy of Gerry Lee).

Figure 2.4. Ashdown heathland (source: Sussex Wildlife Trust).

has termed the wildwood.[16] On higher parts a more open scrub cover of oak and birch would be found; and in the lower stream valleys there would be alder.

However, the present vegetation cover is relatively modern and is anthropogenic, a human-induced plagioclimax maintained by activities such as grazing, turf cutting, wood-cutting, bracken and heath collecting, and burning. Ashdown is as a result known for its lowland heath, defined as 'ericaceous dwarf-shrubs growing at low altitudes, below 800 feet in acidic, nutrient-poor, mineral soils ... [with] wet heath and valley mires.'[17] But Ashdown, unlike some other southern heathlands such as Blackdown, is also characterised by a continuum of vegetation from valley bog to dry heath, due mainly to its variability in altitude, microclimates and human interference. It is the largest continuous area of lowland heath, semi-natural woodland and valley bog in South East England.

This heathland habitat, now greatly prized, comprises several vegetation types. Ashdown was recognised in 1953 as a Site of Special Scientific Interest (SSSI) based on a number of different factors. Although the area of heathland had declined there remained extensive areas of dry heath: complex associations of common heather or ling (*Calluna vulgaris*), as, for example, between Camp Hill, King's Standing and towards Gills Lap; and bell heather (*Erica cinerea*) on drier or freer draining areas, such as sandy banks, edges of tracks or free-draining disturbed soils. Heather grows through four phases: pioneer, building, mature and senescent, and Ashdown's biodiversity would be best served if all four phases were co-present because each offers different ecological possibilities.

On damper areas, cross-leaved heath (*Erica tetralix*) becomes dominant with deer grass (*Trichophorum cespitosum*). Purple moor grass (*Molinia caerulea*) grows in transitional acidic areas or where there is a perched water table, and there is a greater proportion of this plant than on many heaths, such as those on the Greensand or Bagshot Beds. Along the stream beds one might find tussocky *Molinia*, rushes (*Juncus*) and ferns. There are also iron-rich alder and sallow flushes reflecting the complex underlying geological structures. In the acidic valley bottoms are also found *Sphagnum* bogs with moss, together with heather and cross-leaved heath and sundews, bog asphodel (*Narthecium ossifragum*), common cotton grass (*Eriophorum angustifolium*) and local specialities such as marsh gentian (*Gentiana pneumonanthe*), 'one of the jewels of the southern heathlands', white beak sedge (*Rhynchospora alba*) and the rare marsh clubmoss (*Lycopodiella indundata*).[18]

Importantly associated with the heather is dwarf gorse (*Ulex minor*) and common gorse or furze (*Ulex europeaus*), the most common form of gorse, which can grow several feet high. Large stands are to be found at Greenwood Gate and in ditches and roadsides across the Forest. It has in the past tended to block drivers' vision on bends, is a fire risk and as it gets 'leggier' loses some of its wildlife value. But it is high in protein,

---

16 O. Rackham, *Ancient Woodland*, 2nd edn (Dalbeattie: Castlepoint, 2003).

17 Webb, *Heathlands*, 11–12.

18 Webb, *Heathlands*, 123–4. A fuller list of the species to be found on Ashdown associated with the different plant communities is provided in C. Marrable and N. Muggeridge, *Information on Ashdown Forest* (Wych Cross: Ashdown Forest Centre, 2001), 39–45.

Figure 2.5. Bracken near Fords Green with renovated cottage (source: author).

attractive to cattle when the shoots are young, and can be used as winter fodder. It is a fast burner, creating a hot blaze, and has been collected for domestic consumption. There is also the local petty whin (*Genista anglica*). This heathland complex also supports important lichen communities including species such as *Pycnothelia papillaria*.

Also ubiquitous is bracken (*Pteridium aquilinum*), a woodland plant that has spread as grazing and cutting has declined and soil has been disturbed (Figure 2.5). Seemingly indestructible, it is particularly successful, as on other lowland heaths, with rapid vertical and horizontal growth of the fronds suppressing less competitive species, being unpalatable to animals and with energy transferred to underground rhizomes when cut in autumn. Its hugely successful invasive powers enabled it to invade former wooded areas on Ashdown and the process has once again been aided by the recent diminished human activities. The plant pushes through the previous year's litter layer in May, attains full height and dense green canopy by July and collapses as a copper brown colour in October to form the next litter layer. As we shall see in later chapters, the collection of the litter, a precious resource on the Forest, can then begin.

There also remain relict species from the Atlantic period of 5,000 years ago where clearance has not penetrated the deep ghylls. The Ardingly Sandstone provides particular refugia, in the form of some of the oldest and least disturbed woodland in the South East, for these survivors, such as the small leaved lime (*Tilia cordata*), once prevalent over much of the High Weald. Bryophytes and lichens can occur in profusion.[19] The small Forest streams are often lined by alders (*Alnus glutinosa*) with grey sallow (*Salix cinerea*), birch and oak. These ghylls are sheltered from frosts and remain humid in summer, offering the Atlantic microclimate for uncommon bryophytes such as the leafy liverwort (*Nardia compressa*), which otherwise has its nearest locations on Dartmoor or in South Wales, and a range of ferns including the mountain fern (*Oreopteris limbosperma*) and the hay-scented buckler fern (*Dryopteris aemula*), unusual otherwise in South East England.

---

19 P. Sansum, *An Overview of the Character and Ecological Significance of Gill Woodland in the High Weald AONB, including Results of some Preliminary Baseline Surveys for Population Monitoring of Selected Rare Vascular Plants of Gill Woodland*, High Weald AONB Unit Commissioned Report (December 2014).

The damming of streams, digging for marl and stone quarrying have left a legacy of ponds in some areas. Here there may be localised rafts of broadleaved pondweed (*Potamogeton natans*), beds of reedmace (*Typha latifolia*) and water horsetail (*Equisetum fluviatile*), especially in the old marl pits. These habitats support a diverse fauna, including water beetles (*Coleoptera*), a rare midge (*Dixella filiformis*), a diversity of dragonfly and damselfly species (*Odonata*) and the locally uncommon snail *Vertigo substriata*. Some of the ponds also have large amphibian populations, including, importantly, the great crested newt (*Triturus cristatus*).

Ashdown's heathlands and woods support many uncommon beetles, moths and butterflies, including several populations of the rare silver-studded blue (*Plebejus argus*) on the open heath. The bog bush cricket (*Metrioptera brachyptera*) is particularly abundant. Large populations of reptiles, including the ovo-viviparous common lizard (*Lacerta vivipara*) and the adder or viper (*Vipera berus*) occur on open habitats. The complex of vegetation also supports a wide range of breeding woodland birds. Heath nesting birds include nightjar and hobby, and gorse provides an essential habitat for the iconic Dartford warbler (alias furze wren), which has made a recent comeback on Ashdown following the terrible winter of 1962–3 and the frequent Forest fires, but is perhaps helped by global warming. The cutting of rides through the heathland, essentially to restrict the spread of fires, has been of particular importance for various woodland birds, including sparrowhawk. The Forest also provides a refuge for the dormouse (*Muscardinus avellanarius*), which had become scarce over much of the country.

The main types and areas of contemporary heathland on Ashdown are set out in Table 2.2, although it should be remembered that the proportions of heathland types and the ratio of heath to woodland are fluid through time. There were 3,371 acres of heathland calculated in 2008, although other estimates give a figure nearer to 4,000 acres.

Table 2.2. Contemporary heathland on Ashdown Forest.

Type of heathland	Area
Dry heath	741
Wet heath	711
Grass-dominated	573
Bracken-dominated	1203
Gorse-dominated	143
Total	3371

Source: adapted from Conservators of Ashdown Forest, *Strategic Forest Plan of the Board of Conservators of Ashdown Forest 2008–2016* (2008), with author's conversion from hectares to acres.

WOODLAND

Scattered across the heath are to be found many examples of a heath/woodland continuum.[20] This tends to comprise silver birch (*Betula pendula*), pedunculate oak (*Quercus robur*) and Scots pine (*Pinus sylvestris*), in some areas forming secondary woodland and scrub. The most common woodland type comprises birch and oak with a heathland understorey. This can in turn be succeeded by an oak, beech and bramble association, with sweet chestnut, holly and hazel (*Corylus avellana*) shading out the heathland remnants.

Clumps of Scots pine were planted mostly in the early nineteenth century, and there have been subsequent seed dispersals, although this was a native species in the Late Boreal/Atlantic periods (the seventh and sixth millennia BC), with its pollen dominant in Ashdown peat bogs.[21] Today we see pioneer birch and alder developed on mires, as in the one area of alder carr on Kidd's Hill near Newbridge, where coppicing has been instituted since 1993.[22]

Areas of ancient woodland (where woodland has existed continuously since at least 1600) are confined to the ghylls, to wooded banks or to the private enclosed areas of the Forest, such as Pippingford Park, Hindleap or Broadstone Warrens.[23] The older woods are characterised by oak, ash (*Fraxinus excelsior*), field maple (*Acer campestre*) and some birch. At Legsheath and in the Five Hundred Acre Wood great swathes of beech (*Fagus sylvatica*) were destroyed in the Great Storm of 1987, as were many marking the boundaries of the 1693 enclosures. These established woodlands contain bluebell (*Hyacinthoides non-scripta*), bilberry (*Vaccinium myrtillus*), the hard fern (*Blechnum spicant*) and honeysuckle (*Lonicera periclymenum*), with birds-nest orchid (*Neottia nidus-avis*) and violet helleborine (*Epipactis purpurata*) found particularly under beech.

The recording of ancient woodland sites has progressed since the original survey for the Ancient Woodland Inventory (AWI) for East and West Sussex in 1989. For many, the Weald is synonymous with woodland cover, and a 2006 survey showed the number of recorded sites increasing from 949 to 2,323, including an additional 5,000 acres in 1,374 small parcels not previously mapped in Wealden District. The new survey also makes more use of vascular plant 'indicator species' that would have grown in association with older broadleaved woodland, such as holly, hazel and bluebell, all found in over 90 per cent of the surveyed sites, together with wood anemone (*Anemone nemorosa*), which is very slow to colonise new areas, making it a good indicator of the antiquity of a wood.

Revision of the AWI highlighted the problems of classifying woodland sites in historically open areas such as Ashdown and other former Wealden commons and

---

20 For the heathland/woodland continuum – wooded heaths, or heathy woods – see T. Greenaway, P. Roper and K. Ryland, *Wooded Heaths in the High Weald* (Henfield: Woods Mill Record Centre Survey Unit, 2004).

21 G. Christian, *Ashdown Forest* (Forest Row: Society of the Friends of Ashdown Forest, 1967), 6.

22 T. Rich *et al.*, *Flora of Ashdown Forest* (privately printed, 1996), 5; for more on the Newbridge site see J. Thompson, 'Newbridge alder carr. Site survey' (Univ. of Sussex unpub. MSc, 2006). I am grateful to David Streeter for drawing my attention to this source.

23 <http://www.magic.gov.uk/MagicMap.aspx> [accessed 30 June 2018].

hunting forests. Within the revised AWI for Wealden District these habitats were classed as a subcategory of ancient woodland, *pasture woodland*, and defined as a semi-natural habitat that has retained a wooded nature throughout recent history. The Five Hundred Acre Wood is now included, together with small areas of woodland around the Forest edges.[24] The revised Inventory for Mid Sussex covering East Grinstead, West Hoathly and Horsted Keynes includes 607 woodlands not included in the original Inventory, with a total area of about 2,000 acres but very fragmented, the average size of these added woodlands being just under 4 acres.

Much woodland grows on the heavier Wadhurst Clay soils, where pollarded beech are common, especially on seventeenth-century boundary banks where the original trees obtain broad girths. Holly and hazel are again the main understorey species in mixed woodland, and hazel coppice under oak standards can be seen on Wadhurst Clay soils at Cackle Street near Horney Common.

There are nearly 2,500 acres of woodland on Ashdown, including chestnut coppice at Tabell Ghyll, beech near the Five Hundred Acre Wood and mixed woodland at Chelwood Vachery, acquired in 1994. And there are still 1,500 acres of oak woods growing on brown earth soils, which include hazel and chestnut coppice, much merging into the heathland.

HEATHLAND AND WOODLAND

Ashdown's heathland is now highly valued, but, as we have seen, the cessation of Commoners' activities has resulted in fears for its future.[25] There continues to be an invasion by oak, birch and pine scrub, of which secondary birch is most damaging because of its large quantities of dispersible seed and because its leaf litter breaks down quickly to increase soil fertility and microfauna such as earthworms, not wanted for heathland. With decreasing acidity, heather loses its competitive advantage, grasses increase and tree seedlings survive. Eventually a shaded tree canopy coalesces and soils veer towards acid brown earths. One study, comparing aerial photos of about one-third of the Forest between 1947 and 2002, showed that scrub and successional birch had increased from 7.1 per cent to as much as 49.2 per cent of the study area. By extrapolation it has been calculated that, if left unchecked at this rate of change, the open Forest would be completely wooded within 75 years (Figure 2.6).[26]

---

24 S. Westaway, *Weald Ancient Woodland Survey: A revision of the Ancient Woodland Inventory for Wealden District, East Sussex* (High Weald AONB, 2006); S. Westaway, M. Grose and P. McKernon, *A Revision of the Ancient Woodland Inventory for Mid Sussex District, West Sussex* (High Weald AONB, 2006). The 1989 inventory for East and West Sussex was updated between 1995 and 1999, and the 2006 surveys add in the additional Wealden District and Mid Sussex District data. The findings remain provisional as this is an ongoing process with both additional woodland and losses being noted.

25 See pp. 18–19.

26 J. Douch, 'Reconstructing Change in Vegetation Distribution on Ashdown Forest since 1947' (Univ. of Sussex BSc project, 2002).

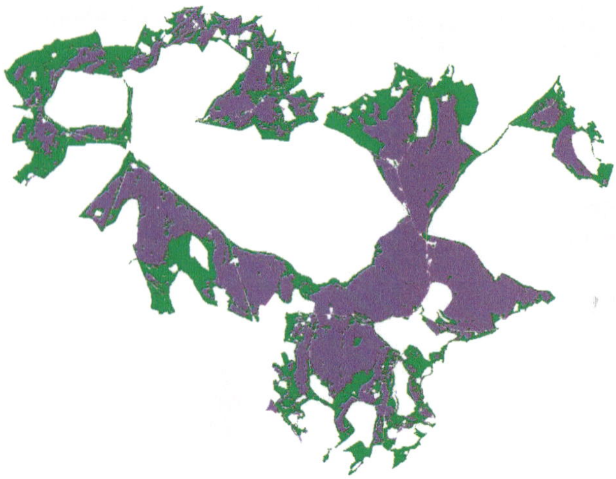

Figure 2.6. Main heathland (purple) and woodland (green) on the Forest (2005). White areas are private and excluded (source: Ashdown Forest, *Strategic Forest Plan 2008–2016*, 5).

An oral history project undertaken in 2005 elicited some thoughts of residents on the changing pattern of woodland:

> You could stand down at Nutley on the village green and you could look right across that valley to, up to ... the next brow ... and there was no trees in between, just grass and a bit of heather, a few gorse bushes, but no birch trees as there is now. You try and look across there now ... you can't see any of the houses. But you could look down and see all the houses, with their gardens with their hedges all around them, the chickens running round ... .
>
> If you look from Friends Clump up the top there, you could look right down into Old Lodge Bottom, not a tree or anything in sight, it was very clear and ... looking back ... down towards the south ... that was the same you could look for miles and there wouldn't be a tree in sight.[27]

And from the Crowborough side of the Forest:

> ... when you came to what is now the Horder Centre, that was completely open, no trees at all; it was all heather ... when you came to the edge of the top of Bunkers Hill you would look down and you ... could actually see the mill pond.[28]

Traditionally scrub birch would have been cut for besoms and the seedlings destroyed by grazing animals. Although cutting is now regularly undertaken, if the scrub has attained five years old the huge amounts of seed means that the same state will soon again be attained.

---

27 ESBHRO, ACC 9997. Roy Pipe, quoted in 2005 from the Ashdown Forest Oral History Project 2005–6.

28 ESBHRO, ACC 9997. Ashdown Forest Oral History Project, Larry Chappell and Len Wickham, quoted in 2005.

Furthermore, there are equally difficult invasions of bracken and purple moor grass. The expansion of bracken at the expense of heather has long been a major source of concern. This was increased by the Commoners' past practice of frequent burning, intended to promote grassland for their cattle, but which resulted in a loss of soil nutrients, the initiation of soil erosion, the destruction of surface organic soil layers, a drastic modification of the microclimate above the soil surface and the spread of unwanted birch as a coloniser. The impact of fires on Ashdown's vegetation has varied according to the temperatures reached and the intensity of burning. Wind direction is also crucial, and controlled burning against the wind, or 'back burning' can result in the complete eradication of all above-ground vegetation. It can take up to two years for *Calluna* to reassert its dominance after a fire and purple moor grass, together with other grasses, may instead become established in place of heather.

Thus, the combined actions of burning, grazing and cutting have prevented woodland regeneration, reduced plant homogeneity and encouraged podzolisation of the soils on the Forest, giving rise to the extensive heathland so characteristic today. The brown forest soils which originally covered so much of the Weald have been degraded as a result largely of human intervention over many generations, and the exercising of the Commoners' rights of grazing, litter collection, estovers and so on must be seen in this longer term as contributing to the impoverishment of the Forest soils. In contrast to the surrounding farmland and in the enclosed parts, there has been no attempt to increase the quality of the Ashdown soils on the open Forest by liming, marling or fertilising.

The grazing activities of different animals on Ashdown should also be remembered. Sheep and horses nibble and graze close to the ground, thereby creating a short sward. Sheep are selective grazers, taking new shoots and flowers from many plants, including sedges and rushes. Horses prefer grasses, seedlings and the tips of branches. Cattle prefer longer grasses and sedges, but will also tear leaves from trees. Cattle, sheep and horses will feed on *Molinia*. Pigs are root feeders, disturbing the topsoil, and will eat acorns, beech mast and even bracken rhizomes in autumn. Deer will feed on scrub, heather, saplings or pastures, although they prefer woodland cover. But sheep have historically been excluded from Ashdown because they compete with fallow deer for pasture, although flocks were being turned out in the nineteenth century, and the rules governing sheep on the Forest have gradually been relaxed such that today, as we shall see in Chapter 12, they are used to help manage the vegetation mosaic. These different foraging characteristics have enhanced biodiversity, as do the trampling and the stocking densities permitted.

THE LANDSCAPE

Seen in the long term, Ashdown is a transient landscape, and shares the catastrophic loss of heathland experienced across most of Western Europe.[29] This is, of course, part of a wider loss:

---

29 P. Dolman and R. Land, 'Lowland Heath', in W. Sutherland and D. Hill (eds), *Managing Habitats for Conservation* (Cambridge: Cambridge University Press, 1995), 267–91.

> England's natural capital – the elements of the natural environment which provide valuable goods and services to people such as clean air, clean water, food and recreation – is in long-term decline. Successive 'natural capital deficits' have built up a large natural capital debt and this is proving costly to our wellbeing and the economy. If economic growth is to be sustained, natural capital has to be safeguarded.[30]

More specifically, Ashdown now faces several environmental challenges common to the High Weald more generally and including climate change, with the impact of rising temperatures and extreme weather events; pressure from recreation; soil erosion on open heathland requiring soil conservation; pollution by nutrient, pesticide and fine sediment runoff into ponds and streams; and non-native invasive species in water bodies.[31]

The basic Forest landscape is, then, a mosaic, a reflection of topography, geology, soils and climate, modified to lesser or greater degrees by human intervention. The rest of this study will henceforth focus on the latter, while paying all due respect to the influences of the encompassing environment – actors on a stage but in which each impacts actively upon the other. The environment of Ashdown has offered both difficulties and opportunities to humans, who have worked with its possibilities but also transformed it through their working and daily lives.

---

30 Natural Capital Committee, *The State of Natural Capital: Protecting and Improving Natural Capital for Prosperity and Wellbeing* (DEFRA: Third Report to the Economic Affairs Committee, 2015), 2.

31 High Weald AONB, *High Weald AONB Management Plan 2019–2024 Consultation Draft* (June 2018), 22.

CHAPTER 3

ASHDOWN BEFORE THE FOREST

> The Weald was of old a desert and a waste, and neither planted [cultivated] nor peopled, but filled only with herds of deer and droves of swine.[1]

IT IS PERHAPS WORTH stating one obvious point: before the medieval period there was no distinction between the area that eventually become Ashdown Forest and the areas that surrounded it. We might point to its higher altitude and more open vegetation canopy compared with most of the surrounding Weald, but in most other respects it was undifferentiated.

The enormous spans of time with which this chapter deals means that much has to be passed over, but also that we are still at the limits of our present knowledge. The extent to which all historical research findings are provisional is nowhere better exemplified than in considering the landscape of early human activity on Ashdown. We are, of course, dealing with many unknowns. Furthermore, the survival of material remains, to some extent, depends on *what did not happen later*. The fact that much of Ashdown remained unploughed and agriculturally unproductive meant that any relict features stood a better chance of survival. The conservation status of the area has also largely negated significant recent building but this has actually minimised developer-funded archaeology through the application of PPG 16.[2]

One unknown is the date and extent of the clearance of the wildwood of the central Weald by human agency. Indeed, what clearances may have predated their arrival, based on the natural interventions of fire and animal trampling? And we must also note that many early finds on or near the Forest that have been reported may not necessarily indicate any degree of settlement but may have resulted from travellers or itinerant workers discarding objects or simply losing material goods *en route* to somewhere else.

E.C. Curwen held in 1954 that the Weald was unattractive to prehistoric peoples because of the difficulties it posed to farming, and was thus left solely for seasonal

1 T.D.W. Dearn, *An Historical, Topographical and Descriptive Account of the Weald of Kent* (Cranbrook: printed for and sold by S. Reader, 1814), xiv.

2 Planning Policy Guidance 16: Archaeology and Planning (1990). The huge significance for the Weald of such funded work has been shown for the Low Weald at Wickhurst Green in West Sussex (A. Margetts, *Wealdbæra: Excavations at Wickhurst Greeen, Broadbridge Heath and the Landscape of the West Central Weald* (Portslade: SpoilHeap Publications, 2018).

hunting. This built on older views of the Weald as primeval forest, long daunting to human settlement. Edwin Guest's 1849 map, copied in J.R. Green's *A short history of the English people* (1892) (Figure 3.1), shows continuous woodland cover across the central Weald. Robert Furley's *History of the Weald of Kent* (1871) contained a more nuanced map showing woodland cover over the central Weald but with clearances including around the headwaters of the Medway, which may have been the Ashdown area. Nevertheless, he believed that 'for many a century it [the Weald] was a desert and a waste, neither planted nor peopled, and filled only with wild animals … .'[3]

Later research has demonstrated that such views of an impenetrable woodland are incorrect. Indeed, the current view is that, with some exceptions, early people lived everywhere, whether downland, woodland or heath. From the mid-1960s local resident Fred Tebbutt, walking on Ashdown and surrounding farmland, noticed flint flakes. He concluded that some degree of settlement, if only temporary, had occurred, and therefore that the prehistoric central Weald had been far from deserted.[4] In what follows we build on Tebbutt's insight.

THE PALAEOLITHIC AND MESOLITHIC

We can now catch mere glimpses of early human connections with the area that was to become Ashdown Forest. Palaeolithic finds at present are dubious; most distribution maps can only refer to the general area of a find spot, but, for example, there are none shown on a map of 2003 by Holgate.[5] The absence of commercial gravel working around Ashdown, compared with sites nearer the coast, has certainly limited the chances of finds here, and for much of the Upper Palaeolithic (40,000–10,000 years ago) the area would have been little more than treeless tundra, often permanently frozen below the ground surface and relatively uninviting even for hunting or gathering. During the so-called Late Glacial Maximum *c.* 18,000 BP, with much of northern Britain covered in ice, sea levels were as much as 300ft lower than those of today and cold winds swept across 'Northsealand'. Periods of warmer (interstadial) and colder (stadial) climate followed.

---

3 E.C. Curwen, *The Archaeology of Sussex* (London: Methuen & Co., 1954); R. Furley, *A History of the Weald of Kent*, I (Ashford: Henry Igglesden, 1871), 3 and map inserted between pp. 27–8; E. Guest, *Origines Celticae* 1 (London: Macmillan, 1883), facing p. 147, based on his earlier map of 1849; J.R. Green, *A Short History of the English People* (New York: Harper & Brothers, 1892), 18.

4 C.F. Tebbutt, 'The Prehistoric Occupation of the Ashdown Forest area of the Weald', *SAC*, 112 (1975), 34–43. Notes on further unpublished finds by Tebbutt are held by the Sussex Archaeological Society. And see also M. Gardiner, 'The Archaeology of the Weald – a survey and review', *SAC*, 128 (1990), 33–53 and A. Margetts, *Out of the Weald, the Secret Weald*, University College London and Surrey County Council (forthcoming).

5 R. Holgate, 'Late Glacial and Post-glacial Hunter-gatherers in Sussex', in D. Rudling (ed.), *The Archaeology of Sussex to AD 2000* (Kings Lynn: Heritage Marketing & Publications, 2003), 29–38.

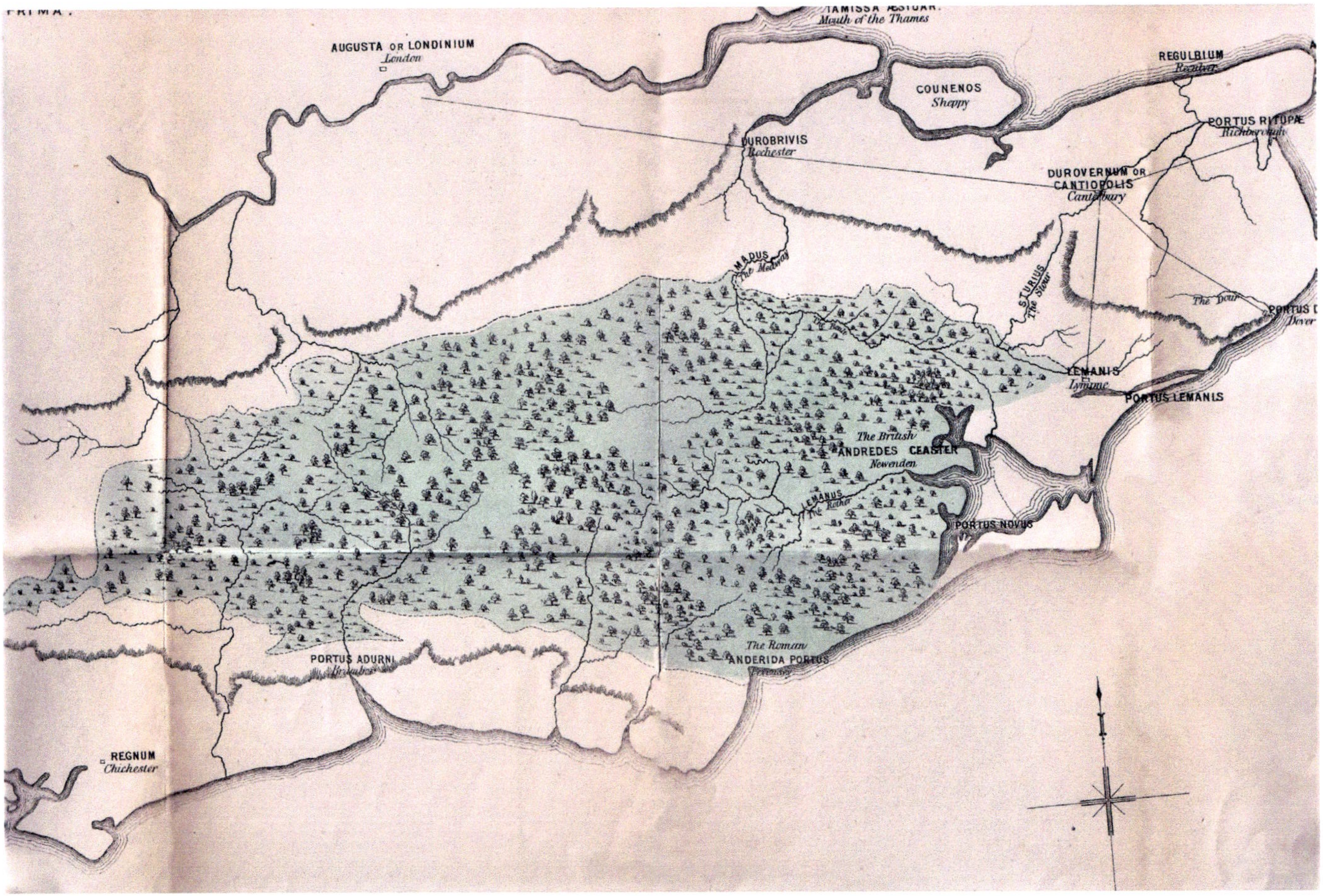

Figure 3.1 (a). Furley's *Anderida* (1871) (source: Furley, *History of the Weald of Kent* I, 27–8).

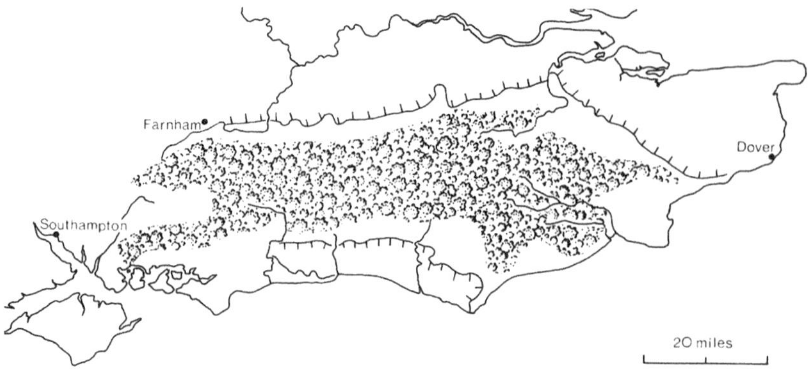

Figure 3.1 (b). The *Andredesweald*, after Guest (1849) (source: Guest, *Origines Celticae*, facing p.147).

One significant find, however, serves as a warning. On higher ground on the Lower Greensand at Pulborough, to the west of Ashdown, a possible hunting camp was established at some point in the early Upper Palaeolithic, offering views across the western Weald.[6] Therefore we clearly cannot rule out future discoveries around Ashdown. Late-glacial hunter gatherers may have visited. But one hindrance to finding evidence is the nature of the modern Wealden landscape. There is little ploughed land, most being under grass, heath or woodland, so ploughing does not bring buried material to the surface. The fields are small and the shaws around Ashdown are daunting to fieldwalkers looking for traces of early presence. Much has relied on opportunistic discoveries, and no distribution map of finds can be regarded as anything more than a temporary indicator.

The so-called Loch Lomond stadial saw a deterioration in climate around 12,000 BP, probably rendering much of what became the British Isles difficult once again for human activity. But with the following amelioration in the climate an initial open environment supporting horse and reindeer may be assumed, but with an advancing woodland cover over the Ashdown area. On Ashdown's sandier soils scrubby juniper, birch and pine would be early colonisers, followed by hazel. Southern Britain now gradually became a more hospitable environment, allowing the development of a mixed deciduous woodland as we enter the Holocene, the current geological period.

Again, however, there is little evidence for early Mesolithic activity in the central Weald. The earliest post-glacial flint assemblages, including the small, thin-bladed flints, or microliths, that are taken as indicating bow and arrow hunting, have been

6 R. Jacobi, N. Debenham and J. Catt, 'A Collection of Early Upper Palaeolithic Artefacts from Beedings, near Pulborough, West Sussex, and the Context of Similar Finds from the British Isles', *Proceedings of the Prehistoric Society*, 73 (2007), 229–326; M. Pope, R. Dinnis, A. Milks, P. Toms and C. Wells, 'A Middle Palaeolithic to Early Upper Palaeolithic succession from an open-air site at Beedings, West Sussex', *Quaternary International*, 316 (2013), 14–26.

dated to *c.* 9,800 BP.[7] Until recently, the assumption has been that Mesolithic communities were engaged in exploiting local resources by hunting and gathering, but with little significant lasting impact on their surroundings. It was held that the wildwood began to be cleared during the Neolithic from about 3,500 BC. But it is now postulated that localised clearances were made by Mesolithic hunter-gatherers, migrating seasonally across a land bridge with the Continent that became, from *c.* 8,000 years ago, the English Channel. At that time the ice sheets were melting sufficiently for the old 'Channel River' to become inundated, at first seasonally and later permanently. A newly cleared landscape would offer rich browse and grasses, attracting animals and thereby facilitating hunting. But Franz Vera also believed it possible that clearings in the woodland may have been opened up by large herbivores, enhanced by falling trees and lightning strikes, allowing grassland and thickets to be 'naturally' present. Thus a 'savannah' landscape was perhaps emerging with groves, glades, grassland and thicker woodland, and with heathland a natural component on the poorer, more heavily grazed soils. In this scenario heathland is therefore initially a natural feature, not anthropogenic.[8]

In fact, as far back as 1975 Tebbutt was writing of the Ashdown area:

> One cannot easily escape the supposition that a great deal of forest clearance took place, even in Mesolithic times, perhaps by fire or ring barking, and perhaps to increase the feeding grounds for wild grazing animals.[9]

Arguments by analogy are contentious, but we may note here the findings of a paper relating to vegetation change on the Lower Greensand soils of the far western Weald. Here Mesolithic clearings were seen as significant but only temporary, with deciduous and pine stands growing again thereafter. The true development of heathland seems to have occurred from the later Neolithic onwards, and particularly during the Bronze Age.[10] Conversely, it may be that we have also underestimated the permanence of settlement resulting from initial seasonal camps. Intimacy with the landscape must now be assumed and many of the finds around Ashdown are in locations offering viewpoints across lower land, essential for hunting. A site at Broadstone with extensive views to the north, for example, yielded *c.* 500 flints, including microliths, a tranchet arrowhead, cores, scrapers and flakes.[11]

---

7 Holgate, 'Late Glacial and Post-glacial Hunter-gatherers', 29–38. Note that the carbon-14 dating method assumes 'Present' (as in BP) to be 1950.

8 F. Vera, *Grazing Ecology and Forest History* (Wallingford: CABI, 2000). But for doubt that large herbivores could create and maintain such clearings see F. Mitchell, 'How open were European Primeval Forests? Hypothesis testing using Palaeoecological Data', *Journal of Ecology*, 93 (2004), 168–77.

9 Tebbutt, 'Prehistoric Occupation', 36.

10 J.A. Groves *et al.*, 'Long-term Development of a Cultural Landscape: the Origins and Dynamics of Lowland Heathland in Southern England', *Vegetation History and Archaeobotany*, 21 (2012), 453–70.

11 Tebbutt, 'Prehistoric Occupation', 39.

Figure 3.2. Rock shelter site: Hermitage Rocks, High Hurstwood, Buxted (source: Peter Anderton).

As yet the evidence is poor, but an important development relating to this period in the central Weald has been the study of river valley deposits, material eroded from upstream which may contain Palaeolithic and Mesolithic finds, and which also offer evidence for deliberate woodland clearance, a purposeful manipulation of the environment, perhaps by burning.[12] The Ouse valley was found to contain as much as six feet of alluvial deposits traceable directly to the localised removal of vegetation cover in the upper Ouse during the Mesolithic period, in the seventh and sixth millennia BC. At this time pine and birch, as early colonisers in the warming post-glacial climate, were being replaced by mixed deciduous woodland, with lime becoming dominant.

There is certainly evidence of later Mesolithic activity in the Weald. There are numerous rock-shelter sites beneath the massively jointed Tunbridge Wells Sand, especially the Ardingly Sandstone, associated with large amounts of microliths and the debitage of microlith production, which are interpreted as short-stay hunting camps within enveloping woodland of oak, birch, yew (*Taxus baccata*), hazel and some beech. From here groups could hunt for wild ox (aurochs), red and roe deer and wild pigs, and forage for hazelnuts and berries.

Such locations to the north of Ashdown include Harrison's Rocks, High Rocks (where charcoal has been dated to *c.* 4,500 BC) and Bowles Rocks, all with Mesolithic

[12] P.A. Mellars and S.C. Reinhardt, 'Patterns of Mesolithic Land-use in Southern England: a Geological Perspective', in P.A. Mellars (ed.), *The Early Post-Glacial Settlement of Northern Europe* (London: Duckworth, 1978), 234–94.

flint assemblages. Tebbutt found a thin scatter of flint from no fewer than 58 locations on and around the Forest.[13] Other nearby sites include the Hermitage, High Hurstwood (Buxted), where over 4,000 pieces were found, together with a sandstone hearth with charcoal that was carbon-dated to around 5,500 BP (Figure 3.2). Elsewhere there have been finds at Rocks Wood, Withyham and Rocks Field, Uckfield.[14] Large assemblages of microliths, cutting blades and scrapers had been made from flints coming from either the North or South Downs.

There is a cluster of Mesolithic flint finds around Camp Hill, but there are many isolated finds right across the Forest. For example, a Mesolithic knife and a small flint tool were found in 1971 near Yew Tree Farm east of Nutley, a tranchet axe was found to the north-west of Old Lodge, and flints have been reported from Lower Parrock in the Medway Valley. At Kilnwood Farm (Hartfield) a sandstone hearth with axe and flint flakes, previously attributed to the Neolithic, is now interpreted as Mesolithic.[15]

There are also streamside locations across the Weald associated with a broader range of implements, including both microliths (frequently Horsham Points) and tranchet adzes, possibly representing longer-stay camps. It may well be that the base camps were on the Lower Greensand, Chalk Downs or Coastal Plain, and all the signs still point to transitory Wealden occupation by small mobile hunting or hunter-gatherer groups, although the true extent of sedentism remains arguable at present as new discoveries and interpretations are published. But the fact that the flint materials were derived from the chalk downland does reinforce the linking of territories between the Downs and Weald at this time.[16]

We can therefore effectively gainsay any notion that this area was avoided in prehistory. The sheer number of Mesolithic flint finds, which can only be a small sample of the total, shows the area to have been a well-known hunting location, with rock shelters indicating at least temporary settlement. This was a long-established way of life, and it did not suddenly come to an end with the advent of farming practices. Rather, the Ashdown area probably continued as the hunting and gathering grounds for people who were otherwise becoming familiar with the basic principles of farming. This accounts for the intermixture of Mesolithic and Neolithic flintwork found in the Ashdown area by Fred Tebbutt, and for the presence of Neolithic pottery at High Rocks, where Mesolithic microliths have also been found.

---

13 C. Greatorex and M. Seager Thomas, 'Rock Shelter Stratigraphy: Excavations at Eridge', *SAC*, 138 (2000), 49–56.

14 R.M. Jacobi and C.F. Tebbutt, 'A Late Mesolithic rock shelter site at High Hurstwood, Sussex', *SAC*, 119 (1981), 1–36; A.F. Harding and J. Ostoja-Zagorski, 'Excavations in Rocks Wood, Withyham, 1982', *SAC*, 125 (1987), 11–32; M.F. Hemingway, 'Preliminary explorations at the Rocks, Uckfield', *Sussex Archaeological Society Newsletter*, 31 (1980), 209–10; M.F. Hemingway, 'Further explorations at the Rocks Early Mesolithic site, Uckfield, East Sussex', *Sussex Archaeological Society Newsletter*, 34 (1981), 243.

15 ESHER, 'Yew Tree Farm, Meso flints', <https://www.thekeep.info/collections/getrecord/ESHER_MES4584#sthash.QMOiBW0a.dpuf> [accessed 18 March 2019].

16 Holgate, 'Late Glacial and Post-glacial Hunter-gatherers', 33.

THE NEOLITHIC AND BRONZE AGE

By *c.* 4,200 BC the beginning of a recognisable Neolithic tradition in Sussex can be seen. New ideas about the domestication of animals and growing of crops are currently believed to have been developed through acculturation, or from a small elite group, rather than by any mass influx of farmers from the European mainland. But the process of 'neolithisation' is still poorly understood. The domestication of cattle may have been the first element, to be followed somewhat later by the tending of crops. The large number of polished stone axes found in the Weald testify perhaps to intensified activity, sending vast quantities of soil downstream as the woodland cover was partially cleared.

The story of the Sussex Neolithic has predominantly been concerned with the South Downs, with its ritual enclosures such as Whitehawk or Cissbury, pottery, long barrow territorial markers and flint mines. But the old Wealden hunting and gathering surely persisted, although any such activity around Ashdown remains elusive. More detailed investigation may yield more information. For example, there was the reuse of the Mesolithic rock shelter at High Rocks, where a sherd from an Early Neolithic round-based bowl has been thermoluminescence dated to 3324 ± 375 BC. There were a few flint implements, two leaf-shaped arrowheads, a scraper and pottery sherds, but no evidence of flint working, so we must again assume a temporary hunting camp, possibly with food transport or preparation indicated by the pottery.[17]

Tebbutt's valuable survey of Ashdown recorded many Neolithic flint scatters. Near the Isle of Thorns, at Warren Farm, Crowborough, and near King's Standing, Beaker-period barbed and tanged arrowheads were found, the latter also with a few flakes. Long-term continuity in this area is also exemplified from Woodland Farm, Maresfield, where Mesolithic, Neolithic and Bronze Age flints have been recovered.[18]

This burgeoning list of finds for Neolithic Ashdown contrasts somewhat with a surprising lack of real evidence when we move into the third millennium BC and encounter the Bronze Age. The transition was far from uniform and we can again envisage a hunter-gatherer Wealden lifestyle virtually unchanging for thousands of years, and perhaps low-level seasonal exploitation. But metal now appears alongside curved knives, barbed and tanged arrowheads and the innovative tall, narrow-necked and highly decorated Beaker pottery. The latter appears in Britain from *c.* 2,500 BC and is found beneath round mounds. Again, the downland provides the best environmental conditions for archaeological discovery, and it is from there that comes evidence of timber longhouses, together with domestic apparatus and the growing of emmer wheat and barley. These circular mounds, appearing during the Neolithic, become the main shape of such mounds from *c.* 2,500 BC, contrasting with those that had been primarily linear in the Neolithic. They appear to represent individual human burials, and again are best known from the South Downs and Lower Greensand ridge.

---

[17] P. Drewett, D. Rudling and M. Gardiner, *The South-East to AD 1000* (Harlow: Longman, 1988), 46–7.

[18] Tebbutt, 'Prehistoric Occupation'.

Nevertheless, there are clues from the Weald. Bronze Age finds have been listed by Grinsell and Tebbutt, including palstaves from Horsted Keynes, Buxted, Mayfield and on the Forest itself. One possible example of a round barrow on Ashdown, much damaged, comes from near Londonderry Farm, Nutley, on a low ridge.[19] Other potential round barrows were recorded by Tebbutt on a ridge to the west of Duddleswell and again to the south of Gills Lap at Peat Lump Hill.

There is more evidence of Bronze Age activity from the St Leonard's Forest area west of Ashdown. Bronze axe heads, bronze tools, numerous finds of flint implements and flakes and a rich assemblage of 'Wessex culture' goods, including arrowheads, a dagger and a Beaker, also pointed to sites in the area ranging from Mesolithic to early Bronze Age, as well as Roman coin and pottery.[20] The central Weald may have been marginal, but was certainly not *terra incognita*. The finds of bronze flat axes certainly indicate the continuing exploitation of the woodland resources of the Weald.

From about 1,700 BC the better soils were being cultivated outside the Weald in organised field systems alongside round houses, crafts, crops and animals. If some farming was taking place in the Weald, together with hunting and gathering, its permanence is questionable, as soils would rapidly have become exhausted. We can perhaps only note the clearance of woodland more generally in South East England, and in particular the decline of lime (*Tilia*), with lime pollen declining during datable Neolithic and Bronze Age periods (about 2,500 BC to about 1,000 BC). Other species were also affected, notably oak, within areas of mixed woodland such as Ashdown. On some Wealden sites the decrease in lime was followed by increases in beech pollen, but this is not consistent, and it is equally likely, especially on sandy soils, for *Calluna* together with *Betula* and *Pteridium* to enter the picture more strongly. But openings for agriculture, and early recolonisation by birch, may thus be indicated.[21]

THE IRON AGE

Evidence for human presence is far stronger for the Iron Age, when there may have been a relatively well-settled landscape. Trade and exchange were now well developed and metal implements gained a status equal to that of the older flint tools. Hunting and gathering in the relatively well-wooded central Weald remained important. The analysis of deposits in the upper Ouse valley at Sharpsbridge, between Newick and Uckfield, suggests woodland clearance during the Neolithic, but particularly during the later Bronze and early Iron Age periods. Tree pollen was gradually replaced by

---

19 ESBHRO, 'Londonderry Farm, ? Barrow', <http://www.thekeep.info/collections/getrecord/ESHER_MES4573> [accessed 11 September 2021]; Tebbutt, 'Prehistoric Occupation', 42.

20 T. Honywood, 'Discovery of Flint Implements near Horsham, in St. Leonard's Forest', *SAC*, 27 (1877), 177–83; S.G. Beckensall, 'The Excavation of Money Mound', *SAC*, 105 (1967), 13–30. I am grateful to M. Weir-Wilson for drawing my attention to these references.

21 M.J. Grant, M.P. Waller and J.A. Groves, 'The *Tilia* decline: vegetation change in lowland Britain during the mid and late Holocene', *Quaternary Science Reviews*, 30 (2011), 394–408; Groves *et al.*, 'Long-term Development of a Cultural Landscape'.

that indicating weeds and cereals, demonstrating cultivation in the area, and even soil deterioration, as shown by the presence of *Calluna* pollen.[22]

A possibly more permanent lifestyle on Ashdown is indicated by several enclosures and possible field systems. Ivan Margary reported on what he believed was an Iron Age cattle enclosure or agricultural settlement south-east of Gills Lap, and again at King's Standing, with associated pottery, and he is quoted as saying that he could detect very faint traces of lynchets to the east of King's Standing.[23] And a system of cultivation furrows with a superimposed undefended enclosure with pottery relating to the later Iron Age was found at Chelwood Gate, the erroneously named 'Danes Churchyard', with evidence of iron smelting activity and late pre-Roman (Eastern Atrebatic) pottery. The site was abandoned before AD 43.[24] At Maresfield a field system and enclosures were orientated north-west–south-east and north-east–south-west, on the same axis as the better-known field system at Ripe, to the south of Ashdown.[25] And further indications of a population here at this time are provided by a late Iron Age coin hoard (*c.* 50–20 BC) from Camp Hill (Maresfield), originally identified as Saxon by Gideon Mantell after their discovery by a peat digger in 1822, but now reinterpreted as pre-Claudian in date.[26]

Of course, the very term 'Iron Age' reminds us of the significance of the Weald at a time when advances in metalworking made the Wadhurst clay iron ore nodules highly significant. There were sandstone blocks that could be used for making furnaces and plentiful timber for fuel and building, and iron-making may have been initiated here by the third century BC, if not earlier, and certainly by the mid- or late first century BC.[27] As corroboration, we have Julius Caesar's account of the maritime area of Britain, possibly the south-eastern Weald, in *De Bello Gallico*, which records outputs of iron, 'but the quantity of it is small'.[28] The Greek Strabo recorded iron among the exports from Britain: 'it bears grain, cattle, gold, silver, and iron. These things, accordingly, are exported from the island, as also hides, and slaves, and dogs … .'[29]

---

22 Scaife and Burrin, 'Floodplain Development', 1–10.

23 Cited as a personal communication in Gulley, 'Wealden Landscape', 483.

24 I.D. Margary, 'A Celtic enclosure in Ashdown Forest', *SNQ*, 3 (1930), 71–2; Margary, 'King's Standing, Ashdown Forest', *SNQ*, 3 (1930), 72–6; N. Wickenden, 'Excavations at Chelwood Gate: Interim Report', *Sussex Archaeological Society Newsletter*, 48 (1986), 467; *idem*, 'Final interim report', *Sussex Archaeological Society Newsletter*, 54 (1988), 11; Historic England, 'Chelwood Gate Enclosure', <https://historicengland.org.uk/listing/the-list/list-entry/1423862> [accessed 11 April 2019].

25 Margetts, *Wealdbæra*, 124.

26 Portable Antiquities Scheme Unique ID: IARCH-E681EF; P. de Jersey, *Coin Hoards from Iron Age Britain*, British Numismatic Society Special Publication (London: Spink & Son, 2014).

27 H. Cleere and D. Crossley, *The Iron Industry of the Weald* (Leicester: Leicester University Press, 1985), 53–6, and 2nd edn with added bloomery sites, including more than 40 of Iron Age or Roman date around Ashdown (Cardiff: Merton Priory Press, 1995); M. Russell, *Prehistoric Sussex* (Stroud: Tempus, 2002), 114; J. Hodgkinson, *The Wealden Iron Industry* (Stroud: The History Press, 2008), 28.

28 Julius Caesar, *The Gallic* Wars, <http://classics.mit.edu/Caesar/gallic.5.5.html> [accessed 12 August 2021].

29 *The Geography of Strabo*, Book IV, Ch 5, <http://penelope.uchicago.edu/Thayer/E/Roman/Texts/Strabo/4E*.html> [accessed 12 August 2021].

A late Iron Age hillfort, perhaps more accurately described as a promontory enclosure, is located at Philpots against Ardingly Sandstone rock cliffs and with traces of Mesolithic flint working, but little else beside a vallum and gateways. There are two others of similar ages at High Rocks and Saxonbury, north-east of the Forest, that yield evidence of ironworking and may well have been administrative or protective centres for the industry.[30] At the former there is also evidence of arable fields and associated earthworks. The univallate Saxonbury camp has yielded mid–late Iron Age pottery with an underlying, possibly Bronze Age, rectilinear field system, but no definite Romano-British material.[31]

ROMAN ASHDOWN

The most significant early survival in the Ashdown area is the Garden Hill fort, identified in 1968 by Fred Tebbutt, who found early Romano-British material in a trial excavation.[32] Following on from this, the Garden Hill Excavation Group reported a scatter of flint indicating a probable presence in the Neolithic or Bronze Age, but two round houses are attributed to the late pre-Roman Iron Age. This area was then developed into a univallate hillfort enclosing about seven acres within a roughly rectangular shape, with stone-revetted and palisaded defences dated by buried pottery to the first century AD, and with an inturned entrance on the eastern ramparts, possibly part of defences against the Romans after AD 43 (Figure 3.3). There are steep natural drops to north, east and south and good views towards both North and South Downs. These earlier foundations were then further developed during the later Romano-British period. Timber buildings were erected at the east end of the hillfort, containing mainly late first-century and second-century locally made pottery, nails, roasting and smelting furnaces, a forging hearth and associated iron slag of the mid-first century. There was also a rectangular building with two verandahs and a four-post structure on the same alignment, both probably of the first century; a timber building and a small and poor stone bath-building with a tiled roof of the second century that copied the more sophisticated Roman originals; and other undated postholes. Other finds here were few, although a rare complete (but broken) pane of Roman window glass was uncovered.[33]

It is possible that Garden Hill was primarily an administrative base from which local small-scale bloomery sites were operated in the first and second centuries. A Roman road (see below) runs to the north of the site, facilitating the export of material from here. During or soon after the end of the second century AD the bath and adjoining

---

30 D. Lea and J. English, 'Analytical Survey and Landscape Context of Saxonbury Hillfort, Rotherfield, East Sussex', *SAC*, 153 (2015), 29–40.

31 For these hill forts and Garden Hill see G. Lock and I. Ralston, *Hillforts: Britain, Ireland and the nearer continent: papers from the Atlas of Hillforts of Britain and Ireland Conference, June 2017* (Oxford: Archaeopress, 2019), <https://hillforts.arch.ox.ac.uk> [accessed 17 August 2021].

32 C.F. Tebbutt, 'Garden Hill Camp, Hartfield', *SAC*, 108 (1970), 39–49.

33 J.H. Money, 'The Iron-Age Hill-fort and Romano-British Iron-working Settlement at Garden Hill, Sussex: Interim Report on Excavations, 1968–76', *Britannia*, 8 (1977), 339–50. Excavations and reports continued from 1969 to 1982.

Figure 3.3 (a). Garden Hill, general view of excavations, showing the bath house on the left and timber building on the right (source: A. Meades/Wealden Iron Research Group photo archive).

structures fell into disuse, and there is evidence of squatters moving in. By the early third century the site seems to have been in decay, but a more substantial occupation possibly persisted somewhat later. A cart track on the north-east corner and a deep disused post-medieval stone quarry destroyed part of the defences.

Here, then, is a site transitional between late Iron Age and Romano-British activity demonstrating a continuity of occupation on Ashdown. And it heralds a phase that has left far more material remains. The Ashdown area may not have been part of a postulated Roman Imperial Estate, which may have existed in the eastern High Weald, with a 'command economy' devoted primarily to the production of iron and metal materials for the fleet, the *Classis Britannica*, based at Dover.[34] Instead this area may have been exploited by private enterprises and was clearly of some importance for iron production (Figure 3.4).[35]

Among the key sites is a first-century bloomery with a domed shaft furnace and large amounts of slag at Pippingford Park, immediately to the south. This was probably linked by an ancient trackway to Garden Hill, representing another transitional Iron Age/Romano-British site, as also did the nearby Cow Park site, with its three

[34] The controversial idea of the eastern Weald as a Roman imperial estate is dismissed in M. Millett, 'Roman Kent', in J.H. Williams (ed.), *The Archaeology of Kent to AD 800* (Woodbridge: Boydell, 2007), 178.

[35] I am grateful to Jeremy Hodgkinson for providing the data for this new map of Roman ironworks around Ashdown.

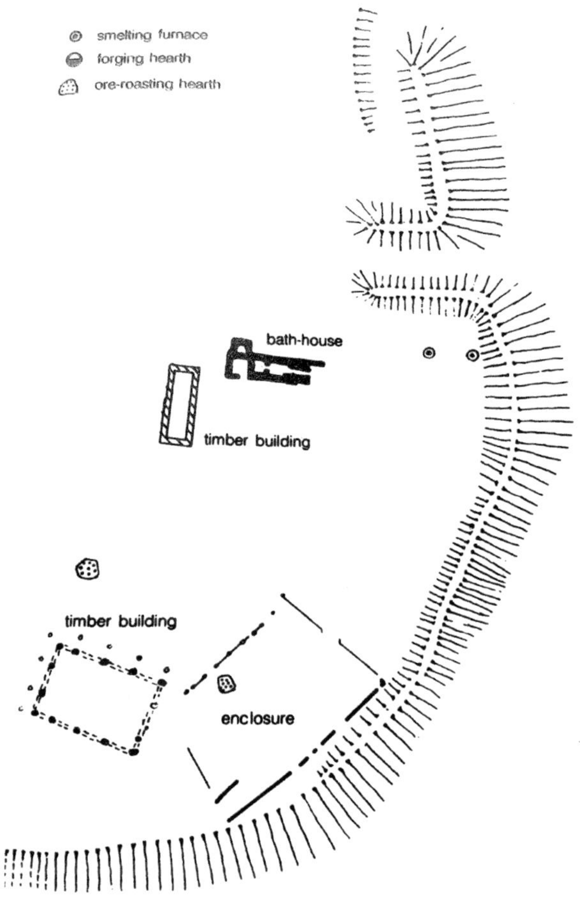

Figure 3.3 (b). Garden Hill, the layout of the Roman site (source: Cleere and Crossley, *The Iron industry of the Weald*, 77).

bloomeries.[36] Pottery at the Pippingford site was dated approximately to AD 43–65, as was a bronze brooch.[37] At Little Cansiron Farm, Hartfield, a substantial area of Roman late first-century/early second-century ironworking, accompanied by iron slag and charcoal (showing use of birch and oak), also had an adjacent tile manufacturing site, with buildings adjacent to the M14 Roman road. Archaeomagnetic dating of the tilery gave a magnetic date of AD 100–130, a date supported by the pottery found on site. Some of the tiles have designs combed on them, and some show dog, cat and red deer imprints.[38] Tiles from here were transported as far north as the Roman villa at Beddington (Surrey) and southwards to the villa at Beddingham.

---

36 C.F. Tebbutt, 'The Excavation of Three Roman Bloomery Furnaces at Hartfield, Sussex', *SAC*, 117 (1979), 47–56.

37 C.F. Tebbutt and H. Cleere, 'A Romano-British bloomery at Pippingford, Hartfield', *SAC*, 111 (1973), 27–40.

38 D. Rudling, 'The Excavation of a Roman Tilery on Great Cansiron Farm, Hartfield, East Sussex', *Britannia*, 17 (1986), 191–230. There is, as yet, no evidence for fallow deer around Ashdown at this time, although there were managed herds at Fishbourne early in the Roman era (R.

It is now clear that by the late first century and into the second century the Ashdown area maintained a busy industrial output, employing many workers, with the most abundant coin finds being those of Hadrian (reigned AD 117–38).[39] The activities of the Wealden Iron Research Group (WIRG) have constantly uncovered new sites in the area and refined thinking on others.[40] At least 246 bloomery sites have been identified as Romano-British in the central Weald, including Ashdown sites at Oldlands (Maresfield), Ridge Hill (East Grinstead) and Blacklands, and a potentially very large site at Great Cansiron (Hartfield), where slag and blackened soil extend over several acres, together with open-cast quarries and cinder-lined trackways, indicating a large operation close to the Roman M14 road (see below). Pottery associated with Roman ironworking at Ridge Hill has been dated to the end of the third century; and there was also pottery from the second century here, and at two other sites in East Grinstead.[41] The site at Oldlands was the first such site to be discovered and published in 1862 by the Rev. Edward Turner of Maresfield, and revealed extensive evidence of Romano-British industrial activity, again close to the M14, with a horizon of bloomery slag over a substantial area associated with stone-built structures, inhumation burials and a coin sequence extending from Nero to Diocletian.[42] There are also many smaller, domestic-scale bloomery sites scattered around the area.

Analysis of charcoal from these sites indicates a promiscuous use of woodland, but with oak very predominant, supported by birch and hazel and with varying amounts of ash, beech, hornbeam, hawthorn and elder. The extent to which trees were being managed remains debatable, and much use was made of secondary regrowth, although Rackham believed, on slim evidence, that a form of coppicing was in use in the Weald at this time. Nevertheless, the impact was considerable and we must suppose that the area was heavily depleted, as was the Wealden iron-producing area to the east. If the calculation made for the latter area also holds for Ashdown (i.e., 84 tons of wood to make the charcoal for one ton of iron), with six main ironworks there with an output of 550 tons per year between AD 120 and 240 then about 75,000 acres of woodland would be required, about 9 per cent of the entire Weald.[43]

Wadhurst Clay iron nodules were mined here, usually from pits or opencast; there were trackways linking the mines, the ironworkings and the better roads. The extent to which such activity was permanent is, however, very uncertain. The lack of traces of permanent shelter at Cow Park, for example, suggested that working was a summer activity, and the tilery at Hartfield may also have been somewhat ephemeral.

Madgwick, N. Sykes, H. Miller, R. Symmons, J. Morris and A. Lamb, 'Fallow Deer (*Dama dama dama*) Management in Roman South East Britain', *Archaeological and Anthropological Sciences*, 5 (2013), 111–22.

39 J. Kaminski, 'The Environmental Implications of Romano-British Iron Production in the Weald' (Univ. of Reading unpub. PhD, 1995), 339–40.

40 See, for example, J. Hodgkinson, 'Field Notes: Ashdown Forest Iron Sites – updated locations', *Bulletin of the Wealden Iron Research Group*, 28 (2008), 3–4.

41 Gulley, 'Wealden Landscape', 474.

42 Kaminski, 'Romano-British Iron Production', 1–5, 325–6.

43 Rackham, *Ancient Woodland*, 108–9; Cleere and Crossley, *The Iron Industry of the Weald*, 36–8.

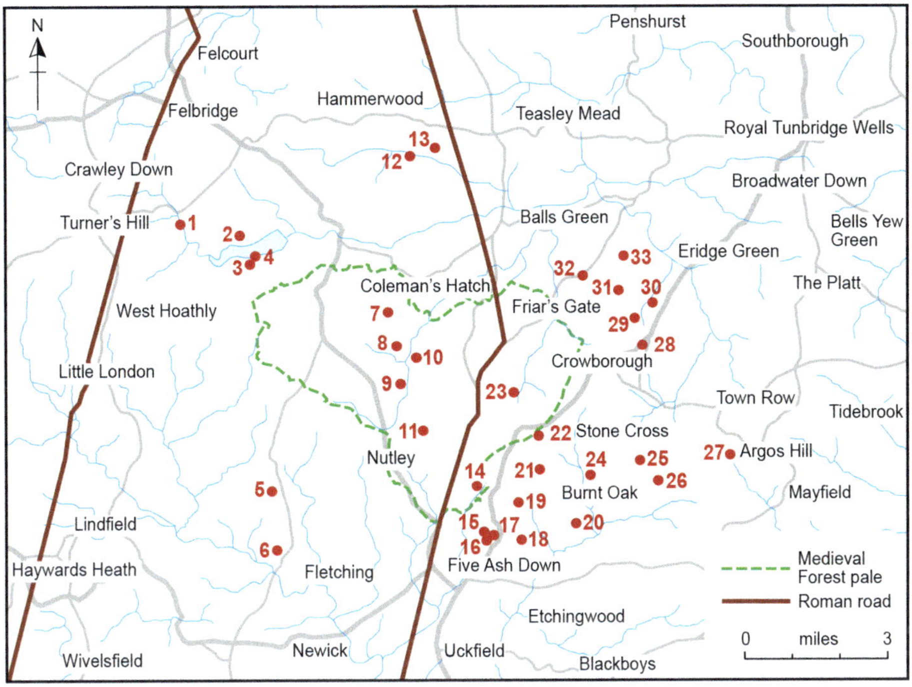

Figure 3.4. Roman ironworking sites around Ashdown (source: Jeremy Hodgkinson). Ashdown, not created at this time, is outlined for context.

1. Ridge Hill
2. Standen
3. Charlwood Farm
4. Walesbeech
5. The Toll, Heaven Farm
6. Coleham
7. Garden Hill
8. Pippingford Park
9. East Wood
10. Cow Park
11. Misbourne 1
12. Great Cansiron, Blacklands
13. Little Cansiron
14. Oldlands
15. Hendall Wood 1
16. Hendall Wood 2
17. Hendall Wood 3
18. Front Wood, Bevingford 1
19. Tilgate, High Hurstwood
20. Morphews
21. Greystones Farm
22. Newnham Park, Chillies Farm
23. Crabtree Farm 4
24. Brook House, Oxley Wood 1
25. Stumletts Pit Wood, Owlsbury Farm 4
26. Limney Farm 1
27. Castle Hill, Home Farm
28. Stone Cross
29. Hodges Wood 1 (Cage Wood)
30. Walnut Tree Field, Renby
31. Minepit Wood / Orznash 2
32. Bingles Farm
33. Rocks Wood

Figure 3.5. The M14 agger at Roman Road car park (source: author).

The pattern of Romano-British settlement in the south-east generally is still focused on the extra-Wealden and coastal areas, where villas and agricultural activity were more prominent. Agricultural activities may have been of secondary importance around Ashdown, although a first-century AD corn-drying kiln from Uckfield does perhaps indicate some farming activity, and the logistics of feeding the ironworkers would seem to demand some local production. At Park Farm, Maresfield there is evidence of ditches demarcating fields, cereal growing and crop processing as well as iron production.[44] To date, though, no large Roman villas are known from the Ashdown area.

The other important landscape elements here at this time were the roads, closely aligned with the development of the iron industry and the corn-producing villas and farms of the South Downs. Two of the key roads stretching south from Southwark skirt Ashdown. The 'M' designations for these roads follow the convention of naming them after I.D. Margary, whose early work on the roads has remained extremely important. The M150 road through Ardingly and what would become Haywards Heath and Burgess Hill ran southwards to join the Greensand Way (M140) at Hassocks and enabled iron products to be transported from the sites at Ridge Hill and Walesbeech.

The M14 road ran from Edenbridge through Holtye – where a 270-yard section of the *agger*, with wheel marks and two ditches on either side, was acquired by the Sussex Archaeological Society in 1939 – and through Galleypot Street to enter what was to become Ashdown Forest at Chuck Hatch, aiming for the heights at Camp Hill and then leaving the Forest via Duddleswell, Fairwarp, Lampool, Flitterbanks and

44 ESBHRO, 'Park Farm, Maresfield: I.A-Rom Features and Finds', <http://www.thekeep.info/collections/getrecord/ESHER_MES25879> [accessed 9 March 2019].

Maresfield Park.[45] Although the road here was partly obliterated by tank training in the Second World War, a section is preserved at Roman Road car park, where the *agger*, of sandstone, is clearly visible and the side ditches are 60 feet apart on either side (Figure 3.5). The road joins the Greensand Way near present-day Lewes, taking iron products from Oldlands and Great Cansiron. Much of the slag otherwise missing from the Cow Park bloomeries was probably incorporated into this road.[46] This process was repeated elsewhere in the area, at, for example, Holtye. The huge amounts of slag incorporated into these roads must indicate a very large output of iron here over a substantial time period.

Across the Weald there were also many older trackways, running broadly east–west with the grain of the topography, and some were incorporated into the Roman road network. The M148, for example, branched from the M14 at Galleypot Street and ran near Colemans Hatch towards Wych Cross and beyond to West Hoathly and Selsfield Common, where it intersected with the M150, and then on to Turners Hill and Horsham.[47] A well-used ridgeway also ran eastwards from Uckfield along the High Weald to the Rye area.

This Roman road and trackway system, allied to the ironworking, tile-making and administrative centres such as Garden Hill, ensured that the area was more populous and more integrated into wider systems of governance than ever before. It was, of course, an exploitative relationship, shipping away large quantities of iron in semi- or finished form, and which depended upon the charcoal, sandstone and clay.

There is, as yet, little evidence that the iron industry continued beyond the mid-third century. At a macro-level the overall decline in Roman control in the fourth century would certainly have impacted upon the sites more generally and closure would have been complete by the end of the fourth century, with both the military and civilian market for iron having dried up. The impact of deforestation and over-exploitation of the ore is uncertain, but closures then followed, although some smaller sites may have continued working, such as that at Oldlands.[48]

ASHDOWN IN THE EARLY SOUTH SAXON ERA

The years following the collapse of Roman rule in Britain have long intrigued historians. The transformation from *Britannia* to an island of Germanic and Scandinavian kingdoms is known to some extent, but the local detail is missing. And this is, of course, the case for Ashdown. Did the lives of Wealden people continue much as before, amidst the turbulent events around them? We do not know to what extent the increasing numbers of Germanic raids impacted on this part of the Weald, nor precisely what might have been the impact of a Fort of the Saxon Shore being

---

[45] I.D. Margary, *Roman Ways in the Weald*, 2nd edn (London: Phoenix House, 1949), frontispiece and 124–64; A. Vincent, *Roman Roads of Sussex* (Midhurst: Middleton Press, 2000).

[46] Tebbutt, 'Three Roman Bloomery Furnaces', 47–56.

[47] I.D. Margary, *Roman Roads in Britain*, 3rd edn (London: John Baker, 1973), 520.

[48] P. Brandon, *The Kent and Sussex Weald* (Chichester: Phillimore, 2003), 39–42; Cleere and Crossley, *The Iron Industry of the Weald*, 57–60.

built by AD 293 at Pevensey (whose Latin name, *Anderitum*, means 'great ford'),[49] nor how much Roman infrastructure survived the transition to a Saxon *Andredesleage*, 'the wood of Andred'. The latter, *Andred*, now appears very early as the Saxon name for Pevensey, and by the early eleventh century a grant of land by King Cnut to Archbishop Aelfstan refers to 'the famous wood *Andredesweald*'. The place-name *weald* has been re-examined by Della Hooke, who notes the word as a Saxon-Kentish variant meaning wood-pasture, open pasture and heath. The name 'Weald' itself was recorded in a Kentish charter of 939.[50]

We do know, however, that the Romano-British economy was breaking down from the middle of the fourth century, with towns and some villas showing signs of severe dilapidation, and that during the early fifth century the final withdrawal of Roman forces was completed. The ensuing century is mostly lost in some turmoil, as Romano-British leaders struggled to continue a Romanised lifestyle while lacking the necessary resources. We certainly do not know the extent to which the Romano-British population of the Ashdown area was diminished or was replaced by early Saxon settlers, in part or whole. There is no evidence for either a wholesale extirpation of the Romano-British from the Weald or for them mingling with newer arrivals. And, in landscape terms, we do not know either whether any Wealden Romano-British fields continued to be worked by the new arrivals, as seems to have been the case elsewhere in Britain.[51] Instead, we might assume that this marginal area reverted to an unknown extent during the Saxon period to woodland, scrub and heath.

One key problem is that Saxon archaeological material in the central Weald is at present almost absent, and the remains of the extensive Romano-British iron industry were seemingly deserted. As is well known, Saxon buildings were primarily wooden and their pottery was friable and stood little chance of leaving traces in these acidic soils. As a result, our knowledge is always going to be conjectural to some extent. But the Roman economy was certainly interrupted and pressure on marginal lands was released, so that the need to colonise the more demanding soils of Ashdown was lessened.

The arrival of the Saxons in Sussex is generally traced archaeologically to the mid-fifth century, perhaps a decade or so later than in Kent.[52] In the *Anglo-Saxon Chronicle* the story begins in 477 with Aelle and his sons, who landed at *Cymenesora* and 'killed many Britons and drove some into flight into the wood which is called *Andredeslea*'. Here the Weald is being mentioned as a refuge. Later came a battle at the unknown

---

49 R. Coates, 'Anderida: not the Roman name of Pevensey', *SAC*, 129 (1991), 250–51. And for the Weald as a place-name, *idem*, *Preparatory to a Dictionary of Sussex Place-names* (2017), <https://core.ac.uk/reader/323894005> [accessed 12 September 2021]. The British form was *Anderitu*, Latin *Anderitum*, Saxon *Andred*.

50 Brandon, *The Kent and Sussex Weald*, 51; D. Hooke, 'Old English *wald*, weald in Place-names', *Landscape History*, 34 (2013), 33–49.

51 A. Reynolds, *Later Anglo-Saxon England: Life and Landscape* (Stroud: Tempus, 1999), 22; H. Hamerow, *Rural Settlements and Society in Anglo-Saxon England* (Oxford: Oxford University Press, 2012), 144–7.

52 M.G. Welch, *Early Anglo-Saxon Sussex*, BAR British Series 112 (Oxford, 1983), 259; Millett, 'Roman Kent', 183.

Mearcredesburna in 485 and then Aelle's massacre of the defenders at *Anderitum*, here now called *Andredescaestar*, in 491. If these events have any historical validity – and the dates have been called into question – it is suggested that this was actually a rebellion by Saxon mercenaries, already ensconced in Sussex, against their sub-Roman paymasters. But it is generally accepted that a nascent South Saxon kingdom, or kingdoms with rival kings, was established in the downland area and in the lower river valleys during the sixth century. But none of these early events are known to have impacted directly on the Ashdown area and indeed, for the early Saxon period (*c.* 410–*c.* 650) we have no direct information for the area whatsoever at present.[53]

One type of evidence does help, however, in reconstructing something of the Ashdown area at this time. Place-names are extremely useful in deciphering more about the process of colonisation during the Saxon centuries. But we must enter the normal caveats, since this work is a minefield for the unwary, and we owe our knowledge of these important clues to the trained linguists. Names change through time and modern names may be subtly or radically different from those bestowed by the Saxons. Many occur in written form only centuries after their supposed formation. The original meanings of many and the chronology within which they are found are still disputed by specialists. With these provisos we can tentatively press on, in the knowledge that place-names can reveal something of the history of the central Weald.

The earliest Wealden settlements seem to be indicated by place-name elements ending in *–ham*, *-ingham* or *-ingas*, elements now considered to be used only in the sixth and seventh centuries and being the earliest place-name elements located beyond the initial Saxon pagan burials outside the Weald. These place-name elements were concentrated on the Wealden margins, with only six *-ham* names and only one – Fletching, 'the people of *Flecci*' – of the earliest *-ingas* names being well within the Weald. In an influential analysis of these early names Dodgson has also included place-names adjacent to the Ashdown area, such as Ardingly, Tillinghurst (Ardingly), Bevingford and Etchingwood (Buxted), Bletchingley (Rotherfield), Chiddingly (West Hoathly) and Ruttingham (Fletching). Additionally, we have the place-name element *-hamm*, relating either to the hamlet of Blackham or to Blackham Court (Withyham), an early Saxon ending associated with an enclosure or a shelf of ground – here overlooking a headwater of the Medway.[54]

Several other place-name elements denoting colonisation in the eighth and ninth centuries include *-tun*, *-ingtun*, other combinations of *-ing*, as found in Dallingridge (Forest Row) or Etchingwood (Buxted), and later *-ing* elements themselves. Such elements did not, however, offer clues to the landscapes or economies of those who settled in such places. Furthermore, these names are relatively rare in the Weald.

---

[53] As seen in the distribution maps and text in Welch, *Early Anglo-Saxon Sussex*, and in Drewett *et al.*, *The South-East to AD 1000*, 255.

[54] J.M. Dodgson, 'The Significance of the Distribution of the English place-name in *-ingas*, *-inga-* in South-east England', *Medieval Archaeology*, 10 (1966), 1–29; *idem*, 'Place-names from *hām*, distinguished from *hamm* names, in relation to the settlement of Kent, Surrey and Sussex', in P. Clemoes (ed.), *Anglo-Saxon England 2* (Cambridge: Cambridge University Press, 1973), 1–45.

However, the area's wooded nature was manifest in a variety of early place-names, such as *hyrst* (as in Buckhurst, *Bocherst* 1199, 'Beech-wood hill'), *-grafa* and *-wudu*. Later place-name elements such as *-leah* reflected husbandry clearings and natural openings, particularly for pastoralism. And we have *-falod* (fold, small enclosure or stock pen) and *-wic* (a farm, often a dairy farm). These latter two elements were more widespread than *-leah* names in the Weald as a whole, although *-leah* did occur on the High Weald; *-wic* was found throughout, but *-falod* was almost wholly confined to the western Weald. We also have Fairlight (near East Grinstead and just north of the Forest) which in earlier versions is *Ferlaga* (1086) or medieval *Farnlehe*, 'bracken clearing'.[55] There is also *-worth*, a private enclosure, as in the early forms of Pippingford, *Pippingworth* (1352), 'Pippa's enclosure', or Cotchford, *Cocheworth* (1265), possibly 'enclosure by the thicket'. Another element, *-feld*, as in Sheffield, Lindfield or Rotherfield, may connote a secondary phase of clearing for open land by groups already settled in the area.[56]

There are more: another place-name element taking us back to the very early days of Ashdown colonisation is *saengat*, a burnt clearing, as seen in Saint Hill, 'singed or burnt hill' (East Grinstead), and St Ives, *Saynt Tye* (1590), a 'burnt clearance' (Hartfield). The *-teag* element is an enclosure, as in *Branbertie* (Brambletye 1086).[57] The element *stede*, stead, a place or location, is also found around Ashdown, as at East Grinstead, *Estgrenested* (1271), 'green place', Walstead (Lindfield), *Walcanstede* (765), 'Walca's place' and Horsted Keynes, *Horsted* (1086), 'a place where horses are kept'. The latter's second element derives from William de Cahaignes, tenant from the time of Domesday Book. One place-name possibly suggests a conflict: north of Splaynes Green, Fletching, lies Lower Flitteridge Wood and Flitteridge Farm, from *(ge)flit*, 'strife, dispute' and *hrycg*, 'ridge', so perhaps named after a ridgetop dispute between two neighbouring swine-pastures?[58]

What of the name Ashdown itself? There seem to be two possibilities: that the element *-aesc*, ash tree, is involved, plus *-dun*, hill; or that the name is derived from a personal name, *Aesc* or *Aesca*. Mawer and Stenton favour the second of these, and refer to M.A. Lower in noting that ash trees are uncommon here. They certainly favour more calcareous soils, generally avoiding acidic and infertile soils. But they are nevertheless widely distributed, so the first explanation should not be completely ruled out, especially as ash leaves are known from elsewhere to have been used shredded as browse for deer, possibly from pollarded trees.[59] Together with oak, the ash tree might also better withstand the pressures of grazing animals, and where older woodland has survived in Ashdown's ghyll bottoms it is characterised by oak, ash, field maple and

55 A. Mawer and F.M. Stenton, *The Place-names of Sussex*, 2 (Cambridge: Cambridge University Press, 1930), 332.

56 P. Brandon, 'The South Saxon *Andredesweald*', in P. Brandon (ed.), *The South Saxons* (Chichester: Phillimore, 1978), 152.

57 Mawer and Stenton, *Sussex*, 2, 328; Brandon, 'The South Saxon *Andredesweald*', 148.

58 D. Hooke, *The Landscape of Anglo-Saxon England* (London: Leicester University Press, 1998), 79; Mawer and Stenton, *Sussex*, 2, 346.

59 O. Rackham, *The Ash Tree* (Toller Fratrum: Little Toller Books, 2014), 36–8.

some birch. Charcoal from Roman ironworkings, as we have seen, also included ash.[60] Dodgson certainly gives the place-name Ashurst (West Grinstead) as *æscett*, 'Ash wood or Ash copse', with *æsc* as the element here, as also Ashurstwood, Forest Row.[61] In Kent we have Ashdown, again given as *æsc den*, 'Ash pasture', a pasture belonging to Lenham, Teynham and Ospringe.[62]

Richard Coates suggests that

> *Ashdown* raises a dilemma. There is no formal reason to prefer the personal name explanation or the tree explanation. The existence of the former Ashdown in Berkshire makes the personal name explanation less likely – it would be an oddly coincidental duplication. We have to bear in mind that OE *æsc* also meant 'spear', but whether that is likely to appear in the genitive case in a name like this is a moot point, and what it might denote in the landscape is another. We can obviously rule out the other sense, 'rowing-boat, war-boat', on hills. OE *æscen*, the adjective corresponding to the tree-word, appears with *dūn* 'down, hill' in Ashdon (Essex) and Ashendon (Bucks). Of recent scholars, Margaret Gelling inclines to the tree explanation, and A.D. Mills and Victor Watts go so far as to derive the Sussex name from the adjective on the basis of two thirteenth-century forms which actually have medial *en*. The more forms with *-en* appear in the Sussex documentation, the safer this explanation becomes.[63]

So the name persists as something of an enigma, with experts uncertain about its meanings and derivation. We might note that early medieval forms of the name are variable: *Hessedon* (about 1200), *Essendon* (1207), *Ashendon* (1234), *Assesdune* (c. 1275), *Essedon* (1284) and *Asshedo(u)n* in the high Middle Ages.[64]

Gulley suggested, based on the lack of the earliest place-name elements in the central Weald, that permanent agricultural colonisation proceeded inwards from its margins, perhaps along early transhumance routes, from the more fertile scarpfoot areas of the North and South Downs, where prehistoric, Romano-British and the earliest Saxon settlements and centres had been established.[65] Such a diffusionist concept is not accepted uncritically, however, since there are actually some early place-names at the very centre of the Weald and it may be that we should rather see a gradual infilling and regulation of the Wealden commons in the later Saxon centuries, hence the need for charters to specify boundaries of holdings.

60 Rich *et al.*, *Flora of Ashdown Forest*, 5; E. Straker and B.H. Lucas, 'A Romano-British Bloomery in East Sussex', *SAC*, 79 (1938), 224–32.

61 Rackham, *Ancient Woodland*, 174, 210; J.M. Dodgson, 'Place-names in Sussex: the material for a new look', in P. Brandon (ed.), *The South Saxons* (Chichester: Phillimore, 1978), 67–8; D. Hooke, '"Wealdbæra and Swina Mæst": Wood Pasture in Early Medieval England', in S. Turner and B. Silvester (eds), *Life in Medieval Landscapes: People and Places in the Middle Ages (Papers in memory of H.S.A. Fox)* (Oxford: Windgather Press, 2012), 32–49.

62 A. Everitt, *Continuity and Colonization: the Evolution of Kentish Settlement* (Leicester: Leicester University Press, 1986), 327.

63 Richard Coates, personal communication, April 2015.

64 Coates, *Preparatory to a Dictionary of Sussex Place-names*.

65 Gulley, 'Wealden Landscape', 445–6.

As yet, we do not know how the territorial organisation developed, but we do know that at some point before the Norman conquest the area of modern Sussex was divided into formalised units later referred to as hundreds and rapes (the latter being territories where a leader exerted authority through juridical or military force), but it is uncertain where their boundaries lay or how many there were at this time. However, these rapal divisions would have had centres in the southern part of Sussex with outliers in the Weald.

SAXON ECONOMIES AND ECOLOGIES

The light woodland with open heathland on the higher land was possibly indicated by the admittedly later Hodore Farm (Hartfield, 1274 *Hodore*, 'heathy bank'), or Hoadley's Farm, north of Crowborough (1287 *Hodleg* 'heath').[66] With the seeming collapse of the iron industry (although this is not yet proven; see below), the most likely use of the area would now have been a reversion to deep ancient traditions of the grazing of animals and the exploitation of timber. The area would almost certainly have been grazed communally with cattle and pigs from the constituent settlements of the rapes, with little need to define territories very precisely. In time this would become necessary, as populations and animals grew in number, culminating by the later Saxon period in swine pastures appurtenant to individual manors (or estates, in their Saxon terminology) and linked by droveways.[67]

The droving routes around and across the Forest have not so far been definitively traced but the ancient routeways, prior to any later turnpiking, can be reconstructed (Figure 3.6). It is incontrovertible that the Ashdown area was used for swine pastures and other resources by the established centres in Kent, Sussex and Surrey.[68] The swine pastures certainly left their names all around the edges of what was to become Ashdown Forest. The place-name element *-denn* is commonly used to denote such pastures, although it is more associated with the Kentish Weald, rather than Sussex, where the equivalent term seems to be *-falod* (fold), at least in West Sussex.[69]

Nearer to Ashdown, Danehill was *denne* (1279), a swine pasture appurtenant to Selmeston and Tarring Neville, near the downland. In Withyham we have Birchden, near Groombridge (*Bircheden* (1203) 'swine pasture overgrown with Birches'), and in

66 Mawer and Stenton, *Sussex*, 2, 367, 372; Dodgson, 'Place-names in Sussex', 67.

67 For a general view of Saxon farming see D. Banham and R. Faith, *Anglo-Saxon Farms and Farming* (Oxford: Oxford University Press, 2014); for ancient commoning practices (though not Ashdown) see S. Oosthuizen, *Tradition and Transformation in Anglo-Saxon England* (New York: Bloomsbury Academic, 2013).

68 D. Chatwin and M. Gardiner, 'Rethinking the Early Medieval Settlement of Woodlands: Evidence from the Western Sussex Weald', *Landscape History*, 27 (2005), 31–49. For a reconstruction of the possible routeways into the High Weald see also <http://www.highweald.org/images/droving_map_highweald_rc.jpg> [accessed 5 January 2020.] I am very grateful to Dr Andy Margetts for his help with Figure 3.6.

69 P. Brandon, *The Sussex Landscape* (London: Hodder, 1974), 72; Brandon, *The Kent and Sussex Weald*, 45.

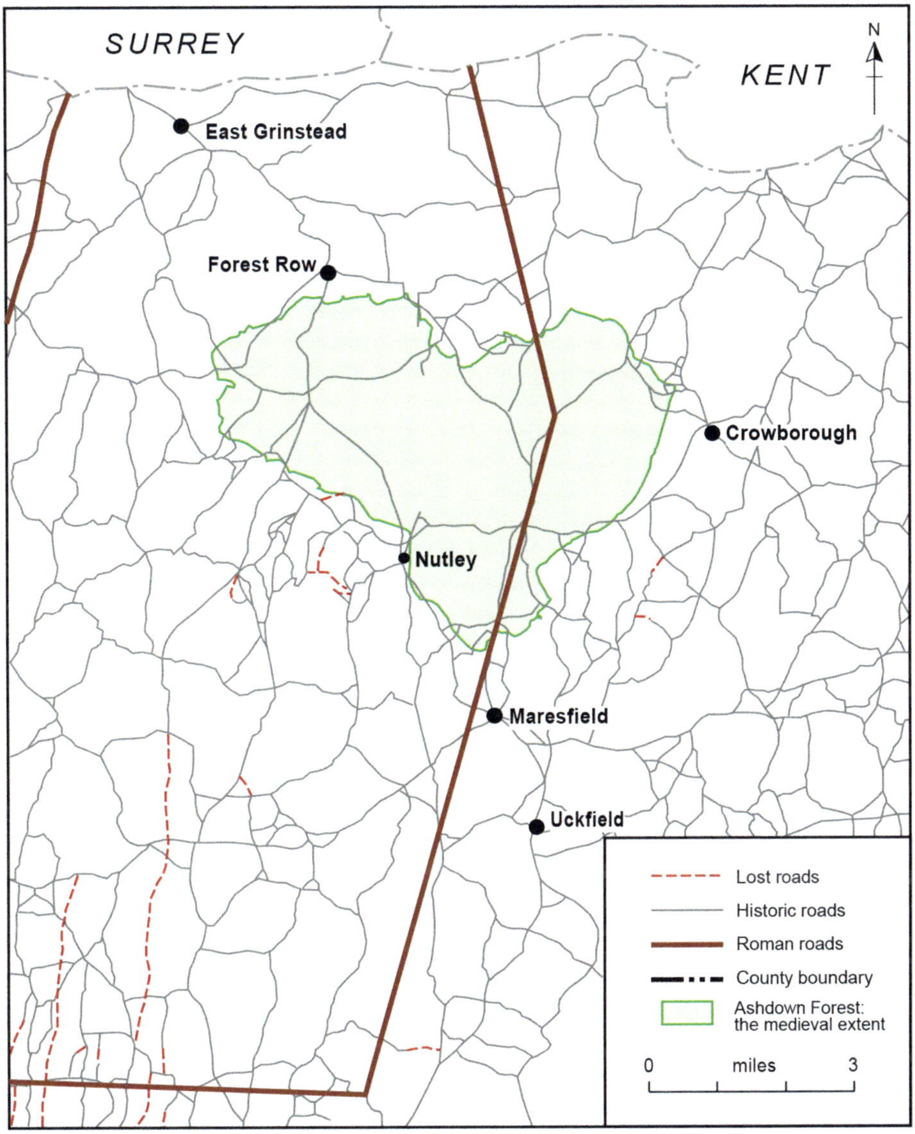

Figure 3.6. Ancient routeways and droving routes (source: based on original research by Dr Andrew Margetts).

East Grinstead we have Hackenden (1300 *Haghindenne*, 'Hacca's swine pasture') and Hazleden (1086 *Halseeldene* [sic] 'swine pasture where hazels abounded').[70] The western edge of the Ashdown area seems to be a dividing line between eastern groups using the element *-denn* for their swine pastures and those to the west using *-falod*.

70 Mawer and Stenton, *Sussex*, 2, 332, 335, 370.

The routes by which the *-denns* were reached by the *drofmen* have been plotted meticulously by Witney for Kent, by Brandon and Margetts for parts of Sussex and by Blair for Surrey.[71] The old Roman ways may have still, in part at least, been usable; indeed, it is likely that the droveways and connectivities had been used for centuries before the Saxon arrival, a theme explored for the Weald by Martin Bell.[72] Small hamlets and isolated farms emerged at this time near these droveways, and in time churches, followed by markets and built-up areas, developed: Ditchling, Shipley and West Grinstead, for example, lie on such routeways, possibly offering intermediate sustenance for the drovers and their swine heading to the main pastures. It should therefore be noted that these outlying smaller settlements are earlier in date than the later villages, a common finding for the Weald. But the patterns of roads, clearings and settlement were fundamentally shaped by the needs of transhumance in the pastoral Saxon economy.

The fattening of swine during autumn on beech mast and acorns was by far the most important economic activity around Ashdown during the Saxon centuries, although the preferred pastures lay among the denser oak and beech woods of the Low Weald, also closer to the manorial centres. But we have Buxted, *Boxted* (1199), 'Place of beech trees'. The true extent of *-denn* activities is unfortunately lost to us, as is the true extent of woodland. Questions remain: why would swine, the 'walking wealth', be driven at the equinox in mid-September, distances of ten miles or more, for six weeks' pannage, only to be driven back again on the Feast of St Martin (Martinmas), thereby losing much of the benefit of the pannage fattening? Why were so many in such hostile and remote environments as the central Weald? And when did the pastures grazed in common change to belong more specifically to a named parent holding? In addition, it now seems possible that the dominance of swine in animal numbers around Ashdown may have been overemphasised, as cattle bones from this period account for as much as 45 per cent of the total bone assemblages in the High Weald, and cattle droving routes may have been as significant as those for swine.

Details of swine totals are lacking, although it seems likely that the parent manors may have frequently had very high numbers. The domesticated swine at this time were actually very similar in appearance to their wild cousins (*Sus scrofa*), with whom they probably intermixed, and who were long-haired, long-legged, razor-backed and slow maturing. Wild pigs survived in some areas of England until the sixteenth century and the place-name element *-denn* could mean a wild beast lair as well as a swine pasture.[73] The importance of swine pannage is hard to overstate: from the manorial

71 Brandon, *Sussex Landscape*, 73; K.P. Witney, *The Jutish Forest: a Study of the Weald of Kent from 450 to 1380 AD* (London: Athlone Press, 1976); A. Margetts, *The Wandering Herd: the Medieval Cattle Economy of South-East England c.450–1450* (Oxford: Oxbow Books, 2021), 81–107; J. Blair, *Early Medieval Surrey: Landholding, Church and Settlement* (Gloucester: Sutton and Surrey Archaeological Society, 1991).

72 M. Bell, *Making One's Way in the World: the Footprints and Trackways of Prehistoric People* (Oxford: Oxbow Books, 2020).

73 Gulley, 'Wealden Landscape', 446–65. With no natural predator, by 2013 there was a breeding population possibly numbering 1,000 in Kent and Sussex after escapes from domestic

demesnes but also from the smaller pig producers, later evidence shows that payments for pannage could raise large sums. Many manors required payment in kind of a pig in some ratio of the total, sometimes 1 in 7, or perhaps 1 in 10, with smaller producers as a whole sending more swine than the larger institutions.[74]

Now for the first time we encounter written evidence in the form of the Saxon charters that demonstrate grants of Wealden outliers to southern centres. A charter of *c.* 765–71 by King Ealdwulf to Hunlaf of 16 hides to found a monastery (sometimes rendered as a minster), possibly at Stanmer, or more probably at South Malling, mentions locations near Ashdown, at Burleigh, Chiddingly, Buxhalls, Paxhill and Lindfield, with the second, third and fourth of these as swine pastures for Stanmer. It may be that the mentions of the *pastus porcorum* have been added at a later date, but the charter remains a significant indication of the Saxon use of the Ashdown area.[75] Burleigh is almost certainly at Turners Hill, Margary seeing the Roman road go close to Burleigh House farm. Chiddingly (*Citangaleahge*, 'the clearing of Citta's group') is Chiddingly Wood in West Hoathly; Buxhalls (*bocgeselle* 765, the early *gesell* element meaning a shelter for animals or herdsman's hut 'by the beech tree') is in Lindfield and in the Hundred of Burleigh Arches, as is Paxhill. It also included Philpots (older *baere leah*, a swine-pasture clearing) now in Lindfield.[76] The charter also mentions the unknown '*wæsteat*' and the clearing of the *Halfocungas*, which were said to be to the east of Philpots – that is, towards Ashdown.[77] And Bletchingley Farm (*Blec(c) ingeleghe* 1279), east of Rotherfield, may well commemorate a similar arrangement of an outlying swine-pasture for either East Blatchington or Blechingley in Surrey. In the will of Alderman Aethelwold in 946 or 947 he bequeathed land at *Æscesdune*, a holding presumably within Ashdown.[78]

There is at present no evidence for any widespread continuation of the Romano-British ironworking here. We have to wait until the Middle Saxon period for any signs of the industry, and then, so far, from just one known site, at Millbrook, near to the Millbrook East car park off the present A22 road. It was a simple smelting furnace, reused for a small number of seasons during the ninth century (dated by radiocarbon

captivity and there are reported sightings from the Ashdown Forest area (BBC News 22 February 2013; *The Times*, 27 July 2015).

74 S. Harvey, 'Domesday England', in H.E. Hallam (ed.), *The Agrarian History of England and Wales II 1042–1350* (Cambridge: Cambridge University Press, 1988), 127–31.

75 E.E. Barker, 'Sussex Anglo-Saxon Charters, I', *SAC*, 86 (1947), 42–101; S.E. Kelly (ed.), *Anglo Saxon Charters: Charters of Selsey* (Oxford: Oxford University Press, 1998), 6. The charter now exists only as a thirteenth-century copy.

76 Mawer and Stenton, *Sussex*, 2, 272, 365–6; Margary, *Roman Ways in the Weald*, 106–7; R. Coates, 'Corrections and Additions to Sussex Material in Later Volumes Published by the English Place-Name Society' (unpub. paper, Univ. of Sussex, 1986), 12.

77 Barker, 'Sussex Anglo-Saxon charters I', 87.

78 E.E. Barker, 'Sussex Anglo-Saxon charters III', *SAC*, 88 (1949), 58–9; Mawer and Stenton, *Sussex*, 2, 377, give East Blatchington or the Surrey town; E.E. Barker, 'Anglo-Saxon charters Part II', *SAC*, 87 (1948), 115 offers only East Blatchington, as does Brandon, 'The South Saxon *Andredesweald*', 146. Blair (*Early Medieval Surrey*) makes no mention of an outlying Sussex pasture for Blechingley.

and archaeomagnetic samples to AD 800–835) and far less sophisticated than its Roman predecessors. There were two hearths, postholes and a slag dump. The charcoal was made from mature oak. There were sherds from a pottery cooking pot and a sandstone hone found on the site. Its technology closely resembled that of earlier Germanic ironmakers in the Netherlands and northern Europe outside the Roman empire.[79] It may well be, therefore, that there was a considerable hiatus in ironmaking with the departure of the Romans, and that the Millbrook site represents a quite different tradition of working. As such, it is an important site in British technological history. Another bloomery furnace to the east of Ashdown at Long Gill, Mayfield, has very broad radiocarbon dating from AD 315 to 785, but lacks other associated datable material. We then have to wait again until the late eleventh century for a mention of an ironworks, *una ferraria*, this time documented in the Domesday Book as being near East Grinstead. It is possible that this was at Lavertye, an outlier of Ditchling.[80]

THE LATER SAXON CENTURIES

By the eighth century it seems that some detached but appurtenant locations were sufficiently acknowledged as to have their own names. It would not be too difficult also to envisage that the swineherds would have cleared and fenced some land for temporary cultivation and grazing. On a droving route from the scarpfoot region into the High Weald, the vaccary at Trubwyk, a Wealden outlier of the Manor of Plumpton or Plumpton Boscage, had an open cattle shed and a hedged enclosure to protect cattle from wolves or thieves. This lay within the bounds of the present Bolnore village, offering grazing land and possible access to summer agistment on the open land of Ashdown.[81] Such protection around Ashdown is evidenced by the place-name *wulfpyte*, a wolf-pit, now Woolpack Farm, near Sheffield Forest. And in due course the swineherds may have brought families and erected more permanent dwellings within woodland clearings. By this time, then, a settlement pattern and rural economy was beginning to emerge. By the ninth century it would be possible, perhaps, to imagine a lightly populated but fully agrarian landscape on the inhospitable soils of Ashdown. Vaccaries or beccaries have not received much archaeological attention, but the vaccary at Trubwyk is also complemented by those nearer the edge of Ashdown, such as Faulkner's Farm (Hartfield) and Chelwood Vachery, with its distinctively oval-shaped enclosure surrounded by the Forest waste.[82]

Nevertheless, the primitive state remaining over much of the Weald during the eighth century is suggested in the *Anglo-Saxon Chronicle* account of King Sigeberht

---

79 C.F. Tebbutt, 'A Middle-Saxon iron smelting site at Millbrook, Ashdown Forest, Sussex', *SAC*, 120 (1982), 19–35.

80 Hodgkinson, *The Wealden Iron Industry*, 36–7.

81 A. Margetts, 'The Hayworth: A Lowland Vaccary Site in South-East England', *Medieval Archaeology*, 61 (2017), 117–48.

82 C.F. Tebbutt, 'A Deserted Medieval Farm Settlement at Faulkner's Farm, Hartfield', *SAC*, 119 (1981), 107–16.

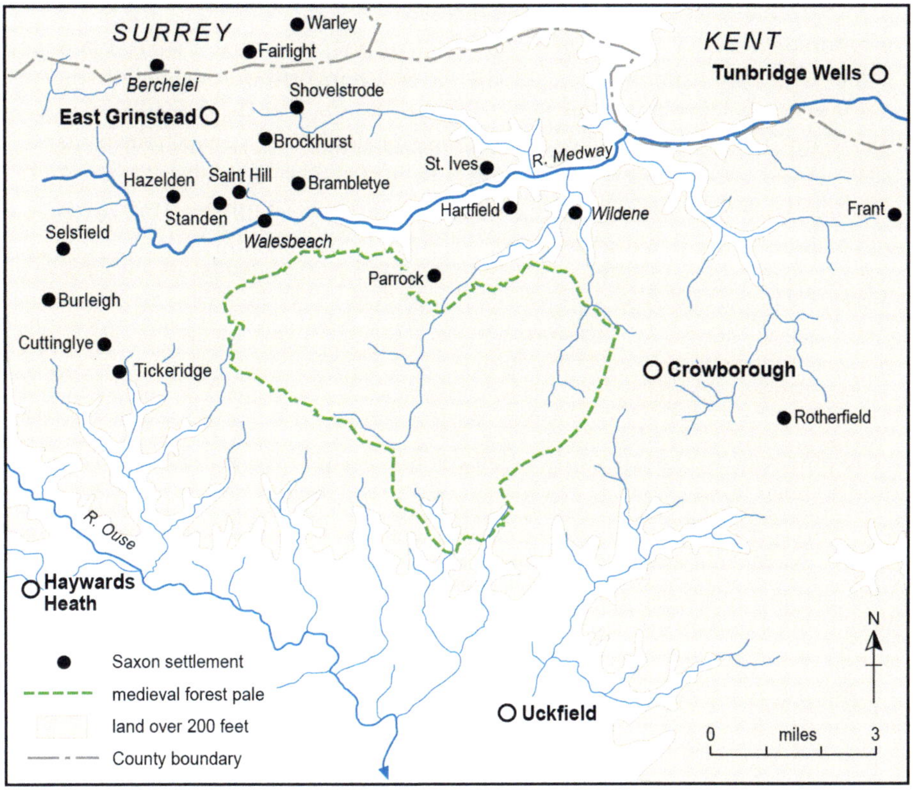

Figure 3.7. Saxon settlements in the upper Medway Valley (source: author, based on Brandon, *The South Saxons*, 144).

of Wessex, who succeeded his kinsman Cuthred in 756 but whose short and bloody reign was ended abruptly by Cynewulf (his successor) and the councillors of Wessex, the Witan, who 'drove him away into the Weald [where] he lived until a herdsman stabbed him at the stream of Privett'. The commentator Leonce de Lavergne translates the herdsman as 'un porcher'. Elsewhere Cuthred is described as being driven 'into trackless parts of the place called the Weald', or *Andred*, and murdered by a swineherd.[83]

Certainly, the Saxon era is crucial to any understanding of the settlement pattern and economy of the Ashdown area (Figure 3.7). Brandon offers a vivid picture of the early Saxon colonists in what he calls the 'frontier region' of the upper Medway valley around the fringes of Ashdown:

83 Léonce de Lavergne, *The Rural Economy of England, Scotland and Ireland* (trans. from the French) (Edinburgh: W. Blackwood and Sons, 1855), 203; Prosopography of Anglo-Saxon England, <https://pase.ac.uk/jsp/pdb?dosp=VIEW_RECORDS&st=PERSON_NAME&value=1776&level=1&lbl=Sigeberht> [accessed 12 September 2021]. Privett's river was in eastern Hampshire, near Petersfield, where Sigeberht ruled for a short time after being deposed.

> Here we shall never be far from civilisation, for along the banks of this now lovely basin settlers had scattered their habitations uninterruptedly on the natural 'shelves' midway up the valley slopes ... *Wildene* (probably Withyham), *Apedroc* (Parrock), Brambletye, Hartfield, Standen ... If we now pass on into West Hoathly and its contiguous parishes there are a half dozen place-names of such great antiquity as to imply that the Saxon was amongst the first pioneers up in these hills, stringing out his fences and building some form of house and barn in the woods. The unfolding drama of Saxon forest clearance is illustrated by such names as Selsfield (open land marked by a hall)... a number appear to have been attached to farms and freshly-turned furrows rising into the view of an observing drover or shepherd, doubtless casting an uneasy eye on these ominous harbingers of a new way of life. ... rough surroundings, the loneliness, the dull thud of an axe denoting the building of some sort of a house over the hill.[84]

One aspect of the later Saxon settlement process was the founding of churches, although much remains obscure in the absence of documents such as those for Kent, which greatly amplify the numbers of churches recorded in Domesday Book. The church at Rotherfield was probably founded in the eighth century, and those at Worth and Horsted Keynes show Saxon details, with the imposing Worth initially built using Tunbridge Wells Sandstone. Near-contemporary sources also note churches by the end of the eleventh century at Ardingly, Balcombe, East Grinstead and West Hoathly.[85]

The extent to which churches around Ashdown were endowed with land to supply their priests with resources is at present conjectural, as is their spatial control. While many parishes seem to have been created in southern Sussex by *c.* 1100, the situation in the Weald is less well understood. Populations here may instead have still been served from minsters outside the Weald, but the fact of there being so many Wealden churches by the eleventh century does at least indicate a population sufficiently large, although perhaps scattered, as to merit their building. It certainly was not the case that churches in this area were built within existing settlements: on the contrary, we might envisage them initially standing somewhat isolated, with buildings accruing around them only later to form villages.

THE EVIDENCE OF DOMESDAY BOOK

The achievements of the Saxon settlers around Ashdown can be caught, to some extent, in the extraordinary source that is Domesday Book. By the late eleventh century the Weald is believed to have had the largest concentration of woodland in England, perhaps covering about 70 per cent of its area.[86] But the full picture cannot be readily deduced from Domesday because of the manner in which the document was compiled, being less a gazetteer of settlements and more a taxation document for William I

---

84 Brandon, 'The South Saxon Andredesweald', 155–6.

85 G. Mayhew, *The Monks of St Pancras: Lewes Priory, England's Premier Cluniac Monastery and its Dependencies 1076–1537* (Lewes: Lewes History Press, 2014), 25.

86 Rackham, *Ancient Woodland*, 124–6.

and his Norman overlords. The manor (*manerium*) was the essential building block for Domesday, but the scribes did not differentiate between the principal settlements, the demesne headquarters and their outlying component Wealden pastures, a pattern replicated for Kent and Surrey. Thus, for example, the place-name elements mentioned in Domesday Book incorporating *-tun* (farm, or even nucleated village) are conspicuously absent from the Weald, but instead clustered around its edges in Kent, Surrey and Sussex.[87]

What light, then, does Domesday shed on the Ashdown area? First, we might notice the fact that the boundaries between Sussex, Kent and Surrey appear to have been largely fixed by this date and follow well-defined natural features: either ridges of higher land (as between Sussex and Surrey) or river valleys, such as the Medway (between Sussex and Kent). Ashdown itself might be expected to have provided the boundary with Kent or Surrey, given its location astride the watershed of higher ground. But somehow the South Saxons had pushed northwards beyond Ashdown and established a shire boundary, perhaps as early as the mid-eighth century but possibly in the ninth or even tenth. Some of the outlying swine pastures, the dens (or denns) belonging to manors on the Kent side of the boundary were actually cut off, thus indicating that the linkages pre-dated the county boundary.[88]

As noted above, the origin of the rapes has long been a subject of controversy and perhaps should not detain us too long, except to note that some form of territorial division, running north–south, pre-dated the Norman conquest and was subsequently modified after 1066, including during the 20 years before Domesday was completed. John Norden commented that 'Every of the Sussex rapes hath its castle, forest and haven, or in tyme past had'.[89] Ashdown was at the northern end of the Rape of Pevensey. In the rape, as in Sussex more generally, Saxon lords were now to be replaced by Normans, with military considerations uppermost: protecting the sea routes to Normandy and defending against Danish attacks or Harold's sons. All of the rape of Pevensey, except for that held by the church or by the king himself, was made over speedily to William's half-brother Robert, Count of Mortain in southern Normandy and significant enough to be recorded on the Bayeux Tapestry, who had contributed many ships to the invasion.[90] The rape included most of the Ashdown area except for West Hoathly and Horsted Keynes, which were in the neighbouring rape of Lewes. Bishop Odo of Bayeux, also William's half-brother and earl of Kent, had acquired the manor of Rotherfield from Earl Godwin but had in turn forfeited it to the king. The latter's possessions across the country at this time surpassed those of any future monarch,

87 V. Watts, *The Cambridge Dictionary of English Place-names* (Cambridge: Cambridge University Press, 2004), lxii.

88 M. Gardiner, 'Economy and landscape change in post-Roman and early medieval Sussex, 450–1175', in D. Rudling (ed.), *The Archaeology of Sussex to AD 2000* (Kings Lynn: Heritage Marketing & Publications, 2003), 155–6; Chatwin and Gardiner, 'Rethinking the early Medieval Settlement of Woodlands'; Harvey, 'Domesday England', 118.

89 Cited in F. Kitchen, 'Cosmo-choro-poly-grapher: an analytical account of the life of John Norden 1547?–1625' (Univ. of Sussex unpub. DPhil, 1992), 97.

90 J.A. Green, *The Aristocracy of Norman England* (Cambridge: Cambridge University Press, 1997), 27–9.

although they were actually rather more limited in Sussex. Instead, land was held by his trusted supporters, who in turn generally sub-leased their vast lands among their own followers. Odo, for example, had over 150 sub-tenants. The most valuable estates, nearer the coast in the case of Pevensey rape, between Lewes and Pevensey, were retained in hand.

Within the rapes were the hundreds. As a unit of administration and taxation these were also well established by this time, with some 60 in Sussex, and with meeting places for the court, to be attended by the freemen of that Hundred. The nominal 100 hides, an early Saxon measure of productivity and for taxation, was actually rather smaller in Sussex than elsewhere. The Ashdown area had its hundreds, although they may have been at a fairly primitive stage: Domesday notes the hundred of *Grenestead* (East Grinstead), together with the long-established Rotherfield and also Hartfield. To the south lay the divided hundred of Ristone (later Rushmonden), with land in Lewes Rape at Birchgrove and Horsted Keynes but also in the rape of Pevensey at Sheffield and Fletching.

For the most part, the smallest unit recorded in Domesday Book was the manor, about which more detail is given in the following chapter. Some manors were multiple, or 'federative', with specialised detached portions, and were sometimes intersected by rape boundaries, something normally entered into Domesday Book. Although the *caputs* of those recorded in Domesday are located primarily outside the Weald, we can glimpse something of the distribution of settlement in East Grinstead, separately recorded in Domesday somewhat fortuitously because the area was later transferred from Lewes to Pevensey Rape for administrative convenience. East Grinstead Hundred was actually sliced through by the boundary with Lewes Rape, leaving only Brambletye within Pevensey Rape in 1086. There were 36 named locations, including Brambletye, together with 13 others (one un-named) belonging as outlying swine pastures to parent manors in Lewes Rape. They include Brockhurst (*Biochest* in Domesday, badger), Fairlight, Hazelden, Shoverstrode (*Calvrestot* or *Celrestius*, from Old English *scylf-strod*, 'marshy ground by the slope') and Walesbeech (*Waslebie*, derived from a personal name, *Hwæl*, hence '*Hwæl*'s valley'). The town of East Grinstead was a later foundation within that hundred, probably founded by the Aquilas in the early part of the thirteenth century on the site of the hundred meeting-place, possibly where the later Sackville College was built.[91]

The hundred of *Hertevel* (Hartfield), with 31 households, included an un-named holding formerly an outlier of Rodmell, which being in the lands of the de Warennes, was now reallocated to the count of Mortain. The entry for *Hertevel* reads:

> In HARTFIELD Walter [of Ricarville] holds 1 hide from the Count. It answers for as much. Karli of Norton held it as freehold. Land for 3 ploughs. In lordship 1½ ploughs; 6 villagers and 2 smallholders with 1½ ploughs. 1 mill at 4s and 350 eels; meadow, 3 acres; woodland, 5 pigs from pasturage. [Value] before 1066 and now 40s; when acquired 20s.[92]

[91] M. Leppard, *A History of East Grinstead* (Chichester: Phillimore, 2001), 4–5.

[92] Opendomesday.org/book/sussex/12/

The hundreds of Hartfield and Rushmonden were also subdivided between the rapes of Lewes and Pevensey and after 1066 many manors had also been broken up and rearranged. So, for example, holdings belonging to the count of Mortain in Hartfield Hundred had parent holdings outside the rape. Thus *Apedroc* (Parrock, an enclosure) was returned in Lewes. It was said never to have paid tax (meaning that it was taxed with its parent manor outside the rape), had formerly belonged to the wealthy Queen Edith, wife from 1045 of Edward the Confessor, and its woodland returned 40 pigs and 12s.There was also mention of a hall here, formerly belonging to Earl (King) Harold, and now belonging to the count. Hartfield itself and Withyham were returned in Pevensey.[93]

If we can place any reliance on the figures for households, the broad conclusion is that population densities were far lower in the Weald than in the surrounding regions of the South East.[94] But population estimates from Domesday are difficult. The few mentions of ploughs and plough teams for the area do indicate some farming operations, but the evidence is slight. There are no explicit references to the standard Domesday entry for land under the plough, *terra est n carucis*, but there was, for example, half a plough-land at Birchgrove (*Bontegrave*, Bunt's woodland'), but no mention otherwise of plough-teams.[95]

Equally frustrating is the lack of evidence in the entries for Pevensey Rape for what have always been seen as the main Ashdown activities at this time: the pannage of pigs and the supply of woodland products. Woodland was returned in terms of the numbers of swine it could feed, or the renders from the swine fed in that woodland, the usual formula for woodland being *silva de n porcis*. But again this was applied to the parent manors, giving little clue as to the actual woodland cover on Ashdown. But single totals for swine grazing in woodland and also on grass are given for Hartfield, Brambletye, Fletching, Sheffield and Shovelstrode. Thus for Sheffield, *De silva et herbagio xxxii porci*. The un-named holding in East Grinstead had '6 swine from wood and herbage'.[96] Compared with elsewhere in England, it is also surprising that no swineherds are enumerated in the Sussex Domesday folios. Devon, for example, recorded as many as 370.[97]

There is no mention of the name of Ashdown itself in Domesday.[98] A park is mentioned at Rotherfield, but there are otherwise mentions of parks only in western

93 J. Morris, *Domesday Book 2: Sussex* (Chichester: Phillimore, 1976), 21d; and end maps of Hundreds.

94 D. Haselgrove, 'The Domesday Record of Sussex', in P. Brandon (ed.), *The South Saxons* (Chichester: Phillimore, 1978), 217.

95 H.C. Darby, *Domesday England* (Cambridge: Cambridge University Press, 1977), 90; S.H. King, 'Sussex', in H.C. Darby and E.M.J. Campbell (eds), *The Domesday Geography of South-East England* (Cambridge: Cambridge University Press, 1962), 407–82.

96 King, 'Sussex', 445–6.

97 Harvey, 'Domesday England', 62.

98 Of the English forests, only the New Forest is recorded as such in Domesday Book (J.A. Green, 'Forest Laws in England and Normandy in the Twelfth Century', *Historical Research*, 86 (August 2013), 416–31).

Sussex. Likewise, no royal hunting forests are mentioned in Sussex apart from Dallington, but this was strictly speaking a chase belonging to the count of Eu, and becoming known as Dallington Chase. However, Frankish kings were developing royal forests from the seventh century onwards, with hunting laws and royal rights preserved, and pre-Conquest elites on both sides of the Channel held hunting in high regard. It might be possible, therefore, that the Ashdown area, once part of a communal space, was perhaps to some extent used for this purpose by the eleventh century, probably driving deer with dogs, in the Saxon fashion. It did have the appropriate environment: open heaths with some tree cover, and thus we have Hartfield, *Hertevel*, 'open space frequented by harts', but as yet we lack real evidence for Saxon hunting here.

ASHDOWN: PRESENCES AND ABSENCES

For so much of the millennia covered in this chapter there are more absences and questions than presences and answers. The suspicion must be that, after centuries of near-avoidance, a community did emerge during the later Iron Age and Romano-British period that supported ironworking and developed small hamlets and field systems to supply local demand. By the Saxon period there is enough evidence to suggest a three-part transition. First there was the primitive Weald, with resurgent woodland and scrub and with swine, deer and wolves, a regional intercommoned wood-pasture. Secondly, it became a space used more exclusively for transhumance by the Pevensey rapal community. And, thirdly, much became appurtenant to individual manors, as demarcated in the Saxon charters and partly revealed in Domesday Book. Jones suggests that 'An image emerges of the forest as a vibrant, locally self-determined entity, part of the interlinking communal resources within discrete territories reaching back, perhaps to small kingdoms and concepts of folk-right.'[99]

But Ashdown was effectively bypassed by many of the great themes of the Saxon years. We do not, for example, see the conversion, by perhaps the 'long eighth century' (670–840), of the older prehistoric and Roman *blockflueren* fields into the well-known open and strip fields of Midland, or indeed southern Sussex landscapes. Nor is there evidence of *Wandersiedlungen*, the 'Middle Saxon shift', as farms and settlements gradually shifted to new, adjacent sites, either organically or as planned villages, that appeared elsewhere in the South East.[100] Instead there was an increasing density of small hamlets and isolated farms that might eventually give rise to a focused village and church and within new parish units, although so much of this, effectively a fourth transition, had to await the post-Domesday period.

[99] G. Jones, 'A "common of hunting"? Forests, Lordship, and Community before and after the Conquest', in J. Langton and G. Jones (eds), *Forests and Chases of Medieval England and Wales c.1000–c.1500* (Oxford: St John's College Research Centre 2010), 53.

[100] See Hamerow, *Rural Settlements and Society in Anglo-Saxon England*, 144–62; H. Hamerow, 'Settlement Mobility and the "Middle Saxon Shift": Rural Settlements and Settlement Patterns in Anglo-Saxon England', *Anglo-Saxon England*, 20 (1991), 1–17. For a Sussex example of settlement shift see G. Thomas, *The later Anglo-Saxon Settlement at Bishopstone: a Downland Manor in the making*, CBA Research Report 163 (York, 2010).

CHAPTER 4

ASHDOWN EMERGES AND THE LANDSCAPE FILLS UP, 1086–1485

The journey to the margin.[1]

FOLLOWING THE CONQUEST, ASHDOWN Forest truly emerges, with documentation that allows us far better insights into its ownership, use, landscape and administration. We also can intertwine its history with the activities of those who were active on the national stage.

In the last chapter we noted land mentioned as being within *Æscesdune* in the mid-tenth century. If this was Ashdown, it is the earliest mention. But the first undisputed mention is at some point between 1107 and 1115, being the earliest and latest dates of Henry I's issue of writs from Westminster, when he confirmed that monks could continue to use a road across *Essessdone*:

> Westminster
>
> Notification by Henry I to R[obert] son of R[alf] de Hastinges, D[rew] de Pevenesel, and all justiciars, &c., of Sussex: that he confirms to the monks of St. Martin's, Battle, and their men, all their roads throughout his lands, particularly that from Battle to Hastings, the roads round Battle, the crossings over (*ultra*) Winchelsea, and their road over (*super*) Ashdown (*Essessdone*), as in the reigns of William I, William II, and in his own.[2]

This immediately raises some points of interest. First, there was therefore a named entity known as Ashdown or *Essessdone*, at least by *c.* 1100, and that it was sufficiently well-known for it to be specifically mentioned; secondly, an acknowledged route ran across it during the reign of William I, being possibly the 'old road', the Roman road

[1] R. Faith, *The English Peasantry and the Growth of Lordship* (London: Leicester University Press, 1997), 208.

[2] C. Johnson and H.A. Cronne (eds), *Regesta Regum Anglo-Normannorum 2, 1100–1135* (Oxford: Clarendon Press, 1956), no. 1670. For the rights of the men attached to Battle Abbey see E. Searle, *Lordship and Community: Battle Abbey and its Banlieu 1066–1538* (Toronto: Pontifical Institute of Mediaeval Studies, 1974), 44–68. Battle Abbey was dedicated to St Martin; Pevensel was an alternative name for Pevensey.

from London to Lewes, or perhaps one of the subsidiary hollow trackways; and, thirdly, the monks whose interests were being upheld were from Battle Abbey, founded in the 1070s in the rape of Hastings.

THE FOREST OR CHASE OF ASHDOWN

Whether or not the Ashdown area was an elite hunting ground during the Saxon centuries, the development of their later equivalences in England begins with decrees made by William I. The new Norman rulers had at least six hunting forests in their duchy by the year 1000, with evidence for one at Jumièges by 716. But it should be noted that dukes, counts and lesser landholders all held lands referred to as forest, and that these areas were important for woodland products, pannage and grazing rights as well as for hunting, which might actually be protected inside enclosures or parks within the forest. A balance had to be struck between hunting, grazing, woodland products and wood provision, with the role of the forester important as a land manager.[3]

It is usually suggested that chases were forests that had passed out of royal control and were thus no longer subject to forest laws.[4] Instead, the chase had its own administration, although the suite of officers had functions that actually paralleled those of royal officials in their forests. Offences against vert (vegetation cover for the game) and venison were carefully reported through woodmote courts and occasional inquisitions, which were superimposed on the pre-existing manorial or hundred courts, or, in this case, courts assigned to the rape of Pevensey. They did, however, escape the perambulations due under the Charter of the Forests (1217). There were no royal forest eyres held in Sussex after 1212, and no royal forests in the county after that date.[5]

But it is clear that such chases, when held by the king, would be used for hunting in the same fashion as a royal forest. The use of sparrowhawks is recorded in 1287; Edward I's huntsman was recorded using dogs in pursuit of game on Ashdown in 1303 and Edward granted the right to Thomas Peynell to hunt the badger, fox, wild cat and hare with his own hounds in Ashdown.[6] We might note here that Horsted Keynes' Norman parish church is dedicated to Saint Giles, patron saint of woodland, whose symbol is a deer.

---

3 Jones, 'A "common of hunting"?' 59; Dolly Jørgenson, 'The Roots of the English Royal Forest', *Anglo-Norman Studies*, 32 (2010), 114–28.

4 J. Langton, 'Medieval Forests and Chases: another Realm', in J. Langton and G. Jones (eds), *Forests and Chases of Medieval England and Wales c.1000–c.1500* (Oxford: St John's College Research Centre 2010), 22–4.

5 David Crook (personal communication) maintains that Ashdown was never subject to forest law and thus never legally a forest. There were absences from the late twelfth- and early thirteenth-century eyres, bringing amercements for non-attendance from the men of Hastings, Chichester and Arundel rapes, but there is no mention here of Pevensey Rape. For the absence of royal forests from Sussex see M. Bazely, 'The extent of the English Forest in the Thirteenth Century', *Transactions of the Royal Historical Society*, 4 (1921), 140–63.

6 Christian, *Ashdown Forest*, 5.

Ashdown also now became a park, with a boundary or pale to retain deer and to demarcate its sphere of control.[7] Similarly, by 1262 the parks of Bletchingly and Waterdown Forest in Rotherfield were impaled as hunting grounds, but with supplementary income from pannage, pasture and timber sales.[8] The area of the enclosed park at Ashdown was a substantial 13,477 acres, with a perimeter length of around 23 miles. Created within a wave of park formations in the twelfth and thirteenth centuries, it was referred to in 1283 as the 'free chase and warren of Essedon', but on passing to the Duchy of Lancaster in 1372 it became known as Lancaster Great Park.[9]

So, a chase, a warren, a park. The pale certainly existed by 1273, when a custumal noted the services of *claustura* with a constant charge for expensive repairs, offset to some extent by local sales of birch and other wood as firewood. This was followed by a record of repairs to the pale in 1282. Perhaps all we can conclude at present, although more detail is given below, is that the enclosure had taken place by *c.* 1270.

THE GOVERNMENT OF ASHDOWN IN THE RAPE OF PEVENSEY

To begin, we should briefly trace the successive holders of Pevensey Rape and the Forest at its northern extremity after 1066, in order to place Ashdown in its broader context. The rape was governed by a succession of Norman and Angevin landholders. As we have seen, following the Conquest William I granted the rape to Robert, count of Mortain, one of about 200 substantial tenants-in-chief. Excluded from the grant were Rotherfield, the largest manor entirely within the Weald, which stayed with the king; Malling, the archbishop of Canterbury's manor, centred on Malling, near Lewes, and running north-eastwards to Wadhurst and the Kent border; and Alciston, given to the new abbey at Battle (Figure 4.1).

Robert of Mortain died in 1095 and was succeeded by his son William, who, loyal to the duke of Normandy and in attempting to repel the invasion of Normandy by Henry I, was made prisoner at the battle of Tinchebrai in 1106. Imprisoned for life, he thereby forfeited all his estates. Henry then split the Honour of Pevensey (most of Pevensey Rape), with some going to endow the Honour of Leicester, but keeping the Hundred of Eastbourne in hand and bestowing the rest on Gilbert de Laigle, from southern Normandy, grandson of a high-ranking nobleman killed at the Battle of Hastings. The Honour stayed with this family for the next 125 years, notwithstanding hostilities with the crown on occasion, and became known as the Honour of the Eagle, Latin *Aquila*; Norman French *l'aigle*.

The Norman and English components of King John's realm fell apart in 1204. The then Gilbert de Laigle (Gilbert IV), left with the choice between his English and Norman holdings, apparently left for Normandy without leave in October 1203 to face a threat

[7] V. Blandford, *Tracing the Pale of the Ashdown Forest Deer Park* (Chris Butler Archaeological Services, 2012), 24. I am indebted to Vivienne Blandford for an early sight of her report. The following account of the pale owes much to her research.

[8] Gulley, 'Wealden Landscape', 392–3.

[9] P. Brandon, 'The Common Lands and Wastes of Sussex' (Univ. of London unpub. PhD, 1963), 101.

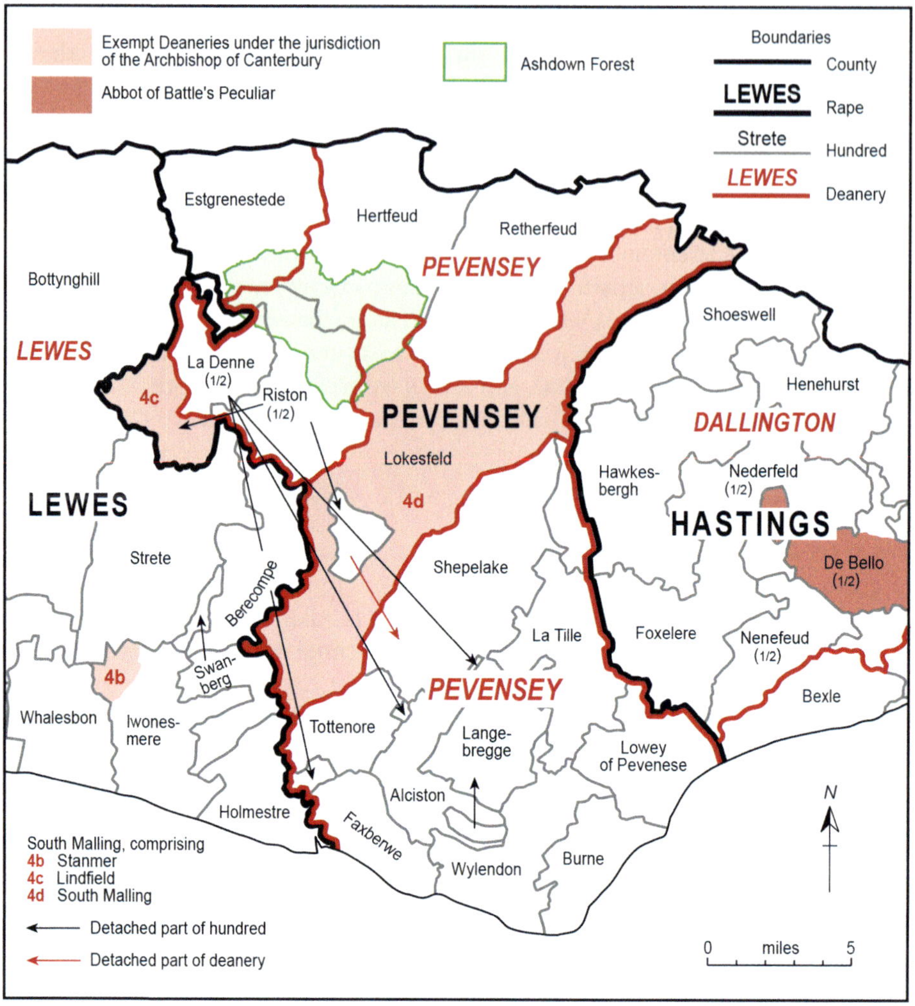

Figure 4.1. The rape of Pevensey *c.* 1278 (source: C. Adams, 'Medieval administration' in K. Leslie and B. Short (eds), *An Historical Atlas of Sussex* (Chichester: Phillimore, 1999), 40 (detail)). The neighbouring rapes of Lewes and Hastings are also shown, and the location of Ashdown Forest is marked. The hundred names are those recorded in the Hundred Rolls of 1278.

to his estates from the French king. His English property was then taken into the king's hands. But he returned to England around 1215 and repossessed his family lands, joined the barons opposing John after the signing of Magna Carta and fell in with the young Henry III following John's death in 1216. A listing that year of the constituent properties of the Honour included specified lands at Maresfield, Grinstead and, in another early mention, *Essesdon*.

Gilbert IV, the last of his line, began work on Michelham Priory in the south of Pevensey Rape in 1220 (it was chartered in 1229), and also rebuilt the manor house at Maresfield. He died in 1231 without heirs, and in 1234 the Honour was granted to Gilbert Marshall, earl of Pembroke. But in 1240 Gilbert was forced to surrender the town of Pevensey itself to Henry III after incurring the king's displeasure, and the Honour of Pevensey then went in 1242 to the influential Peter of Savoy (*c.* 1203–1268), uncle of Queen Eleanor of Provence, wife since 1236 of Henry III.

Although much embroiled with European politics and only an occasional visitor to England, it may have been during Peter's tenure that the park was first enclosed with a pale. In August and September 1263 deer were ordered by the crown to be taken from Peter's Ashdown Forest, although admittedly this is by no means conclusive evidence of a pale. His record was not unblemished, however. He had previously been accused in 1243 of wasting the woods in Worth Forest, while the rape of Lewes had been assigned to him during his custody of the lands of John de Warenne, then a minor. Home Wood (Worth) was the scene of much felling of oaks by 'Sir Peter de Sabaudia' and his bailiffs, selling wood to all comers and sending 60 oaks to Maresfield Park, together with four mill staves and beams. Four more of 'the finest mill staves' were also sent to Maresfield from Cuckfield Park (in Lewes Rape), together with another oak.[10]

Peter died in 1268 and the Honour passed to Eleanor of Castile, who became the first of the English queens, as Edward I's consort, to hold Ashdown Forest. She is recorded as holding Ashdown in 1275, following her coronation the previous year. At her death in 1290 it was recorded that she held the hundreds of East Grinstead, Hartfield and Rushmonden, with only the Forest area in Buxted parish not being hers, being part of the archbishop's manor of South Malling.[11]

Although periodically held by the king or by Queen Eleanor, and later Queens Margaret, Isabella and Philippa, Ashdown remained a chase but retained a close association with the crown as a hunting ground, and it was actually regarded as a royal forest until 1268. Following Eleanor's death the Honour stayed with Edward I until 1299. At St Radegund (August 13) 1302 the king requested venison from Ashdown, requiring the sheriff of Surrey and Sussex to oversee its transport.[12]

In 1299 Edward I, then aged 60, married for the second time: Margaret, the young daughter of Philip III of France, was his new queen and, as with Eleanor, the Honour of Pevensey was bestowed upon her. Edward II came to the throne in 1307 and in 1318 the Forest and Honour of Pevensey were now granted as dower to Isabella (the 'She-Wolf of France'), his adulterous queen and daughter of Phillip IV of France, although she was stripped of all her lands on falling into disgrace in 1330.[13] In that year Edward

[10] Christopher Whittick (personal communication): the depredations also appear to have included woods in Surrey and in Reigate Park, although the jury were barred from entry to the park by Peter's bailiffs (TNA, C 145/16 Chancery, Miscellaneous Inquisitions 1243). Peter of Savoy held the Lewes lands, including Worth Forest and Cuckfield Park, from 1241 to just after 1256 (*VCHSx* 7, 'The Rape of Lewes', 1940), 1–7).

[11] L.F. Salzman, 'The Hundred Roll for Sussex, Pt II, Rape of Pevensey', *SAC*, 31 (1943), 35–54.

[12] TNA, SC 1/12/202. i. Edward I to John de Fulham, his 'valet' venison from Arundel forest. Dated at St Radegund. ii. Memorandum of similar letter to John Lovel: venison from Ashdown forest [dated 22 September 1302].

[13] A. Crawford (ed.), *Letters of the Queens of England 1100–1547* (Stroud: Sutton, 1997), 81–6.

III (reigned 1327–77) successfully staged an attack upon Roger Mortimer, the queen's lover who, with her, had assumed control of the kingdom, and in 1331 the Honour went to Edward's queen, Philippa of Hainault (then still only 17), and she held it for many years, together with numerous other possessions. Again in 1330 we find Daniel de Burgham, from Kent, a loyal supporter of Edward III against Mortimer, petitioning parliament to be restored as bailiff of Ashdown Forest and the Park of Maresfield, to which he had been appointed by Isabella and which she had granted to Bartholomew de Burwash while Burgham was in prison in Scotland. In January 1332 Philippa was granted a weekly market for life in her manor of Maresfield, and in 1350 a fair on the eve of St George (April 22) at Nutley chapel. On her death in 1369 the Forest again reverted to the crown.

In 1370 John of Gaunt, duke of Lancaster and third son of Edward III, was granted the 'free Chase of Ashdown with rights and liberties pertaining thereunto' and in 1372 he acquired the rest of the Honour of Pevensey in return for the Honour of Richmond, granted to the duke of Brittany. Control through the Duchy of Lancaster ensured stronger administration, following a somewhat lax intervening period. As well as Pevensey itself, the grant included the manor, church and free chapel of Maresfield and the Priory at Withyham.[14] In 1377 the town of East Grinstead was also added. Thereafter Ashdown became known as the Great Park of Lancaster, and in 1372 Gaunt referred to it as 'nostre park de Asshedoune' and as the 'chase of Lancaster Great Park'.[15]

Many medieval royal forests covered farmland rather than waste, and there was a necessary distinction, also recognised on Ashdown, between the use of some of the land for farming, which was the prerogative of the landholder, and the use of the land as a whole for hunting, which was reserved and administered for the crown. The holder of the land could profit from agistment and from manorial income. By 1382 it would appear that John of Gaunt had also acquired the hunting rights, since, as we shall see below, he was claiming (and winning) damages for loss of game on Ashdown against the local tightly knit gentry, led by Sir Edward Dallingridge.

At John of Gaunt's death in 1399 the Forest was left with his third wife, Katherine Swynford, and on her death in 1403 it reverted to her son, Henry IV, although the Duchy of Lancaster continued its administration. But with the uniting of duchy and crown Ashdown could, therefore, again be considered a royal possession.

During all this period the rape held its own courts with its own sheriff, coroners and other officials. There were two courts: one at Ripe serving the six southern hundreds and one at Duddleswell serving the three northern hundreds of (East) Grinstead, Hartfield and Rushmonden (Riston). The hundreds covering Ashdown and its constituent parishes were East Grinstead (covering East Grinstead parish and the later Forest Row); Hartfield (Hartfield and Withyham); Rotherfield (Rotherfield and later Crowborough); Buttinghill (West Hoathly); Danehill Horsted (Horsted Keynes and later

[14] R. Somerville, *History of the Duchy of Lancaster* I *1265–1603* (London: The Chancellor and Council of the Duchy of Lancaster, 1953), 52.

[15] Brandon, 'Common Lands and Wastes', 101.

Danehill); Rushmonden alias Riston (Fletching and Maresfield); and Loxfield Dorset (Buxted). The courts heard various causes from tenants of the rape. At Duddleswell we hear that in 1274 an Inquisition Post Mortem (IPM) was held on the death of Jordan de Sackville III (d. 1273); and in 1285 an IPM on the death of Isabel de la Hay states that her manor of Brambletye was held of the Honour of Aquila.[16]

THE DEVELOPMENT OF MANORS AND PARISHES

The period from *c.* 1000 to *c.* 1300 saw many of the older tenements grouped into new manors, sometimes by grouping holdings formerly appurtenant to *caputs*, the manorial centres in the more southern parts of the rape. A network of parent and subordinate manors with scattered outliers and complex jurisdictions now developed around the Forest. Often the first glimpse of this process comes in the Tudor period or even later, but it was clearly a medieval development. It is important also to emphasise that manorial holdings here were scattered, rarely forming a ringfenced block. Thus, the scattered dens of the manor of South Malling were being grouped into South Malling Lindfield manor, and run quite separately, at least from the Tudor period. The manor of Walstead was in turn a subinfeudation from South Malling Lindfield, with scattered properties in Lindfield, Worth, Ardingly, West Hoathly and Horsted Keynes. Imberhorne manor was a substantial collection of holdings originally belonging to the Cluniac priory of Lewes, and may itself have originally been part of the manor of Balneath (Barcombe).

Much more important for Ashdown Forest were the manors of Maresfield and Duddleswell. The former was considered a royal manor but both manors exercised an element of control over the Forest, dealing with assarters and overseeing game rights. Duddleswell's holdings tended to encircle the Forest. Inheritance here was by primogeniture, at variance with the older custom of Borough English, by which the youngest son inherited.[17] In fact inheritance around the Forest varied from manor to manor: the custom at Maresfield was for assart land to be inherited by the eldest son or daughter, although other yardlands were governed by Borough English, as also was the case in Horsted Keynes and Imberhorne. However, at the royal manor of Rotherfield 'farthing' and 'cotman' lands went to the youngest son, but 'farthing' lands, in the absence of a son, would be divided among all the daughters while 'cotman' land would go to the youngest daughter. Assart land here was inherited through primogeniture, but the widow was entitled to a dower on 'farthing' and 'cotman' land 'during [her] chaste widowhood'.[18] The division of lands among co-heirs could result in complex

16 I am indebted to Christopher Whittick, formerly of ESBHRO, for his unpublished paper 'The Honour and Barony of Pevensey, otherwise the Honour of the Eagle' (June 1998). See p. 58 for the retention of Brambletye in Pevensey Rape in 1086.

17 I am grateful to Heather Warne for making her unpublished work on manorial tenements in the Ashdown area available to me.

18 G.R. Corner, 'On the Custom of Borough English, as existing in the County of Sussex', *SAC*, 6 (1853), 164–89.

land-holding patterns, involving even the subdivision of individual fields, thereby rendering them still smaller.[19]

Alongside the development of the manors, the parochial pattern was also now emerging. The first point to note is the unusually large size of the parishes here, a consequence of poor land and a scattered population, when compared with the far smaller parishes outside the Weald. This meant that several older manors or parts of manors could be encompassed within the boundaries of a parish.[20] The second point is that there is no relationship between the parish boundaries and the pale of the Forest. If the pale had been established before the parish formation one might have expected some degree of relationship, but none exists. Therefore, it can be argued that parish formation pre-dated that of the thirteenth-century pale, but was developed after 1100. More generally it is thought that by the end of the twelfth century the parochial system had been extended over most of the country.[21]

The Forest's pale actually takes in parts of six older ecclesiastical parishes: Hartfield, East Grinstead, Maresfield, Withyham, Buxted and a small part of northern Fletching (Figure 4.2). All six run up onto the Forest and it may well be the case that the original 'communia' or early manors in these locations retained these marginal lands for commoning prior to any parochial formation. Such regional intercommons frequently saw the interlocking of later parish boundaries, as in Norfolk, where Birtles has identified as many as eight or ten parish boundaries meeting on several heathlands prior to their later enclosure. Such interlocking probably originated to ensure that each community had a share of grazing and woodland resources, as also found in the Kentish Weald.[22] A well-used trackway that ran from the pale at Reads Gate through Five Hundred Acre Wood to Greenwood Clump and on to King's Standing actually forms part of the parish boundary between Hartfield and Withyham. That the parish boundary makes use of such a routeway must indicate that the route was older than the parish boundary, and may be a further indication of the ancient origin of many of the pack-horse routes across the Forest.[23]

---

19 For the landscape effect as revealed through Historic Landscape Classification see N. Bannister, 'Field systems in the High Weald', <https://landscapesforlife.org.uk/about-aonbs/aonb-case-studies/high-weald-field-systems> [accessed 16 September 2021]. Unfortunately, this 2017 report, for the High Weald Partnership and Historic England, includes no case study from the Ashdown area.

20 The large Wealden clay parish of Rudgwick to the west of Ashdown had parts of at least 16 manors within its boundaries (D. Chatwin, *The Development of Timber-framed Buildings in the Sussex Weald* (Rudgwick: Rudgwick Preservation Society, 1996), 7.

21 N. Pounds, *A History of the English Parish: the Culture of Religion from Augustine to Victoria* (Cambridge: Cambridge University Press, 2000), 3.

22 Everitt, *Continuity and Colonisation*, 284, S. Birtles, '"A Green Space Beyond Self-Interest": the Evolution of Common Land in Norfolk, c.750–2003' (Univ. of East Anglia unpub. PhD, 2003), 26–30.

23 I.D. Margary, 'Pack-horse Routes across Ashdown Forest', *SNQ*, 8 (1940), 36. He mistakenly refers to the gate as Chuck Hatch, although the accompanying map shows the route clearly to be from Reads Gate. The route now forms part of the Wealdway.

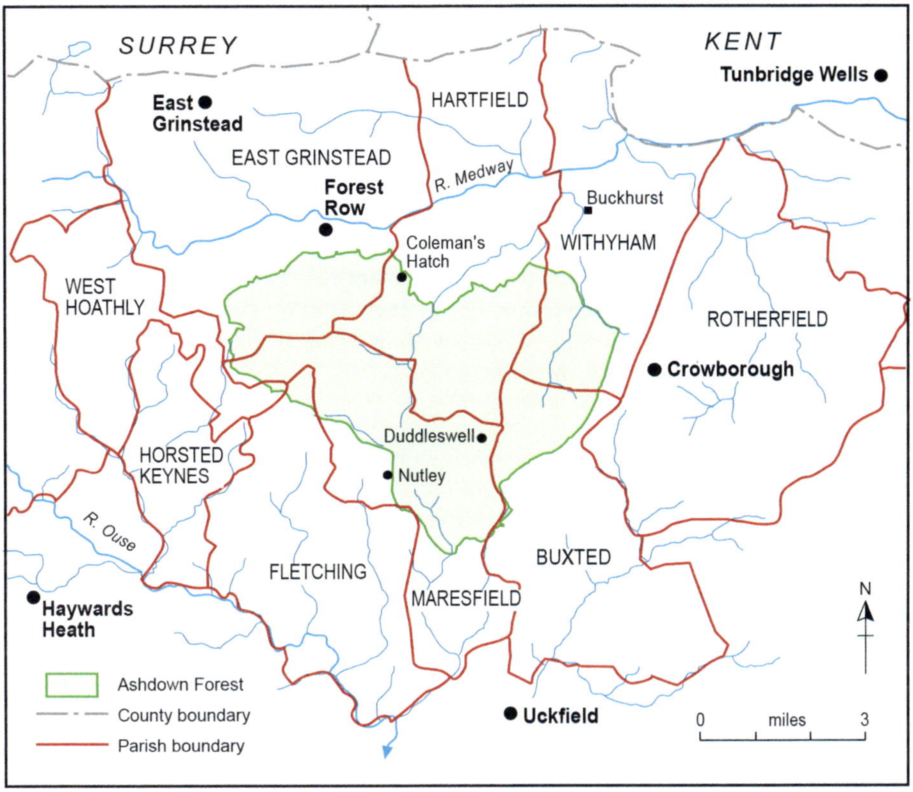

Figure 4.2. Ashdown and its parishes *c.* 1880 (source: author).

In addition to the parishes touching on the Forest, there were others nearby that had intimate associations with Ashdown and are therefore considered in this volume. These are West Hoathly and Horsted Keynes to the west and Rotherfield to the east, the latter being a large parish that later calved off Crowborough. In later years, too, Danehill became a civil parish taking land from Fletching and Horsted Keynes, and Forest Row grew to independent status within the large older parish of East Grinstead.

Associated with the parishes there are early signs of religious activity: the Domesday entry for Brambletye mentions a priest. The place-name for Press Ridge Warren is derived from *Preosta beorg*, 'Priests' hill', possibly also indicating an ecclesiastical presence in the Ashdown area.[24] Robert, count of Mortain, gave the manor of Withyham and the hamlet of Blackham to the priory of Mortain at some time before 1086. The manor was usurped by Walter de Richardeville but restored to the monks about 1095, and further confirmed to them by Robert's son William in about 1100. A single monk appears to have been in charge here and was referred to as prior of Withyham by 1249. In 1325 the monks of Mortain still had property in Withyham and in 1370 were holding

[24] Coates, 'Corrections and Additions to Sussex Material', 5.

the manor and advowson. A return of alien religious houses mentions that the prior of Withyham was an Englishman but had no fellow monk. The grant made by Edward III in 1372 to John of Gaunt included the advowson of this alien cell, which survived for another forty years before being suppressed with the other alien houses in 1413. It was then given first to the New Priory of Hastings and afterwards to King's College, Cambridge. The manor was referred to as Monkenecourt in a list of possessions of alien priories in the 1430s. The adjacent farm is now a Grade II listed medieval building with a nearby moated site and adjacent fishponds.[25]

During the thirteenth century church building moved on apace to give a network of churches and smaller chapels around the Forest edge. The church of St Mary at Hartfield was built at the edge of the then linear settlement and is largely of the thirteenth century. From 1318 it was held from Wilmington Priory, to the south. An earth and timber motte and bailey castle was built *c.* 1300 to the north of the church, presumably by the Sackville family, as a defensive measure against threats along the Medway.[26] St Andrew and St Mary the Virgin (originally dedicated to Our Lady St Mary) at Fletching has a Norman tower and much thirteenth-century work. There is a brass to a member of the recalcitrant Dallingridge family and his wife, *c.* 1380, probably depicting Roger, Sir Edward's father. The church was appropriated by Michelham Priory *c.* 1399.[27]

The original church at Withyham, St Michael and All Angels, built by 1291, was later destroyed by fire in 1663. It may be that a church stood here even earlier, however, also belonging to Withyham Priory, although Withyham manor itself seems to have belonged to Warminghurst Priory by 1085.[28] The church of St Bartholomew at Maresfield has a round-headed lancet of *c.* 1080 in its south wall and the shape of its north doorway could be Norman. It was not mentioned in the 1291 taxation of Pope Nicholas IV although possessing a grant of market rights by 1332.[29]

The seigneurial church of St Denys in Rotherfield, with early pre-conquest foundations and donated to St Denys near Paris in 792, is also mainly of the thirteenth century and, with wall paintings of *c.* 1300, remains a good example of a medieval church. The proprietary church of St Margaret's, West Hoathly, was made over to Lewes Priory by Ralph de Cheyney (alias Keynes), for the soul of Ralph his father, and confirmed to them by William de Warenne between 1091 and 1098. The nave is late eleventh century, part of a small Norman church, and the south arcade and font are late twelfth century. The chancel, with an extension and a south chapel, also dates from the thirteenth century. The aisle was rebuilt in the fourteenth century and the

---

25 W. Page (ed.), 'Alien houses: Priory of Withyham', in *VCHSx* 2 (1973), 123–4; TNA, SC 6/1127/10: Ministers' and Receivers' accounts: Manor in Withyham called Monkenecourt (Possessions of Mortain Priory): Sussex.

26 C.F. Tebbutt, 'Castle Field, Hartfield (TQ 481 361)', *SAC*, 118 (1981), 377–8.

27 M. Jurkowski and N. Ramsay with S. Renton, *English monastic estates, 1066–1540: a list of manors, churches and chapels part 3*, List and Index Society Special Series 42 (London, 2007), 497.

28 TNA, SC 6/1127/10.

29 N. Rushton, 'The Parochialisation of Sussex, 1000–1086–1291', in K. Leslie and B. Short (eds), *An Historical Atlas of Sussex* (Chichester: Phillimore, 1999), 36–7.

tower dates from *c.* 1400. Lewes Priory also held the rectory here as a manor, as well as Imberhorne manor and the churches at East Grinstead and Horsted Keynes.[30] St Margaret at Buxted was also a mid-thirteenth-century foundation and within the great estate of the archbishop of Canterbury. Much of the church remains unaltered, although the tower is a later addition. As we shall see, its singular location in Buxted Park relates to the third Lord Liverpool's decision to allow the village around the church to deteriorate.

The churches themselves were normally established with glebe land to provide for the priests, and this land too could become a manor. Thus, the manor of Dewland, for Rotherfield church, comprised the block of land at Rotherfield, but also detached areas further afield in Wadhurst parish and a subinfeudated and subsidiary manor at Frant.[31]

Chapels were also a landscape element at this time, often built for private seigneurial use, as, for example, at Old Buckhurst for the Sackville family, or when travel along Wealden lanes to reach parish churches was almost impossible. Many can only dimly be seen. A chapel of ease was built at 'Notlye' in about 1230 by Gilbert Laigle IV but its precise location is unknown, although it was said to be 'beside Fair Place' – possibly to the east of the road running between Chelwood Gate and Nutley. And ruins in Chapel Wood, on the road to Chelwood Gate, were to be found within living memory as reported by Hussey in 1852. Sixty acres of land were also given, which Turner placed at Press Ridge to the north. A font from here was recovered and placed in Maresfield church. A chapel belonging to Buxted provided the origins of Uckfield church.[32]

Where churches were established, they might provide a focus for the isolated holdings and hamlets and markets might soon appear; where these were fairly frequent, more buildings might grow and thus a settlement develop. The parish church of St Swithun in East Grinstead was known to have existed by 1106, and its dedication suggests a pre-Conquest origin, when it would have served the whole area, possibly as a minster church. The town developed at a hilltop crossroads around the church from the thirteenth century, achieved borough status in 1235 and had a functioning market by 1247.[33] Other nearby markets and fairs were granted in the thirteenth century at Horsham, Uckfield (developing as a nucleation of shops and housing at this time) and Cuckfield, and in the fourteenth century at Lindfield and Rotherfield (Table 4.1).[34]

---

30 *VCHSx* 7 (1940), 171; Jurkowski *et al.*, *English monastic estates*, 497. All these churches are Grade I listed by English Heritage.

31 C. Pullein, *Rotherfield: the Story of some Wealden Manors* (Tunbridge Wells: Courier Printing and Publishing Co., 1928), 85, 88.

32 Rev. E. Turner, 'The Free Chapels of Maresfield and Dudeney', *SAC*, 9 (1857), 41–4 is now largely discredited; Mawer and Stenton, *Sussex*, 2, 352. Chapels had neither fonts nor cemeteries.

33 M.J. Leppard, 'Church origins in East Grinstead', *SAC*, 142 (2004), 148–9.

34 Gulley, 'Wealden Landscape', 418, 521–3.

Table 4.1. Markets and fairs around Ashdown Forest.

	Market	Fair date	Fair detail
East Grinstead	1247 by Peter of Savoy	25 July (St James the Apostle) from 1247	Henry III to Peter of Savoy
Frant	1297 to be held at the manor of Wm de Brunfeld	1 November (All Saints) from 1297 at Brunfeld's manor	Henry III to Peter of Savoy
Maresfield	1332 by Edward III to Queen Philippa for life	24 August (St Bartholomew) from 1301; 23 April (St George) 1350	Edward I, to be held at the king's manor Edward III at chapel of Nutley
Rotherfield	1318 by Edward II to Hugh de Audele the younger and Margaret, his wife (the king's niece) and her heirs – at the Manor	9 October (St Denis) from 1376	By the recently deceased Edward le Despenser
Uckfield	1220 by Henry III to archbishop of Canterbury, to be held at the manor	3 May (Inventory of the Cross) 1378	By Richard II to the archbishop, to be held at the town

Source: S. Letters with M. Fernandes, D. Keene and O. Myhill, *Gazeteer of Markets and Fairs in England and Wales to 1516*, List and Index Society Special series 33 (pt 2) (London, 2003), 349–55.

The substantial investment in church building and extensions in the thirteenth century underlines the growth of population and settlements around the edges of the Forest, albeit at a relatively late date compared with many areas in the South East. The entrances into the Forest were themselves sometimes foci of settlement, albeit on a small scale. Clusters of buildings grew up at Chuck Hatch (*le Chukhach* in a minister's account of 1467), at Colemans Hatch (*Colmanhacche* in 1495), probably named from the Coleman family noted in an assize roll 1279 and lay subsidy roll of 1327, and Chelwood Gate (*Chellworth Gate* in 1564).

More substantial and of greater later importance was the linear settlement of Nutley (*Nutleg* in 1249, 'nut-tree clearing'), threading in parallel with the pale along its south-western edge.[35] Here was a chapel, and a pinfold to where stock, turned out on the Forest contrary to manorial custom, could be driven. And mentioned somewhat later is Forest Row (*Forstrowe* in court rolls 1467), growing from a line of cottages on the north-western edge of the Forest.

[35] Mawer and Stenton, *Sussex*, 2, 350.

Travel between these settlements and across the Forest might still use the line of the old Roman roads, together with the drove roads, both offering strong north–south links. But also important were the more sinuous unmetalled pack-horse routes, hollow ways and trackways that had developed around and through the Forest (Figure 3.6). Wandering animals and the rutted nature of the tracks caused wider or sub-parallel tracks to develop, but encroachments onto these routeways were frequently reported, and local courts dealt with complaints about bad roads, obstructions by pits or buildings, unscoured ditches or fallen trees. The later use of eight-ox wagons to transport heavy goods such as timber or iron certainly made many routeways difficult to negotiate, especially in winter.

Analysis of the royal itineraries for John, Henry III and Edward I respectively demonstrates that during December and January they restricted their travel mainly to south and south-eastern England, especially favouring the drier chalk and limestone roads and avoiding the more difficult Weald. The Ashdown area was conspicuously devoid of visits.[36] Nevertheless, Maresfield was later frequented by royalty and messages went regularly from there. In 1324 Edward II wrote letters from his hunting lodge in Nutley, including one to the pope. The following year we find Edward journeying through the district: on 21 September at Withyham; from then to 3 October at Maresfield; and on October 3–4 at East Grinstead.[37] The beasts being hunted then were fallow deer (both buck and doe), which were probably hefted to Maresfield Park, together with fox, pine marten and the native roe deer.[38]

ASSARTS, PANNAGE AND COMMONERS

What of the uses prevailing before the chase was proclaimed and the pale raised? Here was the hard-won farmland over which the hunts now ran, the swine pastures belonging to distant manors, the new holdings and small clusters of huts, with families who had already established their own versions of rights to common resources. This was to be a running sore which frequently festered and erupted, as our later narrative will show.

In the 240 years after the Norman Conquest England's population probably trebled. In the 'journey to the margin' new families came to live around Ashdown, with land being assarted for their houses, land and cattle.[39] The easier availability of land, as

36 For the royal itineraries see J.E. Kanter, 'Peripatetic and Sedentary Kingship: The Itineraries of the thirteenth-century English Kings' (Univ. of London unpub. PhD, 2011).

37 E. Hallam, *The Itinerary of Edward II and his Household, 1307–1328*, List and Index Society 211 (London, 1984).

38 In the early medieval period there was a decline in the native red deer, and fallow deer, reintroduced in Anglo-Norman England, became the main deer species on the Forest. The small native roe deer ceased to be a forest beast by a decree of 1339, so receiving no further forest protection (J. Tsouvalis, 'Socialized Nature: England's Royal and Plantation Forests', in I. Cook, D. Crouch, S. Naylor and J. Ryan (eds), *Cultural Turns/Geographical Turns* (Harlow: Routledge, 2000), 294.

39 Faith, *English Peasantry*, 208.

indifferent as it was, favoured earlier marriage, hence larger numbers of children, and the fringes of the Forest now became more heavily populated. There is also evidence of ironworking, as bloomery production appears to have been revived; and trades and crafts can now be discerned. And all these developments were happening, in a sense, regardless of any overall control being exercised from above.

Assarting, the reclamation of small patches of land by individual families, was certainly important at this time. By *c.* 1290 the rental for Maresfield manor shows that about 50 small holdings had been recently established on Ashdown and neighbouring wastes such as *Mallingstone*, around present Mallingdown Farm, probably named from a boundary stone marking one of the manors of South Malling. Only four exceeded 50 acres in size and 35 were less than 20 acres. In addition, another 144 acres were held in small parcels of land as 'new assart demised from the Forest of Ashdown'.[40] The Hundred Roll for Sussex records encroachments on Ashdown at this time: Geoffrey 'le Dykere' was noted as making an encroachment 'for the last four years, by what warrant they know not'.[41] These thirteenth-century incursions were joined by cottages and other assarts, described as 'new' in the fourteenth century, possibly amounting to another 290 acres and noted in 1338 for the manor of Duddleswell.

There was evidently a great hunger for land at this time, despite its frequently uncertain quality around Ashdown. This advance of cultivation into the previously large areas of 'waste' around the Forest also advanced manorial control. Was this resented? We do not know. But Ashdown was now administered primarily from the two manors of Maresfield and Duddleswell, although there were also assarts and unlicensed purprestures, for which fines were paid to the manor of Endlewick (in Arlington and Wilmington). Purprestures were hedged or paled enclosures obstructing the free passage of deer, and presumably also made for purposes of farming.

The area of Ashdown as an open communal landscape will therefore have shrunk. And it should be remembered that it was just at this time that the pale was being erected around it. Was this a defensive measure to attempt to preserve land from the assarters who were reclaiming land, legally or illegally? Did the creation of the pale cut down the amount of available land for cultivation and grazing? There certainly seems to have been no slackening in the demand for land: the *redditus assisae* (the rents of assize paid by villeins to the manor) more than doubled for the manor of Rotherfield between 1262 and 1346 as land was settled around Alchorns and Rotherfield Park. There was even partitioning of the larger holdings and the leasing of land from within the Rotherfield manorial demesne. What was referred to as 'the assart of Crowborough' was poor land taken up by existing customary tenants or free tenants and precisely recorded by the manorial officials.

In part of Withyham parish, on the flanks of Crowborough Beacon and running northwards to Hendal Bridge on a tributary of the Medway, the suggestion of a degree of systematic planning of virgates belonging to the archbishop of Canterbury's vast

---

[40] Mawer and Stenton, *Sussex*, 2, 346; P. Brandon, 'Medieval clearances in the East Sussex Weald', *Transactions of the Institute of British Geographers*, 48 (1969), 135–52.

[41] Salzman, 'The Hundred Roll for Sussex, Pt II, Rape of Pevensey', 35–54.

Malling estate has been made. The resulting fields here are to some extent more regular and are bounded by long sub-parallel hedgerows delineating the virgates. No precise date for this development has been hazarded, beyond the possibility of an early thirteenth-century layout. Such planned or coordinated activity in this area has been uncovered rarely, but it would imply once again that demand for farmland was growing apace, even on such unpromising land.[42]

There was also an outer ring of commons outside the pale, almost certainly at one time part of a larger Ashdown, and which could be regarded as purlieus. The original size of the Forest may have been as much as 20,000 acres, and, following the creation of the pale enclosing about 14,000 acres, this left some 6,000 acres of common land beyond. Although this had also been affected by encroachments, assarting and imparking, there still remained about 1,000 acres of these commons by the mid-sixteenth century. Approximately clockwise, these were: Forest Row Green, Quabrook Common, Coleman's Hatch and Chuck Hatch Greens, Crowborough and Five Ash Commons, Lampool Common, Horney Common, Bunts Grove (Birchgrove) Common, the northern end of Chelwood Common and Stumblet (Stumblewood) Common, and also Pilt Down and Mallingsdown, further to the south-west.[43] The rights exercised on these external commons may have relieved some of the pressure on those whose rights were now compromised by the impaling of Ashdown.

We are fortunate that there are some surviving extents and ministers' accounts dating to the end of the thirteenth century for the Forest.[44] We also know rather more about the tenures by which Forest land was then held. In common with many other woodland and marginal areas, the majority of tenures were free: in the case of the manor of Maresfield over two-thirds were free tenures of some kind. Thus, the keepers held land in the hereditary tenure of sergeanty (2.7 per cent), with the eldest son succeeding; others held by paying a fixed rent in socage (27.7 per cent); there were *liberi homines* (11.6 per cent); and those holding by assart tenure for a money rent only (26.8 per cent). Only about 20 per cent held land for which they owed services. One service at Maresfield was that of ward: watching and guarding the Forest. In practice there might be complex mixtures on any one holding or for any one individual, but the overall picture of a relatively free peasantry compares, for example, with Bishopstone (15 per cent free), Willingdon (20 per cent) or with the Sussex coastal plain, where labour services as components of rents remained important.[45] For Maresfield we can also estimate holding sizes: 42 of the 85 holdings for which a size was determinable

[42] M. Gardiner, 'Planned Medieval Land Division in Withyham, East Sussex', *SAC*, 123 (1985), 109–14.

[43] Brandon, 'Common Lands and Wastes', 102 and TNA, DL 42/112 'Survey of Pevensey, Langney, Willingdon, Seaford, Duddleswell in Ashdown forest, Ashdown forest, East Grinstead etc.' (4 August 1564).

[44] Ministers' Accounts are financial statements from crown officials (Ministris) appointed to manage estates in the king's hands. The extent was a listing of the manorial lord's entitlements to rent and services.

[45] B.M.S. Campbell and K. Bartley, *England on the Eve of the Black Death: an Atlas of Lay Lordship, Land and Wealth, 1300–49* (Manchester: Manchester University Press, 2006), 264–5.

were less than five acres (48 per cent); 23 per cent were between 11 and 20 acres and just 5.4 per cent greater than 20 acres.[46]

An extent of *Essedune* in 1273–4 revealed that by this time there were about 300 Commoners on Ashdown, with 208 being 'old customers' paying a fixed rent of 2d per adult pig for their commoning, together with hens and eggs.[47] These customary tenants, living in the manors of Maresfield and Duddleswell on the edge of Ashdown, were entitled to enter the Forest with carts to carry away estovers – that is, windfall wood (*boscum fractrum*) – between Hockday and Michaelmas within their 'communia', except where winds had torn it up by the roots, in which case it belonged to the king. So Commoners could take windfall wood, heather, brushwood, furze and broom for domestic fuel and depasture all those cattle that could be maintained on their holdings over the winter, although between Michaelmas and Martinmas the mast had to be preserved for pannage. Sheep and goats could not be turned out because of competition with deer for herbage, but Commoners' swine could be grazed except for the fence month, the four weeks at midsummer 15 days before and after St John Baptist (24 June), when does were dropping fawns. The extent also allowed that 'if it be necessary for the improvement of their common pasture they may burn all the aforesaid' (heather, brushwood and so on), an early reference to what much later became a contentious practice.[48]

We also have some ministers' accounts from 1291 onwards for Maresfield manor and the Forest, revealing that there were now also 58 'new customers' paying ½d for each beast. The following year we also hear of the *extranei*, about 20 in number, who paid 1d per beast, and who occupied land in manors other than Duddleswell or Maresfield.[49] Ashdown was now clearly heavily grazed by the Commoners' cattle, especially in the South Ward, possibly numbering between 1,900 and 2,000 head, and this over and above up to 1,000 deer. In addition, and without the need to pay, the priory of Wilmington had the right to graze 36 oxen on Brownknoll (near Crowborough) during early summer, and the chapels at Maresfield and Nutley also had rights to put beasts onto the common.

These large cattle numbers necessitated some degree of supervision. Manorial control over farming operations is generally thought to have been far less forceful in the Weald than elsewhere in the South East, but some arrangements were in place. Thus, commoning involved the payment of *lep* and *haforf*. The former was a fine imposed on those who depastured cattle in the mast period, clearly analogous to the *lefgavel* in the Kentish Weald and similar to the payment of Danger (*dangerium*) for damage to

[46] R. Parsons, 'Status, Occupations and Landholding. The Emergence of an Independent Peasantry in the High Weald of East Sussex in the Late Thirteenth century. The Evidence of Ashdown Forest and its Region' (Univ. of Leicester unpub. MA, 1996), 44–63.

[47] TNA, SC 12/15/46: Ashdown (Essedune) Forest: Extent' 2 Edw 1 (20 November 1273–19 November 1274).

[48] Brandon, 'Common Lands and Wastes', 103–05; J.E. Small, 'A Review of Ashdown Forest and the Common Rights Thereon', *SAC*, 126 (1988), 157. Hocktide was celebrated on the second Monday and Tuesday after Easter.

[49] TNA, E 101/136/19: Particular Forests (south of Trent): Accounts relating to Ashdown forest. 24 Edw I (1295–6) and SC 6/1027/20–22; SC 6/1028/1–2 (Maresfield Manor 1290–1303).

pannage areas by the destruction of herbage or mast-yielding woodland.[50] *Haforf* was probably a payment for depasturing those cattle that could not be accommodated on the holding over the winter months.

By the late thirteenth century receipts from pannage were the largest source of manorial income, continuing the pre-Conquest situation. But the numbers of customary tenants were relatively small and completely overshadowed by the numbers of *extranei*, accounting in the 1290s for generally between 2,000 and 3,000 pigs being sent to the Forest, with, again, a large proportion in the South Ward. The employment of *drofmanni* would have remained important. The pannage receipts can be estimated from the accounts of William de Bergham, Roger de Tukestone and William de Gulderegg, successive master foresters, ministers' accounts and Exchequer pipe rolls (Table 4.2). In 1292 Gilbert Brouning of Heselwode (East Grinstead) accounted for 746 grown pigs and 485 yearlings, all belonging to *extranei*.[51]

Brandon hypothesised that such large numbers resulted from the activities of grazier–butchers and purveyors of salted pork, since Richard Pycham and William Browning, who paid pannage in 1291, also appear in the 1296 Subsidy Roll as being taxed at the port of New Shoreham, and may have been provisioning traders or Edward I's naval victuallers.[52] So swine continued to play a large role in the Forest economy, despite the increases in assarting at the expense of woodland, a greater arable area and growing numbers of cattle. Sometimes numbers of swine to be commoned were limited: the abbot of Bayham's right to grazing in Ashdown was restricted to 20 swine. In 1341 the grazing rights there of parishioners in Withyham were valued at 17s, those of Maresfield at 20s and those of Hartfield at 40s. At Maresfield 20s is given as equal to 20 pasture animals and 24 pigs, which, being a tithe, gives totals of 200 animals and 240 swine being grazed.[53]

Attempts to produce arable crops, almost entirely oats, were, to a limited extent, helped by the use of marl. Land at *le Strode* (East Grinstead) was leased in 1273 on condition that the tenant marl all that part of the acreage not already marled within five years. Small marl pits were dug across much of the Forest, often in new assarts, and much in the south. The benefits were variable as most of this marl, as we have seen, did not contain calcium carbonate, but marled arable at Buckhurst was valued at three times that of unmarled land in 1273, 80 acres of marled land here being valued at 6d an acre, while 100 acres lying unmarled was worth only 2d. There were also 50 acres of heath and broom worth just 1d per acre and the herbage at Newenham Park was worth little because of its proximity to Ashdown Forest, whose bailiffs allowed deer to escape into this neighbouring park.[54]

---

50 F.R.H. Du Boulay, 'Denns, Droving and Danger', *Archaeologia Cantiana*, 76 (1961), 75–87.

51 Michael Leppard (personal communication), and see his 'Pigs and Dens in East Grinstead and their parent Manors', *East Grinstead Society Bulletin*, 118 (Spring 2017), 5–9.

52 Brandon, 'Common Lands and Wastes', 107.

53 Gulley, 'Wealden Landscape', 319–20; Bayham was also restricted at certain times to the felling of one oak and a beech.

54 J. Irons, 'Aspects of the Impact of Man on the Historical Ecology of Ashdown Forest, Sussex before 1885' (Univ. of Sussex unpub. DPhil, 1982), 290; Gulley, 'Wealden Landscape', 343–4; TNA, C 133/8/7 Details from the IPM of Sir Jordan de Saukevyll alias de Sakevile 1274.

Table 4.2. Pannage receipts 1284–97.

Date	Rent from customary tenants	Rent from strangers: *extranei*	Consolidated total rents
1283			£7 2 3½d
1284			15s
1285	£317s 1½d	£18 12s 4½d	
1286		£2 13s 3½d	
1287			£1 12s 3½d
1288	£3 10s 0½d	£11 4s 5¼d	
1289	£3 0s 6d	nil	
1290	£1 11s 8d	nil	
1291	£1 3s 10d	£28 1s 6d	
1292	£3 3s 7d	£25 5s 5¼	
1295	1s 7d		
1297	£3 4s 2d	£17 0s 11½	

Source: Brandon, 'Common Lands and Wastes', 106.

The structure and practices of agriculture were now beginning to resemble those of later periods: small family-held farms, primarily pastoral, with a reliance on the common resources of the Forest. For many, agriculture might be combined with other seasonal or part-time activities, such as woodcutting. Although there is no evidence for an export trade in timber from Ashdown, and while haulage costs were high, there were demands for firewood and timber for Pevensey Castle: in 1317 100 oaks were requested from Ashdown, hinting at a still relatively extensive woodland cover. Repairs in 1300–03 at the castle had already included the supply of timber for the hall beams and 7,000 lathes from Maresfield Park. In 1285 Grenehurst (Buxted), part of the archbishop of Canterbury's great manor of South Malling, reported 35 acres of coppiced wood, indicative of some attempted conservation at this time. But by 1309 there were reports of 'depredations' in Ashdown's woodland, although in 1350 ten wagonloads of beech faggots left the Forest.[55]

Occupational by-names indicated trades and crafts: Eudon le Vyxhunte (hunting); Walter le Warrener (indicating the thirteenth-century increase in rabbit warrens on the Forest); Coleman and John le Colyere (the iron industry and charcoal burning); Turner, Cooper and Pollard (wood products); and William le Weuere (weaving). In 1377 the Poll Tax recorded three carpenters, a cooper and a woodman among the 15

[55] Gulley, 'Wealden Landscape', 300–1; there is no mention of the area supplying timber for export through Winchelsea or other ports in R.A. Pelham, 'Timber exports from the Weald during the Fourteenth Century', *SAC*, 69 (1928), 170–82.

artisans at Blackham (Withyham).[56] Here was 'a flourishing of private skills, trades and landed enterprise behind the façade of manorial routine'.[57]

Other by-employment included the mining of iron ore, charcoal burning and perhaps smelting and smithing. The iron industry, seemingly moribund during the Saxon period, was now revived, with its own demands on the landscape for fuel and building materials. Prior to the late fifteenth century the techniques employed in iron production remained similar to those used in the Romano-British bloomeries, and indeed the differentiation of medieval and earlier sites is difficult without pottery analysis. Until relatively recently it was believed that there had been a hiaitus between the Roman and post-medieval ironworking in the Weald, with the Millbrook Saxon site a rare outlier. But evidence is now accumulating to suggest that there was medieval activity. Iron mining near Sharpthorne, West Hoathly, for example, has left mine pits with fills containing part of a tree trunk and worked timber radiocarbon dated to AD 1220 ± 80 and 1120 ± 75 years. This is interpreted as indicating medieval mining, presumably feeding one or more nearby bloomeries. In 1262 there was a dispute over the profits of an 'iron mine' between Agnes Malameins and Isabel de Aldham on the Lavortye estate, possibly even near the *Ferraria* of Domesday Book.[58]

The expansion of agriculture into the Ashdown area would have generated some demand for farm implements. There are references to purchases of iron from Rotherfield by the early fourteenth century for the repair of carts or ploughs. Paternoster Wood near Upper Parrock (Hartfield), just outside the pale, has yielded a scatter of ironworking sites, many with associated medieval pottery, together with the site of cottages with much late twelfth-century to early sixteenth-century pottery, bloomery slag, nails and clay daub, possibly indicating a small settlement of ironworkers. But the site appears to have been largely abandoned thereafter, probably as the demand for iron products from the Weald dried up.[59] Pottery found with slag at Chandler's Farm (Hartfield) has been dated to the twelfth century. The site at Newbridge, more normally associated with early Tudor blast furnace development, has also yielded bloomery slag and may have been the site of a medieval water-powered bloomery or hammer forge. At Withyham *c.* 1320 John de Lynleghe had a forge and accompanying stock. His possessions here and in Hartfield, Capel (Kent) and 'Tendale' had been seized among those of other 'contrariants', opponents of Edward II. At Rotherfield the inclusion of iron-tipped arrows in rents in the early fourteenth and fifteenth centuries was perhaps significant, with John Inchrede paying 'a barbed arrow, value 1d' for his house and land

56 M. Gardiner, 'The Geography and Peasant Rural Economy of the Eastern Sussex High Weald, 1300–1420', *SAC*, 134 (1996), 125–39.

57 Parsons, 'Status, occupations and landholding', 28–40; F.R.H. Du Boulay, *The Lordship of Canterbury: an Essay on Medieval Society* (London: Nelson, 1966), 185; K.P. Witney, 'The Woodland Economy of Kent, 1066–1348', *Agricultural History Review*, 38 (1990), 20–39; Gulley, 'Wealden Landscape', 299, 318.

58 Hodgkinson, *The Wealden Iron Industry*, 37–48; TNA, JUST 1/912a, m. 17d. Sussex Eyre 1262.

59 A. Sapoznik, 'Rural Industry and the Peasant Agrarian Economy: a study of the Iron Industry in Medieval England', in J. Bowen and A. Brown (eds), *Custom and Commercialisation in English Rural Society* (Hatfield: University of Hertfordshire Press, 2016), 36–50.

in 1428, and duchy ministers' accounts for Maresfield in the 1380s and much of the fifteenth century also making reference to this practice.[60]

So evidence mounts up. An investigation at Crowborough in early 2014 in advance of development revealed firm evidence of medieval ironworking, including smelting and smithing hearths. Spreads of slag contained medieval pottery sherds of the thirteenth to fourteenth/early fifteenth centuries. Nearby is a fifteenth- or sixteenth-century bloomery site in Minepit Wood with roasting and smelting furnaces, a timber-framed building and much slag with fifteenth-century charcoal, primarily from oak and beech but also using birch, ash and poplar. By 2008 36 medieval ironworking sites had been identified in the Weald as a whole, but, given the difficulties of dating, there may be many more awaiting identification.[61]

THE PALE

Having looked at the rape for which Ashdown was the 'waste', and something of the economic activity in the area, we can now turn to the topography of Ashdown's boundaries. To demarcate the park a deep internal ditch was dug and a prominent earth bank built on its outer side, topped with closely spaced tall pointed oak stakes, often with railings, and sometimes live wood. Holly is now found in many places on the earthwork, and may have been planted, with hawthorn, as a living stockproof hedge or fodder supply. Where the holly is dense today it unfortunately still obscures the pale in lidar survey as well as at ground level (as near Oldlands), but its presence also helps to discern the conjectural line of the pale elsewhere. The surviving sections of the pale are found in areas unsuited to later development, in roadside wastes and where sections are used as modern property boundaries (Figure 4.3).

Even with their prodigious leaping abilities deer would find it difficult, though not impossible, to surmount both the ditch and the bank with stakes and were thus corralled. But deer from beyond the pale could more easily jump inside, thereby rendering the pale a mechanism for gaining deer but not losing them. The pale also included 'deer-leaps', or salters, enabling deer to enter or re-enter the park but, again, not escape from it. These may have simply been a raised gap in the bank with a pit inside the bank, or a stone or wooden barrier. The leap may also have incorporated some fencing to funnel the deer towards it (Figure 4.4). The name Hindleap (*hindehlype*) may therefore refer to a deer-leap.[62] In some parks, too, including Ashdown, an area of about 20 feet called a free-board was cleared outside the bank for ease of access to repair the pale, but also to detect escaped deer more easily. At Ashdown, there were also internal boundaries forming compartments to manage the deer and to keep them

---

60 Cleere and Crossley, *The Iron Industry of the Weald*, 89–91; TNA, DL 29/441/7082 Sussex: Lordship of Pevensey and others: ministers' accounts; Pullein, *Rotherfield*, 85, 88.

61 Sussex Archaeological Society, *Archaeological Round Up* (October 2014); J.H. Money, 'Medieval Ironworkings in Minepit Wood, Rotherfield, Sussex', *Medieval Archaeology*, 15 (1971), 86–111.

62 M. Derbyshire, 'Old-style Deer Leaps', *Rural History Today*, 19 (July 2010), 4–5.

Figure 4.3. The Ashdown pale at Legsheath. Modern fencing tops the medieval pale, with the internal ditch clearly to be seen (source: author).

out of wooded areas, perhaps coppiced woodland or rabbit warrens. Thus, there is a reference from 1282 in the Duchy of Lancaster records for a new pale in Costley Ward.

Because of the variations in topography, the Ashdown pale was inconsistent in its morphology, and in some places may have only consisted of a fence or stone wall. It was often placed along higher points, breaks of slope and overlooking streams and ghylls, obviating the need for the ditch and bank altogether. To the north-east of Marden's Hill, near St John's, is a steeply incised ghyll where the natural slope has been cut back to provide an almost perpendicular face between six and nine feet high, dropping to the valley floor some 50 feet below: a formidable boundary. This is also the location for the aptly named Watergate (alias Newnhams). At Millbrook, above Nutley, the stream was also a boundary, and possibly only a modest fence was required here. A double deterrent is found near Chelwood Gate, where there are two bank and ditch arrangements, although the internal bank may have been for stock management purposes.

The pale was costly to build and maintain; the Duchy of Lancaster records contain references to the condition of the pale and the need for repairs, with wood from the Forest sold to finance them. In 1401 wood worth £40 was sold out of the 'Chace of Ashendon [to] expend the same in the repair of the fence of that Chace and of the ponds of Maresfield'. As well as being practical, however, it was a symbol of elite power. It was not, however, totally a symbol of exclusion, as gates and hatches punctuated the pale at fairly regular intervals of one or two miles. Hatches (OE *hæcc*) were high half-wicket

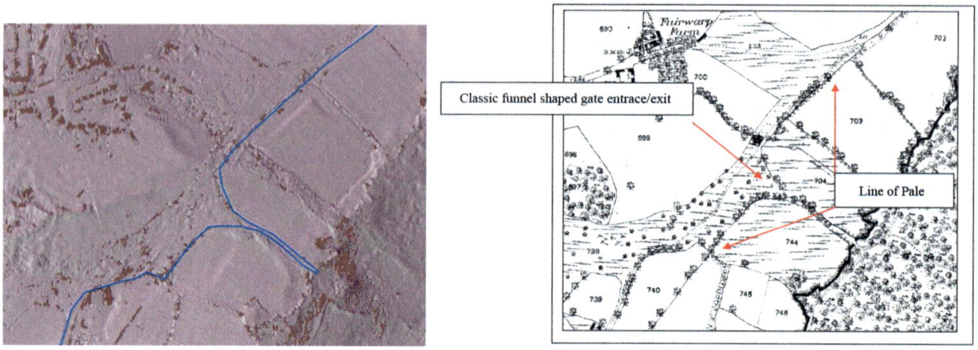

Figure 4.4. Line of pale at Fairwarp, with funnel-shaped entrance at Hendall Gate (source: Blandford, *Tracing the Pale*, 32). Lidar image compared with the first edition Ordnance Survey 25-inch (1:2500) map. The numbers within each field relate to a separate area book.

gates with a bar above to prevent deer from escaping. These were more normally used for access on foot, whereas a gate gave access for horses and possibly carts. The entrances are listed in Chapter 6, with information taken from sixteenth- and seventeenth-century surveys. Some gates dropped out of use early, such as Brikhouse/Brikhost gate, near the southern entrance to the Forest. Deep in the beech woods in the north-western part of the Forest is Mudbrooks Gate, where numerous hollow ways from the Forest converge, leading to a stream and crossing point. These tracks fan out across the Forest, either as footpaths or pack-horse routes. Wheeled traffic, except for small carts, would have been rare over the rough ground. The historical and spatial interrelationship of track, pale and gate is interesting: were the gates/hatches new creations when the pale was built? Or were they placed where pre-existing routeways approached the new pale? The section from Highgate at Forest Row via Wych Cross to Chelwood Gate was part of the important route from London to Lewes, being continued as a hollow way across Chailey North Common to the south of Ashdown.[63]

The north-western edge of the Forest respects the boundary of Pevensey Rape in part, reminding us that Ashdown was originally in the larger Forest of Pevensel (Pevensey), covering much of the northern part of the rape. The boundary also had to respect older parks and enclosures, and so deviated around Buckhurst, Newenham Park (later Oldlands) and the land belonging to Michelham Priory at Chelwood Vachery, a very early enclosure that had its own ditch and bank to keep it separate from the deer park. Originally referred to as Windebeche, the priory had been granted this land some time before 1287, and the name was first recorded in 1292, when we hear of 'Nicholas ater Vacherye'.[64] Norden's map of 1595 and Speed's of 1610 show Buckhurst and Newnd (Newenham), which have a common boundary with Ashdown, together with Maresfield to the south (Figure 4.5). In the Duchy of Lancaster archives there

63 Margary, 'Pack-horse Routes', 34–40.

64 Irons, 'Aspects of the Impact of Man', 237.

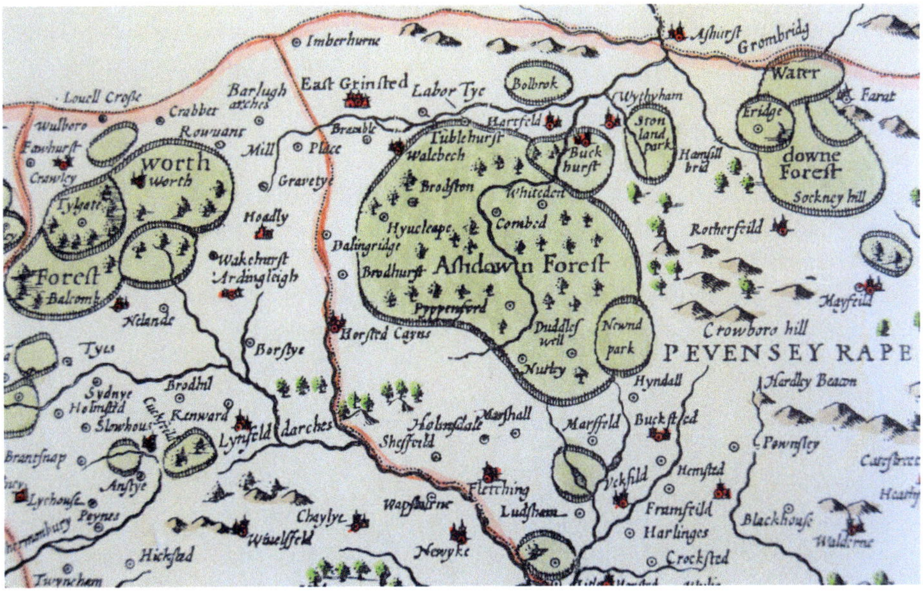

Figure 4.5. John Speed's map of Sussex, 1610 (detail) (source: West Sussex Record Office, PM 892, with thanks to the County Archivist).

is a record for making the new park pale of Maresfield in 1282, so it is arguable that Ashdown was enclosed shortly before Maresfield Park. The manor house at Maresfield probably stood south of the present church.

The shape of the Ashdown boundary has had a lasting effect on the landscape. Although dictated in part by existing routeways, in turn it has shaped the course of some later minor roads and footpaths and defined the line of property boundaries. The ditch of the pale has been utilised to build minor routeways, as at Old Forge Lane to the east of Horney Common, and even part of the main A22, at Horney Common itself. Another example is the location of the modern Crow and Gate pub, the name derived from its location at the Crowborough Gate. The rear boundary of the pub garden is the pale, and this continues northwards, forming other property boundaries, reaching to about four feet high in places. The Crow and Gate building therefore sits at an odd angle in relation to the A26, being aligned to the pale rather than the road.[65]

The pale was therefore a substantial landscape feature, not easily moved. However, some exceptions to this may exist: nine acres at Newbridge Furnace and 14 acres of Newenham Park, opposite Poundgate, where it was recorded in 1650 that some land had been taken out of Ashdown Park. The park pale clearly once included the land on which the furnace and park stood. On the other hand, although primarily *in situ*, it is unlikely that much of the pale has survived unaltered, the least modified stretches being those forming field boundaries, in roadside waste that has reverted to woodland, as at Chuck Hatch, and where slopes have made it difficult subsequently to

65 Blandford, *Tracing the Pale*, 47.

use the land for any other purpose. Vivienne Blandford, concluding her survey of the pale, noted 'the sheer effort and ingenuity involved in the creation of this boundary'.[66]

Parks such as these were very numerous, with about 100 in the Weald alone by *c.* 1300, and Sussex had one of the highest densities of parks in England.[67] Ashdown was one of a chain of such landscapes, which also included St Leonard's Forest, Worth, Waterdown, Dallington and the Frith, in Kent. Here lords could establish 'living larders', rent out herbage and manage timber; but above all the park was a place of pleasure and social status in its provision of exclusive hunting.[68] Venison was commonly given and received as a mark of lordly recognition or at feasts, and the main animals for which these parks were created were fallow deer, which were generally carefully managed. They had been reintroduced *c.* 1100, but whether before or after the Conquest is not yet known.[69]

ASHDOWN'S ADMINISTRATION

By the 1290s the Forest was divided into three wards for administration: Waldhatch (the West Ward), Heselwode (also Costley) and Lampole (South Ward) (Figure 4.6).[70] Such an arrangement, probably dating back to the building of the pale 20 or so years before, would have been similar to that of the New Forest. Aves and Woodmote courts were possibly held in each ward: aves courts fell on the first Tuesday after All Saints Day with the purpose of collecting rents and arbitrating on pannage and agistments, 'wherein they presented abuses of customs, incroachments, spoyle of game or woods, and pay [sic] the aves money for the year past';[71] three weeks later Woodmote courts heard serious offences, such as poaching or encroachment. Courts Baron, with a homage (jury of more substantial landholders), were held every three weeks to deal with issues of land use and customary tenants' rights.

---

66 Blandford, *Tracing the Pale*, 74.

67 L.M. Cantor and J. Hatherly, 'The medieval parks of England', *Geography*, 64 (1979), 75. But, presumably because of its name, Ashdown is not included as a park in L. Cantor, *The Medieval Parks of England: a Gazetteer* (Loughborough: Loughborough Univ. of Technology, 1983), 74–7.

68 Jones, 'A "common of hunting"?' 28, 53.

69 J. Birrell, 'Deer and Deer Farming in Medieval England', *Agricultural History Review*, 40 (1992), 112–26; N. Sykes, G. Ayton, F. Bowen, K. Baker, P. Baker, R. Carden, C. Dicken, J. Evans, A. Rus Hoelzel, T.F.G. Higham, R. Jones, A. Lamb, R. Liddiard, R. Madgwick, H. Miller, C. Rainsford, P. Sawyer, R. Thomas, C. Ward and F. Worley, 'Wild to Domestic and Back Again: the Dynamics of Fallow Deer Management in Medieval England (c. 11th–16th century AD)', *Science and Technology of Archaeological Research*, 2.1 (2016), 113–26. I am grateful to Naomi Sykes for early access to this publication.

70 Where the three wards meet, south of the Pippingford boundary, is still known as the Three Wards.

71 F.E. Sawyer, 'Sussex Folk-lore and Customs connected with the Seasons', *SAC*, 33 (1883), 237–60. The New Forest was, contemporaneously, divided into six walks, later nine, each with its own warden or keeper, and this also applied to Ashdown (E.M. Yates, *The Landscape of the New Forest in Medieval times*, Kings College Univ. of London Occasional Paper 24 [London, 1985], 12, 23.

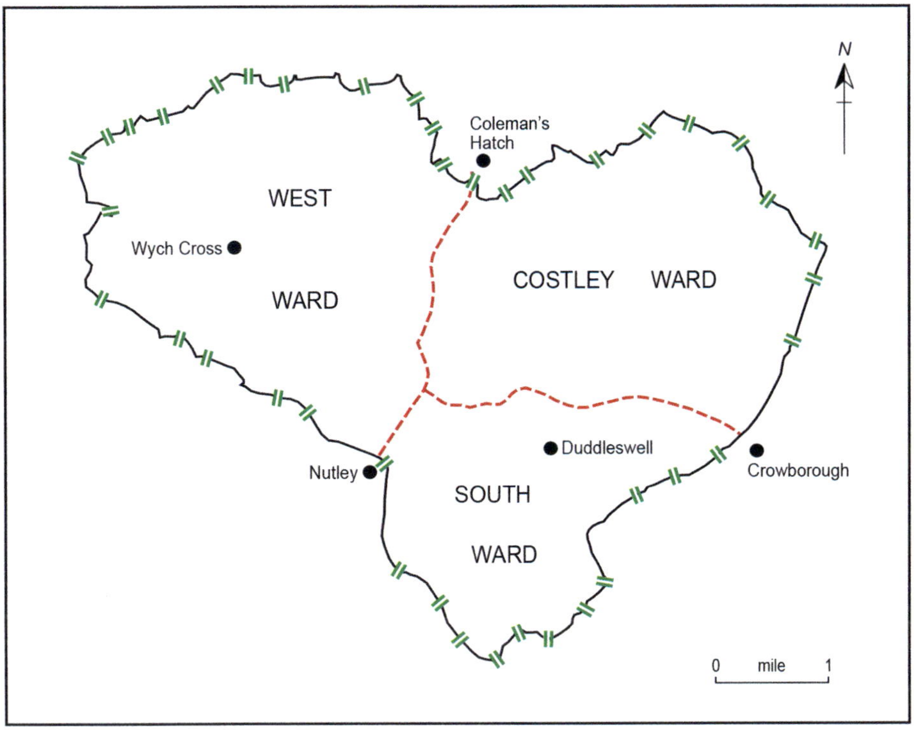

Figure 4.6. Forest administration (source: author).

Within the wards there were further subdivisions of walks and lodges. In 1419 the pale in 'Westwardes' was broken down and totally decayed, as were five wooded gates, to the extent that 'all the Kings beasts' escaped into the surrounding wood pasture. There were deer houses next to gates and beasts found depastured illegally were impounded and removed. Some place-names – Prickets Hatch (either *pricket* a young deer or *pricker*, a forester) and Poundgate – may indicate this use of land.[72] Much effort would continue to be expended on fencing to retain the agile deer within the park.

The central part of the Forest, around the keepers' lodges, was the area reserved for the deer, described as *in defenso*. The lodges were the Old, New and Duddleswell lodges, with the first-named often the location for the Woodmote courts, but other lodges seem to have been in existence by the early fifteenth century.[73] From here the management of the deer and their integration with other affairs could be overseen. In 1273–4 a master forester, who was paid 8d per day to cover himself, his man and his horse, was assisted by eight serjeants or foresters, who may have had to do practical work, for which they were paid 1d per day. Agistment cattle were also admitted for the

[72] D. Hooke, 'The Woodland Landscape of Early Medieval England', in N. Higham and J. Ryan (eds), *Place-names, Language and the Anglo-Saxon Landscape* (Woodbridge: Boydell, 2011), 143–74.

[73] Brandon, 'Common Lands and Wastes', 101–2.

lord's profit, and the lord was also entitled to the first lay of mast before Martinmas, after which it was open to the Commoners. Clearly a system of joint exploitation was therefore now in place that acknowledged the rights of both the lord and Commoners.

Under the Duchy of Lancaster a tighter control over the Forest was exercised, with an army of officials wielding supervisory powers.[74] In 1372 master forester John Seyncler was to deliver 'ij daymes de fermeson deinz nostre park de Asshedoun' – two does at the closure of the hunting season – or 'une bisse et une dayme de fermeson' or 'xx bisses de fermeson et ycelles transmettez a nostre chastel de Hertford devant le fest de Noel prochain'.[75] A bisse was a hind or roe deer, and twenty were to be delivered to Gaunt's castle at Hertford before the following Christmas. In 1374 it was 'deux deymes de grece' – does in grease or fat and fit to kill. Poaching by others was deemed a grave offence. We also find, to the prior of Michelham in February 1375:

> Purce que trové est par enquest que le priour de Michelham et touz ces predecessours priours de la dite maison du temps de leur fundacion ont ehu et joié deinz nostre chace de Asshedoun pasture a sessant grosses bestes, et pannage a cent porcs, et maeresme pur leur edifices et et [*sic*] claustures sanz rien paier, c'estassavoir par an sept cheisnes et deux fages.[76]

Michelham Priory, from the time of its foundation, was found to have enjoyed in the chase of Ashdown, pasture for 60 cattle and pannage for 100 swine, as well as free building and fencing materials in the form of seven oaks and two beech trees annually.

Ashdown provided regular income to the duchy, as well as a means of honouring Gaunt's obligations. By November 1372, therefore, Gaunt commanded the 'warden of our chace of Asshedown' to allow Sir Thomas Ludlow from the Parrock 'within our said chace' common pasture throughout the year 'for all manner of their beasts', free pannage for their hogs and the freedom to pitch 'hog stys' wherever appropriate. Ludlow, a justice sitting in neighbouring Sussex, as well as Surrey and Kent, could also take one oak and four beeches twice a year under the supervision of the bailiff – all to be enjoyed 'peaceably without impeachment'.[77] In 1379 Gaunt sent John Culpeper, created master forester of Ashdown for life in 1375, specific notice informing him that he had pardoned the trespass and taking of venison by the chaplain of his chantry at Maresfield, John de Wollaston, 'perensi qil se port bien et loialment devers nous en tout temps avenir', that is on the basis that he will continue to behave loyally towards us at all times in the future. But in 1386 Roger and Walter Prynkell and others were outlawed in a case brought by Gaunt (referred to as King of Castile by his 1371 marriage until the title was revoked in 1388) for poaching at Maresfield, a case that seemingly rumbled on through

---

74 For the many officials see Somerville, *History of the Duchy of Lancaster* I, 97 *ff.*

75 S. Armitage-Smith (ed.), *John of Gaunt's Register* II, Camden 3rd series XXI (London, 1911), 1071, 1084, 1103.

76 Armitage-Smith, *John of Gaunt's Register* II, 1647. *Sessant* = 60; *Maeresme* = Anglo-Norman for construction timber or wood; *Cheisnes* = oaks and *fages* = beech.

77 ESBHRO, U of Sx Spec Colls, AFC I.

the 1380s.[78] A continuing succession of local landed families provided the administrators of the Forest, at least in a titular sense: the influential John Pelham was 'rider and ranger' in 1413, Richard Lewkenor was master forester in 1486.

CONTRACTION IN THE LATER MIDDLE AGES

In terms of the impact on the landscape, we should see the period from *c.* 1086 to *c.* 1300 as one of pressure to create footholds in and around the edges of the Forest, and indeed the creation of the Great Park itself, with its pale. Such was the landscape impact in many such marginal areas at that time. But the period from *c.* 1300 to *c.* 1500, by contrast, saw a lessening of demand for land and pannage: it was a time of contraction that may have begun nationally with short-term food shortages and higher mortality from the 1290s.

We can assume that hunger for land around Ashdown slowed or even came to a halt with the advent of the poor weather, agricultural crisis and famine of 1314–22, the Black Death in 1348–9 and recurrent outbreaks of plague for over 20 years thereafter. Such calamitous happenings and attendant vagaries affecting farming probably impacted forcefully in this marginal area. Wet weather and disastrous crop yields meant that less grain was available to sow for the following harvest, thereby prolonging the difficulties. Assarting stopped on manors such as Withyham, and from the middle of the fourteenth century a long period of reduced activity ensued, lasting long into the fifteenth century. Instead, woodland and scrub reasserted themselves.

Around Ashdown perhaps as many as one-third of the new assarts remained untenanted after the Black Death, although the fifteenth-century *decasus* (a list of properties for which no rents were collected) in Maresfield manor covers very few properties.[79] The loss of perhaps 40–50 per cent of the English population presented an unimaginable catastrophe. By the late fourteenth century the ministers' accounts for the duchy show that pannage activity was greatly reduced, but in exceptional years receipts still remained high from the depasturing of pigs belonging to the *extranei*. The Sackville manors away from Ashdown sold wool from large flocks of sheep at this time, but not from around the Forest.[80] Instead, the depressed situation by the later fifteenth century meant that the average annual number of cattle grazing Ashdown was only about half that of the thirteenth century.[81]

Some dilapidation may also be surmised from the fact that in November 1374 and again in February 1375 Sir Thomas Travers, chief forester, and John Edwards, 'nostre Seneschal de Marfeld', were being enjoined by the duchy to repair the watermill at

---

78 E.C. Lodge and R. Somerville (eds), *John of Gaunt's Register* I *1379–83*, Camden 3rd Series 56 (London, 1937), 1149; TNA, E 101/510/28, Account of Walter Prynkell, outlawed at the suit of John, duke of Lancaster, lands in Sussex. 7 Rich II (June 22 1383–June 21 1384); Somerville, *History of the Duchy of Lancaster* I, 56.

79 Brandon, 'Medieval Clearances', 141–3.

80 N. Saul, *Scenes from Provincial Life: Knightly Families in Sussex 1280–1400* (Oxford: Clarendon Press, 1986), 133.

81 Brandon, 'Common Lands and Wastes', 105.

Maresfield using wood from Ashdown and Maresfield Park. By February 1380 the mill was still in need of repair and the prior of Michelham, Gaunt's receiver, was instead sent to repair the sluice and bridge there and to buy suitable stones for the mill, under John Culpeper's 'surview and control'.[82]

On the Forest another possibility presented itself, as previously intimated: the expansion of rabbit warrens, typical of areas reverting to scrub or early woodland cover. Henry III certainly had a coneygarth at Guildford by 1235, at a time when their meat was considered a luxury. Only later did they become food for the poor.[83] John de la Parocke claimed warren at Parrock by grant of Henry in 1253, as recorded in the Hundred Rolls, but in turn he was accused of encroaching upon the road at Parrock 'to the extent of two perches, to the injury of the whole neighbourhood'.[84] We have already noted that in 1283 the Forest was referred to as the 'free chase and warren of Essedon', and in that year a commission investigating poaching from the warren of Dowager Queen Eleanor reported its findings: the beasts and fowls of the warren were said to be hare, conies, pheasant and partridge. A royal coney warren at 'Gardine Hill', Hartfield, was said to encompass as much as 96 acres. More investigating commissions similarly reported incursions to warrens in the reign of Edward III.[85]

Within the warrens were pillow mounds with surrounding ditches, constructed to ensure drier conditions for the rabbits and to allow warreners to catch them, often with ferrets.[86] The mounds were generally about 150 yards long, six feet high and 20 feet across (Figure 4.7). In the early fifteenth century we learn of a rabbit warren on an assart at Nutley, offering some return by leasing land for warrens.[87] Such documentary evidence is helpful but archaeological investigations have revealed more pillow mounds than were once believed, although they are difficult to date without identifiable material. Place-names are a pointer, but cannot always be allocated to the medieval period. Thus we have Crowborough Warren, Hindleap Warren, Press Ridge Warren, Wren's Warren and Broadstone Warren from later periods on the Forest. Lidar investigations have actually doubled the number of known pillow mounds on the open Forest, and revealed one group of pillow mounds at Chelwood Gate whose outer boundary makes use of the Forest pale. This newly discovered warren includes the usual straight mound, but also circular and semi-circular constructions and a double mound.[88]

At Tabell Ghyll, on land formerly belonging to the Pippingford estate, similarly unusual pillow mounds have been discovered; one is L-shaped and three more are C-shaped. The mounds, 70 and 109 yards long and with deep ditches to the uphill sides, have not been dated but yew trees of considerable age have colonised one.

---

82 Armitage-Smith, *John of Gaunt's Register* II, 1569, 1634; Lodge and Somerville, *John of Gaunt's Register* I, 241.

83 E.M. Veale, 'The Rabbit in England', *Agricultural History Review*, 5 (1957), 85–90.

84 Salzman, 'Hundred Roll for Sussex', 40.

85 Christian, *Ashdown Forest*, 12–13; Irons, 'Aspects of the Impact of Man', 176.

86 T. Williamson, *Rabbits, Warrens and Archaeology* (Stroud: Tempus, 2007).

87 Brandon, 'Common Lands and Wastes', 109.

88 C. Butler, V. Blandford and A. Locke, *Ashdown Forest Historic Environment Resource: a LIDAR-Enhanced Archaeological Survey* (Chris Butler Associates, May 2011), 16, Fig. 13 and Plate 6.

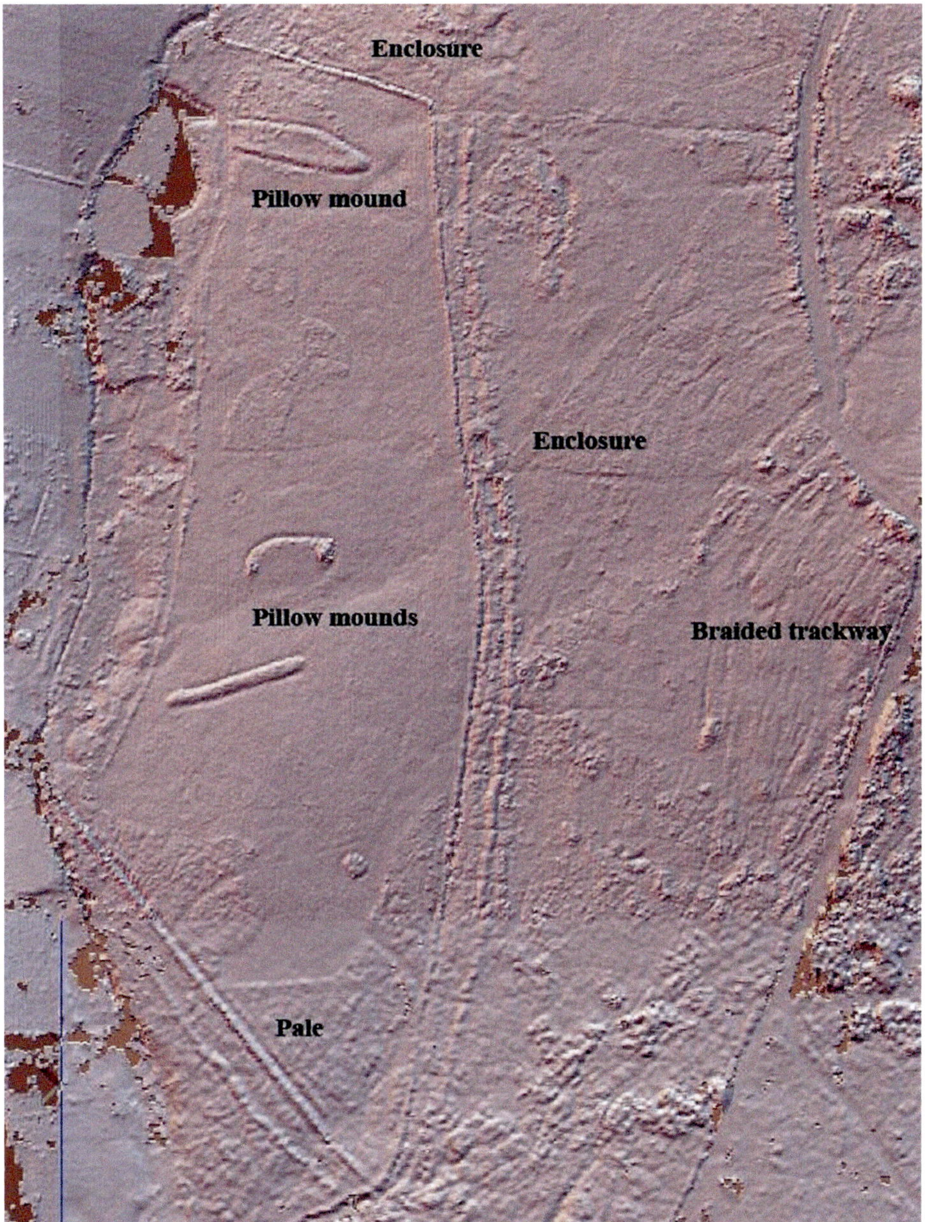

Figure 4.7. Pillow mounds near Chelwood Gate. Lidar image, with the mounds and the line of the pale running from north-west to south-east (source: C. Butler, V. Blandford and A. Locke, *Ashdown Forest Historic Environment Resource: a LiDAR-Enhanced archaeological Survey* (Chris Butler Associates, May 2011), Fig. 13).

Ashdown was clearly an important area for rabbit farming, and Sussex as a county, together with eastern England, experienced a disproportionately large number of both recorded rabbit theft investigations and numbers of free warren licences during the medieval period.[89]

THE SACKVILLES AND THE STRUGGLES FOR CONTROL ON THE FOREST

Sharing a boundary with the Forest were the lands of Buckhurst and the Sackville family, a family whose relationship to Ashdown was often close, sometimes more distant, and whose descent and significance through to the end of the medieval period should now be outlined.

Herbrand de Saucqueville sur Scie, from the Scie valley near Dieppe in Normandy, founded the dynasty. In 1086 he held the estate of Fawley in Buckinghamshire. His son Robert and wife Letitia expanded the family's interests into Essex and Suffolk. But it was Robert's eldest son, Jordan II (d. *c.* 1175), of the Buckinghamshire branch of the family, who pushed the family's interests into the Ashdown area through his marriage to Ela de Dene, heir to Buckhurst, other smaller manors, including Chalvington to the south, and the patronage of Otham, the Premonstratensian canons' house near Hailsham. In about 1208 the canons moved to Bayham, near the Kent border, and the Sackvilles of Buckhurst were buried in the choir there until the mid-fifteenth century. The abbey was suppressed in 1525 and the estate leased by Henry VIII to Anthony Browne, Lord Montague.

Although they could not aspire to the ranks of the Warennes in Lewes Rape, or the FitzAlans at Arundel, the Sackvilles were now well-endowed with lands and influence within Sussex. Sir Jordan de Sackville III (d. 1273), who supported Simon de Montfort against Henry III, was captured at the Battle of Evesham and forced to redeem his forfeited lands, straining the family's resources. His son Andrew I (d. *c.* 1297) was active in Edward I's military service in Wales and Gascony, as, in turn, was his own son Andrew II (d. 1316), who distinguished himself in the Scottish campaign. This Andrew's park at Buckhurst was forcibly entered and damaged in 1301 by poachers who carried away deer and game.[90] After his death Andrew III (*c.* 1306–1369), a minor, was married to his protector's daughter Joan de la Beche, who eventually inherited the manors of Chiddingly, Claverham, Arlington and Waldron, making the Sackvilles now one of the richest landed families in east Sussex.

Andrew III continued his family's military tradition with involvement in the early campaigns of the Hundred Years War, being present at Crecy. But he was also a good administrator, serving the earls of Arundel and Salisbury, Roger Mortimer, earl of March, and Hugh, Lord Despenser, as well as being appointed sheriff of Sussex and

89 D. Gould, 'The Distribution of Rabbit Warrens in Medieval England: an East–West Divide?' *Landscape History*, 38 (2017), 25–41.

90 Rev. C.N. Sutton, *A Short History of Withyham and Buckhurst* (Tunbridge Wells: A.K. Baldwin, 1893), 9.

Surrey (1366–8) and being three times MP for Sussex. On his death his estates were settled on a bastard heir because his two sons had predeceased him. This heir was Thomas Sackville V (1342–1432), whose succession by 1377 was later challenged unsuccessfully by Sir Thomas Sackville of Fawley, of the senior branch of the family who had inherited directly from Herbrand. In the 1390s he was elected three times to parliament, but in 1400 he was arrested under suspicion of involvement in the 'Epiphany plot' against Henry IV, although he was later released. Thereafter he seems to have put his sometimes turbulent past behind him. He was sheriff of Sussex (1406–7) and in 1418 and 1421 was a Commissioner of Array. When he died in December 1432 he left 100 marks to Bayham Abbey, where he was buried.[91]

ANARCHY ON THE FOREST

In the years following the Black Death there are signs that lordly control over Ashdown was contested. Thomas Sackville V, coming to the lordship of Buckhurst by 1377, five years after John of Gaunt's control over Ashdown had been confirmed, strengthened his position in Sussex by an alliance with the locally powerful Sir Edward Dallingridge (*c.* 1346–1393), whose daughter Margaret he married. In the 1380s he gave his backing to Dallingridge, who was then conducting a campaign of intimidation against the Duchy of Lancaster's officials on Ashdown, and in 1384, with his father-in-law, suffered brief imprisonment for his actions, receiving a royal pardon the following year.

Opposition to the duchy's rule came from some local manorial lords who considered themselves exempt from Gaunt's administration, especially those whose manors had been allocated to the Honour of Leicester. Sir Edward Dallingridge, who later built Bodiam Castle, was lord of Bolebrook, Sheffield and other manors, as well as master forester for Hugh Lord Despenser's chase at Rotherfield, and was loyal to the earl of Arundel. Another powerful family, the De la Warrs, originally from Gloucestershire, who had built up influence in the Midlands and locally through a series of judicious marriages, also resented the Lancastrian intrusion into Pevensey Rape, and worked closely with Dallingridge to subvert Gaunt's officials. The De la Warrs were now firmly established in the second tier of Sussex elites, holding land that included the lordship of Fletching.[92]

Gaunt thus faced considerable opposition from three powerful families around Ashdown: the Dallingridges, Sackvilles and De la Warrs.[93] But there had been ill-feeling long before. In July 1314 an oyer and terminer case had been brought by the Dowager Queen Margaret against the brothers John and William Dallingridge alleging that they

---

91 Nigel Saul, 'Sackville family (*per. c.*1066–*c.*1477)', *ODNB* (2004); <http://www.historyofparliamentonline.org/volume/1386-1421/member/sackville-sir-thomas-ii-1432> [accessed 15 June 2019].

92 K. Fildes, 'The Baronage in the Reign of Richard II, 1377–1399' (Univ. of Sheffield unpub. PhD, 2009), 209–48, 291.

93 For more detail see S. Walker, *The Lancastrian affinity 1361–1399* (Oxford: Clarendon Press, 1990), 117–41.

> had entered her free chase in Ashdown Park at Maresfield and her free warrens there and elsewhere, hunted therein without licence, took deer, hares, partridge and pheasants, felled trees and fished in her stews, carrying everything away.

This had arisen from a complaint made by Reyner le Brom, parson of Hartfield and from another influential local family, who had petitioned the king that 'there has been waste and destruction of the woods newly made by the Dallingridge brothers, foresters of the chase to the great damage of the king' He requested that the king 'order certain of his privy justices with Sackville [Sir Andrew Sackville II] who is aware of the matters, as there is evidence if the justice come, but if they don't come, the Dallingridges have procured juries for these trespasses and others to the disinheritance of the King'. He hoped that Sackville and the king's men would visit 'and view the damage and waste which has been made in the same place, and diligently enquire by whom it has been done, and return the inquisition before the King'.[94]

Men from around Ashdown also ventured further afield in illicit deer-taking. In 1375 members of the Cat family from Rotherfield, together with others from Frant and Withyham, broke into the archbishop of Canterbury's park in Wadhurst, taking deer, wounding others with arrows and severely beating the park keeper and his servant.[95] Subsequently John Dallingridge's grandson Edward (*c.* 1346–1393) took a lead in attacking the authority of Gaunt. In June 1377 Ashdown was illegally hunted and lands in Fletching and East Grinstead despoiled. In June 1379, and again in February 1380, warrants were issued to Peter atte Mersh, Gaunt's steward in Sussex, the prior of Michelham, his receiver, and chief forester John Culpeper, giving instructions to pay the bishop of Chichester the tithe for the pannage of Ashdown to which he was found to be entitled by the jury at a Duddleswell court.[96] In March 1380 underwood on Ashdown was fired and on Good Friday 1381 it was claimed that Dallingridge's men drove off livestock belonging to Gaunt at Fletching. John Delves, Gaunt's Lancastrian feodary in Sussex, was also attacked at Ringmer in June of that year.

This period of anarchy, coinciding with the Peasants' Rebellion of 1381 against the unpopular Gaunt and other royal officials, seems to have witnessed some decay on the Forest. Gaunt's closer administration through his own officials also undoubtedly came as an unpleasant shock to many on the Forest, increasing his unpopularity. Steward Sir John Seynclere of Brambletye may have been retaining profits illegally, and his foresters were guilty of illegal hunting and petty extortion. Maresfield manor was in poor condition and receipts from manorial dues were much in arrears.

---

94 TNA, SC 8/36/1783, Petitioners: Reyner le Brom, Parson of the church of Hartfield etc.

95 W.D. Cooper, 'Produce of and Supplies from Sussex', *SAC*, 17 (1865), 115–22.

96 ESBHRO, ACC 1831/1/63, Copy warrants relating to Ashdown Forest for the register of John of Gaunt, 16 Feb 1375, 14 Feb 1380 [copy of 1814].

It was later claimed by Gaunt that Dallingridge and his son-in-law Thomas Sackville, with Philip de Mynstede, who had been sheltered by Dallingridge after a murder, together with John Aleyn, Nichlas Selewyne, John Walsshman, John Kynnere (of Hartfield) and others, had not only disrupted the steward's court but had also 'broke the duke's park at Marsfeld, entered and hunted therein, and in his chace at Asshedon, took his goods and chattels at Eastgrenstede, and deer from his said park and chace, and assaulted his servants there'. It was also claimed that there was general unrest: 'homicides, felonies and other offences in the hundreds of Eastgrenestede, Hertfeld, Rusheyndenn ... Retherfeld', together with 'withdrawals, detention and concealment therein of wardships, marriages, reliefs, escheats etc.'[97]

Profits to Gaunt's courts were reduced at Maresfield and Pevensey, and no courts could be held in Pevensey honour in 1381 because of the general turbulence. That year Gaunt's palace on the Strand in London, the Savoy, was demolished, and many possessions, including many records, were destroyed. Now, hounded by the Wat Tyler rebels, Gaunt chose to placate Edward Dallingridge by appointing him in August 1381 'head keeper of Ashdown chase' (master forester), 'with full power to do all that pertains to the office, receiving the usual fees and wages [and] to hold the office at the will of the Duke'. He also seems to have been permitted to hunt on Ashdown during this period. In October 1382 we find Gaunt's steward, Brook of East Grinstead and Hartfield, together with Dallingridge, Nichole Boille and William de Marschalle being permitted to fell, cut and sell 'mortboys, souzboys et birches', windfall wood, underwood and birch from Ashdown and Maresfield Park for the Duchy. The proceeds from this sale were to be used to enclose the said chase and park under the supervision of John Brook, 'steward in those parts', who also witnessed the order. Thereafter the two men, the military man and Brook, the lawyer, worked together until Sir Edward's death in 1393, although there was another altercation in 1384 (see below). Sir Edward was relieved of the master forestership in August 1383, to be succeeded by Robert de Assheton, 'with full power to do all that pertains'.[98]

The unrest around Ashdown and the animosity towards the duchy's officials continued alongside general disorder in Sussex and elsewhere at this time. And on Ashdown this culminated in an attack on Nicholas Boyle, Gaunt's ranger on the Forest, and the death of William Mouse from Hartfield, an under-forester in March 1384. Furthermore, Sir Thomas Hungerford, speaker of the House of Commons and an elderly but very influential chief steward for the duke's household, was prevented from levying a fine of 10s from Fletching because he was refused access by the villagers. In 1385 we have a record of Sir Thomas holding a tourn at Maresfield. This was part of a circuit to the hundreds of his administration, where he would preside over a full meeting of the hundred court – another annoying interference with the power of the

97 PRO, *Cal. Patent Rolls AD 1381–1385* (London: HMSO, 1897), for 16 and 18 June 1384, pp. 438–9.

98 Lodge and Somerville, *John of Gaunt's Register* I, 1119.

local lords. A jury again conducted an inquest into the pre-Lancastrian right of the bishop of Chichester to tithe of pannage from the Forest.[99]

These disputes also included the interruption of a court held at Hungry Hatch (*Hungerhacche*), south of Fletching, by Gaunt's steward John Brook, which we hear about in June 1384. Dallingridge and his followers compelled Brook, on oath, never to call a duchy court there again. And, at about the same time, Brook and one of the duchy auditors were instructed to inquire into a petition from Dallingridge. When brought to a suit at East Grinstead to answer a special commission of oyer and terminer by Gaunt in June 1384, Dallingridge claimed that it had been Tyler's rebels, gathered together with the intention of killing the unpopular duke's officials, who had attacked Brook in 1381; he had done his best to pacify them, as befitted his influential position.[100] He also claimed innocence of the charges of trespass and hunting offences, the latter because he saw hunting as his lordly prerogative and thus a challenge to his status, but he admitted the disruption of the court at Hungry Hatch, as this was an innovation by the duke and drew suitors away from his own court at Danehill.[101] He also admitted carrying off cattle claimed by the duke at Fletching, since he claimed that they were due to him as forefeited chattels of a convicted felon and thus rightfully his, as lord of the hundred of Danehill Horsted.[102]

Much of Dallingridge's defence centred around the loss of local influence among his tenantry due to the Lancastrian newcomers. But the justices were overwhelmingly drawn from Gaunt's circle and a bribed witness testified to involvement in all the crimes alleged against Dallingridge. The latter was an unruly figure in the court, and he was acquitted only of the charge of burning the duke's brushwood on Ashdown in March 1380 and the attack at Ringmer. He was briefly imprisoned for contempt of court.

Nevertheless, Dallingridge's stand against the duchy enhanced his influence locally, and Gaunt was thereafter less exacting in his control of Ashdown. In 1385 Selewyne, Aleyn, Walsshman and Skynner were pardoned for the death of William Mouse, as was Thomas Sackville for harbouring, aiding and abetting them.[103] In 1387 the court at Hungry Hatch, the scene of Dallingridge's affray, was abandoned by Gaunt and his

---

99 S. Walker, 'Lancaster v. Dallingridge: a Franchisal Dispute in Fourteenth-century Sussex', *SAC*, 121 (1983), 87–94. Thomas Hungerford (d.1398), was Gaunt's chief steward south of the Thames and later south of the Trent, 1375–*c.* 1393 <http://www.historyofparliamentonline.org/volume/1386-1421/member/hungerford-sir-thomas-1397> [accessed 11 March 2022]; Christopher Whittick (personal communication); Walker, *Lancastrian Affinity*, 131.

100 <http://www.historyofparliamentonline.org/volume/1386-1421/member/broke-john-1417> [accessed 2 June 2015].

101 Danehill was *Denne* (1279), later *Denhill* (Mawer and Stenton, *Sussex*, 2), 335.

102 Walker, *Lancastrian Affinity*, 134–5; A. Hicks, 'Hunting, Status and Landscape in Fourteenth-century England: a Dispute between John of Gaunt and Sir Edward Dallingridge', paper given at Australian and New Zealand Association for Medieval and Early Modern Studies third conference 5–8 July 2001, 'Metamorphoses – Peoples, Places, Times' (2001); TNA, JUST 1/947/5; 1/944 General oyer and terminer roll, Sussex, inquiry into felonies and trespasses in Pevensey rape 1384.

103 PRO, *Cal. Patent Rolls AD 1381–1385*, 17 June 1385, at Eltham, p. 590.

revenues from Ashdown fell steeply in this compromise move. Gaunt's third wife, Katherine Swynford, inherited the Forest when Gaunt died in February 1399, together with Maresfield and Withyham Priory, holding these until her death in 1403, when the Forest came under royal control once more.

In 1393 a new power became apparent: John Pelham of Laughton was appointed constable of Pevensey Castle. He took the castle on behalf of Bolingbroke (the future Henry IV), resisting a siege by partisans of Richard II with the shire levies. With Richard's death in 1400, Pelham remained steadfast in his support for the new king, and in return that year he was rewarded for his recovery of Pevensey Castle with its constableship, coupled with a gift of the land nearby that he had previously tenanted. His local standing was further enhanced in 1404 with the stewardship of the duchy's Honour of Pevensey. The following year he was appointed keeper of the New Forest. He was elected to parliament to represent Sussex on five occasions and gained a substantial income from a royal annuity, from many offices and from lands. He was also appointed chief steward of the southern estates of the Duchy of Lancaster in 1405, and in February 1409 the king made him a lucrative grant for life of the lordship of Pevensey, together with all other duchy of Lancaster estates in Sussex. Now, with his estates in Sussex, he became possibly the greatest landowner in the county. In 1411 he became treasurer of the Exchequer.[104] Notwithstanding his national significance, however, in 1418 Pelham was charged with the sale and destruction of timber on Ashdown and in Maresfield.[105] He died in 1429.

The fifteenth century was an unsettled period: wars in France and the internal Wars of the Roses were joined by Cade's rebellion in Kent in 1450. When Edward IV married Elizabeth Woodville in 1464 some duchy lands were settled on her for life, including Pevensey, together with the manors of Willingdon, Maresfield, East Grinstead, Seaford and Endlewick. Edward IV died in April 1483, to be succeeded by Edward V (although never crowned) for just ten weeks. The throne was then 'usurped' by Edward IV's brother, the duke of Gloucester, and Elizabeth was stripped of her lands. Richard III was the last of the Plantagenet kings, and his death at Bosworth Field ushered in a new royal lineage, that of the Tudors.

A FORMATIVE TIME FOR THE FOREST

This area was by no means prosperous compared to many other locations in South East England, having below-average wealth and lower numbers of taxpayers compared with, for example, the downland, the Sussex coastal plain or even Wealden Kent. The 1327 Lay Subsidy and the Nonae Return 1341 both indicate sparser populations and lower average payments in the collection areas of East Grinstead and Maresfield.[106] By the

104 <https://www.historyofparliamentonline.org/volume/1386-1421/member/pelham-john-1429> [accessed 23 June 2015].

105 Saul, *Scenes from Provincial Life*, 71.

106 Campbell and Bartley, *England on the Eve of the Black Death*, 347–8; R.A. Pelham, 'Studies in the Historical Geography of Medieval Sussex', *SAC*, 72 (1931), 157–84. The 1327 tax was

fourteenth century the Weald had become a region of mixed farming with an emphasis on animal husbandry, but the 1334 tax assessments on moveable wealth for the hundreds encompassing Ashdown were still among the lowest in South East England, at about 4s–15s per square mile, compared with 30s–33s on the coastal plain.[107]

Nevertheless, even with a period of stagnation or decline for Ashdown Forest, the medieval centuries were formative. Population and settlements were growing; churches and their parishes had been formed and were becoming the arbiters of acceptable behaviour and local issues; parks and woodland created a landscape mosaic; the old dens and their droveways were losing their importance; assarting had created a now-familiar landscape of small tenements and scattered cottages; and industry offered occupations alongside farming for many. Manorial development had also been gaining a hold, albeit through the process of subinfeudation and often still dependent on older centres elsewhere. But this marginal area was now clearly absorbed through a variety of contacts into the mainstream of south-eastern economic and social life. Of course, for most people, local issues internal to life on the Forest undoubtedly predominated over the concerns of feudal elites.

one-twentieth of the value of goods (persons with values of less than 10s being exempt); the Nonae Returns were of one-ninth of the value of corn, wool and lambs.

107 R.E. Glasscock, 'The Distribution of Lay Wealth in Kent, Surrey and Sussex, in the Early Fourteenth century', *Archaeologia Cantiana*, 80 (1965), 61–8.

CHAPTER 5

SOCIETY AND COMMUNITY ON ASHDOWN FOREST, 1500–1800

Common usage of commons was not a charity for the weakest in the village, it was a resource for almost everyone.[1]

THE ACCESSION OF HENRY VII to the English throne in 1485 ushered in a period on Ashdown for which we have unprecedented amounts of information and much more secondary analysis, quite unlike the preceding centuries. We can therefore gain a clearer idea of the landscape and the main actors and their interactions on or around Ashdown, and we now unfold something of the fortunes of the elite families, the landholders, the farming community and the many poor.

Following his victory at Bosworth the whole Duchy of Lancaster was vested in Henry VII, and the duchy thereafter was merged with the crown, although it continued to keep separate accounts. Following medieval precedent, Ashdown now went to Henry's wife, Elizabeth of York, a procedure also later followed by Henry VIII for Katherine of Aragon, Anne Boleyn, Jane Seymour and Katherine Howard.[2]

It is well known that Henry VII tightened expenditure to rebuild the royal finances: more feudal dues were exacted and attempts were made in 1489 to curtail illegal hunting in Ashdown Forest, where it was also seen that Maresfield tenants claiming common of pasture in Ashdown were misusing the custom to the detriment of the deer. Henry's Scottish campaigns in 1485 and 1496 also had implications for Ashdown, as we shall see, and we also find that Chelwood Vachery, formerly belonging to Michelham Priory, was now absorbed into duchy control.

A hierarchy of officials continued to administer Ashdown. The primary post remained that of master forester, a sinecure for members of the elite who might combine the post with being master of game. In turn, he normally employed a steward, and we find, for example, Thomas Fiennes (Lord Dacre) as steward of Ashdown in July 1519 and other well-connected men involved. George Neville, Lord Abergavenny, master forester 1503–22, was followed by Sir Thomas Boleyn of Hever (d. 1539) and George, his son (executed for treason in 1536); Thomas Culpeper in May 1541; and Sir

[1] J. Neeson, *Commoners: Common Right, Enclosure and Social Change in England, 1700–1820* (Cambridge: Cambridge University Press, 1993), 174.

[2] Somerville, *History of the Duchy of Lancaster* I, 260–86.

Anthony Browne, son-in-law of Sir John Gage, in May 1542. Richard Sackville (1549–53), John Gage (1554) and Richard and Thomas Sackville (1561) continued this line of succession. The Gages continued to exert a major influence in the area and the Boleyn interest was also demonstrated by Thomas Boleyn's lease of the agistment, pannage and rabbits of Maresfield Park for £4 a year.[3] Below the master forester and his steward there would also be a ranger – still a member of the local gentry but in a more 'hands-on' capacity, and normally residing at Old Lodge.[4]

ELITES: THE SACKVILLES

The relationship between the Sackvilles and Ashdown again fluctuated over the course of the sixteenth, seventeenth and eighteenth centuries. They increased their national significance as courtiers and politicians, initially nailing their colours firmly to the Tudor mast, but their Ashdown links became more tenuous, as we shall see, after their move to Knole in Kent. They did, however, continue in their almost hereditary role as MPs for East Grinstead, following on from the first to hold that position, 'John Sakevyle' (1484–1557) in 1541–2.[5] John, also sheriff of Sussex and Surrey under Henry VIII, married Margaret Boleyn (*c.* 1489–1556). By now the Sackvilles were being buried in the church at Withyham, following the suppression of Bayham Abbey in 1525, with John's father Richard (1460–1524) stating in his will that his burial was to be in 'the Lady Ile wher I am wont to sytt'. This, in the north aisle of the nave, became known as the Sackville Chapel or Lady Chapel, and will be described more fully later.[6]

John's son Sir Richard Sackville (1506–66), Anne Boleyn's cousin, enhanced the family's fortunes through family connections and astute dealings in the land market. He was a trusted administrator and a steward of the archbishop of Canterbury's lands in Sussex, and was knighted by 1549. Related to Queen Elizabeth through the Boleyns, he became a member of her Privy Council and MP for Sussex in 1559 and 1563. Although a Protestant, he was also steward for the Catholic earl of Arundel; was escheator of Surrey and Sussex in 1542; and, as we have seen, master forester of Ashdown in 1549. He was appointed as Edward VI's Chancellor of Augmentations in August 1548 until the post was absorbed into the Exchequer in 1554 and joined the commission to value the remaining chantry lands in Sussex. Chantries, serving for the maintenance of prayers (obits) for the dead, were valued at Broadhurst (Horsted Keynes), Rotherfield, Fletching, Withyham and West Hoathly. A chantry in Maresfield had 'the priests felde'. At Hartfield 8lb of wax 'for the light before the image of the

---

3 M. Mate, *Trade and Economic Developments 1450–1550: the Experience of Kent, Surrey and Sussex* (Woodbridge: Boydell, 2006), 174–5.

4 L. Merricks, 'Forest and Waste in Seventeenth century England: the Enclosure of Ashdown Forest, 1600–1700' (Univ. of Sussex unpub. DPhil, 1989), 48.

5 A. Stenning, 'A Return of the Members of Parliament for the County and Boroughs of Sussex', *SAC*, 32 (1882), 159.

6 Sutton, *A Short History of Withyham and Buckhurst*, 39–60. The chapel, badly damaged in the fire at the church in 1663, was beautifully restored. See Figure 5.4.

blessed Mary' was valued. A chantry at Brambletye was also valued, but no incumbent priest seems to have been resident.[7]

Richard's own acquisitions, especially of ex-monastic land, earned him the nickname of 'Fillsack'. In 1564 he acquired valuable land in London, from Fleet Street to the Thames. More locally, however, an attempt to accumulate more land by force at 'Dallyngrigge' (Hartfield) was resisted by Walter Waleys, who took the case to the Star Chamber for adjudication, complaining that Sackville's men had broken in and driven away cattle. He felt that common law would not suffice as 'the said Richard Sakevile is of great might in the said county, so that no officer there dare serve any process against him'.[8]

Evidently, the Honour of the Eagle, most of the old Pevensey Rape, was now being broken up. In 1554 Mary Tudor had granted parts of the Honour, including the manor of Duddleswell and some of the iron and steel forges, such as Stumbletts Furnace (on the borders of Maresfield and West Hoathly), to the Catholic John Gage (1479–1556), Lord Chamberlain of the Queen's Household and Chancellor of the Duchy of Lancaster, in consideration of his good services to the late kings Henry and Edward. Sir John, as we have seen, was also appointed master forester of Ashdown in June 1554.

As a strongly religious conservative family – Sir Edward Gage (*c.* 1500–1569) was organiser of the Marian persecutions and burnings in Lewes in 1555–7 – it is also likely that the Gages would have had little time for the Sackvilles, who were followers of the reformed church.[9] A dispute between the families arose in 1560–62 over the supply of iron ore from Maresfield for Sackville's Sheffield ironworks in Fletching[10] and in 1563 a dispute arose between Sir Richard Sackville and Sir Edward Gage, son of Sir John and lord of the manor of Maresfield. This resulted in what must be the earliest map of Ashdown Forest, made in 1563, describing it as 'all heath and beache woodde', and with Buckhurst described as 'of like ground as the forest save that it is somewhat bettered by industrie' (Figure 5.1). An inset diagrammatic plan shows the manor of Chaunton (Chalvington) and its glebe land, the 'Deane land', a highway, and the hedge causing the dispute. Notes on the plan suggest that it was made by Sir Richard Sackville. Chalvington was an early Sackville possession.

Sir Richard and his wife Winifred had three sons, of whom only Thomas Sackville survived. As well as Buckhurst and other Sussex land, Thomas (1536–1608) inherited land in Essex, Oxfordshire, Yorkshire and Kent, where he was granted the reversion to Knole, though he had to wait nearly 40 years for possession. A grandiose mansion had been planned for Buckhurst but never built, and he actually spent many of his later years in Lewes and planning the move from ancestral Buckhurst to Knole. There is today just the remains of the Tudor gatehouse from the planned layout remaining at Old Buckhurst (Figure 5.2).[11]

---

7 J.E. Ray, *Sussex Chantry Records*, SRS 36 (Lewes, 1931), 183.

8 TNA, STAC 2/23/69.

9 1569 is the date on his wall tomb in West Firle church, and confirmed in ESBHRO, SAS G/19/14 (N. Llewellyn, *East Sussex Church Monuments 1530–1830*, SRS 93 (Lewes, 2011), 382.

10 ESBHRO, SAS G/6/50 Correspondence between Sackville and Gage re Maresfield.

11 R. Sackville-West, *Inheritance: the story of Knole and the Sackvilles* (London: Bloomsbury, 2010), 11–13.

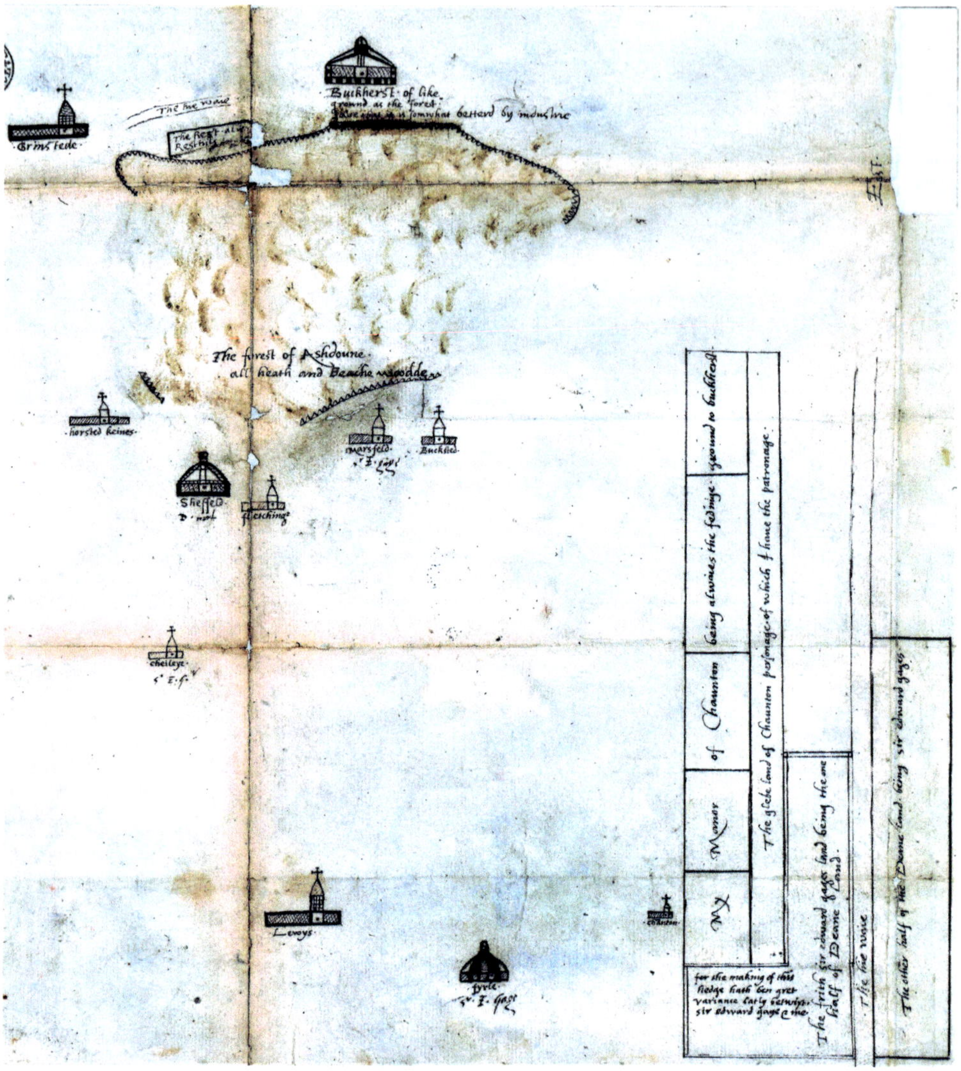

Figure 5.1. The earliest map of Ashdown Forest 1563. 'A platt of the land in contraversie betwixt Sir Richard Sackvil[l]e and Sir Edward Gage' (source: TNA: MPF 1/144, extracted from SP 15/11 [item 118]). 'The forest of Ashdown all heath and beache woodde'; 'Buckhurst – of like ground as the forest ... But bettered by industrie'.

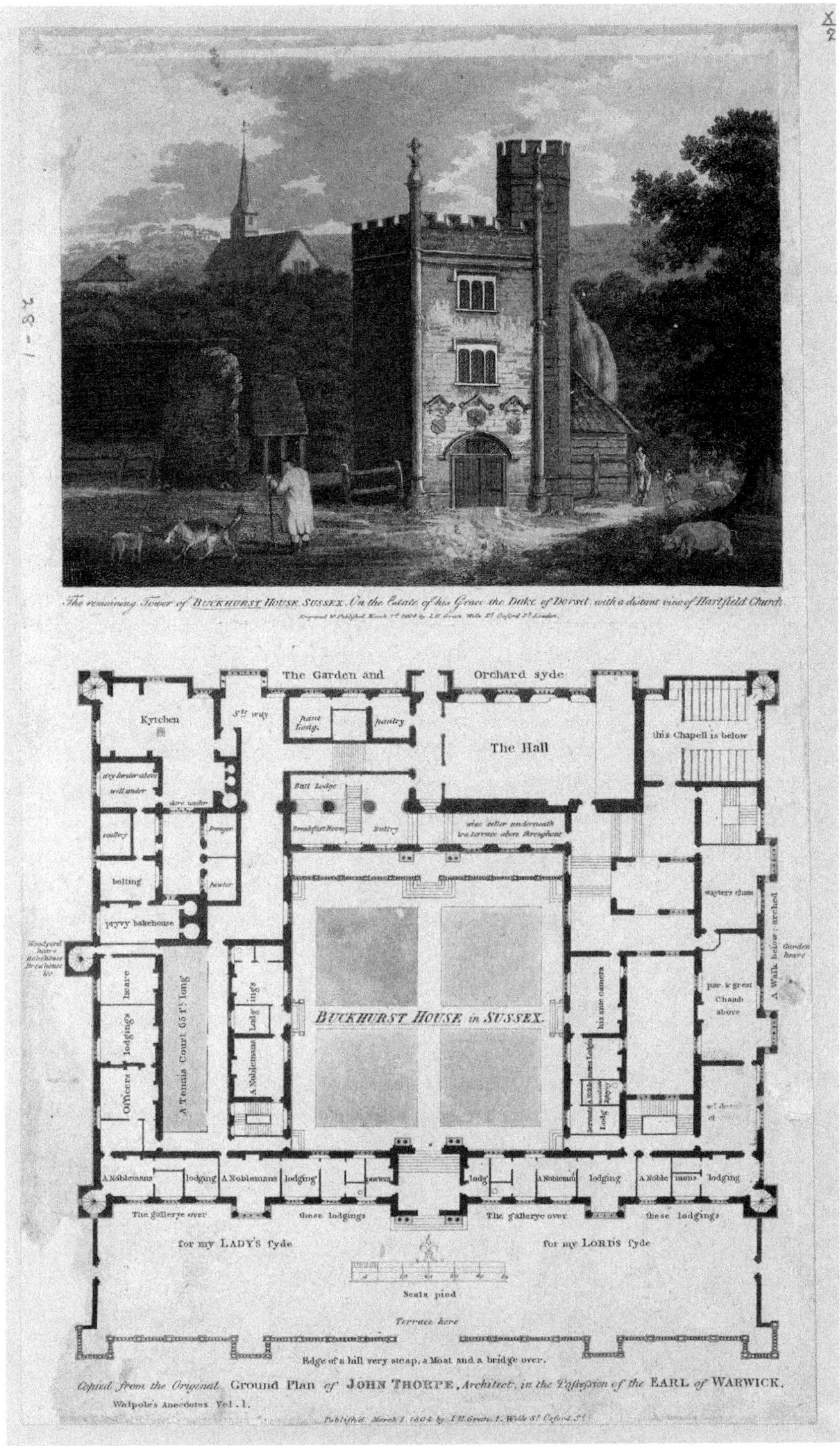

Figure 5.2. Buckhurst: the Tudor gatehouse in 1804 (source: © The British Library Board maps/King George/top 42/28/1).

As courtier, diplomat and MP (again for East Grinstead (1559), where the family owned the majority of the burgage tenements), it was with Thomas that the Sackvilles' ascent into the peerage began. Known also as a poet and dramatist, he married Cicely Baker, of the influential Sissinghurst family, in 1555. Diplomatic successes earned him a knighthood and the title of Baron Buckhurst in 1567, and in 1569 he was appointed a joint lord lieutenant of Sussex, showing himself a moderate Protestant. And he was immensely powerful locally, becoming master forester of Ashdown for life and acquiring wood from the Forest for his iron furnaces. More land in Sussex was now purchased, and he was engaged with such local issues as the drainage of Pevensey marshes. He obtained the manor and advowson of the rectory of Withyham in 1569 from King's College Cambridge, installing a relative as rector.[12] In 1586 he became a privy councillor and, as a trusted advisor to the queen, was appointed Lord High Treasurer in succession to Burghley in 1599. He was created earl of Dorset in 1604.

The influence of the Sackvilles continued through the first half of the seventeenth century, with wealth from property, office holding and their ironworks. Now the undisputed leading family in the area, in 1597 they held 16 manors covering between 16,500 and 17,000 acres.[13] Robert Sackville (1561–1609), briefly second earl of Dorset, survived his father by less than one year, but through paternal influence had previously been MP for Sussex and Lewes. He died at Dorset House (originally Sackville House) in London, but was buried in the Sackville Chapel at Withyham. He had also prospered from trade and again especially from the supply of ordnance. He had married Margaret, daughter (and a suspected Catholic) of the fourth duke of Norfolk, and left money for the building of 'Sackville College for the poor' in East Grinstead, later completed in 1622 using stone from the abandoned Buckhurst project.[14] But by this time the Sackvilles' national ambitions greatly overshadowed their Sussex interests, demonstrated by their move from Buckhurst to Knole, closer to London.

Robert left behind two sons, each inheriting the earldom in succession: Richard, third earl (1589–1624) and Edward, fourth earl (1590–1652). Leading an extravagant lifestyle with a large retinue, Richard was considered 'one of the seventeenth century's most accomplished gamblers and wastrels', sometimes losing at cockfighting with the king.[15] There were repeated infidelities, but he died without a male heir, three sons predeceasing him, and leaving two daughters. He was a leading figure in the Sussex lieutenancy after 1615, attempting to sharpen up the Sussex militia and its musters, which had been neglected. In September 1609 he was granted 'the offices of Master of the King's Forest of Ashdown ... and of the Game thereof; of Steward, &c., of the Honor

---

12 R.B. Manning, *Religion and Society in Elizabethan Sussex* (Leicester: Leicester University Press, 1969), 171–2.

13 E. Straker, *The Buckhurst Terrier*, SRS 39 (Lewes, 1933); Gulley, 'Wealden Landscape', 153.

14 The college includes Lower Tunbridge Wells Sandstone in the walls, Ardingly Sandstone for the dressings and Horsham Stone for the roof (Birch and Cordiner, *Building Stones of West Sussex*, 39).

15 R.M. Cooper, *The Literary Guide and Companion to Southern England* (Athens, OH: Ohio University Press, 1998); K. Hodgkin, 'The Diary of Lady Anne Clifford: A Study of Class and Gender in the Seventeenth Century', *History Workshop Journal*, 19 (1985), 148–61.

of Aquila Castle of Pevensey; also of the office of Feodary of the said Duchy lands in the said county'.[16] As a perquisite arising from his office as master, he held regular hunting parties on Ashdown that included visits from the king. He sold land worth about £80,000 but still died £60,000 in debt.

Richard was also perhaps best known as the first husband of the redoubtable Lady Anne Clifford, long embroiled in legal battles to gain what she considered to be her lawful estates in Yorkshire, who he married two days before his father died. Dorset's death in 1624 made Countess Anne wealthy. However, she encountered 'the envie, malice and sinister practices' of her enemies, among them her brother-in-law Edward, fourth earl of Dorset, who resented her occupying so much Sackville property.[17]

Lacking a male heir, Richard was succeeded by his younger brother Edward, who had been MP for Sussex in 1620 and who dutifully succeeded his brother as lord lieutenant, again enhancing the system of musters and improving the training and equipment. He also became a privy councillor from 1626. An active and effective orator, he was nevertheless in and out of favour at James' court and spent some years on the continent. He was, however, generally favoured by Charles I. But in March 1642 he ceased to attend the Lords, surrendered his commissions of lieutenancy for Sussex and Middlesex and probably joined the king at York during May. Throughout that summer he desperately tried, but failed, to persuade the intractable king to open meaningful negotiations with parliament. Knole was occupied by the army that year, sequestered in 1643 and again occupied in 1644–7. Edward was appointed Lord Privy Seal in 1644, but suffered the murder of his son Edward while a prisoner of the parliamentary army in 1646, and his role in public life thereafter became more intermittent. He never left his London house again following the king's execution, and so had little further real influence on his Sussex estates, which were anyway sequestered and placed in the care of the earl of Pembroke, Lady Anne Clifford's second husband. Driven into debt after compounding, Edward was forced to borrow heavily and in February 1652, shortly before his death, he bemoaned his fate as 'a poore unsuccessefull Cavalier'.[18]

His only surviving son, Richard, later fifth earl (1622–1677), sat in the Long Parliament for East Grinstead but was excluded for royalism, although there is no evidence that he took part in the civil wars. At the Restoration, as we shall see, he was appointed keeper of the Forest but without an absolute grant. From 1670 Richard served, until his death, as lord lieutenant of Sussex and, among other committees, was on the commission for the trial of the regicides.

He died at Knole but had secured the family's failing finances by his marriage to the heiress Frances Cranfield, daughter of Lionel, earl of Middlesex. He seems to have been more financially responsible than many of his predecessors, in 1673 dismissing his then steward Arthur Walker for faulty accounting.[19] Richard and Frances had 13

---

16 *State Papers Domestic* James I, XLVIII, [57a] (1857), 545.

17 R. Spence, 'Clifford, Anne, countess of Pembroke, Dorset, and Montgomery (1590–1676)', *ODNB* (2004).

18 D.L. Smith, 'Sackville, Edward, fourth earl of Dorset (1590–1652)', *ODNB* (2004).

19 D.R. Hainsworth, *Stewards, Lords and People: the Estate Steward and his World in later Stuart England* (Cambridge: Cambridge University Press, 1992), 25–34, 44.

children and the death of their youngest son Thomas, in 1675, is commemorated in the wonderful tomb-chest (1678) by Cibber in the Sackville Chapel at Withyham.

Richard was succeeded by his son Charles (1643–1706), who thus became sixth earl of Dorset in 1677, being already earl of Middlesex through his mother's family. He was known as a poet and courtier, and as a supporter of the invitation to William of Orange became a privy counsellor, Lord Chamberlain (1689 and 1697) and Knight of the Garter (1692). His local interests, however, were quite different from those of his moderate father. He needed income to support his court activities and was less interested in Ashdown's Commoners, becoming a key figure in a scheme to enclose Ashdown in 1679, which, as we shall see, came to nothing.

In turn Charles was succeeded by his son Lionel Cranfield Sackville (1688–1765), seventh earl of Dorset and second earl of Middlesex (Figure 5.3). He was a capable administrator and an enthusiastic promoter of the Hanoverian succession, and was chosen to go to Hanover on the death of Queen Anne in 1714 to inform George, Elector of Hanover, of his accession as George I. Numerous offices followed, from both George II and George III. Created first duke of Dorset in June 1720, he served twice as lord lieutenant of Ireland (1731–7 and 1751–5), but was recalled following involvement in political issues there. He had married Elizabeth Colyear in 1709 and they had six children.[20]

Here, then, was yet another Sackville moving on the national, even international, stage. Although both born and dying at Knole, he too was also nevertheless buried at Withyham. His interest in the Ashdown area fluctuated, but by the mid-1720s he was staying at Stoneland Park and organising the park landscape.[21] As we shall see, he was also responsible for bringing the Forest of Ashdown into the Sackville family by his purchase of the unenclosed parts in 1730.

Lionel was succeeded in turn by his eldest son Charles (1711–1769) in 1765. Following his return from the fashionable Grand Tour in 1731–3, Charles took the family parliamentary seat at East Grinstead, and in the 1740s was MP for Sussex and then Old Sarum. In February 1766 he was appointed privy councillor and lord lieutenant of Kent. He was reputedly somewhat dissolute, but there were no children from his 1744 marriage to Grace Boyle (1723–1763) and he outlived his father by only four years. By this time the Sackvilles possessed a vast spread of land and resources. There was Buckhurst, Newnham [alias Newenham] and Stoneland parks and the lands around Ashdown, including the Five Hundred Acres and the Vachery. In other parts of eastern Sussex there were Broyle Park, several boroughs, more than 20 manors, a share of Lewes Castle and manor, tithes and advowsons, as well as Knole itself, and lands in Essex and the Midlands.[22]

---

20 R.E. Burns, 'Sackville, Lionel Cranfield, first duke of Dorset (1688–1765)', *ODNB* (2004).

21 'Stoneland', <historicengland.org.uk/listing/the-list/list-entry/1000230> [accessed 12 September 2021]. In 1747 the duke was paying tax of £5 2s for the 100 windows ('lights') at Stoneland, about one-seventh of the entire parish tax for Withyham (R. Davey, 'East Sussex Window and House Tax Assessments 1747', <http://www.sussexrecordsociety.org/olb/wtax/> [accessed 9 July 2020].

22 ESBHRO, AMS 5481/1. Marriage settlement for £20,000 ([lease and] release).

Figure 5.3. Lionel Cranfield Sackville, first duke of Dorset (1688–1765). By John Faber Jr, after Sir Godfrey Kneller, Bt, mezzotint, 1731 (the year following his acquisition of Ashdown Forest) (source: © National Portrait Gallery NPG D33101).

The youngest of Lionel's sons, and reputedly his favourite, was Lord George Sackville (1716–1785). A military career was initially successful, but he was disgraced and court martialled for refusing to attack at a crucial moment in the Battle of Minden (1759) during the Seven Years War. Nevertheless, in November 1775 he was appointed Secretary of State for the American Department, at the behest of George III, and was responsible for suppressing the rebellion that had begun at Lexington and Bunker Hill earlier that year. Following the failure to quell the revolt he resigned in 1782, was created First Viscount Sackville, and died in 1785 at Stoneland. His friend Richard Cumberland, the dramatist, who had retired to Tunbridge Wells, described him at Stoneland:

> in his declining years, riding about his estate, followed by an aged groom, who had grown grey in his service, taking an intelligent interest in the welfare of his cottagers and retainers, or in the village church, in quaint Sir Roger de Coverley style, nodding approval of the sermon or rating the rustic choir for singing out of tune.[23]

Because Charles had died childless the title of third duke descended in 1769 to John Frederick Sackville (1745–1799), his young nephew.[24] He was MP and lord lieutenant of Kent 1769–97; ambassador to Louis XVI at Versailles 1783–89, in the years leading up to the Revolution; and Knight of the Garter (1787) and Lord Steward of the Household from 1789 to his death. He is, however, perhaps best remembered for his love of cricket and women. *The Times* reported a match played in the Champs Élysées in 1786: 'His Grace of Dorset was, as usual, the most distinguished for skill and activity. The French, however, cannot imitate us in such vigorous exertions of the body, so that we seldom see them enter the lists.'[25] The third duke was alleged to have had a liaison with Georgiana, duchess of Devonshire, as well as with Marie Antoinette. On his death in 1799 the *Gentleman's Magazine* noted: 'He was not distinguished for mental vigour or literary attainment; but his temper was moderate, and his manners agreeable.'[26]

In 1790 he had married the young Arabella Diana Cope (1769–1825). They had one son, George John Frederick (1793–1815), who succeeded as fourth duke as a boy, but who died after falling from his horse while hunting near Dublin (Figure 5.4).[27] There were also two daughters, Lady Mary (1792–1864) and Lady Elizabeth (1795–1870). Lord George Sackville had married Diana Sambrook in 1754 and their eldest son Charles Sackville-Germain (1767–1843), second Viscount Sackville after his father's death, also now became the fifth and last duke of Dorset. The third duke had left his estates to Arabella, who thereby became a very wealthy widow, controlling both her young son and the Sackville estates. In 1801 she married a friend of her late husband, the diplomat Charles Whitworth (1752–1825), later first earl of Whitworth. He was appointed ambassador to the new French Republic and in 1813 Viceroy of Ireland, becoming a wealthy man in his own right.[28]

Lord Whitworth engaged Humphry Repton in 1805 to advise on improving both the landscape and the house at Buckhurst, for which Repton produced one of his Red Books, *Hints, plans and sketches for the improvement of Stonelands*. Arabella amalgamated Stoneland and Buckhurst Parks and renamed the house Buckhurst Park. In turn she

---

23 P. Mackesy, 'Germain, George Sackville [formerly George Sackville], first Viscount Sackville', *ODNB* (2009).

24 G.F.R. Barker, 'Sackville, Charles, second duke of Dorset (1711–1769)', rev. M.J. Powell, *ODNB* (2004).

25 J. Goulstone and M. Swanton, 'Carry on cricket: The Duke of Dorset's 1789 tour', *History Today*, 39 (1989), 18–23.

26 <http://www.historyofparliamentonline.org/volume/1754-1790/member/sackville-john-frederick-1745-99> [accessed 14 June 2016].

27 His death is marked in Withyham church by a mural monument by Flaxman, one of only three such in East Sussex (Llewellyn, *East Sussex Church Monuments*, 412).

28 Sackville-West, *Inheritance*, 148–9.

Figure 5.4. Monument to Arabella, Duchess of Dorset by Sir Francis Chantrey at the centre of other family memorials (above) and to George, fourth duke (d. 1815) (right) by Flaxman, both in the Sackville Chapel, Withyham church (source: https://www.sussexrecordsociety.org/dbs/esm/church/3020/).

later invited Lewis Kennedy to design new walks and a Swiss bridge and to make woodland improvements. Kennedy had worked in Normandy for the Empress Josephine and, upon his return to England, designed numerous gardens in a Romantic style.[29]

There were no further children. Whitworth died at Knole in May 1825, followed in August by Arabella, who was buried at Withyham (Figure 5.4).[30] Her considerable property (estimated at £35,000 per annum) was divided between her two daughters as co-heiresses, with Knole going to Lady Mary, now countess of Plymouth, and Buckhurst and the Middlesex lands (inherited from the Cranfield family) going to Lady Elizabeth, now Elizabeth Sackville-West, Countess De La Warr, following her marriage to George John Sackville-West, fifth Earl De la Warr (1789–1869) in 1813.[31]

By her marriage to George in June 1813, and following the death of her mother, Elizabeth had effectively finally passed Buckhurst and the manor of Duddleswell out of the Sackville family and into the De la Warr line of succession, where it remained through to the late twentieth century. When the fifth duke died in 1843 George John West, Earl De la Warr, assumed the name of Sackville-West by royal licence.

---

29 S. Daniels, *Humphry Repton: Landscape Gardening and the Geography of Georgian England* (New Haven, CT: Yale University Press, 1999), 268; P. Deitz, *Of Gardens: selected essays* (Philadelphia: University of Pennsylvania, 2011), 248–9. Repton's Red Book, with a preface to Lord Whitworth, bound in red Morocco leather in his usual presentation to his clients, is in Yale University Library (Gen Mss vol. 221).

30 V. Sackville-West, *Knole and the Sackvilles* (London: Heinemann, 1922), 111–220; Llewellyn, *East Sussex Church Monuments*, 411.

31 The deed of partition is at the Kent History and Library Centre U269/E352.

ELITES: OTHER POWERFUL PLAYERS ON THE ASHDOWN STAGE

The Sackville family were by far the most significant in terms of landholding and social capital around Ashdown, thanks in large part to their holding of many different offices of state, their loyalty to different monarchs and judicious marriages and land purchases. But this dominance did not go entirely unchallenged. Other titled families and gentry exerted varying degrees of control, derived from overlapping social circles incorporating patronage and economic, social and cultural capital. We have seen, for example, that the Gage family, although based at West Firle under the South Downs, exercised authority through the manor of Maresfield down to the end of the eighteenth century, and the same distant influence also came from the dukes of Norfolk, who owned the manor of Sheffield in the sixteenth century and, via a dowager duchess, later in the nineteenth century. Later the Shelley family came to dominate much of the Maresfield area.

A further important addition to those bringing wealth and power onto the fringes of Ashdown was the dynamic and resourceful John Baker Holroyd (1735–1821), who purchased Sheffield Park (Fletching) in 1769 from John, second Earl De la Warr. With inherited wealth he employed James Wyatt to rebuild the house in Gothic style during the 1770s and Capability Brown from 1776, and later Humphry Repton, to replace a formal park and redesign the grounds.[32] He was created Baron Sheffield in 1802 and then earl of Sheffield, as an Irish peer, in 1816. From 1794 his second wife was Lucy Pelham, daughter of the second Baron Pelham, later earl of Chichester, and his third wife from 1798 was Anne North, daughter of Lord North, the prime minister 1770–82. With such connections, he became a privy councillor in 1809. George Holroyd (1802–1876), John Baker Holroyd's son by his third marriage, inherited in 1821 as Viscount Pevensey and then second earl.[33] In turn his only surviving son, Henry North Holroyd (1832–1909), became the third and last earl in 1876. When Bateman compiled his lists of great landowners in 1883 the third Lord Sheffield owned 4,537 acres in Sussex, primarily in Fletching, and more than 2,000 acres in Yorkshire and Ireland.[34]

The 1785 Land Tax gives an indication of the extent to which Ashdown parishes were 'close', and dominated by larger landowners, or 'open', where a greater number of smaller landowners existed (Table 5.1).[35] We might first note again the relatively large size of the parishes here, especially Rotherfield and East Grinstead. Except for West Hoathly, all parishes had the presence of a member of the titled gentry. The contrast with the two downland parishes is also pronounced in terms of the numbers

32 N. Antram and N. Pevsner, *The Buildings of England: Sussex: East* (New Haven, CT: Yale University Press, 2012), 389–90.

33 M. Lill, 'George Holroyd (1802–1876), Second Earl of Sheffield' (Univ. of Sussex unpub. MPhil, 2002).

34 J. Bateman, *The Great Landowners of Great Britain and Ireland. A List of all Owners of Three Thousand Acres and Upwards*, 4th edn (London: Harrison and Sons, 1883), 403. The estate also stretched into Chailey, Newick and Lindfield by the early nineteenth century.

35 For the concept of 'open' and 'close' parishes on Ashdown see pp. 217–21.

Table 5.1. Landownership around Ashdown 1785.

Parish and acreage	Total tax collected (£)	Total rental (£)*	Total number of taxpayers	Landowners with annual rents of £50 or more	Top 3 by rental value+	% of total rental by top 3 taxpayers
Buxted (8,961)	266.0.0	1,330.0.0	94	7	Geo Medley 325	39.6
					Mr Oliver 102	
					Lord Pelham 94	
East Grinstead (15,130)	862.13.7	4,313.7.11	181	6	Lord Sackville 322	15.8
					Chas Biddulph 205	
					Rd Austen 155	
Fletching (8,522)	418.17.0	2,094.5.0	98	4	Lord Sheffield 638	55.5
					J Newnham 296	
					Sir Thos Wilson 227	
Hartfield (10,387)	377.12.0	1,888.0.0	63	8	D of Dorset 523	49.4
					L Visc Sackville 226	
					E of Thanet 184	
Horsted Keynes (4,232)	179.14.6	898.10.6	54	5	Visc Hampden 133	43.6
					Rd Wyatt 132	
					Wm Board 125	
Maresfield (8,132)	210.4.0	1,051.0.0	65	3	J Newnham 352	51.9
					Lord Gage 111	
					P Holford 82	
Rotherfield (14,731)	638.17.0	3,194.0.0	169	11	Lord Abergavenny 537	29.8
					J Harrison 247	
					J Newnham 168	

(Table 5.1 continued)

Parish and acreage	Total tax collected (£)	Total rental (£)*	Total number of taxpayers	Landowners with annual rents of £50 or more	Top 3 by rental value+	% of total rental by top 3 taxpayers
W. Hoathly (5,340)	216.11.0	1,082.15.0	76	6	Wm Clifford 209	34.7
					Ellis Bostock 98	
					T.H. Payne 68	
Withyham (8,126)	306.8.0	1,532.0.0	61	4	D of Dorset 650	50.0
					Wm Peerless 60	
					Rev Bale 55	
South Downs parishes for comparison						
Glynde (1,570)	152.15.0	763.15.0	9	2	Lord Hampden 513	77.3
					Thos Hay 54	
					Rev Davies 23	
Stanmer (1,341)	62.4.0	311.0.0	1	1	Lord Pelham	100.0

*Total rental values: calculated by multiplying the collected tax totals by 5, since the tax was levied at 4s in the £ and should be seen as approximate.
+Landowners may have had more than one property within each parish and these have been added together within that parish.
Source: R. Davey, *East Sussex Land Tax 1785*, SRS 77 [Lewes, 1991].

of taxpayers and proportion of rental values accruing from the top three taxpayers, the top Ashdown value being 55 per cent in Fletching compared with 77.3 per cent at Glynde and 100 per cent at Stanmer (Lord Pelham). The domination of the duke of Dorset in Hartfield and Withyham is also noticeable. Of course, since the data is given here by discrete parishes it cannot reveal the true extent of landed property: John Newnham, for example, owned land in Fletching, Maresfield and Rotherfield, part of his extensive landholding around the Forest.[36]

Despite the increasing importance of the estates during the eighteenth century, the manor retained some bureaucratic functions. For example, an Act of 1710 required lords of manors to appoint only one gamekeeper who might kill game as their deputy, and the names of the lords and their gamekeepers were to be returned to a clerk of the peace. Not all the manors recorded were genuinely historical in origin; some were shared, some were farms or even parcels of land, some clearly subinfeudated at an early stage. Nor were all the so-called gamekeepers actually so employed, although some were described as 'menial servant'. Many were in fact gentry themselves, or farmers or even tradesmen who were now being given the right to shoot over the manor's lands. Thus, the reputed 'gamekeeper' for Charles Crawfurd at Mays was William Langridge, an East Grinstead innkeeper. John Frederick, duke of Dorset returned Maurice Halford Barrow (1775–1841), a 'gentleman' just 20 years old, from East Grinstead, as his Imberhorne 'gamekeeper'.

But the extant list of gamekeepers, beginning for Sussex in 1781, again offers a good idea of the main landed gentry around Ashdown (Table 5.2). The dukes of Dorset, Gibbs Crawfurd, John Newnham and Lord Sheffield are all included. The list is repetitive as licences were renewed in later years, although names of owners and 'gamekeepers' did change.[37]

FOREST SOCIETY

When moving on to the more general society of Ashdown, an obvious first question relates to the numbers of people living around the Forest at this time. Epidemic mortality had remained high in the Sussex countryside during much of the fifteenth century and in the rape of Pevensey as a whole the number of men paying the tithing penny in the early 1490s had fallen by one-third from the number in 1449–50, and numbers were to fall again during the poor harvests and years of disease in the 1520s.[38] In 1538 the Maresfield parish register recorded 'In the saide month [October], and the IXth day thereof, I buried a chil of the rayning sickness, whose name was called Parnell Carvell ... I also buried John Hayman, the XIII daye of October, of the rayning sickness'. This was probably the so-called sweating sickness, which visited periodically.[39]

36 See ESBHRO, ACC 3402/8, with details of ownership by the Newnham and Shelley interests.

37 L.F. Salzman, *Record of Deputations of Gamekeepers*, SRS 51 (Lewes, 1951).

38 See M. Mate, *Daughters, Wives and Widows after the Black Death: Women in Sussex, 1350–1535* (Woodbridge: Boydell, 1998), 17–19; Mate, *Trade and economic developments*, 2.

39 R.W. Blencowe, 'Extracts from the Parish Registers and other Parochial Documents of East Sussex', *SAC*, 4 (1851), 243–90.

Table 5.2. Reputed manors, lords and gamekeepers, 1780–96.

Year	Lord(s)	Manor	Gamekeeper	Status (where given)
1781	Duchess of Chandos, Wm Tennant, Jas Norman	Tablehurst	Martin Hoath, E. Grinstead	
1782	Rev Rd Crawley	Rectory of Rotherfield, alias Dewlands	Wm Mepham, Rotherfield	Yeoman
1784	Lord Visc. Sackville	Bolebrook	Hen Gladwell	His servant
1784	J. Biddulph Esq	Brambletye	J. Arnold Jr, E.Grinstead	
		Sheffield Grinstead	J. Bowra, E. Grinstead	Farmer
		Botches [?]	J. Stone, Horsted Keynes	Yeoman
1784	Earl of Abergavenny	Rotherfield	Wm Grinham	His menial servant
1784	Gibbs Crawfurd Esq	Mays	Wm Langridge	His menial servant
		Goddenwicke	Rd Myale, Lindfield	Labourer
		Burley Arches	Wm Young, Worth	Yeoman
1784	Thos Holles Payne Esq	Gravety	J. Clifford, West Hoathly	Yeoman
1784	Geo Medlay Esq	Buxted	Geo Dival, Buxted	Servant
1784	Jas Shard Esq	Parocks	Rd Davis, Hartfield	Gent
1785	Catherine Cranston, widow	Dallingridge	Geo Green, E. Grinstead	Her menial servant
1785	J. Newnham, Esq	Courtland	Walter Dicker, Maresfield	
1785	Lord Sheffield	Sheffield	Thos Vine, Sheffield	Yeoman

1785	Duke of Dorset	Muncklow & Buckhurst	Leonard Hooker, Buckhurst	Yeoman
		Diddleswell, Collinghurst & Alchornes	J. Goring, Rotherfield	His servant
		Buckhurst & Muncklow	Thos Bennet, Withyham	Yeoman
		Michelham Parkgate & Chiddingly	Hen Waller, Michelham	
1785	Warden Sergison, Wm Board, Walter Batchelor, J. Ingram Esq & Wm Clutton, Gent.	Marshalls	Thos Lamon, Brighton	
1785	Rd Wyatt Esq	Cockease	J. Lamb, Horsted Keynes	
1785	J. Lord Sackville, Duke of Dorset	Tarring Peverel	Thos Vine, Fletching	Yeoman
1786	Rt Bostock Esq	Rectory of West Hoathly	Hen Weller, E. Grinstead	
1788	Hen Humphery Esq	Cotchford	Ed. Everest, Hartfield	
1791	J. Newnham Esq	Lavortie	Thos Saunders, E. Grinstead	
1791	Henry Lord Visc Gage	Maresfield	Walter Dicker, Maresfield	Servant
1796	J. Frederick, D of Dorset	Imberhorne	Maurice Halford Barrow, E Grinstead	Gent

Original spellings preserved. After 1785: only included if the manor appears for first time.
Source: Salzman, *Record of Deputations of Gamekeepers*.

However, population levels, calculated by comparing the 1524–5 Lay Subsidy returns and the Hearth Tax returns of 1664–5 or 1670, do then show Wealden population to be growing at this time compared with the lower levels (or even declines) along the South Downs and in non-Wealden Sussex. The area centred on East Grinstead grew by 15 per cent between 1524 and the 1660s, that around Horsted Keynes by 30 per cent. Hartfield actually showed a slight decline of 5 per cent, explained by the presence of 'alien' ironworkers in 1524 who had disappeared by 1660. But, if the latter are discounted, then the populations at Hartfield and Withyham can also be seen to have grown by 10 per cent even between 1524 and the 1539 Muster Returns, when men eligible for military service were assembled for inspection.

This growth was heavily influenced by both in-migration and higher fertility. The possibilities of employment around Ashdown brought migrants in search of work, and work availability also brought a younger age at marriage, with consequent higher fertility rates. Many possibilities for employment offered themselves, particularly connected with the iron industry, but many 'rogues and beggars', wayfarers and vagrants were also attracted to these ironworking districts. Manorial enforcement here was relatively weak and gentry scarce, and there was always the hope of some temporary employment, or at least the possibility of winter warmth in the buildings and near the furnaces and forges. We therefore find 'a travelling woman, dyed at Millplace ffurnace' (East Grinstead) in the early 1660s.[40]

The 1524 and 1670 totals recorded taxpayers, and we also need to allow for levels of exemption through poverty. By 1670, when the Hearth Tax was collected, we have only fragmentary returns available, but these indicate that about one-third of households in Hartfield and Horsted Keynes were exempt, and nearly half those in Fletching. When collections for the relief of Irish Protestants were made in 1642 we similarly find that Hartfield contributed the lowest amounts in eastern Sussex, with 61 out of 104 contributors paying 5d or less.[41]

The subsequent decline of the iron industry, to be discussed in chapter 6, hit hard. Maresfield was severely affected, overwhelmed by its poor at one point and forced to seek help from neighbouring parishes for its 'great and excess charge'. During the first half of the seventeenth century population growth therefore slowed a little: fertility levels remained stable, at between 4 and 5 children per household, but epidemic mortality increased again, plague hitting Fletching in 1603 and what was probably typhoid fever hitting Maresfield badly in 1639. In 1630–31 a hard and exceptionally wet winter led to shortages of oats, difficulties for the poor and unrest at the quantities of grain still being shipped out of Pevensey Rape. Nevertheless, this period remained overall one of population growth and a relatively healthy environment, with fewer

[40] R.P. Crawfurd, *East Grinstead Parish Register, 1558–1661*, SRS 24 (Lewes, 1917), 196. The burial date was illegible in the parish register but was between 15 November 1660 and 2 January 1661. The furnace was to the west of modern Weir Wood reservoir but was discontinued by 1664 (Cleere and Crossley, *Iron Industry of the Weald*, 345).

[41] C. Brent, 'Employment, Land Tenure and Population in East Sussex 1540–1640' (Univ. of Sussex unpub DPhil, 1974), 232–7, 241–55 discusses the nuances of the source material and presents a full analysis of the parishes of eastern Sussex.

carriers of disease passing through, certainly compared with the southern coastal and marshland communities of the South East.

It remained the case that the large Ashdown parishes were more sparsely populated than others in the High Weald. Parts of Rotherfield, Uckfield and, to a lesser extent, East Grinstead were relatively densely populated, but Hartfield, Withyham, Maresfield and Buxted were all below the Sussex average population density at the time of the Compton Census 1676, hardly surprising given the expanse of Ashdown that was within their boundaries.[42] And densities around Ashdown remained among the lowest in Sussex throughout the eighteenth century.

Population growth throughout the later seventeenth and early eighteenth centuries was relatively gentle (Table 5.3). The data for 1676, gathered to determine religious affiliation, are not straightforward, but for the diocese of Chichester, including Ashdown, it appears that only adults over 16 years of age were numbered. Thus a 31 per cent multiplier, as commonly used, has been applied.[43] The figure for Rotherfield seems to be an outlier and possibly scribal error; excluding Rotherfield, population showed a growth of 140 per cent by 1801.

Table 5.3. Ashdown population 1676–1801.

	1676 (with 31 per cent added)	1801 population	Acreage	Inhabited houses 1801
Buxted	556	1,063	8,961	108
E. Grinstead	1,048	2,659	15,130	381
Fletching	590	1,279	8,522	204
Hartfield	440	1,050	10,387	167
Horsted Keynes	328	591	4,232	88
Maresfield	398	960	8,132	170
Rotherfield	3,930	1,963	14,731	269
West Hoathly	393	794	5,340	113
Withyham	187	1,074	8,126	180
Total	7,870	11,433	83,561	1,680

Source: WSRO, Add Mss 30332 parochial totals 1801; A. Whiteman, *The Compton Census of 1676* [Oxford, 1986].

42 M.J. Dobson, *Contours of Death and Disease in Early Modern England* (Cambridge: Cambridge University Press, 1997), 102–10, 124–49, 397–8; D. Martin, B. Short and P. Wilkinson, 'Population in 1676 and 1724', in K. Leslie and B. Short (eds), *Historical Atlas of Sussex* (Chichester: Phillimore, 1999), 66–7.

43 E.A. Wrigley and R.S. Schofield, *The Population History of England 1541–1871* (London: Edward Arnold, 1981), 218. The data is transcribed in A. Whiteman, *The Compton Census of 1676: a critical edition* (Oxford: Oxford University Press, 1986).

One caution has been issued, however: 'the figures calculated from such sources are at best tentative, and at worst downright misleading'. This is especially true where expanses of common land with scattered settlements had attracted many who could slip through the net of those who came to count the people.[44]

The vicissitudes of work availability around Ashdown also meant that many residents looked to seasonal migration for work elsewhere; to the hay harvests, to the downland corn harvest; back into the Weald for autumn hop and apple picking, and perhaps wood-cutting in the winter. Such activities were not, however, always sufficient to maintain a family. Other strategies for survival were adopted. For some the presence of Ashdown Forest behind its pale offered a tempting source of supplementary income, whether through wood-stealing or poaching rabbits or deer. The relative isolation of Ashdown certainly offered possibilities for law-breaking. John Norden, in *The Surveyor's Dialogue* (1607), correlated 'lewdness and rudeness' of the population with the extent of woods and wastes: '*Rudes & refractari Silvicolae* … '.[45]

Indeed, if possible, the area was avoided by respectable people at night. The lack of governance was of particular concern to the Privy Council, and the Justices of Assize concluded their report of October 1587 on the Sussex magistracy by advising that there should be more JPs in Sussex than in other counties: 'for that it bordereth south on the sea and north on the wyld; in which towe places comminly the people begeuen mutch to rwednes, and wyllfulnes'.[46]

Furthermore, in October 1593 the High Sheriff and JPs were addressed in the following terms:

> … whereas we are informed that there are divers lewd people, some of them being Horsemen and other vagrant persons and Rogues, that are gathered together in the Confines where the two Counties of Surrey and Sussex do meet and keeping in the woods do sett upon her Majesty's subjects and such as pass that way, and break also into houses committing also foul disorders, Burglaries and Robberies and gather more disordered people unto them, we canot but marvel yf this information be true, that you the Justices bordering upon the Confines do not take present order to suppress these wicked and lewd persons and therefore we require you without delay to enform yourselves of the truth hereof and that some of yourselves with such forces as you shall think convenient do repair to those places and take such order as they may all or the most part of them be apprehended, committed to prison and so proceeded withall as their most lewd disorders do deserve, least further mischief should ensue of this lewd beginning, wherein praying you to place some extraordinary care we bid you heartily farewell.[47]

---

44 H. Falvey, 'Searching for the Population in an Early-modern Forest', *Local Population Studies*, 81 (2008), 37–57.

45 'Woodlanders are ignorant and stubborn'.

46 W. Durrant Cooper, 'Certificate concerning the Justices of the Peace in Sussex in 1587', *SAC* 2 (1849), 69.

47 J.E. Mousley, 'Sussex Country Gentry in the age of Elizabeth' (Univ. of London unpub. PhD, 1955), 99–101.

Most serious were the 'masterless men' and gangs roaming the Wealden countryside, such as those reported by Sir Richard Michelborne JP for the Horsted Keynes district in 1633: people who he hoped would 'go to the howse of correction there to remayne until further order'. He noted in particular that:

> I sent this 9th of January to the Sessions two women and one yonge fellow the one of the women is called Nan in the Hayre a desperate ruffianly drab she nameth her selfe to be the wife of one Richard Shinn she hath used about Freshfield Forge and Sheffield Forge for the space of this two or three yeares and there hath beene of that company eighteen or nineteen at any one time and they are such as breede a terror in the Country hereabout she may pass for an incorrigible Rouge there must be more severe course taken against her she hath beene twice brought before within these fourteen dayes the first time I sent her away from officer to officer but shee ran away from them and sweares shee will not leave her haunt so George Guilbert who I have apoynted to goe to Lewes with her is able to give good information against her, I pray let him be dispatched as sone as you can.[48]

Parish records, beginning during the sixteenth century, also provide insights into those roaming around Ashdown. The burial records of East Grinstead, for example, confirm the presence of 'waygoers'. Some were unfortunately buried without a name: in November 1575 'a poore Beggar woman'; in October 1578 'One p[e]rson a stranger'; in August 1588 'A poore vagrant woman'; in February 1590 'A poore soldier'; in February 1591 'A poore waygoer; in February 1604 'A poore boye whose name is unknown' and 'Rachel a waygoers child', who at least had a given name.[49]

ORDER AND CONTROL

Around Ashdown, as elsewhere, offences against rural *mores* and public order were dealt with by a variety of courts, as well as more informally. The church still exercised considerable power and the Churchwardens' Presentments detail Ashdown miscreants who had been reported to the archdeaconry of Lewes. A great variety of alleged misdemeanours can be found, ranging from what might today be considered highly intrusive and personal through to concerns about the church fabric, as at Withyham, where it was also reported that a certain pew was to be reserved solely for Rocks Farm.

A visitation on behalf of the archdeaconry in 1686 found the churches here generally in good repair, although needing whiteliming to obliterate unwanted images in the chancel or elsewhere. But there were instances of fallen tombstones, and at Withyham two pits in the churchyard, 'one whereof is very dangerous'. The church at Hartfield, however, was 'much out of repaire. The parish chancel is to be ript. Part of the church adjoyneing to it to be mended. The leads are to be newcast.' Another visitation in

[48] WSRO, Par 379/12/26 Transcripts of extracts from parish and Quarter Sessions records, West Hoathly 1594–1807. A. Fletcher, *A County Community in Peace and War: Sussex 1600–1660* (London: Longman, 1975), 167–8.

[49] Crawfurd, *East Grinstead Parish Register*, 133–44, 153, 170.

1724, Bishop Bowers' survey, inspected the churches and by now most were found to be in decent repair, including Hartfield.

People were regularly presented to the archdeaconry court for non-attendance at church. The wealthy Richard Michelborne (1571–1638) was twice reported for not receiving the sacrament (1587 and 1597) and for allowing the church at Horsted Keynes to fall into disrepair, '*cancella ruinosa*'. Thomas Binnes of West Hoathly was charged in 1597 with 'working on St Luke's Day last' and Thomas Ashbee of Maresfield with 'working his oxen on the day of St Michael' in 1599.[50] Seven East Grinstead residents were similarly presented in 1675.

In 1674 we find two women from East Grinstead presented for each bearing a 'bastard childe', and 'Elizabeth Casselden [Hartfield] hath a sonne that was base borne not long before Easter'. At Fletching a couple 'live as man and wife togeather and refuse to make it known where and by whom they were married', while Thomas Crayford and Elizabeth his wife were presented 'for incontinency before marriage'; and at Horsted Keynes there was a presentation for 'antenuptiall fornicacion'. Here too Mary Vynall was an unlicensed midwife, as were the wives of George Shelley at Maresfield and George Baker, who had been excommunicated. The presentations also included not having banns read, not paying church taxes, failing to have a child baptised and, at Rotherfield, four men talking in the churchyard 'in the tyme of divine service'. In Fletching others were presented for attending services at places other than their parish church. In Hartfield John Farmer and his wife were repeatedly presented as Quakers; twelve others were 'schismaticks' and James Kingsland a nonconformist. At East Grinstead papists, independents and Quakers were presented.[51]

At this turbulent time for religious affiliation the Weald in general was influenced by nonconformist and puritan zeal, perhaps prompted in part by strongly puritan JPs, although around the Forest it was a mixed picture. Richard Gray, rector of Withyham, was viewed by the bishop of Chichester as a troublesome adherent to Catholicism in the 1570s, although he managed to retain his position and was still in post by 1581.[52] The evidence from wills made between 1530 and1559 does little to clarify the picture. It has been said that the Reformation came late to east Sussex, and Ashdown will-makers probably reflected a wider mood in either clinging to the older faith, especially during Mary's reign, or moving to a reformed religion – or, indeed, hovering between the two.[53] Recorded in the Buxted parish registers was one particularly troublesome episode when Walter Cushman, who 'mayntayned divers popish errors, as of the corporal presence in the sacrament, images and other Popish opinions', reportedly 'led his horse up and down in the church, and about the Communion table in the chancel'. This latter action was a common way of insulting members of the reformed churches.[54]

[50] W.C. Renshaw, 'Notes from the Act Books of the Archdeaconry Court of Lewes [1581–1641]', *SAC*, 49 (1906), 49–65. St Luke's Day is 18 October, St Michael's is Michaelmas, 29 September.

[51] H. Johnstone, *Churchwardens' Presentments (Seventeenth Century) Part 2 Archdeaconry of Lewes*, SRS 50 (Lewes, 1949).

[52] Manning, *Religion and Society*, 44–5.

[53] G. Mayhew, 'The Progress of the Reformation in East Sussex 1530–1559: the evidence from wills', *Southern History*, 5 (1983), 38–67.

[54] Blencowe, 'Extracts from Parish Registers', 253.

We should also note that in the reign of the Catholic Mary Tudor three Protestant martyrs from East Grinstead were burnt in the high street in 1556, with two more from Withyham, two from Rotherfield and one from Buxted, altogether being a large proportion of the total burned in Sussex. Although the precise number of such martyrs is disputed, Richard Woodman, an ironmaster from Buxted, has received particular attention. Of the 26 martyrs whose place of abode is known, eight were therefore from the Ashdown parishes, together with nine of the 17 heretics.[55] One from East Grinstead, two from West Hoathly and six from Rotherfield were convicted of heresy. In addition, incumbents were deprived at Buxted, East Grinstead and Rotherfield, and in 1555 Lord Abergavenny had been instructed by the Privy Council to put the inhabitants of the latter parish in order. Rotherfield certainly seems to have been a parish with some adherence to the reformed religion: during Mary's reign the 23 wills made were all overwhelmingly reformist or Protestant, as had been the wills made earlier in the late 1540s by Alexander Hosmer of Rotherfield, later to be martyred with Woodman in Lewes.

By the seventeenth century puritan sympathies might be seen in personal names: we find Absolucon Jenner (1624), Temperance Goffe (1634) and Temperance Shoosmith (1642) from the East Grinstead marriage registers. E.V. Lucas adds Perform-thy-Vowes Seers of Maresfield in 1632 and Kill-Sin Pemble of Withyham. Faintnot Batchellor was assessed for three hearths for the 1662 Hearth Tax.[56]

Incumbents were also sequestered, and during the mid-1640s many suspected monarchist clergy were persecuted and ejected by Presbyterian reformers in the Long Parliament, as recorded in the 'Proceedings of the Committee of Plundred Ministers'. A branch of this committee met at Brambletye, as well as in Chichester, Lewes and Battle, to take depositions. In February 1644 John Large from Rotherfield was examined in his absence but allowed to submit a defence. He described a 'secret plotte and Combination' of some parishioners to eject him because they coveted his incumbency, which was a good living. He was charged with insufficient sermonising but explained that in the short days of winter many found a second attendance at church difficult, as they lived three or four miles away across this large parish, so he issued a sermon of 'neare two hours' for a 'double taste' in the morning! He previously had dealings with Richard Goffe from East Grinstead, serving from 1643, but who was accused of collecting tithes after church on the Sabbath and breaking a cake over a bride's head at a wedding ceremony, an ancient custom now believed inappropriate. Goffe was described as:

---

55 M.J. Leppard, 'Heretics and martyrs in Marian Sussex', *SAC*, 154 (2016), 209–26. For Woodman and the complexities of religious affiliation see P. Quinn, 'Richard Woodman, Sussex Protestantism and the construction of martyrdom', in M. Dimmock, A. Hadfield and P. Quinn (eds), *Art, Literature and Religion in Early-modern Sussex: Culture and Conflict* (Farnham: Ashgate, 2014), 195–220.

56 Fletcher, *A County Community*, 124; M. Burchall (ed.), *Sussex Hearth Tax assessments 1662: Pevesey Rape*, Sussex Genealogical Centre occasional papers 4 (Brighton, 1980); Crawfurd, *East Grinstead Parish Register*, 17–21; E.V. Lucas, *Highways and Byways in Sussex* (London: Macmillan, 1904), 296.

> a common haunter of Taverns and Ale-houses, a common swearer of bloudy oaths, and singer of baudy songs and often drunke, and keepeth company with Papists and scandalous persons ... highly commended Queene Maries time and disparaged Queene Elizabeths, as an enemy to learning ... and that he cared not a figg for the Parliament.[57]

He was ejected in 1646 to make way for Samuel Pretty from Wiltshire, who was promptly returned to that county, to be replaced by the puritan Stephen Watkins, 'Minister of the word', and thereafter by three more puritans, none of whom lasted very long. It would seem that East Grinstead had six ministers between 1643 and 1662.[58]

Bishop Compton's Ecclesiastical Census 1676 was at pains to point out the limited extent of nonconformity. If so, there must have been some hasty revisions of religious feeling among many. But around Ashdown one Presbyterian and five Congregational families at East Grinstead were noted.[59] The latter persuasion also accounted for a family in Hartfield and another in West Hoathly. There were two houses for Quakers in Maresfield. The spread of purpose-built meeting houses for Protestant dissenters followed after the passing of the Toleration Act 1689. It was reported in 1724 that there was a scattering of [Ana]baptists and Presbyterians, but only one 'Romanist' individual in Buxted in another survey of 1717, as well as an 'atheistical'.[60]

Secular matters were referred to justices at the Petty Sessions and Quarter Sessions, or to judges at the Assizes. Many of the Ashdown gentry served as JPs: Sir Henry Compton (Brambletye) until 1644, Anthony Fowle from Rotherfield, Sir Richard and William Michelborne from Horsted Keynes, John Tichborne from Buxted, and John Wilson from Fletching. The Petty Sessions began meeting monthly at Uckfield or Maresfield, gradually taking on more serious cases, such as the preliminary investigation into the death of a Maresfield husbandman who had died after being 'beaten and wounded' in 1657.[61]

The Quarter Sessions did maintain some continuity throughout the Civil War: at the Lewes sessions at Easter 1647 a case was heard concerning Anne Hodely of Rotherfield, who had given birth to a bastard child while travelling in West Hoathly. She was returned to Rotherfield and the reputed father, Richard Wickens, was to pay 12d weekly to the overseers to defray the parish charges.[62] Other cases dealt with in the 1640s included an unlicensed alehouse at Buxted and a beer brewer selling ale to

---

57 F.E. Sawyer, 'Proceedings of the Committee of Plundered Ministers relating to Sussex', *SAC*, 30 (1880), 112–36 and Part II, *SAC*, 31 (1881), 169–200. The quotation referring to Goffe is taken from the alarming *The First Century of Scandalous, Malignant Priests* (ordered by parliament and printed at Blackfriars 1643).

58 W.H. Hills, *The History of East Grinstead* (Lewes: Farncombe & Co., 1906), 77.

59 G. Lyon Turner, *Original Records of early Nonconformity under Persecution and Indulgence* (London: Unwin, 1911); T. McCann, 'Religious Observance in the Seventeenth Century', in K. Leslie and B. Short (eds), *Historical Atlas of Sussex* (Chichester: Phillimore, 1999), 56–7.

60 W.K. Ford, *Chichester Diocesan Surveys 1686 and 1724*, SRS 78 (Lewes, 1994), 41.

61 Fletcher, *A County Community*, 139, 353.

62 B.C. Redwood, *Quarter Sessions Order Book, 1642–1649*, SRS 54 (Lewes, 1954), 127. The following instances are also from the same source.

unlicensed retailers from East Grinstead, where William Burtenshaw, husbandman, was keeping a 'tippling house'.

Sheep stealing was rife. William Beale of Fletching admitted stealing two wethers and a ewe but was 'admitted to clergy and branded'.[63] From Maresfield, William Busse, a collier, was also found guilty of stealing a ewe, and Richard, John and Robert Marchant of 'breaking in and driving off a wether worth 40s'. There were also domestic cases heard: at the Lewes Midsummer sessions 1643 the JPs heard that William Piggott of Hartfield had been accused of being the father of a bastard child born to Elizabeth Gourd, a widow. But she had apparently formerly confessed to other men being the father; nevertheless, Piggott was to pay the quite large sum of £10 to be exonerated and discharged. At the Michaelmas session later that year Elizabeth was sent to the House of Correction in the Cliffe, Lewes, for one year.[64]

For more serious matters, JPs referred cases to the Assizes, held at East Grinstead in March and July, which were visited by judges from London but also by a large cross section of Sussex gentry for the social occasion. In July 1616 judges here found Isaac Taylor, a Rotherfield labourer, guilty of stealing two horses, combined worth £7, for which he was to be hanged. Another labourer, Edmund Harris of Rotherfield, was found not guilty of stealing a brush worth 8d, but was nevertheless sent to the house of correction.[65]

More local offences were also heard by the manorial and Forest courts, which continued to cover infringements well into the nineteenth century, including abuses of Forest custom, the protection of vert and venison, minor infringements and petty theft. Richard Holmes and Nicholas Cade were brought to court for stealing wood in 1613 but were fined 'nil because they are poor'. In 1619 a total of 24 persons was presented for wood stealing, but 19 of these had taken just a quarter of a cartload each and were again fined nothing because they were poor. Widow Ebbes was presented each year from 1617 to 1623 for an illegal encroachment but again fined 'nil because she is poor'. Such repetition demonstrates that, although in theory such cases dealt with 'squatters', in practice the dwellings were permitted to remain by the Woodmote Court, and indeed might be passed down to children, as was the case with the Ebbes family. Widow Ebbes' husband had been presented before 1613 for squatting, but other members of the family later took on the property such that by 1658 it was deemed copyhold of the manor of Duddleswell.

---

63 Benefit of Clergy was the medieval procedure whereby clergy could mitigate their sentences for crimes by reading from the Bible. Later it became available for anyone who could read from the Bible. It was to be applied only once for each person, who was branded on the thumb to prevent another claim. It was abolished in 1827. See H. Gwin, 'Benefit of Clergy', in A. Faulkner (ed.), *Historical Dictionary of Late Medieval England, 1272–1485* (Westport, CT: Greenwood Press, 2002), 52–3.

64 The Cliffe House of Correction functioned from 1610 to 1793, dealing with miscreants from the rapes of Lewes and Pevensey.

65 TNA, Assi 35/58/8: East Grinstead Assizes 8 July 1616.

POOR RELIEF AROUND ASHDOWN

We can now turn to the poorest members of the Forest community. Their numbers do not seem to have lessened. Many found it increasingly difficult to support their families, especially in the harder winter months, and some might be classed as paupers because of age or infirmity. But now the state intervened to some extent to take on the work of the dissolved religious houses, and between 1531 and 1834 there were no fewer than 51 acts of parliament relating to what became known as the Poor Law.[66] Most were concerned with vagrancy, bastardy and, perhaps most contentious, settlement and removal. All affected the inhabitants of Ashdown.

Such issues are significant here because by the early nineteenth century Sussex generally had the dubious honour of having the highest poor relief costs in England, which were especially marked between 1815 and the 1830s during a period of agricultural depression following the Napoleonic Wars.[67] Chronic seasonal and long-term unemployment was exacerbated by discharged militia, the decline in the system of living-in farm servants and, to a more limited extent, the deployment of threshing machines, which took away some winter threshing work.

The important Act of 1601 empowered parish officers, churchwardens and overseers of the poor to raise a parish poor rate from which disbursements might be made to those in need, whether in workhouses, almshouses or their own cottages. But this help was intended for parishioners only, and the Act of Settlement 1662 sought to deal with wandering groups by defining a 'home' parish offering entitlement to relief, to which vagrants could be removed. We have already seen that vagrancy was a particular concern around Ashdown, and now there would undoubtedly be harsh treatment meted out to many. Part of the preamble to the Act of 1662 might have been written with Ashdown in mind:

> ... by reason of some defects in the law, poor people are not restrained from going from one parish to another, and therefore do endeavour to settle themselves in those parishes where there is the best stock, the largest commons or wastes to build cottages and the most woods for them to burn and destroy; and when they have consumed it, then to another parish; and at last become rogues and vagabonds; to the great discouragement of parishes to provide stock, when it is likely to be devoured by strangers.[68]

The system was administered separately in each of the Ashdown parishes until the Poor Law Amendment Act 1834 formally gave control instead to the Guardians of

---

66 W.E. Tate, *The Parish Chest: a Study of the Records of Parochial Administration in England* (Chichester: Phillimore, 1983), 323–7.

67 R. Wells, 'The Poor Law 1700–1900', in K. Leslie and B. Short (eds), *Historical Atlas of Sussex* (Chichester: Phillimore, 1999), 70–71: the poor law records for East Grinstead and West Hoathly are reproduced at <http://www.sussexrecordsociety.org/dbs/pl/> [accessed 22 September 2021].

68 Cited in N. Pilbeam and I. Nelson, *Poor Law Records of Mid Sussex 1601–1835*, SRS 83 (Lewes, 2001), 5.

newly created Poor Law Unions. But prior to 1834 each parish had its own workhouse or poorhouse, imbued with local traditions attaching to poor relief. There was neither any standard type of building around Ashdown, nor any uniform parochial scheme.

The Ashdown poorhouses were scattered around the edges of the Forest. Maresfield's was at Fairwarp Crossroads, where the zealous curate Rev Robert Bingham drew up a list of rules to be observed by the pauper inmates 'concerning church attendance, speaking rights, dress code for adulterous women, representation of the paupers' rights and punishment'. Seeking Lord Sheffield's approval, Bingham pointed out that paupers were neglecting church, that there were 'infamous cases of bastardy in which the women seem to glory as an honour instead of a shame'; and that one pregnant woman in the workhouse 'has had a husband at sea for three years'. Some of the unemployed men were set to work digging stone for the local turnpike roads.[69] Hartfield's workhouse, certainly operating by the 1770s, became what is now the Anchor Inn. A parish farm was also established to provide employment during the 1820s, but a scheme here to break up and enclose some land near Chuck Hatch on the Forest to provide work for the poor was rejected, the parish meeting fearing 'the serious consequences [unspecified] of paying the poor for labour on the Forest'.[70]

East Grinstead's workhouse was leased in 1727 for a three-year period 'to the use of the poor of the parish' and there were proposals in 1735 for farming the running out to Edward Head and William Purchase. They were to pay £300 annually and be responsible for the food, clothing and washing of both indoor and outdoor poor. The parish was to pay expenses for settlement and removal cases and parish officers were to inspect the work that Head and Purchase were to set the poor, such that the two men would make some income. Again, one task for the able-bodied men was the very necessary repairing of the roads.[71]

Fletching's workhouse, replacing an earlier one in the village street that had become unfit for use (Figure 5.5), was opened at Splayne's Green in 1811. On 19 October 1812 89 parishioners were granted outdoor relief in the form of bread, potatoes and cash, and by late December 1818 there were also 77 paupers in the workhouse as well as others receiving outdoor relief. Numbers of inmates during the years to 1820 varied between 60 and 100.[72] In fact control over the running of this workhouse resulted in a disagreement between Lord Sheffield and Sir Thomas Maryon Wilson (1773–1821), seventh baronet of Searles in Fletching. The latter objected to Lord Sheffield's paternalistic or autocratic manner in making appointments of the governor and matron, and he was supported by another influential resident, W.C. Mabbott.

---

69 ESBHRO, ACC 2717/1; SPK 1/98/1; 'Report as to the operation and effects of the Poor Law amendment Act, in the County of Sussex' by William Henry Toovey Hawley Esq, Assistant Poor Law Commissioner (Second Annual Report of the Poor Law Commissioners for England and Wales, June 1836) Appendix B, p. 221.

70 ESBHRO, PAR 360/10/2/1. And see C.J. Griffin, 'Parish Farms and the Poor Law: a Response to Unemployment in Rural Southern England, c. 1815–35', *Agricultural History Review*, 59 (2011), 176–98.

71 WSRO, PAR 348/37/8; 348/37/6/1.

72 ESBHRO, SPK P/4.

The mutual dislike of lord and baronet descended into personal abuse and disputes rumbled on for some years, with intermittent recourse to lawyers.[73]

Wide use was made of allowances to top up wages (the so-called Speenhamland system), usually with some kind of sliding scale of help, as families slipped into poverty in hard times. But this system, held to depress wages, became less popular by the 1820s as costs of poor relief spiralled. An early proposal for an amalgamation of parishes to form a Union under the auspices of Gilbert's Act was put forward *c.* 1810, in recognition of the poverty in the area. This would have included Fletching, Buxted, Maresfield and Horsted Keynes, as well as other non-forest parishes, and would have had a workhouse at Piltdown or on Chelwood Common, but nothing seems to have materialised at that date.[74]

Many Ashdown people endured the most difficult lives and experienced some traumatic situations. We hear from East Grinstead that on 22 December 1699 Mary Wilson:

> who came into West Hoathly being great with child and was delivered there of a male bastard child, has now departed from that parish. As the child has since, illegally, been brought into East Grinstead by the officers of West Hoathly, the child is to be conveyed back to that parish, his place of birth.[75]

In February 1737 it was recorded that William and Mary Best, with two daughters, had arrived in East Grinstead with a settlement certificate and had since had four more children. But William and Mary, with five of their children, including young John, aged two months, were now to be removed back to Snodland (Kent) whence they had come. In February 1746 Richard Feilder, shovelwood maker and turner, and his wife Ann, 'rogues and vagabonds', were to be conveyed out of East Grinstead and into Kent. Removal certificates abound for such unfortunates having to be removed and settled elsewhere. In November 1740 Mary Feast, wife of William, was apprehended in West Hoathly 'as a rogue and vagabond who was last found begging in East Grinstead ... as the place of her legal settlement, birth or parents [sic] abode cannot be discovered, she is to be returned to East Grinstead'. Edward Cranston, JP, ordered the constable of Hartfield in May 1821 'to convey James Starr of Hartfield, shoemaker, to the House of Correction, Lewes, for his refusal to enter into a recognizance as the father of an illegitimate child to be born to Mary Card of Hartfield, single woman'. Similarly, in 1821 came an order from Baden Powell, JP, to the constable of Hartfield to 'convey William Nye to the House of Correction, Lewes to serve three months hard labour for failing to pay £5 12s maintenance for his illegitimate daughter born at the poorhouse to Harriet Fuller of Hartfield, single woman'.[76] And it was not only the poorer members of the

---

73 ESBHRO, SPK P/6–12; H. Rawlings, 'Lord Sheffield's last Battle', *Danehill Historical Society*, 4 (1993), 10–28.

74 ESBHRO, AMS 4899; Gilbert's Act was the Relief of the Poor Act 1782 (22 Geo.3 c.83).

75 Pilbeam and Nelson, *Poor Law Records of Mid Sussex*, 167.

76 ESBHRO, PAR 360/34/2/42. T.C.M. Baden Powell (1767–1844), later High Sheriff of Kent (1832), living in Speldhurst.

Figure 5.5. Fletching's workhouse 1814: by George Holroyd, aged 12 (future second earl of Sheffield) (source: ESBHRO, SPK F1–8. Sketch books of George Holroyd *c.* 1809–15).

community who were involved. In March 1826 we note Ann Miles, with a 'bastard child gone with its mother to Chart Sutton, putative father Capt. Henneker [sic] Esq … Chart Sutton to pay 2/6d per week'. The child was Eliza, entered in the Hartfield registers as 'baseborn daughter' of the unmarried Ann Miles (b. 1804). Henniker (1780–1843) lived at Ashdown Park (High Beeches) and had a distinguished naval career, recorded on a commemorative plaque in Hartfield church. He became a rear-admiral.[77]

And marriages did not always run smoothly. The complexities of the Poor Law system might be illustrated by the case of Elizabeth Chapman in March 1763, involving three of the Ashdown parishes, together with Lindfield, from which John Chapman and his wife Elizabeth had been removed in October 1758.

> Elizabeth Chapman was examined touching the place of the last legal settlement of the reputed father of a male bastard child born to her and of another with which she is now pregnant. Elizabeth Chapman said that she was lawfully married, in the parish church at Lindfield, to John Chapman, who was legally settled in East Grinstead, by whom she has one child named James aged about five years, and that she has not seen her husband for these four years past, and has lately heard his mother say that he was dead. That, about two years ago, she was delivered of a male bastard child in Fletching, since baptised there by the name of George, who is now with her, and that William Weston, now or late of Rotherfield, is the father of the said child. She further said that she is now quick with child and that Thomas Taylor of Fletching, husbandman, is the father of the child [Mark of EC].

77 ESBHRO, PAR 360/12/1 (for Ann Miles).

In April 1763 Elizabeth and son James were removed from Fletching to East Grinstead, as confirmed by Quarter Sessions.[78] So Elizabeth had been removed once from Lindfield to East Grinstead and then again from Fletching to East Grinstead in under five years.

With eligibility for relief varying across the Forest, social practice and local power relations were all-important. Help was sometimes considered too lenient: one West Hoathly churchwarden was called before Petty Sessions for keeping faulty accounts and having relieved the poor in an 'extravagant and unlawful manner'.[79] Common rights were vital to the poor, and from Buxted it was reported that 'a family seated on the common would rather have a cow than 6s a week from the parish'. Indeed, an encroachment onto the Forest with the lord's permission by Wheatley, 'a poor man', was cited with approval; two acres were taken in for planting potatoes and grass for his cow, and a stone cottage was built. Oats would be planted next, and a hog was kept. Wheatley was quoted as saying that this type of land should be broken up by denshiring for potatoes.[80]

Such help might be supplemented from family members and charity, which by now had lost most of its medieval religious connotation. As elsewhere, we find around Ashdown some wealthy residents leaving money in their wills to parish charities.[81] Nicholas Smith's charity ('the Hartfield Dole') was meted out from 1634 to the parish poor each Good Friday outside the church. By custom, following the service the rector and churchwardens walked to what was thought to be Nicholas Smith's tombstone and set out the money, the recipients' names then being called out.[82] Also in Hartfield, Margaret Tufton, dowager countess of Thanet's charity began in 1691 and the Rev. John Grandorge's charity was instituted from 1733.[83] Such help entailed, however, a public acknowledgement of poverty and was only available to those deemed 'deserving'. Thus in 1626 one Hartfield bequest had aimed to 'help the poor and infirmed, but that none of the money should help any persons given to excessive drinking, swearers, pilferers, or persons scandalous, disobedient, or vagrant'.[84] Indeed, the act of giving such assistance calls to mind Thompson's concept of the 'theatre of paternalism', although the offering and receiving of such largesse is fraught with complex intentions and possibly feigned deference here on Ashdown.[85] Dependence was the last resort in a

---

78 Pilbeam and Nelson, *Poor Law Records of Mid Sussex*, 195, 198, 287.

79 WSRO, PAR 379/12/12.

80 A. Young, 'Inquiry by Arthur Young into Propriety of applying Wastes to the Better Maintenance and Support of Poor etc', *Annals of Agriculture*, 36 (1800), 497.

81 See S. Hindle, *On the Parish? The Micro-Politics of Poor Relief in Rural England, c.1550–1750* (Oxford: Clarendon Press, 2004), primarily concerning the poor of Rockingham Forest (Northants).

82 J. Simpson, *The Folklore of Sussex* (London: Batsford, 1973), 112–13.

83 For these and later charities see ESBHRO, PAR 360/24. The pre-1834 Poor Law records are particularly good for Hartfield.

84 M. Pears, *Royal Foliage: a Brief History of Ashdown Forest* (privately printed 1959; reprinted 1983), 40.

85 Thompson, *Customs in Common*, 64.

landscape and society better known for its truculent population than for one securely under elite authority.

For the education of children in the workhouses, charity and parish schools were adopted by almost all the Ashdown parishes at this time. After some delay the Rotherfield workhouse was built and included a schoolroom for a few children by the 1760s. But private charity was more successful here. Sir Henry Fermor, Bt (1667–1734), born at Walshes in Rotherfield, bequeathed money to help the Crowborough area from his fortune made in London, willing:

> ... that my trustees shall lay out any sum, not exceeding the sum of £1,500 in erecting and building a chapel, or church, and a Charity School, in the plainest and cheapest manner, so as it be strong and lasting, for the ease and benefit of the parishes and parishioners of Rotherfield and Buxted, in the said County of Sussex, which Church and Charity School I will shall be erected and built in or near a place called Crowborough and Ashdown Forest, as my executors shall think most convenient, for the use and benefit of the very ignorant and heathenish people.[86]

A further £4,500 was given to purchase land, with the income going towards the maintenance of a schoolmaster and minister, and £3,000 was to be invested in land for the benefit of the poor children to be taught in the school. Sir Henry stipulated that no child should be under seven years of age upon admittance, 'nor proceed in education above the space of four years'. The children were to be taught to read, write and cast accounts, and at no time were there to be more than 40 children in attendance. The school and chapel were completed in 1744 and by 1819 it was reported that there were 30 children from Rotherfield and 10 from Buxted, often walking many miles to attend. But 'the school is always full, and there are many applications for every vacancy'.[87]

Buxted had a school for poor boys and another for poor girls by the 1720s; Fletching had private schools that poor children might attend, with help from Lady Sheffield; Richard Bonner, rector of Maresfield, left a schoolhouse and field in trust for appointing a schoolmaster to teach two poor children. At Horsted Keynes 20 poor children were to be educated, with inculcation with Christian principles uppermost, by a schoolmaster funded from the will of Edward Lightmaker of Broadhurst, who died in 1708. The school appears to have functioned intermittently throughout the eighteenth century but by 1819 there were 12 poor children being instructed.[88]

At Hartfield a schoolmaster was funded from the 1640s will of Richard Rands, supplemented by the earl of Thanet from the rents of Bolebrook Farm from 1725. John Kelton, credited with sketching a map of Ashdown in 1747, was master here for many years before his death in 1777.[89] A schoolroom was built by subscription *c.* 1812. There

---

86 ESBHRO, AMS 5818, Sir Henry Fermor's endowment. Will dated 21 January 1732.

87 Pullein, *Rotherfield*, 369; BPP, *Report of the Commissioners on the education of the poor: Sussex* 1820, 441–2; J. Caffyn, *Sussex Schools in the Eighteenth Century*, SRS 81 (Lewes, 1998), 223–4.

88 BPP *Report of the Commissioners Sussex*, 438–41.

89 See ESBHRO, AMS 4084/3.

was less provision at Withyham, although the overseers were paying for individual children to receive some education by 1744; and at West Hoathly there is a record only of a boys' boarding school by 1752.[90]

AN ASHDOWN COMMUNITY?

Clearly, help for the poorer inhabitants of Ashdown was a community function. Moving forward with the broad idea that 'community' encompasses social and economic relations between individuals and families in a locality and with a sense of belonging, a co-presence, to what extent was there a feeling of community around Ashdown?[91] Were there, for example, any collective rituals binding people together? Were there longstanding networks of interest? While the communal seasonal agrarian practices of common-field parishes are well documented, with strong manorial control in earlier centuries, the degree to which Wealden parishes, in a countryside of hamlets, isolated pastoral farms and cottages, shared any similar sense of communal activity is less well understood.[92]

The location of parish boundaries was certainly a significant feature, determining among other things how the poor would be treated. Although the practice was banned during the Commonwealth, parish boundaries were walked by members of the community and duly recorded – in West Hoathly's case in the parish register for 1788.[93] A large pollarded beech tree on the border between Hartfield and Withyham had HP marked on one side and WP on the other, and the date 1730 carved on it. Walks would presumably have taken place at Rogationtide, accompanied by the required church homily. A mental map and a sense of place and community were thereby upheld by walking the paths, reinforcing boundaries and strengthening the community's identity through this repetition of the traditional act. Here, then, were procession rites with a strong link with Ashdown's landscape.[94]

As elsewhere, there were clearly fractures at many levels, familial and parochial. One piece of doggerel against East Grinstead, presumably alluding to the collapse of the church steeple in 1683 or again in 1684, ran:

> Large parish, poor people:
> Large new church, no steeple.

---

90 Caffyn, *Sussex Schools*, 265, 270.

91 For concepts of community see K.D.M. Snell, *Spirits of Community: English Senses of Belonging and Loss, 1750–2000* (London: Bloomsbury, 2016), 1–28.

92 See Bob Bushaway, *By Rite: Custom, Ceremony and Community in England 1700–1880* (London: Junction Books, 1982). His book focuses on southern England but includes nothing on Ashdown.

93 WSRO, PAR 379/7/8. The boundaries of 1749 were described, together with a perambulation in 1788. And see D. Fletcher, 'The Parish Boundary: A Social Phenomenon in Hanoverian England', *Rural History*, 14 (2003), 177–96.

94 The 14-mile 'beating the bounds' at Crowborough (although not an ancient parish) continues through to the present, as does that at Withyham, last walked in 2017.

And, in Fletching, where a reputation for pilfering, rightly or wrongly, had evolved:

> The people of Fletching
> Live by snapping and ketching.[95]

One obvious source of communal concern would have been the management of common rights over the Forest. It would be an exaggeration to describe the Commoners as a community, as individuals' social status diverged widely. But there was certainly a community of interest in guarding their rights to Forest resources against abuse, and regular meetings of the older manorial and Forest courts provided an ongoing arena for such vigilance. For many the church or chapel services would also have provided the opportunity to meet neighbours (however distant) and discuss matters of mutual interest. In general people felt obliged to attend their parish church on Sundays, although for many this could entail difficult winter journeys in the Wealden mud.

One indication of integration within a community was the extent of intermarriage. As recorded in marriage licences, the number of spouses from the same parish decreased over time. Those for Horsted Keynes, for example, showed a decrease in endogamous (same parish) marriages from 37.8 per cent in 1590–1699 to 23.4 per cent in 1700–1836. Conversely, there was a doubling in the median distance between the addresses of the spouses per decade, from an average of 2.5 miles before 1700 to 5.2 miles after 1700. Greater personal mobility with better communications and improved literacy, allowing regular contact by letter, would both have been influential in this change.[96] The gentry to some extent stood outside this pattern, since marriage alliances, with their attached social and economic implications, might involve more long-distance marriage contracts.

In fact, Ashdown was something of a 'marriage watershed'. Ashdown inhibited, rather than encouraged, courtship. Hartfield parishioners married largely northwards and into Kent and only two extant licences show intending marriages into Buxted and Maresfield to the south. Maresfield parishioners, by contrast, looked firmly southwards; and neighbouring Fletching had only one licence between 1772 and 1837 showing an intended marriage with any of the northern Ashdown parishes. The same avoidance was maintained by Horsted Keynes with parishes north or east of the Forest.

Another signifier of community was the system of living-in farm service. Living-in was a common experience within the life cycle, particularly for young men, who received full board inside the farmhouse in return for farm work, or sometimes, especially for girls, work within the house. Although the hours of work might be long and arduous,

---

95 Simpson, *The Folklore of Sussex*, 152–3. Ketching probably an early form of 'catch', possibly referring to poaching. The original St Swithun, East Grinstead, was struck by lightning in 1683 and its successor was again struck in 1772, with its tower collapsing in 1785. A tower was eventually completed by 1813, being the last element in this Gothic Revival building.

96 B. Short, *The Geography of Local Migration and Marriage in Sussex 1500–1900*, Univ. of Sussex Research Papers in Geography 15 (Brighton, 1981). And for decreasing endogamy through time see P.J. Perry, 'Working Class Isolation and Mobility in Rural Dorset 1837–1936: a Study of Marriage Distances', *Transactions of the Institute of British Geographers*, 46 (1969), 121–41.

and the food and accommodation sometimes poor, in this way neighbours and kin were bound together in a mutually convenient way. The process instilled knowledge of other's affairs and farming procedures, and a sense of belonging to a slightly wider farming community. The extent of living-in fluctuated through time, depending on farmers' calculations of profit and loss concerning board and lodging, the farming family's own age and circumstances and the availability of casual or family labour.

In the 1841 census Hartfield parish returned 188 male agricultural workers, of whom about one-third were living in the farmhouses. Some were put out to service from the workhouse: in March 1827 the Hartfield workhouse contained 39 males, of whom 14 were put out to service, and 19 females, of whom one was put out.[97] John Bedwell from Piltdown (b. *c.* 1804) remembered that 'when I was a child we moved to Nutley for two years then we came to Fletching workhouse for three years. When I was about 12 the parish put me out to service with Mr Cheale at Portmansford Farm in Fletching, one year as under-carter.' He subsequently went on to other farms as a living-in farm servant.[98] A pattern of regular movement for younger farm servants is typical, partly because farmers had no wish to grant a full year's settlement and thus eligibility for Poor Relief. Longer terms of employment were reserved for married and older workers.

Another source of social integration was the coming together of parishioners in scarce leisure time. Beer shops and inns might offer camaraderie, while markets and fairs were occasions to mix and exchange news. At a slightly later date Ashdown villages were also locations for organised music-making. Although the more vigorous church music was to change by the mid-nineteenth century, as 'West Gallery' music gave way to more modern instruments, secular music-making continued and the 'Forest communities of north Sussex ... Horsted Keynes, Nutley, Fairwarp, Turners Hill, West Hoathly, and East Grinstead were veritable hothouses of band activity ... Brass bands seem to have flourished in small-scale communities...'.[99] In Nutley by the end of the nineteenth century we find the 'Ashdown Forest Temperence Brass Band' instituted for the 'cultivation of musical knowledge and abilities of working men'.

In terms of organised sport, hunting was largely confined to the gentry unless others were specifically invited to participate, but cricket had wider appeal. Sussex was an early county for recorded cricket, with several seventeenth-century instances, including one at West Hoathly. And we find in May 1752 two matches played between Withyham and Hartfield during Whitsun week, for 'half a Guinea a man'. The *Sussex Weekly Advertiser* stated that 'it is not doubted but at both matches there will be a great Concourse of People and Exceeding good diversion'. The quality of the cricket is unknown but it should be noted that 'every one of the Gamesters [was] to be upwards of sixty years of age'. Rather later, on 1 August 1796, we find the Fletching

---

97 TNA, HO 107/1115, Hartfield census 1841; ESBHRO, PAR 360/12/12.

98 B. Short, *The Ashdown Forest Dispute 1876–1882: Environmental Politics and Custom*, SRS 80 (Lewes, 1997), 226–7.

99 V. and S. Gammon, 'The Musical Revolution of the Mid-nineteenth Century: from "repeat and twiddle" to "precision and snap"', in T. Herbert (ed.), *The British Brass Band: a musical and social history* (Oxford: Clarendon Press, 2000), 122–54.

XI challenging either the Gentlemen of the South Downs or those of the Weald, the details to be settled at the Griffin Inn, Fletching. And on 29 August a match was to be played on Horsham Common between Horsham and East Grinstead, for £20 a side.[100] Accompanied by extensive betting and refreshments, here were sporting rivalries that would have contributed hugely to a sense of community.

THE DARK SIDE OF COMMUNITY

Of course, life on Ashdown, as elsewhere, was frequently far from idyllic. Glimpses of traumatic episodes were recorded, for example, by the county coroners.[101] Thus in early June 1690 Lewis Andrewes (Hartfield) was walking alone at about 11 p.m. near 'le old bay' in Hartfield when 'by misadventure and against his will he suddenly fell into a deep ditch, whereby he received one mortall bruise on the right side of his head of which he immediately died in the ditch'. His neighbours Thomas and Jane Hobbs, Denham Baker and Edward May were charged at Horsham assizes with his murder, in which Thomas Hobbs struck him with a wooden staff and the others aided and abetted him. However, the grand jury rejected the charge. In mid-December 1691 William Dobson of Maresfield, labourer, murdered William Morris, labourer from Nutley, with a blow to the head with a hoe. At the East Grinstead assizes in July 1692 he was convicted of manslaughter but successfully pleaded benefit of clergy and was branded on his hand and discharged.

A long list of accidental deaths and suicides could also be set out, although in this respect Ashdown was little different from other rural areas. But, more specifically, Mary Roger 'perished with cold' at Rotherfield in January 1758, Ann Watson died on Chelwood Common in November 1801, again 'through the inclemency of the weather on the common', and in August 1759 Richard Finch died from a pistol wound from William Mayor, a Hartfield warrener, although Mayor was later discharged at the assizes. In August 1811 in Buxted Richard Gaston Jr was killed in a fight with Esau Parks, a cordwainer, but Parks was acquitted at Lewes Assizes. There was infanticide and accidental deaths of children, and a death in a hop ground in Maresfield in December 1744 caused by a fowling-piece discharging and killing William Feldwick.

There was a public hanging at Wall Hill, between Ashurstwood and Forest Row. John Beatson and his adopted son William robbed a Brighton mail coach in June 1801, escaping across the fields towards Hartfield with several thousand pounds. They were arrested as far away as Liverpool, brought back to Horsham assizes, found guilty and sentenced to be hanged. As was frequently ordered, the gallows were erected at the site of the robbery and, despite some of the gallows being stolen, the two highwaymen were hanged on 17 April 1802 with an audience of 3,000 interested spectators. Charles Abbot from Kidbrooke wrote to the home secretary requesting that 'the bodies of the highwaymen hung on Wall Hill might be taken down before his wife drove to town'.[102]

[100] T. McCann, *Sussex Cricket in the Eighteenth Century*, SRS 88 (Lewes, 2004), 95, 180.

[101] The following passage is based on entries in R. Hunnisett, *East Sussex Coroners' Records 1688–1838*, SRS 89 (Lewes, 2005).

[102] Pears, *Royal Foliage*, 59.

Poverty could push many into dispensing with legal niceties. As with many rural communities at this time, the distinction between crimes sanctioned by working people, such as poaching or smuggling, or crimes such as theft or murder, which were clearly unacceptable, is an important one. Crimes such as rick burning, machine breaking, animal and plant maiming or the destruction of enclosures were protest actions, which will be considered later. But all of the above were to be found around the Forest and in practical terms one type of crime might shade into another: many a night poacher might be sorely tempted to take a hen or two.

The high food prices of the 1790s, following crop failures in 1794, caused many to look to remoter parts of Ashdown for a chance to boost meat intake, and scarcity was exacerbated by the presence of the military competing for food. Many felt that game was not private property, and while such a philosophy might be found in many sectors of society, more people from poorer families were caught, often hunting at night to provide for their households or for onward sale to urban areas. This could, however, involve armed gangs, and in the depressed years following the end of the Napoleonic Wars there was a considerable increase in such activity.[103]

Class differences had been highlighted by the passing of the Game Act 1671, making hunting the exclusive privilege of the gentry, with a property qualification as a test of legitimacy. The Act was repealed by the Game Act 1831, which ended the property qualification, but it excluded deer and rabbits because these were to be found in 'enclosed' land and thus seen as private property. On Ashdown deer numbers were seriously depleted through the depredations of the seventeenth century, the loss of woodland cover and retaliatory action by farmers whose crops were being eaten. In fact, the Rev. Edward Turner witnessed what he then believed was the last deer killed on Ashdown, in 1808 – a doe that was accidentally sprung by the Hartfield and Withyham Harriers while they were pursuing a hare just below Gills Lap.[104]

Instead, many gangs in Sussex turned to poaching rabbits from warrens, sometimes violently. Not all such poaching was to supply meat for the home pot, for there was also a flourishing black market. The *Sussex Weekly Advertiser* of 8 June 1761 contained the following request:

> Whereas on Sunday night the 31st of May last a rabbit burrow containing seven rods of ground was by some villains destroyed and all the rabbits taken out belonging and in the grounds of Robert Jarrett in Ashdown Forest near Nutley in the parish of Maresfield. Whoever will discover their accomplice or accomplices so that he or they may be brought to justice shall have three guineas reward and his free pardon except the person who did the above damage.[105]

---

[103] P.B. Munsche, *Gentlemen and Poachers: the English Game Laws 1671–1831* (Cambridge: Cambridge University Press, 1981).

[104] Rev. E. Turner, 'Ashdown Forest, or as it was sometimes called Lancaster Great Park', *SAC*, 14 (1862), 62. But see p. 390 for the high numbers of contemporary deer/car collisions on Ashdown.

[105] *Sussex Weekly Advertiser*, 8 June 1761.

In 1723 the notorious 'Black Act' had been passed by the Whig government of George I, bringing legal savagery down upon those poachers who were armed and/or had their faces blacked. The Act deemed it a capital offence, among many activities, if red or fallow deer, hares, rabbits or fish were hunted by persons armed and disguised, especially if caught in the king's forests.[106] Night poachers were particularly punished and deer stealers risked deportation, although the terrorising of gamekeepers ensued as a result. In January 1828 William Vincent of Maresfield was sentenced to seven years' transportation for 'attempted night poaching'.[363] Turner felt that horse-stealers too frequented the Copthorne area, 'the sympathies of the dwellers in that wild locality too generally being with the evil-doer'.[107] One unfortunate death in 1646 saw the tables turned by an animal when John Illman, labourer of West Hoathly, 'was kill'd by a Stag at Gravetye wch was hunted out of Ashdown Forest Sept 14'.[108]

According to Turner 'poaching and smuggling have from time immemorial been foremost among the besetting sins of the foresters'.[109] The French wars at the end of the seventeenth century stimulated import smuggling, and the activities of smuggling gangs have been documented for many parts of Sussex, where there was almost guerrilla warfare between smugglers and government officers. But there is less information on the practice as it touched on Ashdown. However, a gang operated out of Groombridge in the 1730s and 1740s, smuggling tea and brandy through the Forest, where there were certainly many places offering concealment, such as Crowborough Warren, Blacknest, Lampool Farm or Dodds Bank. It is reported that contraband items could be purchased at the Crow and Gate Inn at the eastern Forest edge, and a similar claim was made for other inns, which were the resort 'of various smugglers, poachers, and other lawless inhabitants of the district'.[110] Many Forest inns have been similarly associated with smuggling, such as the Roebuck (Wych Cross), Red Lion (Chelwood Gate), Gallipot [sic] (Hartfield), Hatch Inn (originally 'The Cock', Colemans Hatch) and so on, although the names may have changed over many years. One well-known smuggler from the Crowborough area was 'Duke' Read, a man of formidable stature and strength, who could reputedly carry four small kegs of spirits, a couple thrown over each shoulder.[111]

---

106 E.P. Thompson, *Whigs and Hunters: the Origin of the Black Act* (London: Allen Lane, 1975): C. Griffin, '"Some inhuman wretch": animal maiming and the ambivalent relationship between rural workers and animals', *Rural History*, 25 (2014), 133–60; and Griffin, 'Protest Practice and (tree) Cultures of Conflict: Understanding the Spaces of 'Tree Maiming' in Eighteenth- and Early Nineteenth-century England', *Transactions of the Institute of British Geographers*, 40 (2008), 91–108.

107 Turner, 'Ashdown Forest', 63.

108 'St. Margaret's Church, West Hoathly, Sussex, Burials 1606–1999' transcribed by D.J.H. Clifford (n.d.).

109 Turner, 'Ashdown Forest', 62–3.

110 Leeson Prince, *Crowborough Hill*, cited in M. Payne and L. Batchelor, *Bygone Crowborough* (Chichester: Phillimore, 1987), photograph no. 80.

111 Smuggling, the stuff of local folklore, often lacks historical authentication. But there are many associations in the present landscape, such as 'Smugglers Lane' or the Grade II-listed 'Old Smugglers Cottage', both in Crowborough.

Although the main pack-horse smuggling routes of the carriers from the coast through to London effectively bypassed the Forest, the many hollow ways would have afforded secure passage at night. On occasions processions of smugglers with their goods were witnessed in daylight through villages such as Maresfield.[112] In February 1736 an affidavit was sworn to the effect that on Christmas Eve 1735 a witness saw William Weston, a Rotherfield farmer, with about ten others, being a London smuggling gang, unloading about 7cwt of tea from a vessel at Bexhill. The illicit trade in tea came to dominate, accompanied by brutality on occasion and by complicit arrangements with excisemen in many cases; trials of smugglers at East Grinstead were commonplace. The returns from smuggling, even from helping in a small way by running the contraband, were far higher than labourers could otherwise earn, however. In correspondence between Holroyd and Arthur Young in 1771 it was noted that

> All the lively and able young men are employed in smuggling. They can have a guinea a week as riders and carriers without any risk: therefore it is not to be expected that they will labour for eight shillings a week until some more effectual means are taken to prevent smuggling.[113]

Many were the skirmishes between customs officers and 'owlers'. The leader of the 'Mayfield Gang', Gabriel Tomkins, was captured near Cackle Street, Nutley. An account in the *London Daily Post* of 19 September 1721 described the armed confrontation between grenadier guards and smugglers, beginning near Burwash and continuing with a pursuit across the Weald ending at Nutley, where four smugglers with five horses and equipment were captured and committed to Horsham gaol. Information then given by Henry Kitchell, an East Grinstead warrener, among others, revealed the prisoners to be part of a notorious gang who used the lanes around Ashdown with 20 or 30 horses laden with brandy and other contraband. In October 1783 a conversation was held at Buxted between a gentleman and a group of sailors who had just escorted smuggled goods onto Ashdown Forest and were about to return to their vessels.[114]

SOCIETY, WAR AND THE THREAT OF WAR ON ASHDOWN

The Civil War brought still more men into the lanes and byways: bands of unwilling pressed men or groups of discharged soldiers. At an alehouse in Withyham in 1648 a gang of soldiers 'committed misdemeanours by drawing of swords, cokking of pistols and thretninge words'. One soldier was apprehended and complained that he had been 'set upon by the country people' after revealing that he had marched with Cromwell's colours – probably unwise in the earl of Dorset's sphere of influence.[115]

112 Rev. J.L. Ward Petley, *Maresfield Old and New* (Chichester: J.W. Moore, 1896), 70.

113 A. Young (ed. M. Betham-Edwards), *The Autobiography of Arthur Young* (London: Smith, Elder, 1898), 58.

114 W. Albery, *A Millennium of Facts in the History of Horsham and Sussex 947–1947* (Horsham: Horsham Museum Society, 1947), 462–3, 491–2. And see L. Poore, 'Violence, Authority, Cultures and Community in Sussex and Kent c.1690–1760' (Univ. of Hertfordshire unpub. PhD, 2013), 100–53.

115 Fletcher, *A County Community*, 341–2.

There was no obvious attempt to settle scores or to dispute elite control around Ashdown, or indeed in eastern Sussex more generally at this time, as in some other forest and fen areas, or, for example, in the clothing districts of the West of England, where gentry and authority figures were busy with large national issues.[116] But highways and bridges were somewhat neglected and the Parliamentary Surveys of Ashdown in the 1650s refer to fences being broken, trees destroyed, deer killed and lodges damaged. It is uncertain, however, to what extent this resulted from actual lack of governance by the royalist Dorsets, or whether it was propaganda aimed at Edward, fourth earl, who remained strongly Royalist, as did his brother-in-law Henry Compton at Brambletye.

However, armed conflict largely bypassed Ashdown: no battles or skirmishes are recorded, these being confined to western Sussex. There was a confrontation at Muster Green, Haywards Heath in December 1642, when a band of Royalist conscripts was defeated, and a rising of country people at Horsham in July 1648, again suppressed by Parliamentarian forces. The gentry of eastern Sussex were broadly sympathetic to the Parliamentarian cause and the ironworks were a major resource to be guarded against Royalist interests. But many here were 'neuters' and enthusiasm for recruitment for either side ran low: 'Ancient' Streater was involved in a riot at West Hoathly fair in May 1643 while trying to recruit volunteers, and the head of his drum was beaten in.[117]

In the early months of the conflict Parliament passed a resolution to the effect that all 'Malignants' supporting the king would have their property sequestered, and additionally in 1644–5 that landowners could free their estates by compounding for a sum of money, overseen by the County Committee, which had been appointed in 1643 to administer income, defences and sequestrations. Sussex Royalist sympathisers, such as Dorset and Compton, duly conformed.[118] But they and their families, together with many others of Ashdown's elite, heartily welcomed Charles II on his arrival.

During the succeeding century, as we shall see, Ashdown was only indirectly connected with warfare through the vicissitudes of contracts for the produce of its iron industry. But Sussex was in the front line of anxieties about a French invasion during the Revolutionary and Napoleonic wars between 1793 and 1815. While Ashdown was at some distance from the coast, there were impacts. One was that the area around Copthorne – noted, as we have seen, for its lawlessness – was somewhat strangely chosen as the place of assembly for gentry families resident in Pevensey and Lewes Rapes, in order to provide safety in the event of invasion.[119] And a government advisor assumed that one of the locations that would be chosen to make a stand against the French would be Wych Cross. Understandably, there was much consternation.[120]

[116] See B. Sharp, *In Contempt of all Authority: Rural Artisans and Riot in the West of England, 1586–1660* (Berkeley: University of California Press, 1980).

[117] C. Thomas-Stanford, *Sussex in the Great Civil War and the Interregnum 1642–1660* (London: Chiswick Press, 1910), 64.

[118] Thomas-Stanford, *Sussex in the Great Civil War*, 143–7.

[119] Turner, 'Ashdown Forest', 63.

[120] For Wych Cross see J.H. Rose and A.M. Broadly, *Dumouriez and the Defence of England against Napoleon* (London: J. Lane, 1908), 218–19.

Figure 5.6. The military camp near Duddleswell 1793 (source: TNA: MPH 1/407 [extract]).

A second and more immediate impact was from the large numbers of troops stationed in Sussex at this time, with East Grinstead almost becoming a garrison town. There were men from the regular army, the militia (only full-time in war, and then to serve only in England and to keep the internal peace, especially in Sussex against smugglers) and volunteers, who would fight in their localities in the event of invasion. Troops were marched all over Sussex responding to invasion threats, and at one point in 1793 were camped on Ashdown. A map of their disposition shows the location of the field kitchens to provision a camp of 7,000 horse and foot soldiers, with 12 regiments commanded by the duke of Richmond (Figure 5.6). Arriving on 4 July they remained for a week before marching to Brighton, where they were inspected by the Prince of Wales. The remains of the kitchens are in two blocks, discernible on lidar images as a line of small, closely spaced mounds: one group near Camp Hill, and a second nearer to Friends Clump.[121] In Nutley High Street 'the Barracks' was a group of cottages used as officers' billets during this period, but conditions were less good for private soldiers, such as the unfortunate Edward Stibs, who 'died through the inclemency of the weather' at Nutley in January 1805.[122]

From the late 1790s the threat of invasion had grown. The government called for churchwardens and overseers of the poor to produce data to help assess the amounts of moveable and permanent goods that would be involved in 'driving the country' in the event of invasion: that is, evacuating much of Sussex and Kent. The results became known as the civil defence and livestock returns, and they shed some light on the Ashdown response to the French war.[123] In fact, by 1803–4 evacuation was seen as impractical, indeed cruel to local people, whose crops and mills would be burnt, their elderly carted away, and who would be threatened by famine. However, enlistment and recruitment were hastened and armed associations formed in many parts of Sussex. Beacons were readied at Crowborough and West Hoathly in preparation to signal the French arrival. In 1803–5 the invasion scare reached its height, and an act of parliament now sought to identify those men who could be co-opted into a defence force.

All able-bodied men between the ages of 17 and 55 not already serving in the army were listed (Tables 5.4 and 5.5). There is superb detail for four Ashdown parishes, giving an idea of the range of occupations for just under 1,000 men, although we should remember that the distinction between a servant and a labourer was by no means clear-cut, and in fact those compiling the list for Rotherfield did not use the term 'servant' at all. It also omits men aged over 55, who would still have been active in the workforce. Nevertheless, the occupational structure is much as expected: in this agrarian economy many of the servants were probably living-in, and adding labourers and servants together

---

121 TNA, MPH 1/407, Plan of the camp in Ashdown Forest, Sussex, showing the positions of the regiments; I.D. Margary, 'Military Field Kitchens of the Eighteenth Century', *SAC*, 103 (1965), 60–66; Butler *et al.*, *Ashdown Forest Historic Environment Resource*.

122 Hunnisett, *East Sussex Coroners' Records*, 45.

123 ESBHRO, L/C/G/3/1–2 Proceedings under the Defence Acts 1801–06; and for a full analysis see R. Pearce, *Facing Invasion: Proceedings Under the Defence Acts 1801–05*, SRS 99 (Lewes, 2019); G.H. Kenyon, 'The Civil Defence and Livestock Returns for Sussex in 1801 with Particular Reference to the Returns for Kirdford Parish in 1798', *SAC*, 89 (1950), 57–84.

gives an overall total of 67 per cent of the working population for the four parishes. The only other significant group was farmers (14 per cent). Some men declared that they were already enrolled in the voluntary infantry or cavalry. Men with two or more children would be exempt, but a significant proportion were unfit. The family of the Moons, Rotherfield weavers, were described as either 'infirm' or 'dumb'.

In fact, by no means all the men listed in the returns declared themselves willing to volunteer. The figures were somewhat variable: in the four Ashdown parishes as many as 8.3 per cent were unfit, leaving 75.7 per cent of the men declared willing to serve, and just 1.7 per cent already volunteering in one capacity or another. Of the latter, 11 were volunteers for the infantry and three farmers from Hartfield (Thomas Spencer, Steven Waters and William Woodhams) for the more prestigious cavalry; two more would act as guides through the Ashdown area, and one Henry D'Oyley, gentleman, was described as 'belonging to the line'.[124]

Table 5.4. The militia lists 1803.

Occupation	Buxted	Hartfield	Rotherfield	Withyham
Labourer	86	79	249	87
Servant	42	50		63
Farmer/farmer's son	31	16	54	37
Carpenter	7	6	4	1
Cordwainer	7	4	3	
Shoemaker	1		4	7
Blacksmith/farrier	5	7	5	5
Wheelwright	5	6	6	2
Butcher	2	1	2	
Baker				1
Miller	2	3	2	3
Bricklayer	1		2	2
Mason/stonemason			1	2
Thatcher	2			
Sawyer			2	1
Tile maker				1
Plumber		2		
Tailor		4	2	3
Weaver	2	2	8	
Cotton spinner	1			
Glover			1	1

[124] Henry D'Oyley was the youngest son of the rector of Buxted, one of five brothers, and became a major general.

Collar maker		1		
Fellmonger			1	
Turner				3
Hoop maker			2	
Chair maker			1	
Clocksmith			3	
Carrier/higgler	1	2		
Shopkeeper/victualler/grocer	4	1	4	
Hawker			1	
Publican/innkeeper	1		4	
Bailiff	2			
Gamekeeper	1			
Gardener	2			3
Groom				1
Shepherd				1
Footman				1
Clergyman	1			1
Gentleman	3		1	
Surgeon	1			
Schoolmaster			2	
Other	3		1	
Total	213	184	365	226

Source: https://theweald.org/Militia.asp?So urceld=Bu1803Dir; Ha1803Dir; Ro1803Dir; Wi1803Dir.

Table 5.5. Breakdown of the 1803 militia lists.

	Buxted	**Hartfield**	**Rotherfield**	**Withyham**	**Total**
Unfit to serve	21 (9.3%)	9 (4.9%)	36 (9.9%)	16 (7.1%)	82 (8.3%)
No declaration of willingness to serve	37 (17.4%)	45 (24.5%)	85 (23.3%)	73 (32.3%)	240 (24.3%)
Volunteers etc.	7 (3.3%)	6 (3.3%)	2 (0.5%)	2 (0.9%)	17 (1.7%)
	5 infantry; 1 guide; 1 military	3 Infantry; 3 vol cavalry	2 infantry	1 guide; 1 infantry	
Total men listed	213	184	365	226	988

Source: http://theweald.org/HCensus2.asp?HMatch=HMi03.

The North Pevensey Legion was raised in 1803 under Lord Sheffield, who declared that this part of Sussex was full of 'a bad breed of Smugglers, Poachers, Foresters and Farmers' servants, who, in the case of confusion, are more to be dreaded than the march of a French army'. By the end of the year he had 1,000 men from all social classes. A company of skirmishers was formed from 'Smugglers, Poachers (whom it may be necessary thus to occupy that they may not take a worse course) and unsightly men (with whom the Farmers would not Chuse to rank)' because 'the worst looking and worst made men often make excellent skirmishers'. The senior officers were Lord Sheffield (Colonel), Major Edward Cranston and the Hon. Charles Abbot (Lt-Col.), who raised a troop of 46 men from the Forest Row area.

The most prestigious of the volunteers were the gentlemen and farmers of the yeomanry cavalry. One such troop was raised in 1794 by John Trayton Fuller at East Grinstead; another was raised by William Sewell, referred to as the Lindfield, or West Hoathly or East Grinstead troop; and a Forest Row cavalry troop existed from 1794 to 1802. Lord Sheffield was a member of the organising committee from 1794. The North Pevensey legion was disbanded in 1806 but was anyway subject to apathy as the crisis subsided after the Battle of Trafalgar and with Napoleon's decision to move his troops away from the Channel coast to face the Austrians. Lord Sheffield now described his legion as 'generally undisciplined and insubordinate, and which, on the slightest dissatisfaction or caprice, might vanish in an instant'.[125] No doubt Ashdown residents were relieved when the militia were disbanded, as they had acquired a reputation for vandalism, poaching and wood theft.

A third, and perhaps inevitable impact, was on society and especially local women. Uniforms might dazzle and scandals erupt, and it was said that 'The militia's reputation, after these scandals, would be more about the risks it posed to English ladies' virtue than the threat it made to Frenchmen's lives.'[126] Thus in October 1805 Amelia Budgen, 'singlewoman', declared 'that Lewellyn Thomas, Sergeant in the Monmouth and Brecon Regiment of Militia, is the father of a female bastard child born to her on 18 July last, at her mother's house in West Hoathly'. Thomas was to pay £2 18s for lying-in and 2s a week; the mother to pay 6d a week.[127] More happily, when Sarah Bridger of Buxted, spinster, married Andrew Alchorn of Isfield, labourer, in August 1803, he was returned in their marriage licence as being in the militia. Their son Andrew, the first of seven children, was born in November that year, so she was six months' pregnant on her wedding day – not an altogether unusual state of affairs, but could it be in part put down to his military bearing?[128]

[125] A. Hudson, 'Volunteer Soldiers in Sussex during the Revolutionary and Napoleonic Wars, 1793–1815', *SAC*, 122 (1984), 165–81; C. Sebag-Montefiore, *A History of the Volunteer Forces from the earliest times to the year 1860* (London: Archibald Constable and Co. Ltd, 1908), 263–5.

[126] T. Fulford, 'Sighing for a Soldier: Jane Austen and Military Pride and Prejudice', *Nineteenth Century Literature*, 57 (2002), 156.

[127] WSRO, PAR 379/34/4/1; Pilbeam and Nelson, *Poor Law Records of Mid Sussex*, 329.

[128] E.H.W. Dunkin and E.W.D. Penfold, *Marriage Licences at Lewes, 1772–1837, A to L*, SRS 25 (Lewes, 1917).

IMPROVED ROADS AND THEIR IMPACT

A recurring issue concerned the state of the highways around the Forest. The local JPs were kept busy dealing with encroachments onto the highways, and the ways in Withyham were especially troublesome. Withyham's inhabitants were indicted at the Lewes Easter 1642 sessions for 'not repayring severall highways' and three men appeared to pay a fine. At Midsummer 1644 William Constable 'made oath that the highway leading from Leighgreene to Forestgate – "januam Forest de Ashdowne" is repaired and is discharged'.[129] The same issue necessitated a parish representative attending the Lewes sessions in Easter 1645 to attest that his hedges and ditches along the same road were now repaired. But the issue arose again at the Lewes Epiphany session in 1648/9, together with the requirement to repair a bridge in Withyham, and again at the Lewes Midsummer and Chichester Easter sessions in 1649. Similarly, the road from Holtye Common to East Grinstead was 'in decay' in 1643. Clearly the roads to the north-east of Ashdown leading to the Forest were in a poor state.

To help remedy the appalling state of the roads, some private donations were forthcoming. The will of John Baker of Withyham in 1555 specified that 'I gyve and beqweth to the mendynge of hye ways within the paryshe of Wethyham £6 13s and 4d to be delyvered to the hands of Richard Grey clerke Parson of Wethyham within two yeres next after my death.' Other wills made similar provision.[130] John Pope of Fletching in his 1547 will bequeathed 'XXs to the hye ways between Prykars [Prickets] Hatche Gate and Clapwater Bridge, to be paid out of my land, the Vyere after the decease of me and my wiff'. In 1596 Richard Leche likewise bequeathed 'threskore cowdes to the amending of the ways leading from Godstone to Lewes ... '.[131]

But even by the eighteenth century pack-horse and horseback traffic was the most that could be hoped for during much of the year and wheeled traffic drawn by oxen would often find itself in difficulties. As Defoe had noted in 1725 of the Weald in general, 'the corn is cheap at the barn because it cannot be carried out; and dear at the market because it cannot be carried in'.[132] Travel could be hazardous: William Cranston wrote to John Collier on 15 October 1745 'by that time I got to Maresfield could scarce see the church, but however, being well acquainted with the road, I ventured over the forest, upon which I found a great fogg and bitter dark, but surmounted all difficulty, and reached Eastgrinstead before 8'.[133] Dr John Burton confided to his journal in 1752 'There did not appear to be public ways, but bye-roads, or rather the haunts of kine, for the oxen had left everywhere deep marks of their cloven feet.'[134]

---

129 'januam Forest de Ashdowne' – at the entrance to the Forest.

130 R. Garraway Rice, *Transcript of Sussex Wills – Vol IV. Racton to Yapton*, SRS 45 (Lewes, 1940).

131 R. Garraway Rice (ed. W.H. Godfrey), *Transcripts of Sussex Wills Vol II*, SRS 42 (Lewes, 1937), 188; Rev. S.D. Wilde, 'Fletching, Parish and Church', *SAC*, 4 (1851), 232–42.

132 D. Defoe, *A Tour through the Whole Island of Great Britain* (Harmondsworth: Penguin, 1971; first published 1724–6), 441.

133 R. Saville, *The Letters of John Collier of Hastings 1731–1746*, SRS 96 (Lewes, 2016), 309–10.

134 BL, Add MS 11,571, 'A Traveller's reveries, or Journey through Surrey and Sussex ... by John Burton, M.D. 1752 [transcript].

But now a new landscape feature was appearing. Turnpike Trusts were being established, since the arrangements by which parishes maintained their own roads were clearly breaking down as roads were made execrable by the heavier passage of people, oxen with waggons, and especially Wealden iron and timber tugs. In 1718 an Act was passed creating a turnpike trust over the road from London to Croydon and Kingston, and following a petition from Sussex it was shortly afterwards extended to East Grinstead. By April 1723 a turnpike was approved to extend the road further onto Highgate, thereby reaching the edge of Ashdown and giving improved access to East Grinstead from the south.[135] By the 1780s the turnpike had been extended again to Wych Cross, meeting another turnpike that had been constructed from 1752 between Wych Cross, Uckfield and Lewes. And in the same year construction had begun on a road offering a more direct route from Wych Cross to Lewes via Offham. Wych Cross, with a toll house, had now become an important turnpike intersection (Figure 5.7).

Most of the turnpikes around Ashdown were products of the second half of the eighteenth century and many were concerned to convert older driftways and narrow lanes into wider roads suitable for carts. In 1766 a turnpike was opened from Tunbridge Wells through Friar's Gate, King's Standing and Duddleswell to Maresfield (now the B2168), with the fairly standard justification that the road 'is impassable for Wheel Carriages in the Winter Season, and very dangerous for Persons travelling in Carriages, or on Horseback, at all Times … '. A second Act in 1788 authorised the turnpiking of the road that branched off from this route at Florence Farm, near Groombridge, to run through Hartfield, Withyham and Forest Row to East Grinstead. In addition, during the 1760s turnpikes reached from Tunbridge Wells to Crowborough Gate and Rotherfield, and from Bromley to Hartfield and Beggars Bush. By the 1770s, therefore, Ashdown residents had decent access both to Tunbridge Wells and beyond to London, and south to Lewes and thence to Brighton and the coast.[136]

However, these turnpike roads, especially on pre-existing routes, were not universally popular, and neither were their tolls, with many finding ways to circumvent the toll houses.[137] Some minor changes were made later: the steep hill at Danehill (across Dennwood common) was bypassed in 1818 and the road became a quiet residential area.[138] And not all the roads were necessarily kept in good repair, as hard stone could not always be found – indeed, flint had to be procured at considerable

---

135 Hills, *History of East Grinstead*, 155; B. Austen, 'East Sussex Turnpikes of the High Weald', *Sussex Industrial History*, 43 (2013), 31–47.

136 Taken from G.D. Johnston, *Abstract of Turnpike Acts relating to Sussex* (transcribed by R.B. Snow from an original 1947 typescript in Sussex Archaeological Society Library).

137 G. Joan Fuller, 'The Development of Roads in the Surrey–Sussex Weald and Coastlands between 1700 and 1900', *Transactions of the Institute of British Geographers*, 19 (1953), 37–49. She states that by 1840 only 22 per cent of Sussex roads had been turnpiked. For one such scheme for a 'free road' to bypass the Wych Cross turnpike by a road from Forest Row to Uckfield see BL Add MS 32,732 Newcastle Papers XLVII (1753).

138 P.G. Lucas (ed.), *Church Lane, Danehill*, Danehill Parish Historical Society (2010), 1–2. The road was eventually crowned with a new church in 1892, thereby becoming the modern Church Lane.

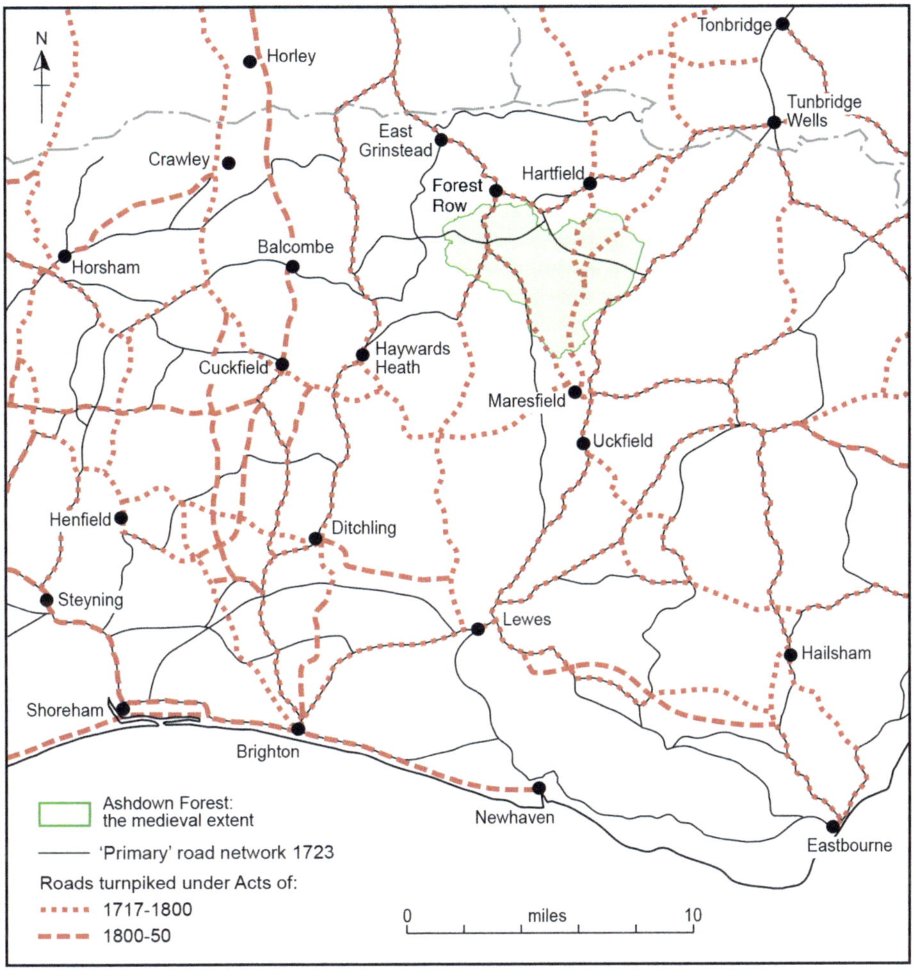

Figure 5.7. Turnpike roads in the Ashdown Forest area (source: adapted from G. Joan Fuller, 'The development of roads in the Surrey-Sussex Weald and coastlands between 1700 and 1900', *Transactions of the Institute of British Geographers*, 19 (1953), 37–49).

cost, even though Ashdown's quarries were at hand. But local residents might also take a hand in improvements: Charles Abbot from Kidbrooke, next to the turnpike road from East Grinstead, certainly took an active interest as chairman of the Sussex and Surrey Turnpike Trust. Under his chairmanship roads were straightened and gradients lowered where necessary, and by the 1820s they were further improved by applying the new macadamised surfaces, with John McAdam being appointed first as an adviser, and then government surveyor, to nine of the Sussex trusts.[139]

139 The turnpike milestone markers along the A22 are claimed to be the longest stretch in the country. And see the evidence of J. Irish relating to the Maresfield road to the *Second Report*

GENTEEL SOCIETY AND THE FOREST LANDSCAPE

Wider and straighter roads with gentler gradients brought coaching traffic and coaching inns to East Grinstead and elsewhere on the fringes of the Forest. Visitors were encouraged by the ability to escape 'the Wen' for greater tranquillity but still be within travelling distance. The area also provided relatively cheap house and land purchases for those requiring access to the facilities at Tunbridge Wells, booming Brighton or administrative and legal business at Lewes. Thus, wealthy families were now arriving in larger numbers around the edges of the Forest.

Possibly for the first time, Ashdown's landscape was also being appreciated for its aesthetic qualities. Its scenery might be viewed with a mixture of delight and horror – both co-existed by around 1800. In 1768 William Gilpin had published his *Essay on Prints*, defining 'the picturesque' as 'that kind of beauty which is agreeable in a picture', and his *Remarks on Forest Scenery, and other woodland Views* (1791) was based largely on the New Forest. Any devotion to heathland was perhaps yet to emerge, but his views were taken up by others and gradually developed into a lexicon of landscape appreciation that was to flourish more particularly in the nineteenth century.

Such new ideas about British scenery filtered into the area, partly engendered by the curtailment of the Continental Grand Tour because of the later eighteenth-century wars. Now parks were being remodelled by 'Capability' Brown at Sheffield Park and by Repton at Buckhurst/Stoneland, Sheffield Park and Kidbrooke. Ashdown House near Forest Row was designed for Trayton Fuller by Benjamin Latrobe in 1793, just a year after he had designed the renovated Hammerwood Park two miles to the north as a mock grand hunting lodge with Greek Doric porticos for John Sperling. These were designed by Latrobe shortly before he left for America where he became a prolific and admired architect.[140]

But class sensitivities were always just below the surface. The tensions and conflicts on Ashdown were not unique. Gilpin certainly recoiled from the lawlessness of many forest inhabitants, and the feelings of some writers, viewing the situation from outside, were epitomised by John St John, whose book of 1787 included the thoughts of John Norden, derived no doubt from his travels as Crown Surveyor of Woods and Forests for James I. For him forests were:

> so ugly a monster as of necessity will breed … more and more idleness, beggary and atheism, and consequently disobedience to God and the King … wherein infinite poor, yet most idle inhabitants have thrust themselves, living covertly without law

of the Select Committee of the House of Lords to examine the Turnpike Returns (August 1833), 34–5. I am grateful to Martin Berry (personal communication) and see his 'The Turnpike Roads of Ashdown Forest: a Forgotten History?' *Ashdown Forest News* (Autumn 2016).

140 A.-N., D. and E. Pinnegar, *Hammerwood Park East Grinstead, Sussex: 1792, first house of Benjamin Henry Latrobe* (East Grinstead: Hammerwood Park, 1992); Antram and Pevsner, *Sussex: East*, 96–7, 414–16.

> or religion ... among whom are nourished and bred infinite idle fry, that coming ripe grow vagabonds, and infect the Commonwealth with most dangerous leprosies.[141]

But Ashdown was changed forever during the late seventeenth and eighteenth centuries in ways both social and topographical. Changes were afoot, too, in how the inhabitants earned their livings, and it is to this theme that we now move.

[141] Hon. John St John, *Observations on the Land Revenues of the Crown*, 2nd edn (London: J. Debrett, 1792), Appendix 2, 290. And see S. Hindle, 'Not by Bread Only? Common Right, Parish Relief and Endowed Charity in a Forest Economy c.1600–1800', in S. King and A. Tomkins (eds), *The Poor in England 1700–1850: an Economy of Makeshifts* (Manchester: Manchester University Press, 2003), 39–75. I am grateful to Steve Hindle for drawing my attention to this quotation.

CHAPTER 6

ASHDOWN'S FOREST ECONOMY

> Tis a melancholy consideration that mankind will inhabit such a heap of dirt for a poor livelihood.[1]

THE YEARS THROUGH TO the end of the sixteenth century are sometimes seen as witnessing the transition from a medieval economy to a more modern agrarian capitalism. While this *might* hold true for prosperous localities, it cannot be read off as a linear progression, nor a spatial one, since the same cannot necessarily be assumed for socially and economically more marginal areas such as the lowland forests and heathlands.

Much of Ashdown would have reverted to birch scrub by 1500 and the early medieval assarts had been hard to sustain. Nevertheless, they remained the most numerous form of tenure: they were in practice secure, held with old-established and unchanging rents, with low entry fines often equivalent to an annual rent or single heriot, such as demanded in Duddleswell manor. But here, as elsewhere, Ashdown's 'peasantry' was itself intensely differentiated: 'different stories for different groups of peasants'.[2] Much depended on the resources offered by the local environment, the consequent seasonality of work, the family and household structure and the manorial institutions.

However, although the land itself was in poor order, by 1500 many of the small severalty farms around Ashdown did have some arable land, generally less than ten acres, primarily for oats, with some wheat and peas. There was a hesitant recovery from the post-Black Death recession. But much was now under pasture, and the enclosed land was still relatively well wooded. By the 1590s the Buckhurst estate, largely in Hartfield and Withyham, and much lying in the more favoured Medway Valley, had 1,329 acres of arable, accounting for just under a third of the farmland. Most fields here were between three and 12 acres in size, commonly with encompassing shaws. There were 1,026 acres of highly prized and frequently subdivided meadow in the Medway Valley, and 2,052 acres of pasture, some stretching alongside roads, as at Lye Green outside Friar's Gate.[3]

1 Lord Chancellor Cowper on the country around Horsham, 1690. R.W. Blencowe and M.A. Lower, 'Extracts from the diary of a Sussex tradesman, a hundred years ago', *SAC* 11 (1859), 183.

2 T. Scott, *The Peasantries of Europe: from the Fourteenth to the Eighteenth Centuries* (Harlow: Addison Wesley Longman Higher Education, 1998), 375.

3 Straker, *Buckhurst Terrier*, 2–29. Straker gives 21 per cent as meadow, 34 per cent pasture, 31 per cent arable, 12 per cent wood and 2 per cent sundry (E. Straker, 'Agricultural History in

With the farms held in severalty, some flexibility was possible. Crops would be taken for a year or so, and then the poor soils allowed to recover once more under grass. A fallow season might be introduced and marling used to ameliorate the soils. Marl's use was now being popularised among literate farmers in Gervase Markham's *Inrichment of the Weald of Kent* (1625), some of which was derived from Fitzherbert's earlier *The Boke of surveying* (1523). Thus, we find the surveyors of the 1650s recommending heavy dressings of marl, the most accessible source of which was the Wadhurst Clay outcrops on the northern (Claypits Gate) and southern (Sweet Mine Pits) edges of the Forest, which were open to Commoners. But pits were also dug somewhat indiscriminately within many fields, since the costs of moving marl any distance were high.[4] In 1875 Topley recorded many old marl pits, some with trees 200 to 300 years old growing in them.[5] Mention was also made of the application of chalk in the 1658 surveys of Prestridge [Press Ridge] Bank and Comedeane Lodge, with its 'killyard', but this would have been an expensive undertaking.

Cattle and pigs were by far the most prominent animals. A few dairy cattle were kept and there were dairy houses and cheese chambers in most Wealden farmhouses. But little left the family holding and the quality of Wealden cheese was never held in high esteem. By the seventeenth century the droving routes from Wales to the Pevensey fattening pastures passed close to the area, bringing lean cattle that might be purchased, and as a result some fattened cattle entered London markets via local fairs or butchers.[6] The drovers would arrive at the November fairs in East Grinstead, and an entry in the *Universal Magazine* for 1760 refers to 'Welch runts, which are brought [sic] up here by the Kentish and Sussex farmers' who would fatten them and sell them to local butchers.[7] The St George's Day fair at Nutley was noted for the purchase of dairy cattle and steers by the 1690s, and farmers travelled to attend the fairs at East Grinstead (now on St Andrew's Day, 30 November) and Rotherfield (9 October).

However, there were no large commercial cattle enterprises, although considerable numbers of cattle might be agisted by the foreign tenants and by the keepers. Sheep were far fewer in number than outside the Weald, but they might be grazed on some of the subsidiary commons around Ashdown. In 1650 a hammerman and collier from Hartfield were charged with stealing sheep from a local weaver at Holtye Common, while sheep kept by a Buxted spinner were savaged on the waste only a few yards from his door.[8] Oxen were the commonest working animals before the eighteenth century,

---

the Hundred of Hartfield', *SAC*, 76 (1935), 172–7).

4 Gulley, 'Wealden Landscape', 144–5, Fig. 17.

5 W. Topley, *The Geology of the Weald (parts of the counties of Kent, Surrey, Sussex, Hants)* (London: Memoirs of the Geological Survey, 1875), 387–8. Topley (1841–94) worked extensively on Wealden geology and its associated economic aspects. See W. Topley, 'On the Agricultural Geology of the Weald', *Journal of the Royal Agricultural Society*, 8 (1872), 262–3.

6 In December 1633 'John, a welch drover' was buried at East Grinstead (Crawfurd, *East Grinstead Parish Register*, 180.

7 I am grateful to the late Michael Leppard for references to the droving trade in East Grinstead.

8 Brent, 'Employment, Land Tenure and Population in East Sussex', 223.

whether for pulling carts or ploughing, being normally worked until about ten years, when they would be fattened and sold.

The Duchy of Lancaster derived a good income from agistment on Ashdown and from its parks at Maresfield and Newnham. But in 1539 a royal commission investigated the management of the Forest and heard complaints that many foreign cattle were being agisted to the detriment of the tenants, and that keepers were withholding profits that should have gone to the king.[9] The keepers were also unpopular during the sixteenth century because of their attempts to enclose areas around their lodges. Each of the six keepers looked after one of the Forest Walks and lived in a lodge within the Walk, such as Comedeane or Broadstone. Their main tasks were the protection of venison and vert, maintenance of the pale, monitoring the cattle commoned and agisted to maintain a balance between the resultant income and browse and cover for the deer, and organising the drifts of illegally depastured animals. There was much room for local disagreements, and the Forest courts frequently dealt with cases of overpasturing of animals, the turning out of pigs during fence month or guarding common rights against non-Commoners. By the 1650s the parliamentary surveyors estimated there to be about 1,000 Commoners' cattle and 700 cattle agisted by the Keepers alone.

In 1564 a detailed report listed 147 Commoners and their payments.[10] Despite the relative freedom from the heavier manorial dues exacted in southern Sussex, there were still regulatory powers exercised by Duddleswell and Maresfield manors. The differentiated system of common rights, described for the medieval period, seems to have persisted. Free tenants, intertenants and foreign tenants were again recognised in an inquisition of the manor of Duddleswell in 1610, which now found 223 Commoners exercising their rights, a 50 per cent increase since 1564.[11] Most of the foreign tenants lived in Maresfield or Fletching. The king's free tenants could have two loads of alder or birch yearly, stakes and binders for enclosing their lands (frith and tenant), and wainbote and ploughbote of beech, birch and alder.

There was certainly now less timber on Ashdown, but Commoners in 1610 were also entitled to two free wagon loads of alder, birch or willow 'agaynste the feste of Crysmase'.[12] All had common of pasture and herbage but only free or intertenants who had houses on their land were allowed estovers. Others also had rights, such as those belonging to the assart holdings that had grown and even given rise in some places to small hamlets, such as Horney Common. But the interpretation of common rights did vary over time: of the free holdings in 1610, only 38 per cent enjoyed rights of herbage. Of the foreign holdings, 21 per cent now apparently enjoyed estovers as well as herbage, although formerly they were only entitled to herbage. There did appear

9 Mate, *Trade and Economic Developments*, 174–5.

10 TNA, DL 42/112, ff.153–64.

11 Gulley, 'Wealden Landscape', 102.

12 Willow, less commercially valuable, was added to the list of estover species during the early modern period, possibly to supplement cottagers' supplies when more traditional species were being commercially sold (Irons, 'Aspects of the Impact of Man', 121).

to be a loosening of manorial restrictions, and there were also reportedly 14 cottages built on the Forest 'without lawful licence'.[13]

As we have seen, given the small size of the Forest holdings many depended on family labour and on the essential common rights, as well as on external sources of income, including day labour elsewhere, such as in the woodlands, where underwood might be gathered and sold locally as firewood, wattles, hurdles or for other domestic use. The 1610 inquisition revealed that nearly 50 per cent of the holdings on the Forest itself were of four acres or less, so it was imperative that those entitled to common rights should exercise them. The most important was the depasturing of cattle and horses, followed by the collecting of estovers (the cutting of wood for fuel and building) and the taking of turf, marl and stone, the latter most commonly for building work. Sheep remained forbidden: a warrant was issued by Richard Sackville in February 1675 to his bailiff Richard Homewood of Maresfield to detain sheep pastured on the Forest, which were to be kept 'until compounded for and payment made'.[14]

As in the medieval period, small tradesmen remained a feature. By the seventeenth century butchers, tanners, glovers and tailors were to be found, together with some clothiers. There were fulling mills at Fletching (1656), Maresfield (two in 1545), Uckfield (1612) and Buxted (1619). The cloth was not of high quality, and numbers of weavers were low. One was George Peene of Withyham, who in 1643–4 purchased 362lb of white wool from sources in Hartfield, Withyham and Robertsbridge. In late 1597 the survey of the manorial lands of Buckhurst revealed that Thomas Tailor, clothier, 'holds by indenture ... the tenement called Newbridge House, and two watermills under one roof, also 6 acres of pasture, and 40 acres of pasture lying within the Forest adjoining the 6 acres, sometime enclosed with a pale'.[15]

Before the end of the fifteenth century there had been manifest concern over timber destruction, even before any over-exploitation of the Forest's timber to provide fuel for the iron industry.[16] Indeed, by this time the Forest was described as barren heath with beech trees and some covert but very little underwood. By the sixteenth century the relatively heavy depasturing, together with the effects of burning the herbage to improve the quality of the grassland and the leasing of parts for rabbit warrens, had already led to clearances of woodland. A custumal of 1520 reported that one-third of the Forest had already been cut down for coaling and that all the birch wood between Nutley and Fairwarp, indeed all the bottom of South Ward, had been felled for the same purposes. At a 'Coorte holden at Notlye' we hear that 'all the byrche wood between Notlye and Fayre Wharpe hath bene felede'.[17]

Beyond the provisioning of elite households or local sales of timber by tenants, there is still little evidence that the area had any regional trading links for wood or wood

13 Small, 'A Review of Ashdown Forest', 162–3.

14 ESBHRO, ACC 1831/1/40.

15 Brent, 'Employment, Land Tenure and Population in East Sussex', 125–6; Straker, *The Buckhurst Terrier*, 16.

16 Brandon, 'Medieval Clearances', 135–53; P. Brandon and B. Short, *The South East from AD 1000* (Abingdon: Longman, 1990), 72–4.

17 Christian, *Ashdown Forest*, 15.

products. However, the woods received increased attention because of the resurgence of the iron industry. Timber was cut for charcoal making, a useful by-employment, and several charcoal-making platforms have been identified on the slopes of wooded ghylls with the help of lidar, particularly near ironworking sites, such as Newbridge or Crowborough Warren.[18] Additionally, oak was much in demand for construction timber and for the pale; beech was employed for panelling, tools and furniture; and there may have been some hazel coppice providing material for wattle and daub.

WEALDEN IRON

The Wealden iron industry, including the initiation and spread of the new blast furnace technology and allied forges (or hammers), is the most closely researched subject in the Ashdown area. The work of many researchers is summarised in the books by Cleere and Crossley (1985), Hodgkinson (2008) and Auty (2019). Because of their assiduous work, and that of the Wealden Iron Research Group, the following account only summarises the industry within or around Ashdown.[19]

The sixteenth and seventeenth centuries were the heyday of the Wealden iron industry (Figure 6.1). The early Tudor kings needed armaments for wars against Scotland and France, which the Weald was able to satisfy, at least in the early years of the industry, together with all manner of iron products. And the Ashdown area was the early heartland of the industry. The resources were readily available locally. There was stone for the furnaces 18–20ft high, and fast-flowing winter and spring streams propelling waterwheels that powered bellows and hammers. A steady water supply was ensured by damming the narrow ghylls with a bay to form hammer ponds; valuable coppice wood was cut on a seven- to 12-year rotation to be coaled on charcoal hearths for the furnaces and forges, and more substantial timber was available for construction. And iron ore was mined nearby, as, for example, at Tugmore Shaw, Hartfield, where 'mine' was extracted from open-cast and bell pits, leaving about 250 pits and saucer-shaped depressions of varying sizes, some redug but many now filled with water, lying close to fields named 'Great Mine Pits' and 'Little Minepits'.[20] But while providing a valuable source of work, the digging of ore was not always seen as a boon: in 1520 customary tenants complained that 'in tyme paste the fforest hathe bene dygged for Irne by the whiche man and beste be in jeopardie'.[21]

---

18 For the contribution of lidar to studies of wooded environments see B. Lennon, *Savernake Forest: Continuity and Change in a Wooded Landscape*, British Archaeological Report 555 (Oxford, 2012).

19 Hodgkinson, *The Wealden Iron Industry*, building upon Cleere and Crossley, *Iron Industry of the Weald* and the work of the Wealden Iron Research Group from 1968, as published in their bulletin *Wealden Iron*, <http://www.wealdeniron.org.uk> [accessed 12 September 2021]. For the continental context of the Wealden iron industry see the late Brian Awty, *Adventure in Iron*, 2 vols, Wealden Iron Research Group (Tonbridge, 2019).

20 'Iron minepits in Tugmore Shaw', <https://historicengland.org.uk/listing/the-list/list-entry/1002230> [accessed 12 October 2019].

21 ESBHRO, U of Sx Spec Colls, AFC I, 10 October 1520.

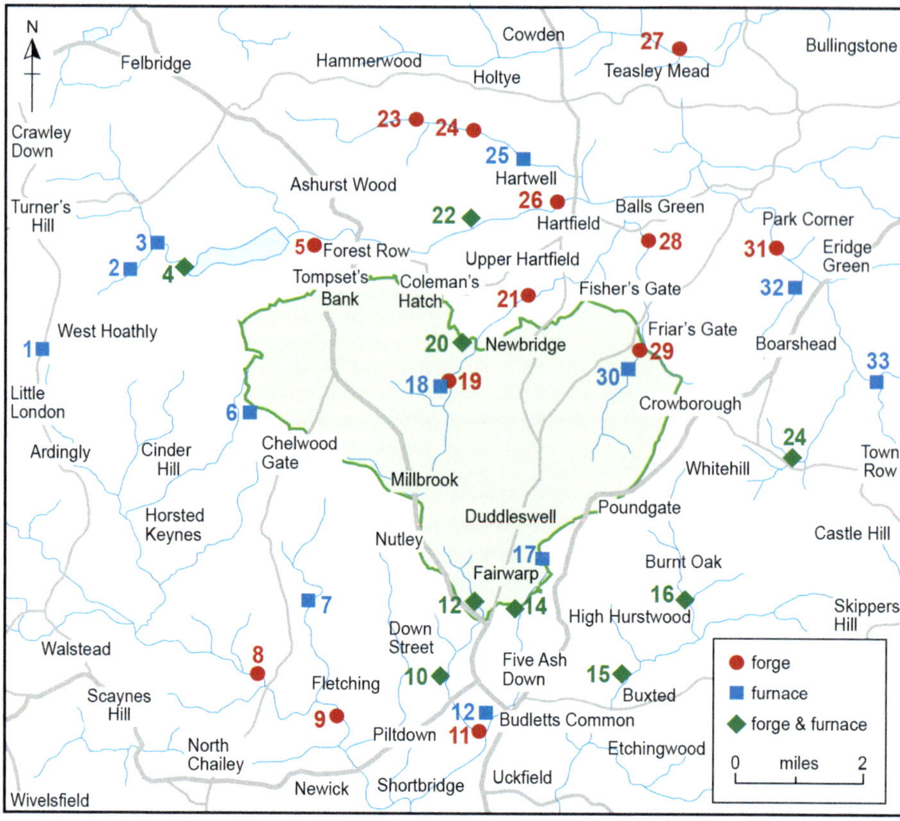

Figure 6.1. Blast furnace and forge sites around the Ashdown Forest. Key: sites where both operations existed are shown, though not necessarily operating at the same time (source: Jeremy Hodgkinson).

1. Chittingly Furnace
2. Gravetye Furnace
3. Mill Place Furnace
4. Stone Furnace and Forge
5. Brambletye Forge
6. Stumbletts Furnace
7. Sheffield Furnace
8. Sheffield Forge
9. Fletching Forge
10. Langley Furnace and Forge
11. Maresfield Forge
12. Maresfield Furnace
13. Marshalls Furnace and Forge
14. Hendall Furnace and Forge
15. Queenstock Furnace and Forge
16. Little Buxted Forge and Furnace
17. Oldlands Furnace
18. Pippingford Furnaces
19. Steel Forge
20. Newbridge Furnace and Forge
21. Cotchford Forge
22. Parrock Furnace and Forge
23. Bower Forge
24. Cansiron Forge
25. Bassetts Furnace
26. Hartfield Forge
27. Pilbeams Forge
28. Withyham Forge
29. Grubsbars Forge
30. Crowborough Warren Furnace
31. Birchden Forge
32. Hamsell Furnace
33. Cowford Furnace

The working period usually began following the harvest and continued into the following spring, when the runoff from Ashdown's headwater streams began to lessen. A sluice through the bay controlled the water for the waterwheel and a spillway allowed excess water to escape. The molten iron from these furnaces included charcoal impurities and was brittle on cooling. It was therefore reworked in a finery forge, or hammer (hence 'the indirect method'), also powered by wheels fed from the hammer ponds, to produce a cast iron 'sow' or 'pig'. These were transferred to a chafery hearth, where a bar of iron was produced. This was sold, to be made into tools, horseshoes, plough shares, wire, firebacks, nails (as at Sheffield furnace), graveslabs (to be found at the churches such as East Grinstead, West Hoathly and Maresfield) and many other domestic or working products. Cast iron cannon and cannon balls were also early products of Newbridge furnace (Figure 6.2).

The earliest known Wealden blast furnace and forge was at Queenstock, worked by 'iernefounders at Buxstede' by the end of 1490 on land in the valley of the river Uck held by the archbishop of Canterbury.[22] It appears that the furnace was out of operation by 1509 but revived again later in the sixteenth century – a common occurrence with the blast furnaces. The site at Newbridge was carved from the Forest, with a lease of December 1494 relating to Duchy of Lancaster mills. The lease provided for the supplying of ironstones, with as much wood and fuel as necessary from the Forest, and with continental workers to be imported at Henry VII's expense. The rent for this site, in production by at least 1497, was to be either £6, or six tons of iron, to be delivered to the house of the goldsmith Henry Finer in Southwark. The initial lessee was Peter Robert *alias* Graunt Pier of Hartfield, a French iron founder, who was replaced after two years by another Frenchman, Pauncelett Symart, who remained for 14 years producing components for ordnance, although he was soon indebted to the Duchy of Lancaster. The works were now said to be in a poor state, and Symart was succeeded in 1512 by the king's gunfounder Humphrey Walker, who produced bullets for the king's ordnance. By 1525 Sir Thomas Boleyn was the lessee but the furnace was reportedly decayed by 1539.[23]

Again, in 1503 we find a lease of six acres and a watercourse to Claude Rombosson, a cutler, and Vincent Breke, steelmaker of London, to construct 'mills and forges, a great hammer and wheels' 'for to make stele and plates for harneys therin'. The latter refers to body armour, and the reference is to the early form of steel making at this site, built by 1509 near Pippingford, and referred to in 1523 when John Glande held a tenement called a 'Forge of Steele'. This was seemingly in production until at least 1550, although once again reportedly ruined in 1539. The forge was said to be 'standing upon the brook of Newbridge and in distance from the said iron mills the

22 B. Awty and C. Whittick, 'The Lordship of Canterbury, Iron-founding at Buxted, and the Continental Antecedents of Cannon-founding in the Weald', *SAC*, 140 (2002), 71–81.

23 Cleere and Crossley, *Iron Industry of the Weald* (2nd edn), 389; B. Awty, 'Henry VII's First Attempt to exploit iron in Ashdown Forest', *Bulletin of the Wealden Iron Research Group* 11 (1991), 11–14. The former Newbridge hammer pond demonstrates the colonisation by alder carr (*Alnus glutinosa*), with sedges and bryophytes on the damp surface (J. Thompson, 'Newbridge Alder Carr. Site survey' (Univ. of Sussex unpub. MSc, 2006), 13 and appendix.

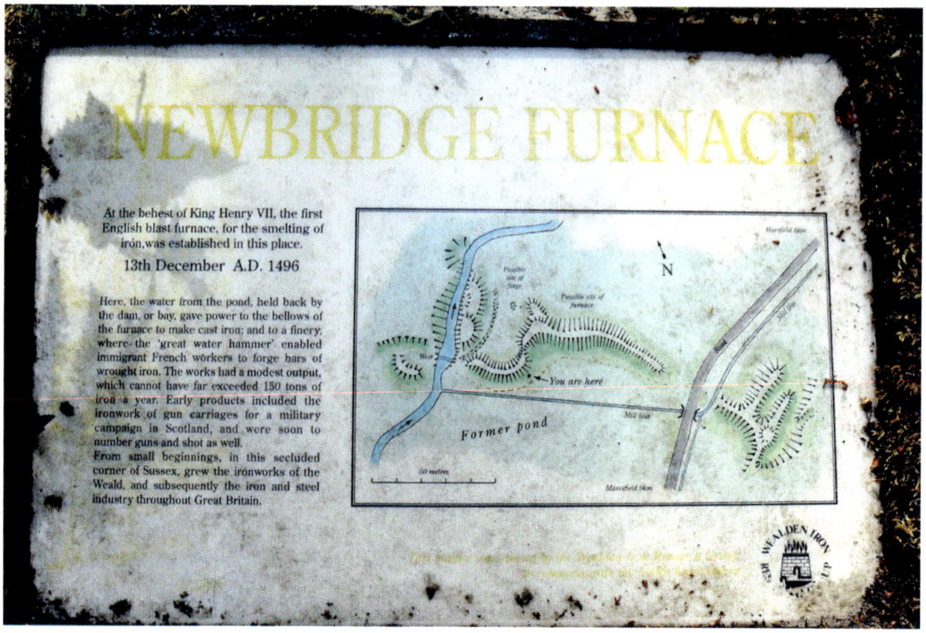

Figure 6.2. Plaque at the site of Newbridge Furnace and the reconstructed spillway (source: author).

space of 2 flight shots or more and yearly rented to the King's Majesty'. The lease again allowed for timber from the Forest for building, ore to be mined and alder, birch and willow to be coaled. The Steel Forge River, between Pippingford and Newbridge, was named after the works.[24]

The earliest skilled workers included many from the Continent, particularly the Pays de Bray. By 1496 the French ironworker John Stiele was at Hartfield forge and there was a prominent nucleus of foreign workers by the 1520s: in the Subsidy Returns for Hartfield in 1524–5 we find such entries as 'Garfysh, a Frenchman', or 'John Myskyn, Frenshman', as well as another twenty noted as 'alyen'. There were altogether 36 aliens in these Subsidy Returns, the highest number anywhere in the Weald. Here were Colne Capyll, Perese Botell and Gwilliam Bristoye, the latter from Glisolles, Normandy, who came to England in 1517.[25] We find in the will of 'Rogerius Machyn parochiae de Rotherfeld' in 1524 the mention of 'All my iron mylles that is to say the iron mill of Brechenden [Birchden], the iron mille of Buksted and the iron myll of Still [Steel]forge'. Machyn was employing four Frenchmen in 1524, probably at Queenstock Furnace. In the Maresfield parish registers we find references in 1551 to 'Buried Peter, a frenchman', in 1557 to 'Buried Peter Gagge, Gallus' or in 1570 'Buried John, a Dutchman'.[26]

These men were the initiators or supervisors of the ironworks, but the need for unskilled labour was met by husbandmen or other local men working seasonally, and there were large numbers of miners, colliers, woodmen, carters and their assistants. During the sixteenth century the coaling operation was also frequently managed by skilled French immigrants. From these early beginnings the blast furnaces and forges multiplied around the Forest, under the supervision by 1541 of William Levett, the king's gunstone maker. At least 36 sites are known in the parishes constituting or bordering on the Forest, including those built for the crown, such as Newbridge, Steel Forge and Stumbletts furnace on Stumblewood Common near Chelwood Gate, the latter built in 1539 but granted by Queen Mary to Sir John Gage in 1553.[27] There were at least four ironworking sites partly extending into the Forest itself: Crowborough Forge, Crowborough Warren, Newbridge and Old Forge.[28]

---

24 Somerville, *History of the Duchy of Lancaster* I, 309; Cleere and Crossley, *Iron Industry of the Weald*, 115; 2nd edn, 391.

25 B.G. Awty, 'The Continental Origins of Wealden Ironworkers, 1451–1544', *Economic History Review*, 34 (1981), 524–39; Cleere and Crossley, *Iron Industry of the Weald*, 122; J. Cornwall, *The Lay Subsidy Rolls for the County of Sussex 1524–1525*, SRS 56 (Lewes, 1957), 130; Awty, 'Aliens in the Ironworking Areas of the Weald: the Subsidy Rolls, 1524–1603', *Bulletin of the Wealden Iron Research Group*, 4 (1984), 13–78. The workers are listed at <http://www.wirgdata.org/searchpro2.cgi?personid=363> [accessed 23 September 2021], and see 548 for Bristoye. The details of the iron workers are to be found in Awty, *Adventure in Iron*.

26 Blencowe, 'Extracts from the parish registers', 247.

27 E. Straker, 'A Lost Tudor Iron Furnace found', *SNQ*, 6 (1936), 217–18; Hodgkinson, *The Wealden Iron Industry*, in Appendix 2 lists Buxted (3); East Grinstead (2); Fletching (3); Forest Row (formerly also East Grinstead, 2); Hartfield (7); Horsted Keynes (2); Maresfield (6); Rotherfield (5); West Hoathly (2) and Withyham (4).

28 Butler *et al.*, *Ashdown Forest Historic Environment Resource*, Fig. 14.

From 1543 complete cast-iron cannons were being manufactured at Buxted under the supervision of the French gunfounder Peter Baude, who was succeeded by the local ironfounder Ralph Hogge. The cannon required a water-powered boring mill to smooth the inside of the barrel. Hogge now played a pivotal role in the Ashdown area. In 1559 he was appointed the queen's 'gonnestonemaker' for life and married Margaret, daughter of Edmund Henslowe, master of the game on Ashdown. In addition to furnaces at Buxted and Hendall, he now added two more, at Marshalls and Langleys (Maresfield), mostly for ordnance and shot. The crown was the main market, but private shipping was also supplied via Lewes.[29]

By 1548 there were said to be 53 furnaces and forges (and possibly some remaining bloomeries) working in the Weald as a whole, and by 1574 there were 50 furnaces alone, together with a similar number of forges. As demand grew, from urban, military, naval and agricultural sources, so more nobility and gentry became involved, especially during the peak of Wealden iron output during the Armada years. Returning to favour under Queen Mary in 1553 following his earlier attainder under Edward VI, the duke of Norfolk controlled Sheffield furnace and forge, but with a change of monarch the Sackville family acquired it and were running the furnace in the 1560s.[30] The Sackvilles clearly saw the industry as of considerable importance and also either ran or leased out Parrock furnace and forge, Maynards Gate furnace and Fletching forge. In the later phases of the industry, in 1644, Sir John Sackville and William White, executors of Sir Thomas Sackville, entered into partnership with the ironmaster Peter Farnden.[31] That many ironmasters were well rewarded for their investments and skill can be seen in part from existing properties around Ashdown (Figure 6.3). Gravetye (West Hoathly) was built in 1597–8 for Richard Infield of Mill Place furnace; the impressive Cromwell House (East Grinstead) was built in 1599 for Edward Payne of the nearby Stone ironworks; and Gibbshaven (East Grinstead) was the home of the Thorpe family, ironmasters at Warren furnace from 1582.

There has been longstanding uncertainty about the impact of the iron industry on Ashdown's woodland. The establishment of the Newbridge furnace and forge and the forges at Hartfield by 1511 clearly impacted locally on the woodland cover. But thereafter successive enquiries investigated the continuing destruction throughout the sixteenth century, including the felling generated by rangers and keepers and also the Sackvilles, the latter on a truly massive scale. Coppicing henceforth assumed more importance to provide a steady supply of charcoal.

Contemporary sources criticised the negative impact, and Michael Drayton had trees cry out against 'these yron times':

29 Hodgkinson, *The Wealden Iron Industry*, 70.

30 M.S. Giuseppi, 'The Accounts of the Iron-works at Sheffield and Worth in Sussex, 1546–1549', *Archaeological Journal*, 69 (1912), 276–311. The Sheffield ironworks were described in a 1549 survey relating to the execution of Lord Seymour who had leased them from Edward VI following the attainder of the duke of Norfolk.

31 Fletcher, *A County Community*, 19.

Figure 6.3. Ironmaster's house, Richard Infield's Gravetye. The surrounding landscaping was created after 1884 when the house and estate were purchased by William Robinson, the well-known landscape gardener (source: author).

> Jove's oak, the warlike ash, veined elm, the softer beech,
> Short hazel, maple plain, light asp, the bending wych,
> Tough holly, and smooth birch, must altogether burn;
> What should the builder serve, supplies the forger's turn,
> When under public good, base private gain takes hold,
> And we, poor woful woods, to ruin lastly sold.[32]

However, the evidence is conflicting. By 1560 the Forest's tree cover extended to about half its area, about 7,600 acres, and in 1561 it was claimed that oaks and beeches on Ashdown were 200 years old.[33] Although much had been cleared before the advent of blast furnace technology, Duddleswell Walk in the South Ward remained well wooded with beech, birch and oak, and Costley Ward had a cover of beech and birch. But the West Ward was only thinly wooded, with apparently much having been cleared for the ironworks as early as 1519. It is certainly likely that the operations around the Forest edge would have depleted the tree cover, as in the case of Hogge's furnaces between Buxted and Maresfield.[34] Commoners brought a complaint as early as 1520 that William Mugglewick had cut down the wood grove on Quabrook Common

[32] Michael Drayton, *Poly-olbion* (printed by HL, 1613), *Song XVII*, 396. And see S. Trevisan, '"The murmuring woods euen shuddred as with feare": Deforestation in Michael Drayton's Poly-Olbion', *The Seventeenth Century*, 26 (2011), 240–63.

[33] Somerville, *History of the Duchy of Lancaster* I, 310.

[34] Hodgkinson, *The Wealden Iron Industry*, 137.

for charcoal and sold it to the Steel Forge.[35] A single furnace might consume 4,000 cords or more annually.[36] The Newbridge works alone needed five and a half loads of charcoal per ton of iron produced and perhaps three colliers with assistants would be required for every furnace and forge. Each load of charcoal required nearly three cords of wood, supplied by the woodcutters, working seasonally at piece-rates.

There was a clear short-term advantage to the ironmasters around Ashdown in clear-felling timber for their fuel supplies, rather than using the products of coppice husbandry.[37] But it would have been recognised at an early date that the conservation of woodland was necessary, especially as shortages meant that the general price of charcoal quadrupled between 1540 and 1600. Almost certainly this would have tempted many to cut cordwood before its generally recognised 16-year cycle was completed, thereby depleting future resources. It also affected supplies of billetwood for domestic fuel, which also correspondingly rose in price, affecting many craftsmen dependent upon its supply.

Tenants complained in 1526 that there was insufficient beech mast for their pigs and that their rights to estovers were being compromised by demands for cordwood and also by the burning of beech for ashes for top dressing heavy soils for tillage. In 1558 a 'waste and spoil' case noted the recent felling of 2,000 beeches, 100 oaks and several thousand birch. There could be bitter disputes over access to wood for coaling, as when William Bassett of Withyham was robbed by Richard Maynard and about 40 men in Buxted in 1591:

> being weaponed with swords, daggers, staves and other weapons ... did pull up a Bridge standing over a River there, beinge the necessarye and usuall carryinge waye of ... coles unto divers of his aforesaid Ironworkes and after this donne, the said ryottous persones upon the said seconde daye of August in riottous and terrible manner did enter ... Buxstede woode ... and there did make an assault upon one William Growte Collyer and servant unto your said Subject and him did beate and greveously wounde and mayme in one of his handes by reason whereof he is not able to worke to gett his livinge and ... did take and carrye awaye Tenne loads of Cole and fowerscore & tenne Cordes of woode ... in waynes and cartes.

The gang returned the following day and carried off another 22 cords of wood and two loads of charcoal. It seems that Sir Thomas Palmer of Angmering had sold the same wood to both men (and incidentally, to a third party also) and that Maynard had already marked the trees to be cut, but that Bassett had then entered and taken it.[38]

35 ESBHRO, U of Sx Spec Colls, AFC I, 10 October 1520.

36 A cord was 8×4×4ft or 14×3×3ft. In general, 3.5 cords made one load of charcoal, weighing about 17.5cwt, and 14 loads of charcoal were needed to produce one ton of bar iron (Irons, 'Aspects of the Impact of Man', 118–21).

37 Brandon, 'Common Lands and Wastes', 116–18. More cases were reported between 1539 and 1545.

38 B.G. Awty, 'A New Forgemaster, William Bassett and an Old Name, Grubsbars, for Crowborough Forge', *Bulletin of the Wealden Iron Research Group*, 9 (1989), 33–7.

From some accounts it seems the impact of the iron industry was considerable. At Sheffield furnace in the 1540s:

> Wood cut by ye cord to be coled to ye furneys 6299 cordes at 3d ye cord ... Cole made by ye lode to ye furneys 2020 lodes at 12d ye lode.'

And for the forge (hammer):

> Wood cut by the cord to be coled to ye hamer 6742 cordes at 3d ye cord. Coles made by the lode to the hammer 1811 lodes at 12d ye lode.

And:

> Dygging of Stone to make the herth of ye furneys 6s Caryage of stone out of ye forest to ye furneys 32s 2d Dygging & caryage of sand to ye furneys 2s 3d & making of dykes to convey ye water to save ye myne in Denewood. In all 50s 5d.[39]

Interestingly the account also features an acknowledgement of negative impacts:

> Recompence made to dyvers persones for their landes lyeng under water at furnes & hamer poundes ... . The tythe of ye proffuytes there growing pd to ye person [parson] of Sheffeld after ye rate of £30 by yere of clere gaine growing of ye said yron myll & for that the groundes of pasture erable & meadow be dygged for myne mych to his losse of ye 10th therof for two yeres at 60s by yere.

An attempt to reconcile reports of timber losses with reports of coppicing and woodland management was made by Brandon. He noted that the Wealden ridges were already severely depleted of woodland by about 1500, and reports of waste mainly date from the sixteenth and early seventeenth centuries, when the iron industry was at its height. In 1585 a statute (27 Eliz. c19) ordered that no new ironworks in Kent, Surrey or Sussex should be built unless their fuel needs could be supplied from the owners' own lands.[40] By the mid-seventeenth century coppice woodlands supplied most of the cordwood. But otherwise regeneration was inhibited by grazing and burning of the heath.[41] It was also the case that the browsing of deer and the building of rabbit warrens would have made regeneration less certain. Where enclosure and management of woodland was possible the cover would have regenerated, as in the Five Hundred Acres. But by 1632 it was claimed that few great trees were left on Ashdown and by 1658 none at all, and illicit cutting of the coppices was still being reported.

But some maintained that the iron industry had actually prevented the complete destruction of woodland, or at least coppices, as grudgingly admitted by Evelyn in his *Sylva* (1664), and also voiced by Andrew Yarranton:

39 Giuseppi, 'Accounts of the Iron-works', 293–4.

40 Gulley, 'Wealden Landscape', 52.

41 Brandon, 'Common Lands and Wastes', 128–9.

> If the ironworks were not in being these coppices would have been stocked up and turned into pasture and tillage as is now daily done in Sussex and Surrey where the ironworks, or most of them, are laid down.[42]

THE DECLINE OF THE IRON INDUSTRY

Significant as it was locally, demand fluctuated and the industry was unstable. Ironworks might close, only to be fired up again when demand improved. And costs of production were exacerbated by the poor muddy roads. Anthony Fowle had a furnace and cannon foundry at Maynards Gate from *c.* 1560, but his forge was at Langleys (Maresfield), six miles away. The teams of oxen employed to move the cannon and other heavy iron products churned up the roads to the detriment of other users, including the timber hauliers. The Sheffield ironworks in the mid-sixteenth century kept 14 oxen for haulage.[43] By 1597 Sussex ironmasters were forced to pay a tax for carrying charcoal, iron ore or iron on public roads in the winter months and had a duty to repair the roads in the summer – another financial penalty expressly aimed at the Weald.[44]

There were also competitors springing up elsewhere in the country. By the early seventeenth century this competition was hitting the production of bar iron and, with a late sixteenth-century prohibition on exporting products still in place, furnaces were soon being laid up. Thus by 1653 Maynards Gate was 'ruined before & so remaine[s]', while those at Maresfield and Buxted were 'yet continued in hope of encouragement'.[45] The forge at Sheffield was 'laid aside & not used' and the 'Messuage called Freshfield Forge or Hammer and the ground … whereon a Forge Hamer or Ironworke lately stood' was noted in a conveyance of April 1677.[46] Although the Dutch wars between 1653 and 1664 provided some respite for ordnance producers, by 1667 only 14 Wealden furnaces were operating and the importation of cheaper Swedish iron ore was also being felt.

In a late development, the wartime conditions of the 1690s, with France again the enemy, led to the establishment by two Londoners of one of the last Wealden ironworks, at Pippingford. An indenture of 16 October 1696 typifies the complex nature of the ongoing speculation. It was between Alexander Staples Jr, Mary his wife, and Alexander Staples the elder, of the first part; John Staples clerk of East Grinstead of the second part; James Hooper Esq. of Middle Temple and Francis Diggs of London, gent. of the third part; Richard Lechmere of London, merchant, of the fourth part; and John Asgil Esq. of Middle Temple of the fifth part.[47] The property was:

---

42 John Evelyn, *Sylva, or A discourse of forest-trees, and the propagation of timber in His Majesties dominions* (London, 1664); A. Yarranton, *England's improvement by sea and land* (London, 1698; first published 1677), 60.

43 Giuseppi, 'Accounts of the Iron-works', 287.

44 Hodgkinson, *The Wealden Iron Industry*, 105–6.

45 J.L. Parsons, 'The Sussex Ironworks', *SAC*, 32 (1882), 19–32.

46 ESBHRO, DYK 784.

47 John Staples, vicar of East Grinstead for 44 years and another member of the Staples family, buried 4 August 1732.

> All that piece of land lately marked out to be enclosed ... upon which said piece of ground there hath been lately erected several houses and buildings in order to the making of iron. To hold to the use of Hooper and Diggs together with houses, bays, water courses, common profits and apparatus ... .

To complete the merchant/ lawyer/ironmaker link here, we might note that James Hooper appears as steward of the court baron of Duddleswell in 1700. By 1717 the ironmaster Charles Manning of Dartford, Kent, was leasing the ironworks at Pippingford to Charles Hooper, with its furnace and boring mill. Hooper was allowed timber to repair the furnace with its gun casting pits for the manufacture of cannon. He died in 1718, some of his gun tackle was sold in 1723 and the works seem to have gone by 1744. Before the closures here, however, a second furnace was constructed alongside the old one, although for the manufacture of pig iron rather than explicitly for ordnance. A faulty cannon was reused in the foundations of the second furnace.[48]

The advent of war in 1739 and again in 1756 once more whetted some appetites. There were forges that continued to operate, such as that at Maresfield by 1737, and there were unsuccessful attempts to revive the industry in the unsettled mid-eighteenth century, including a new furnace at Gravetye and revived furnaces at Warren and Mill Place. But all had disappeared by 1787, and the iron industry here was, to all intents and purposes, dead. Board of Ordnance contracts were switched to other regions.[49]

Where water power had been harnessed, however, there might sometimes be alternative uses. Around Ashdown there were in the sixteenth century, for example, water-powered corn mills. Nutley mill on the Millbrook was one such, and the mill at Newbridge, noted above and still standing today, was mentioned in 1545, continuing through until at least 1695, when repairs there were recorded.[50]

THE EIGHTEENTH-CENTURY AGRARIAN ECONOMY

Ashdown remained an unprepossessing place for ambitious agriculturalists. The ever-present problem of the atrocious roads certainly hindered prospects for a more commercial approach. Although the old-established practice of taking in agistment cattle continued after 1660, up to 3,000 head being recorded, the other ancient practice of swine pannage had gradually died out during the sixteenth and seventeenth centuries. The decline of woodland on the Forest itself may well have been a prime cause.[51]

---

48 C.F. Tebbutt, 'Iron Sites on Ashdown Forest', *Bulletin Wealden Iron Research Group*, 11 (1977), 9–11; Cleere and Crossley, *Iron Industry of the Weald*, 195–9, 259, 350; D. Crossley, 'Cannon-Manufacture at Pippingford, Sussex: The Excavation of Two Iron Furnaces of *c.* 1717', *Post-medieval Archaeology*, 9 (1975), 137; ESBHRO, AMS 683 counterpart lease.

49 J. Hodgkinson, 'The Decline of the Ordnance Trade in the Weald: the Seven Years' War and its aftermath', *SAC*, 134 (1996), 155–67; in October 1761 Thomas Todmans died 'a natural death' at Gravetye Furnace (Hunnisett, *East Sussex Coroners' Records*, 4).

50 Irons, 'Aspects of the Impact of Man', 246–7.

51 Brandon, 'Common Lands and Wastes', 134–5.

In this marginal farming countryside the pattern of landholding remained as complex as during the medieval period.[52] Many held land from different manors; many held by a combination of freehold, leasehold, copyhold and assart tenure; and land was also sub-leased to bring in rental income. There were many transfers between living members of the same family (*inter vivos*), rather than upon death and the payment of heriots.

Although still formally important, in reality, as we have seen, the power of the older manorial courts was in decline and it was increasingly the estate owners who controlled the lives of their tenants. But these were not compact estates: we have already seen how that of the wealthy John Newnham Jr (his father John Sr died in 1764) had been accumulated, together with income from the iron industry, within the manor of Maresfield, but was scattered across several parishes by the late eighteenth century, comprising over 70 parcels of land and cottages.[53] Many holdings were still referred to as assart – Rogers Field was 'a close of assert land' of six acres – heriots and fines were still levied, and both hen rents and broad arrows were due.[54] Newnham's estate also included freehold property: a limekiln at Piltdown; a blacksmith's shop; and 'lands sometime bondlands [land anciently cultivated, as distinct from assart land] said now to be held by lease', such as the 20 acres of Southlands in Maresfield. He also held land technically in the manor of Barkham, a sub-manor of Maresfield, 'by yearly rent of 10d and 2 broad arrows, at 6d each.' There Newnham held four plots of meadow land among the Ouse headwaters, for which he paid an annual quitrent and heriot.

The customs of the manor of Barkham also offer other glimpses into Forest life and indicate how interaction with Ashdown was regulated. In 1786, for example, there was to be no cutting of heath without a licence (stipulated originally in March 1635); no cottages or limekilns were to be erected on the waste without a grant (stipulated originally in April 1678); and tenants were 'not to dig or take way loam, sand and stone' without a licence, nor to 'carry away hay, straw, corn, dung, marl, stone etc from the copyholds without licence and 4d a load paid for acknowledgement, and for wheat straw 6d per load'.

Furthermore, neither copyholders nor customary tenants were to cut timber without licence and 4d was to be paid for each tree to the lord as acknowledgement, since all timber on the waste belonged to the lord. These customs were set down but were intended as a framework within which Ashdown's inhabitants worked, but the many nuances and small transgressions arrived at through decades of actual use meant that the ancient customs set out by court officials did not always dovetail with current and actual practice.

---

52 M. Mate, 'The East Sussex Land Market and Agrarian Class Structure in the late Middle Ages', *Past and Present*, 139 (1993), 46–65.

53 BL, Egerton MS 1967 ff.129–41. The latest entry in the rental is 1794, and the reference to John (Frederick), duke of Dorset (d. 1799) provides a *terminus ante quem*.

54 The broad arrow payment is certainly a throwback. The mark of a broad arrow was still used on trees on common land in colonial North America to indicate that they were reserved for the Royal Navy.

The clause covering the need for licences to cut timber belonging to the lord on the waste was also included within the customs of the sub-manors of Gravetye and Dallingridge, where it was stipulated that tenants had rights of commonage on Ashdown, and repeated on the rent roll of the manor of West Hoathly Rectory in 1788. In the latter it was also stated that '*consuetudinem manerii*' descent of property was to the youngest son (or the son's youngest son).[55]

Probate inventories also help us at this time to examine the agrarian economy more closely. When Widow Walls of East Grinstead died in 1745 her total goods were worth just £59 6s 6d, with the most valuable item being four milch cows worth £13, together with a few other animals. In the ground were four acres of wheat, three acres of oats and seven acres of clover and grass. Her only recorded implement was a cart worth 10s.[56] Other inventories mention butter and cheese, but William Marshall, in his 1798 volume on his travels in the south, related that 'the mystery of cheesemaking is not known in the Weald of Sussex; highly favourable as its soils naturally are to that useful and profitable art'.[57]

Improved grasses were, however, making some progress. As included in Widow Walls' inventory, clover was now established, being mentioned alongside nonsuch grass in a Fletching tithe dispute in 1681. The Fletching blacksmith Nicholas Turner had £1 worth of clover in the ground in 1715, and William Ovenden of Withyham, 'yeoman', died in 1748 with 20 acres of clover on his holding.[58]

There was another innovation: hops were appearing, forerunners of a future regional specialisation in eastern Sussex and Kent. Small acreages were mentioned at Rotherfield in 1593 and Horsted Keynes in the 1640s.[59] In good years they might be marketed outside the area, together with some wheat, but oats and pulses would be consumed on the farm. So here was a partial integration into wider markets, whenever a small surplus might be produced, but without overt dependence on the marketplace. Ashdown was only marginally part of Western Europe's developing commercial economy.

John Hull (almost certainly Hall), 'clerke', brought a tithe dispute in 1670 against John Payne and Jasper Wheeler, who were growing hops in West Hoathly, and two years later he was similarly acting against Thomas Comber, who was sending hops to Jeremy Smyth, 'furnaceman', for drying, and was growing sufficient quantities for them to be taken to London for sale. The West Hoathly parish register for 1708 lists 19 growers in the parish, complete with acreages and the number of hills.[60] Many

55 BL, Egerton MS 1967 'Rental and customs of Barkham Manor; Sussex, 1723, 1786', ff.220–23; West Hoathly Rectory Manor; Sussex: Rental and customs 1788, ff.173–5; Dallingridge Manor; Sussex: Rental and customs: 1785–91, f. 179; Gravetye, co. Sussex; 1755, f. 177.

56 ESBHRO, PBT 1/10/2938B.

57 W. Marshall, *The Rural Economy of the Southern Counties*, vol. 1 (London: printed for G. Nicol; G.G. and J. Robinson; and J. Debrett, 1798), 93, 146.

58 ESBHRO, PBT 1/10/751; PBT/1/1/57/669B; TNA, Exchequer Depositions E 134/33/34 Chas 2, Hil. 6.

59 Brent, 'Employment, Land Tenure and Population in East Sussex', 61.

60 TNA, Exchequer Depositions E 134 21 Chas 2, Trin 8; 24 Chas 2, Mich 9. Depositions in this case *c.* 1668–71 are also at ESBHRO, Add MSS 3303–7. Disputes arose over tithe payments for

eighteenth-century inventories contained references to hops: that of Richard Goodman also included reference to a hop garden, poles and an oast house with its hairs, pokes and three loads of charcoal.[61]

William Ovenden died possessed of two oast hairs and 500 new hop poles, which, including the poles in his hop garden, were valued at £20 4s. Thomas Smythe, vicar of Hartfield since 1716, died in April 1735 with 'three bags and two pockets of hops in hand' worth £15, and poles in his brewhouse chamber. Other than wearing apparel, purse, keys and ready money (£55) and arrears of tithes due (£35), the hops were the most valuable item in his inventory.[62] Thomas Ovenden Jr of Rotherfield, dying in October 1740, left 5,000 hop poles and an oast hair. But he was owed 18 guineas for his hops, which would have been processed by October. He was also owed £18 0s 3d in debts and £13 3s in 'desperate' debts (i.e. debts with little hope of recouping, as opposed to 'sperate' debts) – a reminder of the very common practice in rural society at this time of the lending and borrowing of money between members of the community.

We might also note the continuing role of dual employments around the Forest. Nicholas Birchfield was a carrier from Fletching who died in 1715 and whose inventory was worth the substantial sum of £804 6s 6d, although £210 of this was for debts, of which £50 was desperate debts. His carrier business included a warehouse, seven waggon horses, four mares, one old horse and one colt, three carts and two waggons, six pack saddles and five pairs of harness. But he was also a farmer: the October date of the inventory meant that there was just one and a half acres of turnips in the ground, but there was unthreshed wheat worth £49, oats worth £20 and peas, barley, tares, hop poles (£37) and six bags of hops (£55 5s), the latter giving some idea of the value of this unpredictable crop. In store was hay (£29), flax/hemp (£1 10s) and wood (£5). There were also four oxen, four cows with calves, 19 sheep and nine pigs.[63]

A final example was William Cork of Rotherfield, a cooper, whose inventory of January 1741 referred to his timber hoops and cooper's tools (£7) but was dominated by the value of his farming activities. Alongside 30 acres of sown wheat (£30) he had two oxen (£26) and ploughs, harrows, a cart and waggon, seven cows and four heifers, and steers, yearlings, sheep, hay, malt, oats, wheat and buckwheat. His cooper's trade seems to have been somewhat subsidiary by the time of his death.[64]

---

hops in the Weald. The reference to hills in the West Hoathly register relates to the practice of setting hops to twine around individual poles set in a 'hill' (ESBHRO, PAR 379/1/1, fol. 39, now transferred to WSRO).

61 The poke: a large hessian bag for collecting hops from the hop garden; the hairs were horsehair cloths used in the hop drying process.

62 A. Hughes, *Sussex Clergy Inventories 1600–1750*, SRS 91 (Lewes, 2009), 101–3.

63 ESBHRO, PBT 1/10/714.

64 ESBHRO, PBT/1/10/2849.

'IMPROVEMENT' AND THE DEFENDERS OF COMMON RIGHTS

By the final decades of the seventeenth century the contrasting ideologies of 'improvers' and Commoners were becoming obvious, and were the forerunners of future conflicts on Ashdown during the eighteenth and early nineteenth centuries.

Gordon Mingay's account in 1963 of English landed society in the eighteenth century saw the landed interest as being at the height of its power, which would remain 'virtually unchallenged for the next hundred years ... Land thus dominated Parliament.'[65] This generalisation has been tested in more localised studies and here we can demonstrate that, although the Ashdown elite retained their grip on broad local affairs, their hegemony was by no means accepted by all and would be severely challenged on occasions, something not unique to Ashdown and which was manifest in many other areas with common land.[66]

The eighteenth century brought still more criticism of Wealden farming in general, with farmers more inclined to follow the traditional practices of their forebears rather than share in the developing agricultural progress experienced elsewhere at this time. Much continued as passed down through generations of heath-dwellers within what have been called 'antecedent structures', and any changes would need to be assessed against the 'photosynthetic constraint' of the heathland environment.[67] To quote Dr Burton again: 'surely we cannot wonder if the rust, contracted in this muddy soil should clog the energy of the mind itself ... their greatest pride is to be thought connoisseurs in cattle'.[68]

By the late eighteenth century the economic situation had improved somewhat for the farmers. The new turnpike roads offered better access to growing markets, and the cereal prices commanded during the French wars of 1790–1815 were an inducement to produce more for the market, as were higher prices resulting from poor harvests in the 1790s.

However, contemporary agricultural commentators were generally critical. William Marshall wrote of 'The Forest of Ash Down ... as bleak and barren as the Morelands of Yorkshire or Westmorland', that 'I learnt little that was profitable in this district.' Towards the centre of the Weald, presumably including the farmland around Ashdown, he found '9/10s or 19/20s of cleared lands occupied by arable crops; mostly corn and some raygrass', but the corn 'weak, thin and spindling'. Furthermore, 'even the value of the little wheat that is grown, is much of it sunk in the expense of carrying it to market. What a field for improvement!'[69]

---

[65] G. Mingay, *English Landed Society in the Eighteenth Century* (London: Routledge and Paul, 1963), 10.

[66] See A. Wood, *The Memory of the People: Custom and Popular Senses of the Past in Early Modern England* (Cambridge: Cambridge University Press, 2013) and Neeson, *Commoners*.

[67] T. Williamson, *The Transformation of Rural England* (Exeter: University of Exeter Press, 2002), 21; J. Langton, 'The Historical Geography of European Peasantries, 1400–1800', in T. Scott (ed.), *The Peasantries of Europe: from the Fourteenth to the Eighteenth Centuries* (Harlow: Addison Wesley Longman Higher Education, 1998), 372–400.

[68] BL, Add MS 11,571.

[69] Marshall, *Rural Economy of the Southern Counties* I, 93, 104, 146–62.

Arthur Young noted that 'Ashdown Forest is of great extent – Rutherfield [is] reckoned the poorest parish in the county.'[70] His son, Rev. Arthur Young, writing of the Sussex 'wastes', felt 'it is not a little curious, that such immense tracts of land should be still left in a desert state, when they are every where intersected by turnpike-roads, and in the neighbourhood of such a market as London!' Furthermore, common rights were 'unexceptionably the most perfect nuisance that ever blasted the improvement of a country; and till they are done away, no tolerable husbandry will flourish in those districts where they are in force'.[71]

In another manifestation of the anxieties caused by the French wars, in 1801 fears of food shortages prompted the Home Office to send printed forms to bishops and thence to their parish clergy, reckoned the best source for more information on the state of farming in their parishes. For Ashdown they show that in general the area was growing some combination of wheat and oats but very little barley, except for some small amount in Horsted Keynes, and relatively few peas, beans or turnips (Table 6.1). The returns are often defective, as clergy were greeted with suspicion by many farmers, fearful that their tithe payments might increase.[72]

Table 6.1. The 1801 crop returns for Ashdown parishes.

Parish	Wheat	Barley	Oats	Potatoes	Peas	Beans	Turnips /rape	Total
East Grinstead	1173	48¼	1440	45¼	112	57¼	36½*	2912¼
Hartfield	649	30	632½	10¼	82	46½	24¾	1475
Horsted Keynes	500	100	600	80	50	40	90	1460
Rotherfield	942	35¼	1040	81	95¼	29¼	34¼	2257½
West Hoathly	442½	4½	569	46	50½	25½	21½	1159½
Withyham	569½	20	602	15	93	49	60	1408½

*turnips only

Note: the original 1801 returns also collected information on acreages of tares, cabbage and rye but none were returned for the Ashdown parishes. No returns were sent from Maresfield, Fletching or Buxted.

Source: TNA, HO 67/7 Parish acreage returns: Chichester Diocese.

70 A. Young, 'A Tour in Sussex', *Annals of Agriculture*, 11 (1789), 170–304; quotes on pp. 191, 239.

71 Rev. A. Young, *General View of the Agriculture of the County of Sussex* (Board of Agriculture, 1793; 2nd edn 1813), 187, 464.

72 M. Turner (ed.), *Home Office: Acreage Returns (HO 67), Transcript and Analysis, Pt. 3, Staffs.–Worcs, 1801* (London: List and Index Society, 1983).

Soon after these returns were made it was decided that more information was required, in part to indemnify those who might suffer in any 'driving of the country' in the event of a French invasion. Churchwardens and overseers of the poor were charged with collecting and forwarding this information, offering a large amount of local detail, some of which relating to livestock is listed (Table 6.2). It can be seen, for example, that there were actually more draft oxen than draft horses in Hartfield, Rotherfield and Withyham. There were surprisingly high numbers of pigs and sheep, although little to compare with such downland parishes as Falmer (6,182) or Rottingdean (5,738).

Also included in the returns were the numbers of men able to move waggons, cattle and sheep; the numbers of men to serve in the cavalry or infantry, and how they were to be armed; and those able to bring such things as felling axes or bill hooks. The return also included the names of water and windmills in each parish, the numbers of ovens and the bread that could be produced and any barges or boats that could be commandeered. Of the latter there were six in Fletching and two in Horsted Keynes, all normally plying between Newhaven and Freshfield Wharf.[73]

Table 6.2. Return of livestock in Pevensey Rape under the Defence Acts July 1803.

	Draft oxen	Fatting oxen	Cows	Young cattle & colts	Sheep	Hogs and pigs	Draft horses	Riding horses	Waggons & carts
Buxted	97		168	182	1861	282	144	19	70
East Grinstead	251	50	292	598	1494	902	302	64	338
Hartfield	129	6	178	424	1095	407	120	16	145
Horsted Keynes	74		107	216	375	283	110	9	107
Maresfield	70	7	151	210	779	388	121	23	117
Rotherfield	286	16	343	484	1755	975	276	32	295
West Hoathly*	96		144	274	647	375	124	10	127
Withyham	164	28	217	431	1682	608	162	24	182
Total	1167	107	1600	2819	9688	4220	1459	197	1381

*In the Rape of Lewes

Note: there was no return for livestock from Fletching parish; goats were also to be included but only Hartfield listed 2.

Source: Pearce, *Facing invasion*, 230, 250.

73 ESBHRO, L/C/G/3/2 Proceedings under the Defence Acts vol. 3 (1803–6).

The malign views of agriculturalists such as Young and Marshall were later echoed by travellers such as the oft-quoted William Cobbett. Leaving East Grinstead in January 1822 and heading for Uckfield via Forest Row:

> you cross Ashurst [sic] Forest, which is a heath, with here and there a few birch scrubs upon it, verily the villainously ugly spot I ever saw in England. This lasts you for five miles, getting, if possible, uglier and uglier all the way, till, at last, as if barren soil, nasty, spewey gravel, heath and even that stunted, were not enough, you see some rising spots, which instead of trees, presents you with black, ragged, hideous rocks.' There may be English-men who wish to see the coast of Nova Scotia. They need not go to sea; for here it is to the life. ... From the end of this forest without trees you come into a country of but poorish wettish land.[74]

Nevertheless, Cobbett was forced to admit that such 'wretched tracts and the not much less wretched farms in their neighbourhood breed the cattle that we see fatting in Romney Marsh'.[75]

While Ashdown's agriculture attracted little admiration, the arrival of John Baker Holroyd in the 1760s, as we have seen, did herald another attempt to improve Wealden farming and earned the praise of Arthur Young Sr, who publicised the work being undertaken at Sheffield Place. Holroyd was a regular contributor to Young's *Annals of Agriculture* from the 1780s onwards. There was discussion of the Sussex breed of oxen and their capacity for hard work, although 'as to the tillage of the country, it has not one feature that deserves any attention'. Indeed, the large shaws environing the fields were 'a system of greater barbarity [that] can hardly be imagined'. Holroyd was also interested in the cultivation of cabbage and potatoes, and in under-draining. He was instrumental in establishing the Sussex Society for the Encouragement of Agriculture and Manufacturing (1772) and became President of the Board of Agriculture (1803). Although this was not sheep country, he did keep a much-praised flock (corresponding regularly with John Ellman at Glynde) – although, as he wrote himself, they were 'subservient to the beauty of the park, rather than to a system of the greatest profit'.[76] The Sheffield Park estate contained 25 to 30 farms in the original sale agreement of 1769, to which another ten were added by 1821, most having common rights on the Forest. At the time of Holroyd's elevation to an earldom in 1816 an estate map was drawn up showing the estate, which now covered 4,028 acres.[77] Holroyd also

74 W. Cobbett, *Rural Rides in the Southern, Western and Eastern Counties of England*, ed. G.D.H. and Margaret Cole (London: Peter Davies, 1930), vol. 1, 61.

75 Cited in J.P. Boxall, 'The Sussex Breed of Cattle in the Nineteenth Century', *Agricultural History Review*, 20 (1972), 17–29.

76 Lord Sheffield, 'Communications relative to the circular letter', *Annals of Agriculture*, 14 (1790), 432.

77 I am grateful to Malcolm Lill for information on the earls of Sheffield. For Holroyd's connections with the Youngs, father and son, see J. Farrant, '"Spirited and Intelligent Farmers": the Arthur Youngs and the Board of Agriculture's reports on Sussex, 1793 and 1808', *SAC*, 130 (1992), 200–12.

acquired the advowson of Fletching (1778) and several manors elsewhere in Sussex. The Sheffield Arms inn was begun in 1777 on the turnpike road. In 1808 the Home Farm was also rebuilt, incorporating a 'model farm' design.[78]

Instead of trying to farm these unrewarding soils, some decided to try elsewhere. Richard (1714–1780) and Elizabeth Ellman (née Shoosmith) moved from Hartfield in 1761 to begin farming in Glynde on the South Downs. In the following 35 years two of his children, including the renowned sheep-breeder John Ellman (1753–1832), married Hartfield residents, and Ellmans acted as bondsmen for two further Hartfield marriage licences.[79] In 1783 John Ellman married Elizabeth Spencer from Hartfield, soon after taking over his late father's farm in Glynde. Apparently, some of her relatives around Ashdown had fallen on hard times, and any dowry from her family was returned by Ellman. In his obituary in *The Agricultural and Horticultural Gleaner* we find a fulsome eulogy of his farming:

> Let us point out here how, in agriculture as in breeding, Mr Ellman rose superior to his fellows and the evil examples which surrounded him, surmounting, by perseverance and individual industry, the prejudices, nay, in reference to the period to which we have gone back, we may say the barbarous ignorance of Crowborough and Ashdown Forest.[80]

78 K. Coutin, 'Lord Sheffield's Model Farm', *Historical Farm Buildings Journal*, 4 (1990), 10–15; Young, *General View*, 479–80.

79 Dunkin and Penfold, *Marriage Licences at Lewes, 1772–1837, A to L*; Dunkin and Penfold, *Marriage Licences at Lewes, 1772–1837, M to Z*, SRS 26 (Lewes, 1919).

80 Baxter and Son (Printers), *The Agricultural and Horticultural Gleaner* (Lewes, 1836), 133.

CHAPTER 7

THREATS TO ASHDOWN FOREST

To see the Russet Heath turn'd into greenest Grass[1]

THE FACE OF EARLY MODERN ASHDOWN

THE EARLY MODERN PERIOD allows us a far more detailed insight into the topography of Ashdown. We learn more about the enclosures, about some of the buildings and about the pale itself.

The pale required endless attention, with repairs made by a 'Knockpyn' or Pallister using oak trunks. The location of oaks around the Forest was the subject of an enquiry in 1509, when concern at oak felling was indicated, and landowners were separately listed.[2] In September 1518 comes reference to bargains made for the paling and in 1521 a warrant was issued for the 'Forest of Ashdown to be new enclosed with Pale', presumably replacing the wooden stakes topping the banks, although the cleft oak stakes were not always rotted but might be reused. Copyholders had complained in 1520 that timber had been cut down from their holdings by the Forest officers for repairing the pale, as well as oaks being cut for the repair of Maresfield church and the manorial pound there.[3] In 1605 Thomas first earl of Dorset endorsed a petition from the tenants of Ashdown for a commission to himself to cut timber for repair of the pale 'in order to preserve the game in which the King delights', a reference to the hunting interests of James I. Timber taken for its repair was listed in 1608 but in December 1610 and again in 1613 the pale was reportedly ruinous once more.[4] Furthermore, at this time Richard Holmes 'loosened, pulled down and took divers pales enclosing the Forest aforesaid and burnt the same in his own house'. The fine on this occasion was 5s, suggesting that it was considered a relatively serious offence.[5]

1 R. Weston, *A Discours of Husbandrie used in Brabant and Flanders* (London: printed by William Du-Gard, 1605 [1650]), 27. Weston, a Royalist, travelled to the Low Countries during the troubles of this period, where he encountered different ideas on sheep farming, marling and crops such as clover and turnips.

2 TNA, SP Henry 8, 1/231 f.137; 1/232 f.106.

3 ESBHRO, U of Sx Spec Colls AFC I, 10 October 1520.

4 ESBHRO, U of Sx Spec Colls AFC I, 23–105; TNA, SP 14/13 f.19.

5 Merricks, 'Forest and Waste', 31, 47, 77–8, 97.

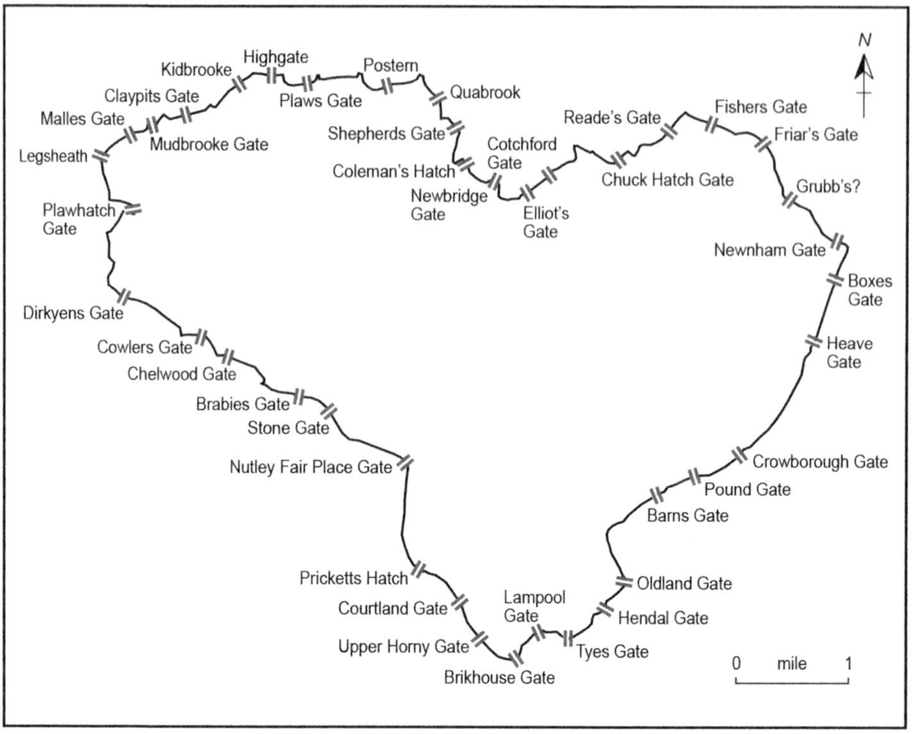

Figure 7.1. Ashdown's pale and gates (source: author, based partly on V. Blandford, *Tracing the Pale of the Ashdown Forest Deer Park* [Chris Butler Archaeological Services 2012] <https://www.ashdownforest.org/enjoy/docs/PaleReport.pdf>).

The 1539 survey by royal commissioners included a perambulation of the pale, from gate to gate, offering thoughts on its effectiveness as a boundary.[6] We now learn that certain sections of the pale were 'in great ruyne and decay', 'leid to the ground', and that the deer 'many tymes goeth out of the same in great daynger never to come home again'. Given their prodigious leaping abilities, the upkeep of both pale and ditch was certainly essential.[7] Between Highgate, Kidbrooke Gate and Legsheath 'the pale is in great ruin and decay. The red and fallow deer break out of this ward, into the sourrounding wood pastures'. The pale on the Crowborough side of the Forest was also ruined 'by reason of burnying of heth upon a common adionyng to the same called Crobarrough Common whereby most part of the said pale is burned'. At one place, near Calcott Gate (Calkett in Table 7.1), a Duddleswell rental of 1564 noted that Richard Martin's cottage and one acre of new assart 'standeth within the pale and

6 Blandford, *Tracing the Pale*, Tables 1 and 2.

7 P. Stamper, 'Woods and parks', in G. Astill and A. Grant (eds), *The Countryside of Medieval England* (Oxford: Basil Blackwell, 1988), 141.

the rest without', which would seem to indicate a decayed or missing, but certainly disregarded, pale at this time.[8]

The 1539 survey is incomplete: there are known gates missing, but it can be complemented by later information from 1564, the 1650s, 1693 and 1880 (Figure 7.1, Table 7.1). The rendering of the gate names over 350 years is of interest in itself. The 1539 survey does yield interesting information when combined with field archaeology: we learn of a forgotten gate and stretch of pale in Steppey Lane, between Newbridge and Chuck Hatch, and another on Nursery Lane, Maresfield, a minor road, where the pale is extant on both sides of the road.[9] The 1539 survey also reveals that at this time at Plaw(s) Hatch there was a deer house offering shelter and feed for deer, and possibly accommodation for hunting parties, but it needed to be 'new sett and more post and rails'.

Old Lodge was centrally placed within the Forest, supposedly to allow the ranger better oversight. In fact, the Old Lodge enclosure descends into Old Lodge Bottom, and the view is not extensive, as anyone looking north from the modern Friends' Clump will confirm. The ranger had managed to enclose about 100 acres adjoining, to the annoyance of the Commoners, but an earlier building had been destroyed by the Earl of Wiltshire, then master forester, in 1541. It was rebuilt, however, with Jacobean flourishes, by the mid-seventeenth century since in 1650 it was described as containing four rooms and offices, so still relatively modest in size.

This was one example of rebuilding at this time: another was an imposing mansion, Brambletye (Forest Row), completed for Sir Henry Compton (*c.* 1584–*c.* 1649) in 1631. Having purchased the previous house in 1619, Compton had it rebuilt entirely of stone in Jacobean style using the proceeds of his office holding, inheritance and marriage contracts. He also held the office of ranger, which he had assumed by 1638. Henry's mother had married into the Sackvilles and he himself had first married his step-sister Cecily, daughter of the second earl of Dorset, and then Mary, daughter of Sir George Browne of Kent. He was MP for East Grinstead on no fewer than eight occasions between 1601 and 1640, was knighted in 1603, was a JP 1625–44 and Deputy Lieutenant for Sussex by 1624. In London he was *Custos Brevium* of Common Pleas, then worth £2,000 per annum. He sided with the king, but died in 1649, leaving a fine for compounding to be paid by his heirs.[10]

The occupiers of the larger gentry houses and the more substantial yeomen farmers on the edge of the Forest might look to renovate in the fashion of the times. Interiors were now modified, with first floors added into the old open halls, staircases (or ladders) inserted, chimneys built and subdivided quarters provided for the family and servants. Separate rooms now might serve as kitchens, butteries, parlours and bedrooms. Outshots or one-storey additions might be erected, but there were few other external changes.

8 ESBHRO, ACC 1831/6: Copy of survey of the manor of Duddleswell, 1564 (made *c.* 1814). I am grateful to Heather Warne for this information.

9 Butler *et al.*, *Ashdown Forest Historic Environment Resource*, Fig. 8.

10 Merricks, 'Forest and Waste', 61; ESBHRO, AMS 3959, p. 16; A. Davidson, 'Compton, Sir Henry (c.1584–c.1649), of Allfarthing, Wandsworth, Surr.; Brambletye, nr. East Grinstead, Suss. and Finch Lane, London', in A. Thrush and J.P. Ferris (eds), *The History of Parliament: the House of Commons 1604–1629* (Cambridge: Cambridge University Press, 2010).

Table 7.1. The gates and hatches of Ashdown Forest 1539–1880.

1539	1564	1650s	1693	1880	modern
Plaws Hatch	Plawhatch Gate	Plawhatch Gate	Plawhatch Gate	Plawhatch Gate	Plawhatch Lane
		Dallingridge Gate			Dallingridge
	Footbridge, Fudbridge or Dirkyens		Fudbridge Gate	Tudbridge Gate	Tudbridge Gate
Paynes Corner					
Calkett Gate	? Cowlers or Collyers Gate	Coolers Gate	Gowlers Gate	Scowlers Gate	Birch Grove House
Chelworth Gate	Chelwood	Chelwood Gate	Churwood Gate	Churwood Gate	Chelwood Gate
	Brabies Gate		Brabeshatch	Brabies Gate	Either side of Chelwood Vachery
Vachery	Stone Gate	Stone Gate	Stone Gate	Stone Gate	
Nutley Strete		? Nutley Fair Place Gate			
	Prickets Hatch Gate	Pricketts Hatch to Millbrooke	Pricketts Hatch Gate	Pricketts Hatch Gate	Pricketts Hatch
? Hervey Gate					
	Courtland Gate	Courtland's Gate	Courtland Gate	Courtland Gate	Courtland Wood
	Upper Horny Gate	Horney Gate	Upper Horny Gate	Upper Horny Gate	Pickets Lane/A22
	Lower Horny Gate	Whitehouse Pond to Horney Common	Lower Horny Gate	Lower Horny Gate	Horney Common

Brykhorst Gate	Brikhouse Gate	Lampert Gate to Whitehouse Pond			?
?Maresfield Gate					
?Fysshepen Brook					
	Lampool Gate	Lampert Gate	Lampool Gate	Lampool Gate	Cophall Farm/B2026
	Tyes Gate	Tyes Gate	Tyes Gate	Tyes Gate	Nursery Lane
	Hendal Gate	Hendal	Hendale Gate	Hendal Gate	Furnace Wood
	Oldlands Gate	Oldlands	Oldlands Gate	Oldland Gate	Oldlands Lane
	Barns Gate	Barnsgate	Barns Gate	Barns Gate	Barnsgate
	Pound Gate	Pound Gate	Pound Gate	Pound Gate	Poundgate
	Crowburrough Gate	Crowborrow Gate	Crowburrough Gate	Crowborough Gate	Crow and Gate pub
	Heave Gate	Heave	Heave Gate	Heave Gate	Heavegate Road
	Boxes	Boxes	Boxes Gate	Boxes Gate	Boxgate/Cooper's Lane
Newnham Gate or Watergate	? Grubbs Gate	? Grubbs		Newnham Gate	Grubs Farm
Friar's Gate	Frayers Gate	Ffrayes Gate	Frayers Gate	Frayers Gate	Friar's Gate
		Ffidges Gate*			Fishers Gate
	Reades Gate	Reades Gate	Reades Gate	Reade's Gate	Neaves Farm
Chuck Hatch Gate	Chuck Hatch Gate	Chuckhatch Gate	Chuck Hatch Gate	Chuck Hatch Gate	Chuck Hatch
	Cotchford Gate				Marsh Green
	Elliot's Gate				Steppy Lane
? Herse Gate					

(Table 7.1 continued)

1539	1564	1650s	1693	1880	modern
Newbridge to Hartfield Gate	Newbridge Gate	Newbridge Gate	Newbridge Gate	Newbridge Gate	Newbridge
Coleman's Hatch	Coleman's Hatch Gate	Coleman's Hatch	Coleman's Hatch Gate	Colemans Hatch Gate	Coleman's Hatch
	Farmers or Shepherds Gate		Shepheards Gate	Shepherds Gate	Holy Trinity Church/ Shepherds Hill
	Quabrook		Qwabrook Gate	Quabrook Gate	Quabrook
	Postern	Postern			?
Plaws Gate	Plaws Gate		Plaws Gate	Plaws Gate	Bank Fm/south end Chapel Lane
High Gate	High Gate	Highgate	High Gate	High Gate	Highgate
	Kidbrooke Gate	Kittbrooke	Kidsbrooke Gate	Kidbrooke Gate	Kidbrooke
	Claypits Gate		Claypits Gate	Claypits Gate	Tylehurst
	Mudbrooks Gate	Mudbrooke	Mudbrooks Gate	Mudbrooke Gate	Footbridge across the Mud Brook onto Forest south of Mudbrooks House
	Mallers Gate		Malles Gate	Malles Gate	Trackway north-south
	Legsheath	Legsheath	Lagsheath Gate	Lags Heath Gate	Legsheath Lane

*Fishers Gate is on the line of the pale but is mentioned only in passing in the 1658 survey. It may have been a private gate into Buckhurst Park.

Note: the survey runs anti-clockwise around the pale, beginning at the western end. It is not always possible to reconcile the 1539 survey with later gate names.

Source: based partly on V. Blandford, *Tracing the Pale of the Ashdown Forest Deer Park* [Chris Butler Archaeological Services 2012], Tables 1 and 2 <https://www.ashdownforest.org/enjoy/docs/PaleReport.pdf> [accessed 2 November 2020].

David Gilbert, yeoman of Withyham, left an inventory in June 1711 that lists buttery, hall and wash house chambers, milkhouse and chambers over the kitchen for beds.[11]

Boyleys (now Boyles) Farm, on a minor routeway south of East Grinstead, was a late fifteenth-century hall house but incorporated late Tudor additions, including bedrooms in the attic space and a seventeenth-century dairy roofed in Horsham stone. Further south along the same road was Walesbeech Farm, now lost beneath Weir Wood Reservoir; it boasted lavish ornamentation on top of the usual medieval hall, solar and buttery, and with upper storey overhangs resembling the imposing west Kent 'cloth halls'. At Hill Place Farm, just west of East Grinstead, the hall was converted before the end of the sixteenth century reusing blackened medieval rafters; and Tilkhurst Farm showed an almost complete hall structure, but with inserted Tudor chimney. At Marshalls manor (Maresfield) the early fifteenth-century timber frame was preserved, but the walls were rebuilt with stone and the roof raised and renewed.[12] Stone was now employed far more in these substantial houses, perhaps tapping into the fine-grained brown and yellow-grey freestone from the communal quarry near Stone Gate. It might be combined with half-timbering, stone foundations supporting timber framing, or plaster or stone rubble infilling.

Nevertheless, many plain farmhouses had changed little from their medieval origins. And humbler cottages were of timber and wattle-and-daub, although oak and other timber was becoming more expensive as its spatial extent shrank. Thatch and ferns helped with roofing, and weatherboarding and tiling were employed to keep the elements at bay. But the spread of these humbler dwellings meant that many Forest courts were concerned with the growing numbers of illegal cottages and small encroachments, which had declined during the late fourteenth and fifteenth centuries but which were increasing again by the seventeenth century. Between 1599 and 1619 there were 21 unauthorised cottages built on the Forest area of Maresfield, and Forest courts continued to hear presentments regarding further encroachments through to the Civil War.[13] The 1650 parliamentary survey recorded encroachments amounting to 185½ acres on the Forest, together with cottages built recently on copyhold land; 22 enclosures, together with 11 ongoing and two new cottages, were reported by 1657.

THE SURVEY OF 1650

By the mid-seventeenth century the possibilities for selling produce in the London markets were burgeoning and, as investors looked to capitalise on the demands for food, a steady literature now urged the enclosing and increased productivity of the commons around London. Indeed, many felt it a moral imperative and uncultivated commons were thought degenerate. Adam More's *Bread for the poor* (1653) inveighed against 'uplandish' commons overgrown with fern and furze.[14]

---

[11] ESBHRO, PBT 1/10/168.

[12] R.T. Mason, 'East Grinstead. Notes on its Architecture Part II. Medieval Farms', *SAC*, 81 (1940), 2–16; Gulley, 'Wealden Landscape', 167.

[13] Brent, 'Employment, Land Tenure and Population in East Sussex', 251–2.

[14] Adam Moore, *Bread for the Poor and Advancement of the English Nation promised by Enclosure of the Waste and Common Grounds of England* (London: printed by R. & W. Leybourn, for Nicholas Bourn, 1653).

This demand for food and 'improvement' was the context for the most important and informative window into Ashdown in the seventeenth century. It followed the execution of Charles I in January 1649 and the passing of the crown lands, including Ashdown Forest, into the hands of parliament. An act for the sale of these crown lands was passed later in 1649 and parliamentary surveyors under the national command of Surveyor-General Capt. William Webb then undertook surveys and valuations, with Ashdown part of this process. The Forest was, however, exempt from the general scheme to disafforest and sell off the royal forests, but instead was to be held as collateral security to meet future army pay arrears. But quite clearly the objective was ultimately to enclose the Forest to ensure its greater productivity. The Forest was to be taken from the crown and put in the hands of trustees, one of whom was William Webb himself.

Indeed, it did seem as if conversion to farmland would surely follow on from two comprehensive surveys. Following the valuation of a few urban properties in East Grinstead and part of Lingfield (Surrey), a 1650 survey of Ashdown began with the primary manor of Duddleswell.[15] The surveyors' task was to assess the manor's resources, including claims to common rights, estovers and the depasturing of cattle. The report of the 1650 survey begins by noting the annual values of quitrents, copyhold rents and income from pannage payable by free and foreign tenants for agisting mares, cattle and swine.[16] It values the Forest drifts and courts and perquisites deriving from the court baron, the aves court, woodmote court and sheriff's tourn court, the latter still held on Berwick Common. Freeholders and copyholders paid ½d for turning out a bullock, 1d for a horse and 2d for any cattle turned out between Michaelmas and Martinmas. Customary tenants also claimed the ancient right to turn out two mares and one colt all summer until Michaelmas if they kept a draught of oxen. The tenants of both Duddleswell and Maresfield manors also claimed rights to turn out their pigs, except in fence month, with the payments according to the pigs' ages, except for 'sucking pigs', which were free. They were also entitled to two loads of birches or alders free at Christmas. Duddleswell copyholders also claimed 'frith and tennett' (coppice wood and dead timber) for fencing their holdings within the park. Finally, 'by their ancient Customalls' they claimed an allowance of 100 loads of marl for 2d, mud for their walls, stone to underpin their houses and fern to cover them.

Foreign tenants also claimed customary rights, but the surveyors were not persuaded. They also reported that Sir Thomas Gage, lord of the manor of Maresfield, had allowed cottages and enclosures to be built within the pale, had cut down timber trees 'about 14 yeares since, but by what right we know not', and was said to be receiving hens and oats as rent for custom in the park from households with no affiliation to the two manors, 'which is very prejudicial'.

---

15 ESBHRO, ACC 1831/1/9. Parliamentary Survey of the Manor of Duddleswell and Ashdown Forest, with Customs (June 1650).

16 A full account of all the 1650s surveys, including transcriptions, is in J.R. Daniel Tyssen, 'The Parliamentary Surveys of Sussex', *SAC*, 23 (1871), 242–313; 24 (1872), 189–287 and 25 (1873), 23–61. Detailed analysis is in S.J. Madge, *The Domesday of Crown Lands* (London: Routledge, 1938). He corrects totals given in Daniel Tyssen for Warren Lodge, Pippingford Lodge and the 1650 survey. For a detailed evaluation of the proceedings see Merricks, 'Forest and Waste', 104–71.

A perambulation of the pale and its gates was undertaken, and an overall value put upon Lancaster Great Park of £2,256 13s 4d for 13,991 acres (Table 7.2). This included the 185½ acres of copyholds within the pale but excluded the Vachery and the keepers' enclosed lands. The survey then described the boundaries of the six walks, noting their lodges and keepers. The walks were Southward or Duddleswell; Pippingford; Hindleap; Broadstone; Comedeane; and White Deane.

Of the buildings, Duddleswell Lodge (worth £15) was described as consisting of hall, parlour, kitchen and other necessary rooms on the ground floor, with four chambers and garrets above. Outside were a barn, stable, garden and about 30 acres of enclosed land occupied by keeper Robert Brooks, who was appointed in 1646 by the fourth earl of Pembroke who had died in January 1650 while master of game. The surveyors also noted that the White House in Duddleswell Walk was in poor repair and the land in need of fencing. Much of the house had been demolished by Sir Henry Compton, when ranger, in about 1638.[17] Old Lodge (Comedeane Walk) was occupied by Henry Ford, described as 'an intruder [who] hath burnt ye pales and fences about ye said land and almost ruined the said dwelling house'. Despite the anti-royalist subtext, the poor state of the lodges was not something attributable solely to Civil War neglect. A brief survey by John Hatton in 1632 had noted that the Forest lies 'all upon Hills and Dales', that the soil was poor and barren and that the lodges on the Forest were all in poor repair. Each lodge had a small garden and paled enclosure but the pales were repaired using quickthorn 'because timber is very scarce thereabouts'; however, cattle and deer were eating the quickthorn away.[18]

The 1650 survey also noted the park's deer, both red and fallow. These were now estimated at about 120 in number and were valued at £120. Woods and underwoods were valued at £620. Attention then turned to cottages and encroachments within the pale. The cottages of William Pollard near Prickets Hatch and Widow Hover in Hartfield were seen as encroachments, but:

> the said Pollerd and Hover ... are poore, especially the said Hover, having three children, and both have been at charges in building the said cottages and fencing ye said plats of ground, therefore ...doe conceive the said widdowe especially is to be pittyed and considered.

Others were not treated so leniently. Eight named occupiers and their holdings were:

> very prejudicial to the said parke, and rather to be pulled downe then continued ... alsoe divers other cottages erected, some of latter yeares, upon small parcells of the lands holden by coppie ... all which are very prejudicial to ye said parke.

The claims, leases, perquisites and fees of Edward, earl of Dorset, were also examined. A critical analysis of his time as master of the Forest and master of the game adjudged his performance as allowing destruction of woodland, pale and fencing around

---

17 See p. 189 for the disrepair of White House (*alias* Whitehouse) at this time.

18 BL, Harleian MS, No. 1579, f.22. The survey was just 2pp long.

encroachments, and allowing income from the Forest courts to diminish through neglect. The report therefore declared his possessions to be void, 'having almost destroyed all the woods and underwoods and suffered ... many of the incroachments and inclosures ... and the Pales to bee ruined and privileges of Courts to be lost'.[19]

The six keepers were stripped of their allowances for winter hay for the game, since the game was now said to have been destroyed; and their ability to take in large numbers of agisted cattle was severely curtailed, since any profit should have gone to the lord of the manor, not to themselves. They had in essence become cattle keepers rather than park keepers. The survey also calculated that 2,783 cattle were turned out by 219 proprietors from nine parishes over an area of 4,522 acres, with John, earl of Thanet, allowed 162 acres for his 100 cattle. Altogether about 2,175 acres had been enclosed in the large blocks within Ashdown's pale by 1650.

A final memorandum attached to the survey, which was presented on 1 June 1650, and which quite clearly sets out the grounds for enclosure of the Forest, reads:

> upon consideracon of all the tenants Cattle ... who have custome, and of all others that claim custome ... and considering the numbers of catle and horses allowed to bee kept by ye officers of the saide Parke ... wee coceive yt a considerble quantity of ground may bee conveniently sett out in convenient places for all that have right of Custome in the said parke wch may be of greater benefitt to ye tenants then what they now enjoy ... and alsoe a considerable quantity disposed of for the use of the Commonwealth.

Less than a month after the survey was completed the Forest was subject to a contract of sale dated 26 June 1650 to John Sanderson for £41,299 17s 4d.[20] This may have been the Col. Sa[u]nderson who had fought with Cromwell and who possibly had the contract on the Forest intended as back-payment for army personnel. However, the Forest remained unsold, probably because Sanderson died shortly after making his will in September 1650.[21]

THE SURVEYS OF THE LATER 1650S

A hiatus followed Sanderson's death until April 1656, when another team was appointed to resurvey the Forest in even greater detail. Again, the intention here was ultimately to enclose the Forest, and again surveyors were to value former royal property, with monies from its future sale allocated to the back pay of army officers and other parliamentary officials. And once more each survey report was endorsed by Surveyor General William Webb.

The surveyors were to 'ascertain the bounds of the forests, enquire into claims, appoint juries for trial of claims, respect all rights existing before 25 March 1641, set

19 Daniel Tyssen, 'The Parliamentary Surveys', *SAC*, 24 (1872), 302–3.

20 Madge, *Domesday*, 240.

21 <https://www.british-history.ac.uk/no-series/cromwell-army-officers/surnames-s> [accessed 10 April 2020].

out a fit proportion of waste ground for common, but allowing no claims by virtue of general words in grants'. They were to discover the yearly value of the lands and timber, take note of the timber, keep a diary of proceedings, employ a clerk, messenger and labourers to carry the chains and other equipment, and take copies of records. A committee of appeal was also established and the commissioners now summoned 'all who had a pretended interest therein to appear by a certain day'.[22]

Essentially the same formula was applied to each site: the surveyors recommended that ditches and quickset hedges be instituted against the Commoners' cattle; that marl be applied to all lands to be ploughed, directing improvers to the locations of the marl; and that the free collection of stone for the improvement or building of dwelling houses be available at Stone Hill quarry, conveniently near the London road and the grounds of Broadstone Lodge.

The surveyors estimated the size of the Forest at 12,596 acres (over 1,000 acres less than in 1650) and gave details of the courts held and the claims relating to the Forest. Furthermore:

> There has been much waste of wood and total destruction of deer, it being almost dispaled. Those who have right of common should be punished for their offences therein by being deprived of it, and those who have no right should be summoned before his Highness and Council.

A schedule of offenders and their trespasses since 1649 was enclosed. The overall value of the Forest was assessed at £1,205 5s, compared with the 1650 value of £2,507. This may have been an accurate assessment if the decay during the 1650s is to be believed, or it may represent the surveyors' difficulties in reaching more accurate figures in either or both of the surveys.

The survey areas, as carefully delineated by Ivan Margary and published in 1940, are set out in Figure 7.2 and valuations, together with the earlier survey of Duddleswell Manor and land at Cotchford and East Grinstead, are set out in Table 7.2. The divisions did not necessarily correspond with the old tripartite wards and walks.[23] For the properties newly surveyed the boundaries were traced, in part being established by dowles, wooden pegs sometimes referred to as cross dowles, and staked out across the Forest. Each memorandum included the opinion of the surveyors on the merits, or otherwise, of the surveyed land. The first of the new surveys to be entered was for Prestridge Bank (Press Ridge Warren) and Footbridge Bank, 'common wast ground', and here they advocated improving the existing woodland, with the rest being used to breed young cattle.

---

[22] *Commons Journal*, 18 June 1657, 562–3. 'An Act for instructions, agreed upon in Parliament... Commissioners for Surveying the Forest of Sherwood ... the Forest or Chase of Needwood etc' <https://www.british-history.ac.uk/no-series/acts-ordinances-interregnum/pp1116-1122>. The officials were also to survey Sherwood, Needwood and Kingswood as well as Ashdown.

[23] I.D. Margary, 'The Parliamentary Surveys of Ashdown Forest – Topographical details', *SAC*, 81 (1940), 136–39.

Table 7.2. The surveys and valuations of the 1650s.

Division	Number in the TNA calendar	Date of 'perfected' survey	Acres	Annual value	Gross value
Prestridge Bank and Footbridge Bank	10	29 March 1658	417.0	73.0	120.0
Come Deane Lodge	11	3 April 1658	1,055.0	102.7	
Warren Lodge	12	25 May 1658	834.3	157.6	100.0
Hindleap Lodge	13	18 March 1657	361.3	63.9	70.0
White Deane Lodge	14	20 March 1657	1,856.0	150.0	
Old Lodge	15	30 April 1658	165.0	20.7	4.0
Broadstone Lodge	16	24 May 1658	1,145.5	249.7	80.0
Pippingford Lodge	17	26 April 1658	725.5	108.0	40.0
Cottesford Mill and Forge*	25	29 Sept 1656	140	65.0	
Duddleswell Lodge & Manor	27	7 July 1658	1,301.0	279.8	233.0
Duddleswell Manor	26	1 June 1650			
East Grinstead	29	30 May 1650			

*Cotchford

Source: Madge, *Domesday of Crown Lands*, 379; Daniel Tyssen, 'The Parliamentary Surveys of Sussex'.

Next came Come Dean (Comden) Lodge.[24] This was quite substantial, set in a 15-acre enclosure and included a 'kell yard' where chalk would be burned for liming. The adjoining attached common was estimated at 1,040 acres, essentially forming the western portion of Costley Ward and including what became Old Lodge Warren and the Five Hundred Acre Wood.[25] Much of the boundary was difficult to determine, other than by a line 'from dowle to dowle through the wast and open land'. Two cottages were 'new erected', one with land 'late intrenched'. Another, lived in by 'Widdowe Greene', was considered fit only for demolition. It then advised that marling and ploughing be undertaken on some of the land, with the remainder to be used for breeding cattle and a sheepwalk. The keeping of sheep is mentioned more than once by the surveyors, apparently disregarding their locally accepted prohibition within the Forest.

Warren Lodge was held from the earl of Dorset 'by pretence of licence', with 86 acres of cony warren at Garden Hill, 'wch paling is all gone and taken away'. The

[24] The name of Come Dean has now disappeared, although it is still marked on the 1693 map of the Forest and on John Kelton's map as Kingslands Lodge (for John Kelton see p. 127).

[25] Margary, 'The Parliamentary Surveys', 136–9.

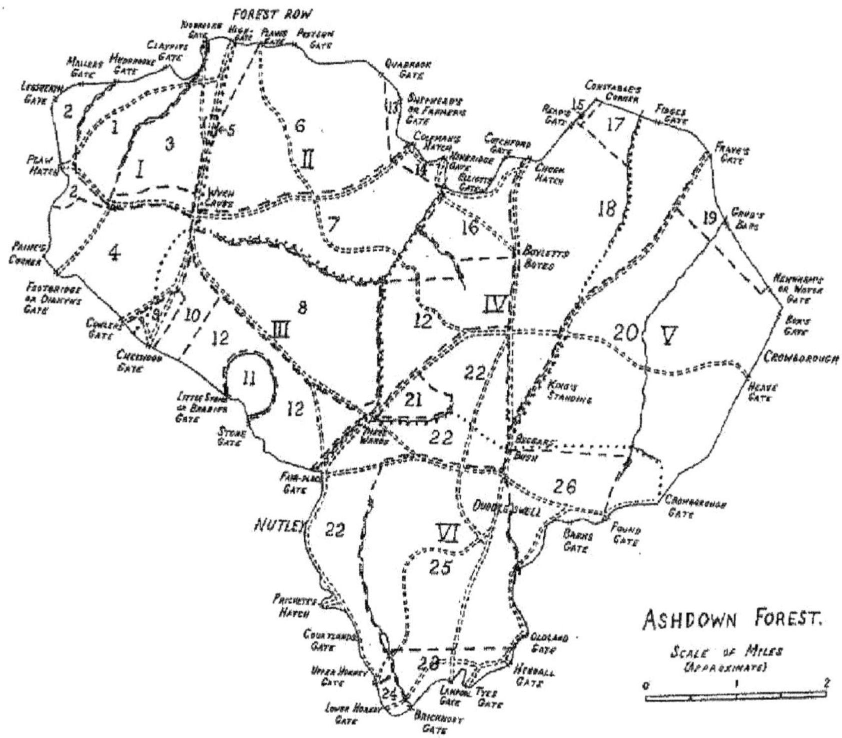

Figure 7.2. Ashdown Forest: the wards, walks and parliamentary survey areas of the 1650s (source: I.D. Margary, 'The Parliamentary Surveys of Ashdown Forest – topographical details', *SAC* 81 [1940], 136–9). Margary's original map, annotated by hand, is reproduced here except that the key in the original is replaced for the sake of clarity. See Figure 7.1 for a clearer version of the names of gates and hatches.

The boundaries of the ancient walks or divisions of the Forest • • • • •

I Hindleap Walk II Broadstone Walk III Pippingford Walk	West Ward
IV Comedeane Walk V White Deane Walk	Costly Ward
VI Duddleswell Walk	South Ward

The boundaries of the Parliamentary survey areas – – – – –

1. Hindleap Lodge
2,3. East Grinstead commoners
4. Footbridge Bank and Prestridge Bank
5. The Stone Quarry
6. Broadstone Lodge
7. Warren Lodge
8. Pippingford Lodge
9. West Hoathly commoners
10. Horsted Keynes commoners
11. The Vachery
12. Fletching commoners
13-16. Hartfield commoners
17. Buckhurst Park
18. Comedeane Lodge
19. Withyham commoners
20. White Deane Lodge
21. Old Lodge
22, 23. Maresfield commoners
24. White House
25. Duddleswell Lodge
26. Buxted commoners

boundary encompassed Newbridge Mill, 'including and taking in all the old hamer and furnace ponds, bayes and places where Iron works have ben heretofore neere to the said mill', and enclosed in all 744 acres. The woodland here was of 'little worth but for firing and [there is] much spoyle and destrucon'. The surveyors advocated mending the fences and digging a ditch around the lodge, that some of the land be ploughed, using marl from pits within the grounds, some kept as rabbit warren, much for woodland, and the residue for rearing young cattle and keeping sheep. The woodland at Hindleap Lodge was also of 'little worth but for fiering', having suffered 'much spoyle and distrucon'. Once again some land should be ploughed, again using the marl, some set aside for wood and the rest again for rearing young cattle and keeping sheep.

White Deane Lodge was a 'capital messuage' occupied by John Palmer, the keeper of White Dean Walk, which occupied the eastern half of Costley Ward. The house was said to be 'all out of repaire'. The common land attached to the lodge covered 1,843 acres, the largest of the individual grounds. Again, the recommendation was for ploughing some, making a sheepwalk of some, and breeding, rearing and keeping young cattle on the rest. There was little woodland.

Next came the more modest Old Lodge, occupied by 'the widow fford'. As in 1650, the surveyors criticised Henry Ford, now 'lately deceased', who 'did in his life tyme intrud into the said house and premises, and utterly destroyed the fences of the said inclosure, as also the ffruite trees and conyes, and alsoe suffer'd the said house to goe much to decay for want of repaire'. Dowles, local names and ghylls set out the boundaries, encompassing 155½ acres, with the timber again 'little worth but for fiering'; in the standard phrase, 'much spoyle and distrucon having ben made'. It was to be converted wholly into tillage with marl from Stone Hill, or from any of the 'Comon Marle pitts, within the said fforest'.

The survey of Broadstone Lodge described 37 acres immediately around the lodge, wherein dwelt Elizabeth Norman, widow of the late John Norman, keeper of Broadstone Walk. The bounds encompassed 1,108½ acres but again the wood was worth very little. A cottage inhabited by Thomas Hover near Quabrook Common had been built by Thomas' father 'under pretence of licence or leave from Sir Henry Compton, of Brambletye, sometimes ranger of the said fforest, chace or parke contrary to the laws in force; the said Hover for the tymes past hath neither paid rent nor done any service for the same'. However, the surveyors judged it fit to stand. Again, the land was to be mostly marled and ploughed, some woodland preserved, and the residue to be for keeping sheep and young cattle.

Pippingford Lodge was occupied by John Franke 'gent', keeper of Pippingford Walk and 'accountable keeper' of the entire West Ward. The house and outbuildings were 'out of repaire'. The common here contained 704½ acres, but again with woodland considered fit only for firing, having suffered much 'spoyle'. Part was to be ploughed and marled, the rest to be for woodland or rearing young cattle. A whapple way was now set out for use through Warren Lodge to Newbridge Mill 'with horse and harness as often as they or any of them shall have occasion'.[26]

Cottesford Mill and Cottesford Forge (Cotchford, Hartfield) were also surveyed, but by a different team of surveyors.[27] Cottesford Forge consisted of five rooms below stairs and three above with outbuildings, orchard, gardens and 12 fields amounting to 140 acres. Of Cotchford Mill it was stated that the premises were the 'discovery' of Capt. Christopher Bodley.

Finally came the substantial Duddleswell Lodge and enclosed grounds, still occupied by Robert Brooks, keeper of Duddleswell Walk and accountable keeper of the South Ward. The survey encompassed 1,203¾ acres, now to be reserved mainly for wood, with some for tillage (to be marled from the common 'great marle pitt' between Nutley and Duddleswell or the Sweet Mine Pitts at Stone Hill) and the residue again being for young cattle and sheep. Significantly, the walk also included Stone Hill, although north of Broadstone Lodge, where stone was available free to Commoners.[28] Trees here were valued together with any trees on the highways belonging to the manor of Duddleswell. There were also two claimed encroachments from the Forest: one of 44 acres occupied by Symon Smyth, gent, and taken into Buckhurst Park; the other of 14 acres taken into Newnham Park held by the widow Mary Middlemore. Both properties were owned by the earl of Dorset, but in neither case did the surveyors see any title to the land, which they nevertheless valued.

They then turned to the commonable acreages allocated to each parish and the cattle numbers turned onto the Forest in 1653–6, detailing the numbers of holdings involved from each parish and the head of cattle, which totalled 2,746 – about the same number as in 1650 (Table 7.3). The number of properties with common rights varied by only one from the total in 1650.

Table 7.3. Cattle commoned on Ashdown and area of allocated common 1653–6.

Parish	Properties	Cattle	Acreage (acres–roods –perches)*	Average cattle per acre (to nearest whole number)
East Grinstead	38	445	723–0–20 in 2 parcels	12
West Hoathly	8	49	79–2–20	6
Buxted	12	220	357–2–0	18
Horsted Keynes	3	88	143–0–0	29

[26] Whapple way – a bridleway for horses only, no carts being allowed.

[27] Much of the document studied by Daniel Tyssen was missing, having been destroyed by damp during the fire at the House of Lords, presumably the fire of October 1834, which destroyed many such records and led to the passing of the Public Record Act (1838) and the formation of the Public Record Office in Chancery Lane.

[28] This yielded a brown and yellow–grey fine-grained freestone (J.L.M. Gulley, 'The Great Rebuilding in the Weald', *Gwerin*, III.3 [1961], 9).

(Table 7.3 continued)

Hartfield	26	384	624–0–0	15
Withyham	13	101	164–0–20	8
Fletching	49	749	1217–0–20	15
Maresfield	69	710	1153–3–0	10
Total	218	2746	4462–1–0	13

*Four roods to an acre, 40 perches to a rood.
Source: Daniel Tyssen, 'The Parliamentary Surveys of Sussex'.

West Hoathly had about 79 acres adjoining Cowlers Gate 'where the said claimants doe most usually bring in their Cattle to the said fforest'. There were just eight claimants. Buxted had its land in one parcel near Poundgate, with 12 properties claiming for 220 cattle. Among the claimants, again, was Mary Middlemore of Newnham Park, claiming for 30 cattle on 48¾ acres, but the largest claim was from Hendall, for 50 cattle on 81 acres. Horsted Keynes, further removed from the Forest pale, had the fewest claimants, just three for 88 cattle on 143 acres near the Isle of Thorns. It should be noted that individuals claimed for pasturage based on properties in different parishes: thus John Mills claimed for properties in East Grinstead, West Hoathly and Horsted Keynes, and in two cases with John Lewkenor. Edward Michelbourne, from the powerful Horsted Keynes family, claimed in partnership with John Newnham and others.

Hartfield was allocated 624 acres in four separate parcels around Colemans Hatch, Buckhurst Park and Chuck Hatch, where 'all the old hammers and furnace ponds bayes and places where Iron works heretofore were'. Of the 26 claimants, the largest was for 100 cattle by John, earl of Thanet, presumably in respect of Bolebrook.[29] Withyham was allocated just over 164 acres in one parcel around Friar's Gate, 'Grubbs Bars' and Newnham Gate.

Fletching claimed the largest area and the most cattle, all in one parcel near the Vachery and Nutley. The largest claimant here was William Wilson, shortly to be created a baronet in 1661, with 55 cattle on 89 acres. The final allocation was to Maresfield, with more claimants than anywhere else, on about a quarter of the entire allocated Forest. Sir Thomas Gage, the third baronet from West Firle, and John Nutt, gent, claimed for 55 each; John Kidder, farming 85 acres held from the manor of Maresfield, claimed for six cattle on nine acres, but few others claimed more than ten cattle.[30] The surveys also allocated 60 acres to Symon Smyth at Buckhurst Park, 'as the said Buckhurste Parke was anciently enclosed according to the old bounds or old banks of the said fforest not according to the late inclosure or incroachment'.

[29] John Tufton, second earl of Thanet (1608–1664), staunch Royalist, had married Margaret, daughter of the third earl of Dorset and Lady Anne Clifford. Living at Wiston House and in Southover, Lewes, he was sheriff of Kent in 1654 and had an interest in Bolebrook through inheritance from the Dorsets, although he probably allowed it to deteriorate to little more than a farmhouse.

[30] John Kidder was assessed for four hearths in the hearth tax of 1662; see Burchall, *Sussex Hearth Tax assessments*. The Kidder family worked Lampool Farm in the sixteenth and seventeenth centuries. Correlating names in the Hearth Tax with those in the Commonwealth surveys would be a worthwhile project.

The survey then recites the highways across and around the Forest before listing the freeholders and copyholders with the names and rents of their holdings. Unsurprisingly, some farms comprised both freehold and copyhold land, and some of the latter was sub-let. Many of these mixed holdings exceeded 100 acres, with the largest reaching over 500 acres.[31] But nearly two-thirds were under five acres, with at least 357 acres of these small holdings located within the Forest pale.[32] Many holdings were fragmented, often with small areas of detached meadowland. In 1597–8 fewer than two-thirds of the farms on the Buckhurst estate had been compact units, with freehold farms less compact than copyholds owing to their greater freedom for sales and leasing.

Further details then follow about Duddleswell Manor. There was still a three-weekly court baron to deal with matters under 40s. The copyholds were old assart land, descending to the eldest son or daughter and paying one year's quit rent for admission together with a heriot of the best beast. Now it was said that the manor had three common marl pits: the great marl pit, the Sweet Mine Pit and a third in Toods Ghyll. But the claim of Daniel Rogers for common rights from the Vachery was now disallowed.

These meticulous surveys, clearly reliant on local knowledge, were finished at a particularly troublesome time. Oliver Cromwell died on 3 September 1658, soon after the last of the surveys had been 'perfected' and signed off. Richard Cromwell renounced power in May 1659 and in April 1660 Charles II was invited to return. Against this background, no action, once again, was taken to follow up the survey recommendations. Valuable as they remain to later generations of historians, the surveys led to no practical action at the time, but if their recommendations had been implemented Ashdown would now be characterised primarily by enclosed farmland or woodland instead of open Forest. Here, then, by the turn of events, was a threat averted.

THE RESTORATION PERIOD

If the surveys of the 1650s are to be believed the Forest was seemingly in a poor condition by 1660, and the Interregnum certainly spelled the end of Ashdown as a deer park. Less fashionable anyway than in previous centuries, deer were now giving way to the requirements of food supply. But threats to the Forest continued. For Ashdown the years after 1660 proved to be ones of legal and social turmoil as repeated attempts to enclose the Forest were resisted by the Commoners.

Just a month after Charles II was received with jubilation in London we find a warrant for a grant to Richard, fifth earl of Dorset, for the office of master of the Forest of Ashdown for a rental of £5, 'with perquisites of courts, underwoods, and coppices, as formerly granted to his ancestors'.[33] However, he succeeded only in obtaining the post of keeper. In November 1662 he tried again, this time for a lease of the Forest for

---

31 Gulley, 'Wealden Landscape', 156–60. Copyholders were generally able to lease out land for 366 days without licence, or for longer periods by permission (J. Whittle, 'Leasehold tenure in England c.1300–c.1600: its form and incidence', in B. van Bavel and P. Schofield (eds), *The Development of Leasehold in North-western Europe, c.1200–1600* (Turnhout: Brepols, 2008), 139–54).

32 E. Straker, 'Ashdown Forest and its inclosures', *SAC*, 81 (1940), 124.

33 *Calendar S.P. Dom.* Vol. XXIX, 5 ff.38, 40 (29 June 1660).

51 years at a rent of £5, 'granted to his late father on rental of 30*l.*, but the cottages are ruined, and the coppices so wasted that there is little covert for the deer'.[34] But he failed again. Other applicants came forward: Sir Paul Rycaut, distinguished authority on the Ottoman empire and secretary to the third earl of Winchilsea, the Ambassador in Constantinople, asked for a grant of Ashdown to his Dutch mother, 'her dower lands being sold to disengage her husband from the debts and losses incurred by his loyalty'. His father, Sir Peter, had supported Charles I but his Kentish estates had been sequestered and 'his houses rifled plundred and his Family turned out of doores, soe that he suffred in his Estate to the value of One Hundred Thousand Pounds'.[35]

But this was to no avail. Instead, a Catholic supporter of Charles II, George Digby, second earl of Bristol – 'one of the ornaments of the court', but a man of 'irresponsible brilliance' – became the master of the Forest. He was given a 99-year lease with the right to all wood and liberty to sell it, and powers to dispark, to agree [negotiate] with those claiming common rights and to disafforest and enclose, so as to allow him to improve it 'and convert the same into tillage', at the rent of £200 a year. In 1662 24 commissioners were appointed to make proposals to improve the Forest and also Broyle Park, north-east of Lewes; these were to be approved by the king and involved dividing the Forest so as to allocate portions for the Commoners in lieu of their common rights. It was said that 'Ashdown and Broyle Park have been laid waste, the deer and other beasts killed and the fences pulled down in the late distractions and the king has disafforested Ashdown Forest.' However, the proposals, which came before parliament in 1663 and 1664, failed to win approval even with modifications reported in the Lords by the fifth earl. The bill returned to the Commons in April 1668, but nothing came of it.[36]

A dispute now arose in 1664 between Bristol and Dorset concerning their respective roles of grantee master and keeper, finally settled by Dorset receiving an £800 lump sum and £100 per annum. But Bristol's attempts to improve the Forest by enclosing portions for sale, despite parliament's reluctance to agree, were frustrated 'by the crossness of the neighbourhood'. His fences were thrown down and the crops he rather ambitiously sowed were trampled by cattle. He would doubtless have concurred with John Aubrey that woodlanders were 'the mean people [who] live lawless, nobody to govern them, they care for nobody, having no dependence on anybody'.[37] At this time the Duddleswell aves court was certainly busy dealing with all manner of transgressions, such as failures to pay custom, pasturing without right and cutting and carrying away wood, although it is not clear whether such issues assumed greater importance at this time compared with others.[38]

---

34 *Calendar S.P. Dom.* Vol. LXIII, 5 (17 November 1662).

35 *Calendar S.P. Dom.* Vol. CXLII. 205 (undated, between October 1665 and July 1666).

36 *House of Lords Journal* 11 (9 May 1663), 520 <https://www.british-history.ac.uk/lords-jrnl/vol11/p520>; *House of Commons Journal* 9 (15 April 1668) <https://www.british-history.ac.uk/commons-jrnl/vol9/p81>, Ashdown Forest.

37 J. Thirsk (ed.), *The Agrarian History of England and Wales IV 1500–1640* (Cambridge: Cambridge University Press, 1967), 111.

38 ESBHRO, ACC 1831/2/15. Presentments of tenants at aves courts for the manor of Duddleswell and Ashdown Forest.

Additionally, Bristol led an expensive lifestyle and was a spendthrift. He failed to develop the Forest, sold large amounts of cordwood and defaulted on his payments to the crown and to Dorset. His cattle were distrained at one point when he owed the king £700. He had also unsuccessfully attempted to impeach his old enemy Clarendon, and he spent two years in concealment to avoid arrest.[39] Instead, from 1664 the Forest was granted to Henrietta-Maria, the Catholic queen-dowager. An act was drafted in 1668 'for the improvement of the forest of Ashdown and the park called the Broyle Park' but with no practical outcome, and she had anyway left for France in 1665, dying there in 1669.[40]

Meanwhile, on the ground itself disputes continued. A memorandum of 1670 concerning Whitehouse Farm noted:

> that Robert, earl of Dorset purchased the house and land to plant furnaces and works for casting iron guns and ordnance, built workmen's houses and made ponds; when the woods in the forest were consumed the works were demolished, the enclosures laid open and the site became derelict, open and common, only the mansion house remaining; so it continued until 1657, when the Forest was appointed to be enclosed ... and it was measured and enclosed; some tenants claimed common in the land as part of the Forest; of late John Hothe, a young man, a tenant of the Forest and living near the house has pulled up some of the fences and committed trespasses, as have others.[41]

It is uncertain to what extent the king's next action was legal. Although Dorset was also still pressing for a lease of the Forest he was again disappointed by its grant to the daughters and trustees of Colonel Washington on 22 October 1673. This was to Sir John Packington, Reginald Graham and George Legg for 31 years for £200 rent, intended as income for Penelope, Katherine, Elizabeth and Mary, daughters of the late Colonel Henry Washington, as reward for the latter's service to Charles I during the Civil War. He had been governor of Worcester but had died in 1664, leaving the daughters and a widow, herself the daughter of Packington.

But, once again, the grant was ineffectual and the rent was not forthcoming because of 'the unfruitfulness of the premises, which will not without much expense be reduced to a condition of yielding any advantage'. Treasurer Clifford had anticipated the problem, noting:

> Lancaster Great Park or Ashdown Forest, co. Sussex, which is a chace belonging to the Duchy of Lancaster, computed to contain about 7,000 acres of waste ground in common to the tenants and inhabitants of several adjoining manors and parishes, which are very many, without whose consent (which will be very difficult, if possible,

[39] R. Hutton, 'Digby, George, second earl of Bristol (1612–1677)', *ODNB* (2004).

[40] House of Lords L/PO/JO/10/1/329/138: Draft of an Act for the improvement of the forest ... (29 February 1668).

[41] ESBHRO, ACC 1831/4/287. Analysis of the title of The Whitehouse in Ashdown Forest, held of the manor of Duddleswell, 1572–1657, and recent trespasses by John Hothe and others.

> to be obtained) an improvement can hardly be made of any part thereof sufficient to bear out the rent of 200*l per an*.[42]

Instead, only a year later in 1674 came Sir Thomas Williams of Elham, Kent (1621–1712), agreeing to pay £1,700 to the trustees of the said children, £1,000 in satisfaction 'of other pretences to the premises' and a yearly rent of £100 1s.[43] Williams was 'Chymical Physician' to Charles II from 1669 to 1689 and MP for Weobley (Herefordshire) 1675–1678.[44] It must be said that his interests were focused far more on the Welsh borders and West Midlands, or on Charles II's court, than on Ashdown.

Williams granted the rent of Ashdown to Charles Sackville in 1677, the year that Charles succeeded his father to become sixth earl of Dorset, but we find Sir Thomas and his son Sir John Williams (1653–1723) attempting to enclose the Forest, aided by their trustee Joseph Fells, a goldsmith, and through the holder of a grant, Alexander Staples, an East Grinstead lawyer, JP and member of the Middle Temple. During the Williams' tenure there were 45 further copyholds granted up to 1693, together with the grant of a lease in 1675 of the Five Hundred Acres to the surveyor Simon Smyth on behalf of the earl of Dorset, followed by its outright sale in 1678.[45] In that same year Crowborough Warren, encompassing the grounds of what had been White Deane Lodge, was leased to Thomas Raymond, including the pillow mounds that were described in 1728 as 'ancient' berries.[46] In 1696 the earl of Dorset paid William Smith £280 for 560 acres, presumably the Five Hundred Acres, which was then enclosed.

Demonstrating the very active land market at this time, the Williams, father and son, had mortgaged the Forest for £500 in 1681, again in 1685 for £1,200, and again in 1688 for £1,500. They then sold the lease for £3,600 in 1690 to Alexander Staples and his younger son, also Alexander.[47] The Williams had previously also leased Hindleap Warren to Staples in 1675, reciting the bounds of the property and indicating the poor state of the pale at this time:

> Beginning at Kittbrookes Gate and so running where the park pale formerly stood to the Claypitts over against the Priory land and from thence by the place where the old pale stood to the lower end of Mudbrookes Gill … [with] full power to cut down and grub up and convert into tillage.

---

42 Report from Treasurer Clifford to the king on the petition of the four daughters of Col. Henry Washington (for a grant of certain lands, part of the queen's jointure) House of Commons Entry Book: January 1673, 16–31 (Calendar of Treasury Books, 4, 1672–1675), <https://www.british-history.ac.uk/cal-treasury-books/vol4/pp37-55> [accessed 30 January 2022].

43 State Papers Domestic Entry Book 26, *f.* 197, 15 August 1675. The King to Sir Robert Carr, Chancellor and to the Attorney of the Duchy of Lancaster, <https://www.british-history.ac.uk/cal-state-papers/domestic/chas2/1675-6/pp243-281> [accessed 30 January 2022].

44 Also described as 'Thomas Williams of Carwardine, co. Hereford, Doctor in physic' (ESBHRO, SAS F1). His baronetcy is of Elham, Kent, his first wife's home (not Eltham, as commonly mis-transcribed).

45 Straker, 'Ashdown Forest and its inclosures', 124–5.

46 The lodge itself had disappeared by 1800 (P. Combes, 'The Crowborough Warren estate, 1809–44', *SAC*, 123 (1985), 215–23).

47 ESBHRO, DLW 283, 286, 293.

Staples the elder had acted as a steward for the Dorsets in the 1660s, and this was part of a package of property speculation that also included 120 acres of Plawhatch from 1682 and which included its demise at one point to Thomas Hobbs, a London 'chiurgeon'.[48] Williams had been a staunch supporter of both Charles II and James II, to the extent of converting to Catholicism. But this backfired badly for him at the Revolution of 1688, and he was ruined. His son Sir John had married an heiress and was cushioned from a similar fate, sitting as MP for Herefordshire in 1701–5.

By contrast, as we have seen, as a supporter of the invitation to William of Orange, the sixth earl of Dorset fared well at court. But his interests were quite different from those of his moderate father. He needed income to support his court activities and was less interested in Ashdown's Commoners. By 1679 another scheme for enclosure was drawn up, with the grantees bringing forward proposals and setting out supposedly generous amounts of land near the pale as compensation for the Commoners. But they rejected the proposal.

It was instead agreed that independent arbiters should make an award, and in April 1680 the influential Sir John Pelham from Laughton and Sir John Fagg from Wiston, the wealthy Presbyterian squire and MP for Steyning, were appointed. They recommended that 5,500 acres be given to the customary tenants in lieu of their common rights. But in June of that year 99 claimants refused the offer and no further action to enclose the Forest was taken throughout the 1680s. Instead, as we have seen, more copyholds were granted during this time.

The Commoners now accused Staples of clearing and selling woodland to the detriment of their estover rights; enclosing for grain, pasture and meadow; establishing coney warrens; agisting sheep contrary to custom; and allowing cottages and other buildings to be erected within the Forest. They continued to graze their cattle where they had always grazed them, even if they were now turning them into fields of corn – and by 1689 any new enclosures were being physically broken down in an open and methodical fashion at the behest of the wealthier Commoners. All appeared to believe that Staples had acted illegally and contrary to Forest custom.

There must also have been doubts in the minds of these potential 'improvers' about the wisdom of attempting to grow arable crops, even if they could erect barriers against the Commoners' cattle. Wheat simply would not answer, and sheep clearly provoked local hostility. Rabbit warrens might have been a more sensible proposition, and the pillow mounds at Pippingford north of Old Lodge may well date to this period.[49] Unsurprisingly, perhaps, many who took on land from the Staples at this time became impoverished, and some later even had to look to their parishes for support.[50] In 1709 Alexander Staples Sr was imprisoned in the Fleet, where he died intestate, in debt or bankruptcy.[51]

48 ESBHRO, SAS PN/272, 273; 733; AMS 364 (Manor of Ringmer 1666); SAS F1.

49 'Pillow mounds on Ashdown Forest', <https://historicengland.org.uk/listing/the-list/list-entry/1002234> [accessed 23 September 2021].

50 ESBHRO, U of Sx Spec Colls, AFC I, 281–6; Irons, 'Aspects of the Impact of Man', 249.

51 TNA, C 7/2/2 Court of Chancery 1710; ESBHRO, AMS 718.

Ashdown Forest had by now survived several attempts at enclosure. The surveys of the 1650s, intended as a preliminary to enclosure, had been halted by the Restoration; the earl of Bristol's attempts in the 1660s were denied by the Commoners, as were the attempts of the Washington trustees, the Williams faction, the sixth earl of Dorset in 1679, and Alexander Staples. The proposal in 1680 to enclose also came to nothing and any further enclosing activity by Staples during the 1680s was physically resisted.

With the further confusion following William III's invasion, and immediately after William and Mary's coronation, we find that during Easter week 1689 many of the fences erected, as described previously, were destroyed. Commoners and their employees attacked both the larger enclosures, such as those near Broadstone Lodge, and the recently granted smaller enclosures in the southern Forest. Grants had been made of land running to an average of about one a year from the manor of Duddleswell from 1660 to 1690. In 1690–91 Staples was reportedly oppressing the Forest at Broadstone 'by putting in great flocks of sheep and hath set up many warrens and made large berries for conies within the same'.[52]

All the holders of such properties, such as John Awcock at Nutley Croft, were associated either directly or indirectly with Alexander Staples. Awcock seems to have been given preferential treatment in paying a lower rent than others nearby, and this would also have caused resentment. Although he held his land there by an ancient copyhold and the family was well known in the area, he had also purchased and fenced Forest land at Nutley Croft in 1671. By 1687 he had been granted several small properties in the area amounting to about 16 acres. But by the end of Easter week in 1689 it was claimed that 'the greater parte of the forest lay open and unenclosed … there were not one thousand acres of the said forest enclosed when the said fences were thrown down'.[53]

The outcome of these more intense attacks was a case brought before the court of the Duchy of Lancaster on 13 May 1689. The complainants were Charles sixth earl, Sir Thomas and his son Sir John Williams, Joseph Fells and Alexander Staples. The defendants were all Commoners: John Newnham and many others, 132 in total.[54] Eventually, in December 1691, it was decided that commissioners should again be appointed to allocate private and common areas on the Forest. An interlocutory decree initially stabilised activities, and then the impressively busy Whig Sir John Pelham, together with Colonel Butler (*c.* 1642–1704), who was active in James II's parliament, and Peter Courthope (1639–1725) of Danny Park, were chosen to adjudicate, together with several local men.[55] Their award, so important for the future of the Forest, was given on 9 July 1693. An accompanying map, surveyed in 1692–3, was made by two

[52] Cited in P. Brandon, 'Warrens and Fossil Fields on Broadstone Warren, Ashdown Forest', *SAC*, 112 (1974), 164.

[53] Merricks, 'Forest and Waste', 262–8.

[54] ESBHRO, AMS 715 Copy Decree of the Duchy of Lancaster Court relating to Ashdown Forest 8 July 1693.

[55] Pelham was MP for Sussex 1660–81 and 1689–98, and deputy lieutenant for Sussex at this time; Butler was MP for Tavistock; Courthope was the patron of John Ray, the naturalist.

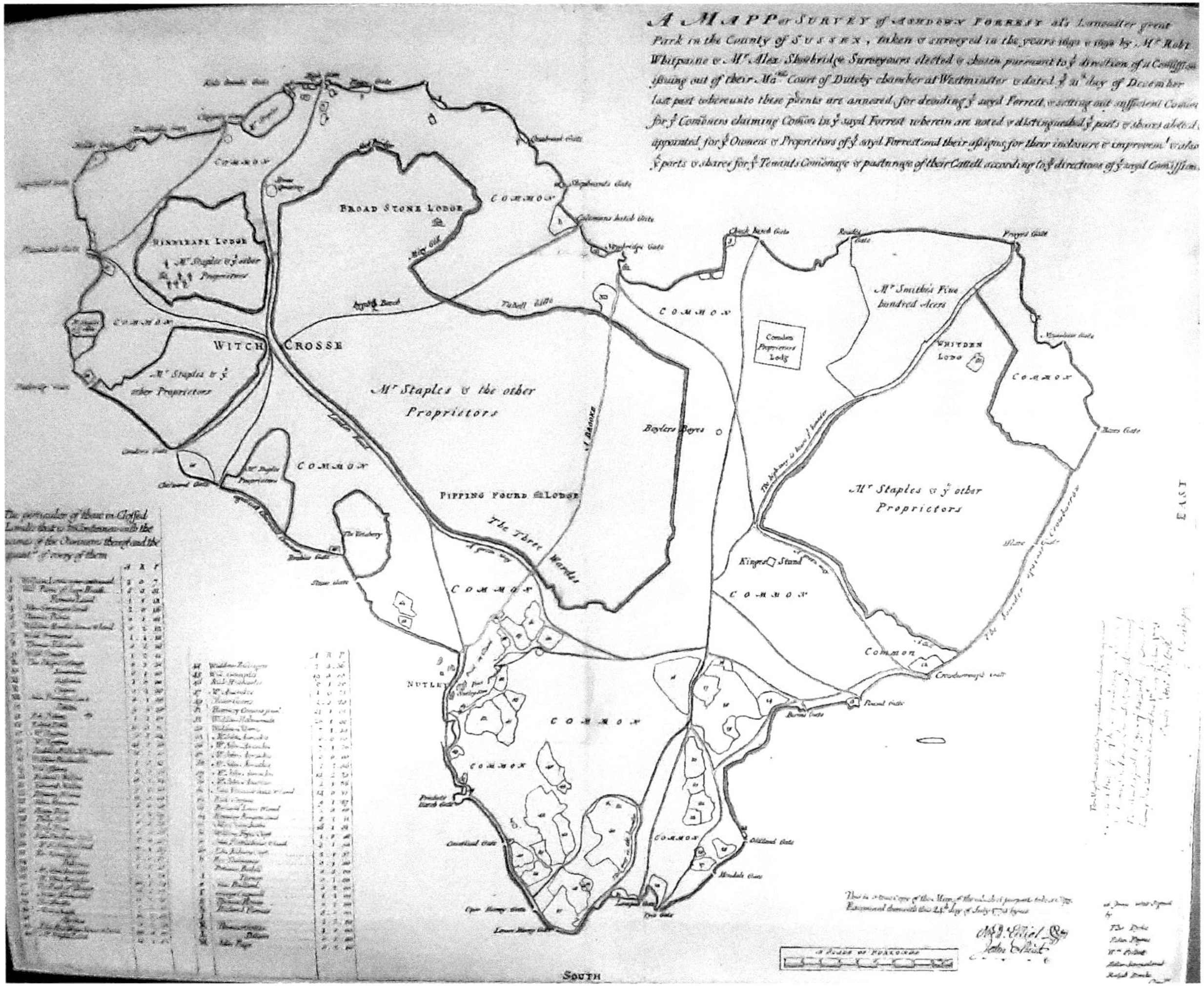

Figure 7.3. The award map 1693 (copy of 1773) (source: unknown, but the original map, too dark to reproduce, is at TNA: MPC 1/74, Map of Ashdown Forest *alias* Lancaster Great Park, Sussex, showing place-names 1693. Another copy is at ESBHRO, AMS 4084/1 but this omits some detail, including the names of proprietors and the acreage of their enclosed holdings).

appointed surveyors, Robert Whitpaine and Alexander Shoebridge (also one of the commissioners) at 4½ inches to 1 mile (Figure 7.3).[56]

All the speculators had struggled to take any profit from the Forest because of either its inherent physical difficulties or consistent opposition from the Commoners. Staples faced increasing resentment as a result of his letting land for pasture for sheep and attempts at wheat growing. Many copyholders and freeholders were also incensed because Staples had enclosed about 1,300 acres in the Forest in the 20 years up to 1691, much being the better land close to the pale previously used by the Commoners, who were now forced to drive their cattle onto and from the commons by roundabout routes.[57]

In the ongoing spirit of 'improvement' many recognised that enclosure was possible, but, mindful of the recent protests, also that adequate land should be reserved for Commoners. In fact, 130 individuals claimed rights on behalf of 215 holdings. Excluding holdings for which no totals were given, Table 7.4 reminds us of the small sizes of these holdings, with over half being under 20 acres. The mixture of medium and small farms around the Forest edges remained typical.

Table 7.4. Sizes of commoners' holdings 1691.

Under 10 acres	69
11–20 acres	47
21–30 acres	21
31–40 acres	16
41–50 acres	0
51–60 acres	11
61–70 acres	9
71–100 acres	9
101–149 acres	8
150–210 acres	6
260+ acres	5
Total	201

Source: Irons, 'Aspects of the Impact of Man', 65.

The commissioners found 'that upon view, perambulation, admeasurement, and survey of the Forest and Great Park, and after due consideration of the evidence … 6,400 acres would in their opinion, be sufficient common of pasture and herbage.' This formed the basis for the duchy decree of 9 July 1693, with the lands to be enclosed

[56] For the decree see ESBHRO, ACC 5411/120 Copies of decrees in the Duchy Court of Lancaster relative to Ashdown Forest, 1691, 1693; for the map see ESBHRO, DLW 465. A photograph of the original presented by Ernest Straker is in the British Library (BL, MS Facsimile Suppl IV W).

[57] Irons, 'Aspects of the Impact of Man', 249.

now set out 'by metes, marks and bounds'.[58] The 6,400 acres for Commoners' use were 'in most convenient places contiguous and adjacent to all several villages towns and farms'. The division in fact closely resembled that proposed by the parliamentary surveyors in 1658 and by Pelham and Fagge in 1679. The commonable parts were denoted formally as those that had existed during the medieval period: primarily just within the pale and on the ridge-top watershed. Staples and his associates were to have the remaining 7,600 acres, of which 'some are new inclosures and some are inclosures ready made'. The lord of the manor retained soil, timber and mineral rights over the commonable area.

This time the Commoners acquiesced. On the unenclosed land they were granted sole right of pasturage and the right to cut two annual loads of birch, alder and willow for each tenement. They also claimed rights to windfalls and roots, although claims to estovers were disputed from time to time, in the interests of woodland cover. Marl could be taken freely and, importantly, the cutting of bracken 'to mend land', for flooring cattle folds and subsequently for fertiliser was included. The award of 1693, as set out in the accompanying map (Figure 7.3), still largely dictates the open land of Ashdown Forest today. Much of the common land is fragmented and interspersed with the many private enclosures, of which the centrally located Pippingford Park was the largest.

This binding settlement recognised 685¾ acres as legal copyhold, much of which had been enclosed prior to the 1650s, as well as land around the park lodges, even though some of it had actually gone out of use by 1693.[59] Altogether about 55 per cent of the Forest within the pale was assigned for 'inclosure and improvement' and any Commoners remaining on such enclosures were subsequently stripped of their rights. Following the decree, in December 1693 the private interests were partitioned between Staples and two Londoners, John Holland and Nicholas Lechmere, the last named taking the lordship of the manor of Duddleswell and all the crown rights to the soil of the open Forest.[60] Staples engaged in further legal action to ratify his proprietorship of his enclosed areas and again pursued mortgages to raise money from his land.[61]

The manor of Duddleswell continued to be the subject of financial dealings and mortgages. By 1727 we find it the subject of deals involving John Crawfurd and his wife Eleanor (née Gibbs). Crawfurd (1694–1762) had moved as a young man from Scotland, where he had been disinherited, and by 1725 was admitted to the Inner Temple and referred to as 'de Ashdown Forest in com. Sussex armiger'. He purchased properties around East Grinstead and by 1720 this included the manor of Duddleswell. From about

---

58 ESBHRO, ACC 5411/120; Turner, 'Ashdown Forest', 35–64.

59 Brandon, 'Common Lands and Wastes', 137–8: Gulley, 'Wealden Landscape', 13.

60 Judge Sir Nicholas Lechmere (1613–1701) was a leading Whig, distinguished lawyer and Worcestershire MP. His grandson, Nicholas (1675–1727), of the Middle Temple, was Chancellor of the Duchy of Lancaster and created Baron Lechmere in 1721. John Holland (1603–1701) was an East India Company merchant and first governor of the Bank of Scotland. His son Richard inherited the Ashdown estate.

61 ESBHRO, AMS 6190/1–2; DLW 466 p. 327; SAS F3.

1733 he assembled the house and estate that eventually became the family residence at Saint Hill (East Grinstead), helped by the Gibbs' inheritance to Eleanor.[62] His son, Gibbs Crawfurd (1732–1793), was, by 1788–92, erecting 'a grand house' to replace the older one. Gibbs had married Anna Payne, last of the Payne family, longstanding residents in East Grinstead, thereby bringing more still wealth to the Crawfurds. They had two daughters, who predeceased them, and two sons: Charles Payne Crawfurd (1765–1814), JP, barrister and paymaster of widows' pensions, and Thomas Gibbs Crawfurd (1768–1832), of Paxhill Park, Lindfield, JP and military officer.[63]

However, on 15 July 1730 Lionel, first duke of Dorset, purchased Ashdown Forest from Crawfurd, together with the (somewhat redundant) Honour of Pevensey, for £1,840, representing 26 years' value of the quitrents of Duddleswell manor.[64] Now the ownership of Ashdown, the manor of Duddleswell and the Honour of Pevensey came to the Sackvilles. The timber at this time was valued at just £210 and there had clearly been much felling by the previous speculators. Crawfurd, probably with the sale in mind, had contracted with at least two local men to take wood: in 1727 with 'Edward Medhurst of Hartfield, carpenter, to cut all the oak trees of four feet and above by girt measure standing on the wastes of the manor of Duddleswell ... with the tops and lops', and at the same time with Richard Game, farmer of Withyham, to cut wood, pollards, holly and whitethorn underwood for three years.[65]

FOREST TROUBLES AGAIN

Following the purchase of Ashdown by Lionel Sackville in 1730 attention turned, once again, to the rights of the Commoners. They remained socially disparate: some from elite society and others solidly respectable yeomen called upon to perform various duties such as attendance at homages or parish offices. But many were hardly more than cottagers with just a few acres of land, and were reliant upon their common rights.

Manorial courts still attempted to maintain some semblance of control over the growth of such holdings. A Duddleswell court baron on 20 May 1765 heard:

> that John Brett a loose idle disorderly fellow hath lately enclosed a parcel of ground and erected a hut or lodge upon the Wast of the lord of the Manor near to the King's

[62] J. Howard and F. Crisp, *Visitation of England and Wales notes* vol. 3 (London, 1898; reprinted 1997), 125. In 1959 Saint Hill, described as one of England's finest sandstone buildings, was sold by the Maharajah of Jaipur to L. Ron Hubbard, the founder of Scientology. Two watercolour sketches are in the British Library: see J. Farrant, *Sussex Depicted: Views and Descriptions 1600–1800*, SRS 85 (Lewes, 2001), 233.

[63] For the wealth of Gibbs Crawfurd see J. and B. Chibnall, *Great Holly Bush Farm, East Grinstead: a Report prepared for the National Trust* (1988), 11.

[64] ESBHRO, ACC 1831/1/101 Account of money paid by the duke of Dorset for the manor of Duddleswell.

[65] ESBHRO, ACC 1831/2/8 Articles of agreement for the sale of oak timber for £352 2s 6d; ACC 1831/3/132 Articles of agreement between John Crawford of the Inner Temple, esq, and Richard Game of Withyham farmer for woodcutting.

> highway leading from Withcross to Nutley and affixed the same to the timber trees belonging to the lord and hath taken to his own use a parcel of Hollys and bushes growing near to the said hutt or Lodge and lopped the timber trees and committed other Wast and Spoil... and that the same hutt or lodge is a great nuisance to all manner of persons traveling that road and therefore the said John Brett is ordered to remove the said Hutt or Lodge and throw out the enclosures again to the Forrest forthwith under the penalty of twenty pounds in default thereof.[66]

Nevertheless, grants of waste continued.[67] Thus we see recorded in a court roll for September 1782 the admission of John Ridley, husbandman, to land 'formerly parcel of the lord's waste (two acres), on which is erected a cottage, at Nutley'.[68] Six enclosures with their origins between 1780 and 1820 typified this on the northern fringes of the Forest west of Colemans Hatch, where holdings taken in during the later eighteenth century were awarded copyhold status during the early nineteenth century. But none was larger than four acres. All had a self-built cottage and a garden,and were enclosed predominantly by holly hedges, with boundaries frequently following small streams.[69]

The agreement of 1693 notwithstanding, the issue of common rights seemingly became an ever more pressing issue to the Sackvilles. They now maintained that, for example, turf was not to be cut and neither would trespass on the Forest be condoned. By 1795 the third duke submitted a case to the eminent lawyer Mr Serjeant Hill, stating:

> The farmers adjoining the Forest, many of whom are Copyholders of the Manor, and as such have right of Commonage, as well as many others who are not Copyholders and have no such right, have for many years past made a practice of committing depredations upon the Forest by cutting and carrying away the heath to the amount of many thousands of loads in the course of a year, by means of which the herbage is not only destroyed, and the tenants who have rights of Commonage prejudiced, but the Lord of the Manor, who is entitled to the timber in the Forest, is much injured, inasmuch as the young oak trees, which may be coming up amongst the heath, are cut down by the scythe, and consequently no timber can ever grow where these cuttings take place. Independently of this injury, the black game which used to abound in this Forest, and which the Duke is extremely desirous of preserving, are by this practice almost extirpated. His grace is therefore determined to put a stop to it if it is possible to do so.[70]

---

66 ESBHRO, U of Sx Spec Colls, AFC II, 39.

67 ESBHRO, AMS 4072 Analysis of Grants of Waste on the Duddleswell court rolls (1604–1878); ADA/ACC 3628, 3628–1 Minute Books Manor of Duddleswell 1745–1805.

68 ESBHRO, SAS AB/824.

69 C. McCorquodale, 'The Landscape of Enclosure on the Fringes of Ashdown Forest from 1780' (Univ. of Sussex unpub. BA dissertation, 2009).

70 ESBHRO, U of Sx Spec Colls, AFC II, 46–7.

However, Serjeant Hill cautioned: 'that if the Commoners had been accustomed to cut heath for estovers as long as any living witnesses could remember, they could not be restrained from doing so'.[71]

A very similar view was given by Sergeant Shepherd to the dowager duchess Arabella in October 1807. Much depended, he wrote, on whether the rights were established by usage exercised within living memory.[72] Nevertheless, the dispute thus being resurrected, the duchess and Lord Whitworth attempted that year to restrict litter cutting by Commoners and non-Commoners alike, but resistance from the influential Charles Abbot (1757–1829), speaker of the House of Commons 1802–17 and later Baron Colchester, saw the restrictions withdrawn (Figure 7.4). Abbot was well acquainted with local issues: in September 1802 he had moved from Bolebrook to purchase Kidbrooke Park. This had originally been redeveloped in 1733–4, using local sandstone, for William Neville, forty-second Lord Abergavenny, who had also benefited from an enfranchisement of over 100 acres on Forest land from the duke of Dorset. But Eridge was always the preferred home for the Nevilles and, after a period when it was leased out, Abbot purchased Kidbrooke.[73] Extensive work on the grounds was carried out by Repton through to 1808, with the Kidd Brook now passing through lakes and cascades. The grounds were then extended by bringing in a large area of heathland at Hindleap Warren in 1809, where an arboretum was begun. George Dance was hired to alter the house itself.[74]

Abbot's interest in Ashdown was double-edged. He may have appeared to support the Commoners, but a letter from him seeking support from Lord Sheffield in September 1808 also reveals another intention, that of pushing forward a Bill 'to divide allot and lay in severalty or inclose Ashdown Forest', with the expense offset by selling 'part of the waste'. Nothing further, however, seems to have come of this proposal.[75]

Abbot certainly made use of the Forest resources in reconstructing his park and gardens at Kidbrooke. George Tester, born around 1785, deposed that from the age of 16 to 22 he had worked for Abbot and then, after a short break as a warrener, for his

---

71 The title of Sergeant refers to elite barristers exclusively entitled to practice in the Court of Common Pleas, and to be appointed as judges there. The title died out from the later nineteenth century.

72 ESBHRO, U of Sx Spec Colls, AFC II, 52.

73 'Abbot, Charles (1757–1829), of Kidbrook, Suss.', <historyofparliamentonline.org/volume/1790-1820/member/abbot-charles-1757-1829> [accessed 15 July 2019]. The design work on the house exterior remains visible, including a large coat of arms of 1818 in Coade Stone (A. Kelly, 'Coade Stone in Sussex', *SAC*, 126 (1988), 186). Kidbrooke is now Michael Hall, a Rudolph Steiner school. See <https://historicengland.org.uk/listing/thelist/listentry/1000305> [accessed 23 September 2021] and E. Byford, 'Kidbrooke over 200 years', *Forest Row: Historical aspects and recollections* 1 (1984), 5–17.

74 The Abbot family held Kidbrooke until 1874, when it was sold to the solicitor Henry Freshfield, later first chairman of the Conservators of Ashdown Forest. The 1874 sale catalogue in the British Library includes a note from the *Daily Telegraph*, 5 November 1874, to the effect that Kidbrooke was sold for £24,000.

75 ESBHRO, SPK 1/88/9. Abbot appeared to have the support of messrs Magens, Cranston and possibly Fuller.

Figure 7.4. Charles Abbot, Lord Colchester, by Charles Picart, after James Northcote, stipple engraving, published 1804 (source: © National Portrait Gallery D7128).

successors. He remembered getting quicks (probably hawthorn) and holly bushes off the Forest for hedging, together with alders, birch, a great deal of turf for lawns and gravel for carriage drives. Large stones, including one of nine tons, were conveyed in a purpose-made truck drawn by seven pairs of oxen. John Gilham had heard that the great stones in the Speaker's cascade were taken from the Forest outside his fence.[76] Abbot, retiring in 1817 from his Speaker's role, remained active. It seems that he had an arrangement with 'old Lord de la Warr' to shoot over the Forest between Kidbrooke Park and Hindleap Warren.

[76] Evidence of George Tester and John Gilham from the notebooks of William Augustus Raper, which will be referred to more fully in the following chapters, and transcribed in Short, *Ashdown Forest Dispute*, 67–8 and 78.

On 25 March 1811 a journey was undertaken 'with the Maresfield tenants over the Forest taking the names of the encroachers'. A list of 51 names was assembled whose 'lands [were] thrown out on [the] forest'. Many of the names were to become familiar in the ensuing century: Tester, Ridley, Hobbs, Carr, Turner, Edwards; even Rev. Robert Bingham was included.[77] But, conversely, by 1816 some enclosures had been made by Arabella and Lord Whitworth in an attempt to introduce more cover for black game (Black grouse, *Lyrurus tetrix*), but the Commoners, at a meeting with Abbot present, resolved that the enclosures be thrown open, but at the same time that litter cutting be restricted. In the same year Whitworth posted warning notices on the Forest against trespassing to cut bushes, turf, peat, heath, fern, furze or grass or to dig stone, mine or minerals. This was probably unsuccessful.[78]

In part the difficulties may have stemmed from the fact that different groups still enjoyed different rights: the tenants of Maresfield and Duddleswell manors, for example, were permitted two loads of wood annually, fallen timber, brushwood, roofing ferns and mud for walls. Pasture was still normally stinted at rates commensurate with the stock overwintered on their holdings. Foreign tenants were still allowed fewer privileges, but there was also a third group: squatters and cottagers around the Forest edges who were non-Commoners but illegally used the Forest. The hasty building of cottages on encroachments certainly continued, and in May 1827 Mr Mannington wrote to the steward of Duddleswell manor:

> Some persons are on the point of building two cottages close by Mr Halfords ... partly in his disputed enclosure ... and partly on the waste. Upwards of twenty men were at work making enclosures on the forest; they say the forest belongs to who could get it ... they are so resolutely determined to proceed, follow what may.[79]

With landowners also encroaching onto the Forest to protect their game and to recoup financial outlays by selling timber and underwood, house and kiln faggots and hop poles, the situation was becoming fraught. And now Commoners were also attempting to enforce their claims to common rights by, for example, litter cutting. By the early nineteenth century the estover species, alder, birch and willow were now joined by hazel, beech, oak and holly in being cut, although the taking of the latter was expressly forbidden by the manor.

John Wheatley (b. 1809) went with his mother to Brabies Hatch to cut brakes and remembered 'an elderly man who used to go about the Forest as a Ranger. I think his name [was] Funnell. He had a gold band round his hat and if he caught people cutting Hollies, Oak or Beech he used to pull them up but he interfered with nobody who cut

---

77 ESBHRO, U of Sx Spec Colls, AFC II, 56.

78 Kent Archives and Local History Centre, U269/E388/1 Copy of legal opinions in connection with a dispute between the lord and commoners over cutting turf 1816.

79 C.F. Tebbutt, 'The One-night House and its Distribution', *Folklore*, 95 (1984), 122. Letter contributing to a discussion of the reputed 'anciently held belief' of being able to continue residence of a cottage if it could be built in a day and smoke came from the chimney by nightfall.

anything else.'[80] Many men interviewed later in the century suggested that there had been more timber on the Forest at the beginning of the nineteenth century than by the 1870s. Benjamin Bridger (b. 1811) remembered as a boy 'fine oak, large old beech pollards, yews and hollies of great age', but thought that much was felled by the poor during the 1830s. An 1815 survey of the timber on the Dorset estates in Sussex noted at least 50,000 thriving standard trees ('tellows') on Ashdown, but clearly timber was being sold off the Forest in some quantity during the first part of the nineteenth century. Holly was being cut for fuel, Christmas decoration, walking sticks and hedging. Henry Heasman (b. 1805) recalled that soon after 1820 'the tenants of Maresfield came and cut the hollies and carried them away and after that everybody cut hollies'.[81]

The ongoing antagonism was apparent to many. James Everest (b. 1800 at Marsh Green) remembered as a boy seeing:

> Wheatley (the father of Ginger) building a house on the Forest just above where Baldwin now lives by Hartfield Grove. Just when he had it finished ready to go in Lord De la Warr had it pulled down. My brother John helped pull it down. He told me he was employed by Lord Whitworth or Lord De la Warr, I forget which. I was looking on and saw it all. I was a young man at the time. Before that time Lord Whitworth used to throw out encroachments. When I was a boy I went round with one of the parties of about 20 men and saw them throw out the enclosures. I remember it was a very cold day and Mr. David Edwards, the Lord's Steward was there on horseback giving the men directions what to do.[82]

Arabella and Whitworth wrote to Augustus Elliot Fuller of Ashdown House (the older name for the house being Lavertye) in October 1822 warning him off hunting with his small pack of harriers and threatening proceedings against him 'as a wilful trespasser'. They then sought legal opinion. John Heaver understood that the outcome was a payment of £50 by Fuller on the understanding that he might thereafter hunt where he liked on the Forest. Henry Heasman thought the sum nearer £100, but with a similar outcome. This action against Fuller was contrary to the understanding expressed in a letter to him by the third duke in 1791 welcoming him to the area:

> I am only afraid the nature of the country will not offer you much sport with your beagles, I beg you will use my manors as your own. I am very anxious to preserve the Black-game on the Forest, and the game in general upon Broadstone Farm. I shall send directions to my keepers to assist you in every way they can.[83]

Fuller was a substantial presence on the Forest, and many local men and boys accompanied him on his hunting expeditions. He was cousin to the eccentric 'Mad Jack' Fuller of Brightling Park, and inherited his estate in 1834. In an 1830 list of holdings

80 ESBHRO, ACC 3715/2, 127–8. Notebooks of William Augustus Raper.

81 ESBHRO, ACC 3715/1 Benjamin Bridger; 3715/2 Henry Heasman.

82 ESBHRO, ACC 3715/3 James Everest.

83 ESBHRO, SAS RF 14/87.

with common rights Fuller was returned with seven such properties, including Lines and Fincham Farms and the manors of Lavertye and Parrock. His mother, the Hon. Anne Fuller, daughter of Lord Heathfield, hero of the Siege of Gibraltar, held another 898 acres of such land.[84]

Colonel Keith Young (1780–1827), JP, of Holly Hill had similarly received a warning letter in April 1814 from Lord Whitworth's solicitor requesting that he desist from sending men to cut litter.[85] After his death in September 1827 he was remembered as a Commoner who had grazed sheep on the Forest with a shepherd, contrary to custom, before levels of sheep stealing reduced his activities.[86] His workers also cut stone for gravel for his coach road and for use in drains, and dug marl and peat (from Newbridge) for his greenhouse heating and for treading into manure in his sheep yards. Some of his tenants offset their rent by collecting litter for him, sometimes using his ox team. Holly Hill was put up for sale in June 1818. The Hartfield Grove (formerly Snow's Hole) portion of his estate was sold to Captain Hall, who had Hartfield Grove House built.

The conflicts over the Forest's diverse resources and their management were complex, but all parties agreed that those without any rights should be excluded. However, the policing role could be arduous and unpopular. Robert Edwards (d. 1831) combined the offices of Duddleswell manorial wood reeve and estate bailiff for Buckhurst until his death. He left fragments of a diary that include references to supervising improvements and new rides in the Five Hundred Acre Wood, which he was engaged in on 18 February 1815, when news came through of the death of the fourth duke. He was thereafter engaged in making arrangements for the funeral and burial in the Withyham family vault. Soon afterwards he returned to dealing with 'depredations' involving poaching and cutting underwood and beech trees. On one occasion three people were 'sent to the treadmill for a month' for cutting bark, and two tramps at Chuck Hatch, suspected of stealing a donkey, were sent to the Cage in Hartfield, pending their appearance before a magistrate.[87] He also organised the cutting and selling of trees from Broadstone and elsewhere, the procurement of Forest gravel and setting it onto the roads, and the leasing of the Buckhurst estate farms. In July and August 1830 he drew up a list of those he considered to be trespassers on the Forest, with their parishes.[88] And so the conflict continued.

---

84 Short, *Ashdown Forest Dispute*, 69.

85 ESBHRO, U of Sx Spec. Colls, AFC II. For offsetting of rent see the evidence of Martha Baker (ESBHRO, ACC 3715/1).

86 ESBHRO, SAS G/ACC/903, 190; PAR 360/33/1/37 and 46.

87 Robert Edwards Sr's extant diary contained extracts for parts of the years 1815 and 1823–4 (H. Cheal, 'A Buckhurst Diary', *SCM*, 12 (1938), 392–5 and 448–50). A brick-built 'cage' in Hartfield was recalled in ESBHRO, AMS 6062/1/1, Notes and Recollections of Alterations and Changes in Hartfield since my Childhood: by George Jenner … .

88 ESBHRO, ACC 1831/3/41.

CHAPTER 8

VICTORIAN ASHDOWN: A CHANGING SETTING FOR AN ESCALATING CONFLICT

> inhabiting the cottages on the borders were turbulent foresters, unfettered by any moral restraints … .[1]

THE GROWING TENSIONS OVER access to Ashdown's resources did not abate with the accessions of William IV or Victoria. The power derived from Empire and a leading role in world affairs hardly masked underlying social and economic problems at the local level. This was certainly true of the burgeoning towns and cities, growing too fast for their administrations to ensure decent living conditions, but it was also true of the countryside. A gap now yawned still further between the lives of those with sufficient incomes to live comfortably in pleasant rural surroundings and those whose incomes were precarious and whose living standards seem to have shown little change for generations.

POPULATION AND SOCIETY AROUND THE FOREST IN THE NINETEENTH CENTURY

To offer a basic overview of population changes around Ashdown we can now turn to the decennial census, begun in 1801, and partly through the prompting of Charles Abbot (Table 8.1). Overall, Ashdown's population nearly trebled by 1911, thanks largely to urban growth at East Grinstead and later Crowborough, which became a separate parish on the edge of the Forest, taking in the western part of Rotherfield. Fluctuations in population numbers sometimes resulted from the temporary migrations of farm labourers seeking work elsewhere, as at Withyham, Rotherfield and Buxted in 1841, the latter parish also seeing a reduction because of the transfer of paupers to the new workhouse in Uckfield. Many rural parishes across England suffered from an exodus of younger people looking for work in towns and, although Ashdown does not really show this, there was something of a standstill or at least only moderate growth

[1] G.G. Jackson and P. Row, *The Sussex Highlands*, 3rd edn, vol. 96 (London: The Homeland Association Ltd, 1929), 24.

between 1831 and 1861. The population boom years between 1901 and 1911 were exceptional, and again largely due to the growth of East Grinstead, Crowborough and now also Forest Row.

The 1831 census abstracts also allow a more detailed demographic analysis than available for earlier periods, showing numbers of families, gender balance, numbers of dwellings and an outline of the occupational structure (Table 8.2). The table emphasises that agriculture remained key to employment, with the result that there were slightly more males in the population. But East Grinstead, with more diverse urban activities, is an exception, with the lowest dependence on agriculture, and a higher number of females, many employed in service occupations. With 152 female servants, the town, together with the large parish of Rotherfield with 192, had by far the largest number around the Forest. By contrast, Maresfield, for example, had 43 and Fletching 52.

The tendency for there to be more families than individual occupied dwellings is an expected feature, multiple occupancy being common. A note in the original 1831 census tables tells us that eight families had emigrated to America since 1821 from Horsted Keynes, although the overall population within the parish had grown over the ten-year period. Conversely, many residents never travelled far from their places of birth at this time. For example, the 1851 census schedules for Maresfield reveal that, out of a total population of 1,805, as many as 1,024 (56.7 per cent) remained living in their parish of birth. Additionally, another 372 (21 per cent) had been born in one or another of the other Forest parishes.[2]

Of course, these abstracts hide much that was occurring socially and economically. Wealthy men moved their families into the area, as had Sir John Shelley (1772–1852), sixth Bt, of Michelgrove in western Sussex. The nucleus of his Maresfield estate had been built up through the marriage in 1769 of John Shelley, fifth Bt (1730–1783), to Wilhelmina (d. 1772), daughter of the wealthy John Newnham of Park House, Maresfield. Sir John changed the name of the house, near the church, from 'The Cross' to Maresfield Park and the family resided there for about 100 years, welcoming such guests as the duke of Wellington in 1819, still triumphant from the Battle of Waterloo, and Queen Victoria in 1845. In 1830 the Maresfield Land Tax recorded Sir John owning 35 properties, by far the largest number in that parish of any individual. In 1832 the estate passed, on the occasion of his marriage, to their son Sir John Villiers Shelley (1808–1867), MP for Westminster 1852–65. But there was clearly a rift between father and son, and in 1842 his father reported to Sir Robert Peel, of whom he was a supporter and friend:

> It is painful to me to mention the cruel conduct of my eldest son, particularly to his devoted mother, but though we have been silent, I fear it is too well known that from the day we signed the marriage settlement by which we made him an absolute gift of Maresfield, he has tried to fasten a quarrel upon us, with the object of avoiding the understood engagement that his home was to be our country house and home ... His determination remains unchanged never to let us enter his doors and the hope which induced us to make him independent is at last completely extinct.[3]

---

2 TNA, HO 107/1640 Maresfield census 1851.

3 <http://www.historyofparliamentonline.org/volume/1820-1832/member/shelley-john-1808-1867> [accessed 23 September 2021].

Table 8.1. Population changes in the Ashdown parishes 1801–1911.

	1801	1811	1821	1831	1841	1851	1861	1871	1881	1891	1901	1911
Buxted	1063	1292	1509	1642	1574	1694	1624	1868	1934	**2039**	2038	1665*
East Grinstead	2659	2804	3153	3364	3586	3820	4266	5390	6968	5180*	6094	**7089**
Fletching	1279	1397	1690	1870	1914	2007	2028	2155	**2213**	1171*	1088	1116
Hartfield	1050	1250	1440	1455	1603	1573	1451	1540	1558	1558	1501	**1628**
Horsted Keynes	591	627	713	782	812	847	790	811	**1149**	792*	888	931
Maresfield	960	1117	1439	1650	1579	1805	1911	1965	2082	2116	2001	**2278**
Rotherfield	1963	2122	2782	3097	3036	3531	3413	4149	4334	**5099**	2823*	2870
West Hoathly	794	840	943	980	1095	1068	1120	1210	**1547**	1442	1446	1522
Withyham	1074	1155	1393	1610	1607	1692	1597	1970	2150	2073	2147	**2479**
Crowborough											3639	**5148**
Danehill										**1214**	1140	1131
Forest Row										2389	2516	**3035**
Hadlow Down												**1043**
Total	11,433	12,604	15,062	16,450	16,806	18,037	18,200	21,058	23,935	25,073	27,321	31,935

*affected by boundary changes (Buxted/Hadlow Down; East Grinstead/Forest Row; Fletching/Horsted Keynes/Danehill; Rotherfield/Crowborough).

Note: bold type shows the parish population maxima during the period.

Source: county of Sussex population tables, TNA: HO 107, census 1851; RG 12 census 1891; RG 13 census 1901; RG 14 census 1911.

Table 8.2. Ashdown 1831: area, population, housing and occupations.

	Area	Population			Houses (inhab)	Families	Surplus of families over houses	Occupations of families		
		total	Males %	Females %				Agric %	Trades etc %	Other %
Buxted	7,020	1642	53.8	46.2	290	317	27	80.4	10.7	8.8
East Grinstead	13,390	3364	49.4	50.6	524	663	139	51.7	29.3	19.0
Fletch	5,830	1870	51.0	49.0	271	343	72	70.3	20.4	9.3
Hartfield N	8,420	437	52.9	47.1	51	80	29	76.3	8.8	15.0
Hartfield S		1018	54.5	45.5	184	218	34	66.5	27.1	6.4
Horsted Keynes	6,210	782	52.6	47.4	116	150	34	66.7	19.3	14.0
Maresfield	7,750	1650	51.9	48.1	233	314	81	64.3	19.4	16.2
Rotherfield	14,710	3097	53.2	46.8	492	535	43	77.9	16.3	5.8
West Hoathly	4,420	980	52.9	47.1	159	177	18	59.9	26.6	13.6
Withyham	9,080	1610	51.0	49.0	284	299	15	70.9	20.7	8.4
Sussex			49.7	50.3				42.6		

Note: Hartfield was divided into north and south sections in 1831.
Source: ESBHRO, LIB/506776 1831 Census: County of Sussex.

When John Villiers Shelley died of gout, 'the old enemy', in 1867, Maresfield Park passed to his widow, Louisa, and on her death in 1895 to their daughter Blanche, who had married Hervey Pechell in 1871. However, Blanche never took up residence and in 1898 the estate was sold to Count Alexander Munster.

Such newer arrivals were in part attracted by the relative proximity to London as well as the availability of medium- or poor-quality (and therefore cheaper) land. Their sources of income were diverse: some, such as the Shelleys, benefited from the old-established processes of marriage settlements and estate rentals; others, such as Charles Abbot, from prestigious office holding or the law; and others were wealthy through London mercantile, colonial and other financial interests. Older sources of income, such as the Wealden iron industry, had now dried up.

Some income was in fact derived from West Indies plantations. Slave owners were wealthy men and, when in 1833 parliament finally abolished slavery in the British Caribbean, Mauritius and the Cape, compensation was paid for the freed slaves. Such compensation came to six men around Ashdown, either directly or through inheritance. Pippingford was purchased by a Jamaican plantation owner, Henry Shirley, but the house was severely damaged by fire on 5 November 1836.[4] The estate was sold to John Gordon of Newton, Aberdeen (1802–1840), who had received compensation for nearly 400 freed slaves in four different plantations in Tobago. Another receiving compensation by way of inheritance for freed slaves was Josias Jackson Jr, son of Josias (1745–1819), who had lived at The Rocks, Uckfield, but who had died on his estate on St Vincent. A fourth was Samuel Jeffries (1745–1819) of Pixton House, East Grinstead, whose three executors received compensation for 187 freed slaves from Jamaica. Jonathan Worrell (1767–1843), who had left Barbados in 1816, was living near Saint Hill, East Grinstead aged 74 with his wife Rebecca in the 1841 census. He was of independent means, part of which must have been derived from compensation for about 140 freed slaves.[5] Finally we might note Augustus Elliott Fuller (1777–1857) at Ashdown House, whose compensation came through his inheritance from his uncle John 'Mad Jack' Fuller of Rose Hill, Brightling. He received over £9,000 for just under 500 slaves from the latter's Jamaica plantations, in turn inherited from John's uncle Rose Fuller (1708–1777), whose fortune had been made from slave-owning and gunfounding.[6]

But such wealthy families were at one end of a spectrum of income and society. In stark contrast, many others struggled to get by and the numbers of paupers in the Weald generally was as high as any comparable area in the country. Poor incomers

---

4 Rev. E. Turner, 'Maresfield', *SAC*, 14 (1862), 145–6.

5 There were altogether 129 entries for individuals with Sussex addresses receiving compensation after 1833. Some, like the Fullers, are well known for their plantation holdings. Many found their way to Brighton at this time (University College London, Centre for the Legacies of British Slavery, <https://www.ucl.ac.uk/lbs/search/legacies (of British slave-ownership)> [accessed 23 September 2021]. I am grateful to Alan Lester for informing me of this source.

6 See D. Crossley and R. Saville, *The Fuller Letters 1728–1755: Guns, Slaves and Finance*, SRS 76 (Lewes, 1991).

still arrived: the West Hoathly poor relief ledgers noted that 'Elizabeth the daughter of John & Ann Stanfield, travellers, was born on Thursday the 7th of April 1831 in Cookham Hovel in the parish of Horsted Keynes'.[7] Undoubtedly there would have been many who felt aggrieved about their living conditions and precarious working lives, but any dissent would have been expressed covertly. However, there had been food riots in Sussex in the 1790s and further unrest in 1800. There were many cases of hunger-related thefts of food – indeed, many householders were forced to take foodstuffs with them to their bedrooms at night. During the French Wars the shortages of provisions and their high costs had stimulated various forms of protest: in March 1796 the Chequers Inn, Maresfield, had hosted a convention of 'working men' in the blacksmith and wheelwright trades; and in February 1801 a contingent from Buxted 'had set out with the determination of lowering the price of provisions'.[8]

Other issues also served to strengthen resistance to traditional authority in this area. The interaction of church and community did not always go well. The aforementioned Rev. Robert Bingham, curate of Maresfield by 1804, was particularly assiduous in checking disorderly alehouses. He denounced John Guy, a Nutley publican, and in September 1810 was instrumental in refusing a licence to Goldspring, who kept the 'Hare and Hounds' 'on a very wild part of Ashdown Forest'. The resulting animosity included letters threatening, among other things, arson. His protection came from Lord Sheffield, who urged him to maintain his stance, although Bingham replied 'My Lord, a person in my situation with a wife and twelve children cannot always act like a man.' On 15 September his stables were set ablaze.

The mood of the neighbourhood was not improved by the actions of Whitworth and Arabella's men, accompanied by Commoners and special constables, who on 2 January 1811 began to dismantle and throw open small Forest enclosures, a process continuing into March, as we have seen. Now letters were sent threatening anyone seeking to evict the families from their enclosures. On 17 January Bingham's house was burnt down, with a fire started in his schoolroom. But suspicion fell on Bingham himself after a young witness claimed that he saw a letter drop from Bingham's possession one Sunday morning, which it was claimed was written by Bingham himself. Addressed to the farmer 'Mr Rt Jinner', it read:

> Murder, Fire, and Revenge
>
> Fifty of us are determed to keep our land or have revenge; therefore pason, churchwards and farmers, your barns and houses shall burn, if you take our land, your lives two shall pay, your sheep we will eat, your oxen we can mame, your stacks shall blaze, Dick you shall be shuted as you comes from market or fares. We are united an are sworn to stand by one another
>
> Fifty Good Fellows

---

[7] WSRO, PAR 379/31/2/6. Note on the front endpaper.

[8] R. Wells, *Wretched Faces: Famine in Wartime England 1793–1801* (London: Breviary Stuff, 2011, first published 1988), 164–5. And for a general view of rural food riots, including from Sussex, see C. Griffin, 'Rural Workers and the Role of the Rural in Eighteenth-century English Food Rioting', *The Historical Journal*, 64 (2021), 1–27.

Bingham was arrested and eventually committed to Horsham gaol to await trial for trying to defraud two fire insurance offices. However, he averred that the fire was started by foresters, rather than by him, and thanks to Lord Sheffield's appearance at his trial as a character witness he was acquitted of arson and attempt to defraud at the March 1811 Horsham Assizes. Lord Sheffield and the bishop of Chichester thereafter corresponded over Bingham's fate and agreed that, although he was zealous in his duties and had started a school under his own roof, he should not return to Maresfield. He left Sussex, dying at North Kelsey, Lincolnshire in 1819.

Bingham was caught up in the malevolent feelings of the foresters towards higher authority. Certainly, the reputation of the foresters within polite society was poor. William Cramp, keeper of the House of Correction in Lewes, spoke of 'The people called Foresters ... a sort of people living on plunder chiefly'. At Bingham's trial a barrister had asked: 'Are not these foresters a very wild sort of people?' Richard Jenner, the Maresfield farmer, replied 'Not more than others; some of all sorts.' The barrister then asked 'Do you mean to say that they are the most moral, decent people you ever knew?' Jenner replied 'I will not say that'.[9]

CAPTAIN SWING 1830

And so a reputation for truculence was handed down through the nineteenth century. Edward Skinner of Ashurst Wood wrote in 1893 that 'In later times the Forest became so notorious as a resort of smugglers and poachers, so much so that the name of Forester was enough to stamp a man as a rogue and a thief.'[10]

But the best-known manifestation of overt protest came with the cluster of incidents known as the Swing Riots, the culmination of many difficult years. The riots were widespread across southern England, beginning in the second half of 1830 in East Kent and moving rapidly across Sussex. The 1830 harvest was meagre, bringing renewed fears of hunger. At Forest Row on the night of 10 July 1830 'a barn and outbuildings, belonging to Rist-Land Farm ... the property of the Hon. Mrs Fuller, it is supposed were maliciously set on fire and burnt to the ground'.[11] On 17 October a barn belonging to a Hartfield blacksmith and assistant overseer was also burnt.

The protests involved a loose gathering of disaffected groups, although farm labourers formed the majority, many incensed by the employment of itinerant Irish harvest labour. There were demands for higher wages; for reductions in tithe payments by small farmers; attacks on workhouses and the overseers of the poor; and calls for the destruction of the hated threshing machines, seen as replacing traditional

9 The verbatim account of the trial, taken from shorthand notes, can be read in *The Trials of the Rev Robert Bingham* (J.M. Richardson, 1811) at <https://archive.org/stream/trialsrevrobert01binggoog#page/n33/mode/2up> [accessed 27 May 2021]; ESBHRO, AMS 6403 Narrative of the ... arrest of the Rev Robert Bingham, curate of Maresfield, 7 Feb 1811; SPK 1/98; Petley, *Maresfield Old and New*, 18–19, 49. Rev. Petley was a curate at Maresfield.

10 ESBHRO, AMS 6403, 39; E.F. Skinner, *A Handbook for Sussex for the Pocket, the Library, or the Tea Table* (London: John Haddon & Co., 1893), 148. I am grateful to the late Michael Leppard for drawing this to my attention.

11 *Sussex Advertiser*, 19 July 1830. From 4 October this newspaper chronicled the events in detail.

threshing by hand during hard winter months. The latter grievance was more muted around Ashdown as there were fewer such machines (although see the visits to Edward Howis below). During what Lowerson referred to as 'bottled-up fury', barns and ricks were fired and threatening letters sent, often signed by the mythical 'Captain Swing'.[12] This was a serious challenge to public order. In some cases, but by no means all, the grievances were met by farmers increasing wages.

On Ashdown during late October and November about half of the incidents were associated with gatherings of disaffected people. The Rev. George Woodward (1750–1837), taking over from Bingham at Maresfield from 1812, received a chilling letter on 13 November:

> Sir, we have enquired into your tithes, and we have determined to set fire to you in your bed if you do not lower them. You receive from Fletching £500 a year, and give your curate only £100 a year, and you starve your labourers that work for you, you old cannibal. You parsons have fleeced the country long enough. Strain, if you dare. You and your daughter shall be burned in your beds if you do. We shall tell all the people in the parish not to pay you, for we are determined and our names are legions of liberty. Deem this as friendly, and consider that we would not burn you up without notice.'
>
> Y.Z.X.
>
> [drawing of penknife, with drop of red ink]

William and Henry Bish – brothers, schoolmasters and 'respectably dressed young men' – were tried at Lewes assizes for sending this letter, but found not guilty.[13]

There were clearly local grass-roots radicals willing to form coalitions to express anger at their living conditions. At Rotherfield and Crowborough the cry was 'Every man should live by his labour'.[14] One enterprise that suffered was that of Edward Howis (1774–1830), whose farm lay between Rotherfield and Buxted. In 1809, as a wealthy London businessman, he had purchased an estate of 1,700 acres at Crowborough Warren with ambitious plans for making farming profitable on these notoriously poor soils. A mixed farm was begun, together with lakes and water mills, the latter to process his own produce along with cereals from the locality, which, in the wartime price boom, had been grown by many small farmers. This large and ambitious

---

12 J. Lowerson, 'The Aftermath of Swing: Anti-Poor Law Movements and Rural Trade Unions in the South East of England', in A. Charlesworth (ed.), *Rural Social Change and Conflicts since 1500* (Hull: Humberside College of Higher Education, 1982), 56.

13 C. Griffin, *The Rural War: Captain Swing and the Politics of Protest* (Manchester: Manchester University Press, 2012), 180; D. Hay, P. Linebaugh, J. Rule, E.P. Thompson and C. Winslow, *Albion's Fatal Tree: Crime and Society in Eighteenth-century England* (Harmondsworth: Penguin, 1975), 315–16. William Bysh [Bish, Byshe], b. 1809, and Henry, b. 1812, were from a large Ashdown family and born in East Grinstead.

14 The classic account is E. Hobsbawm and G. Rudé, *Captain Swing* (Harmondsworth: Penguin, 1973), 67. The reference given in that book has now been updated and is now TNA, TS 11/1007/4051.

undertaking, with many employees and much machinery, included threshing machines which brought Swing rioters to the estate on 15 November 1830. They were met by Howis's son Edward and son-in-law, William Enderby. Howis himself was gravely ill, dying shortly after the incident. The rioters, about 50–60 of whom had assembled at Crowborough Lodge, demanded that the threshing machines be dismantled, to which Edward grudgingly agreed.

The rioters then collected some of Howis's employees and marched on to Withyham through Five Hundred Acre Wood, with the slogan 'one and all, one and all, we'll stand by one another'. They told one of Earl De la Warr's stewards that they were demanding increased wages and reductions in tithes. They helped themselves to food and drink from Withyham workhouse but then dispersed after reinforcements failed to materialise.

William Enderby initiated a prosecution at Lewes Assizes against John Wickens from Crowborough and Richard Hodd. Wickens' brother Joseph was separately charged with obtaining food from Withyham workhouse under false pretences. John Wickens was sentenced to one year in Lewes gaol and Richard Hodd to 18 months. Hodd (b. 1804), a farm labourer from Withyham in the employ of Earl De la Warr, exclaimed at the Lewes winter assizes that 'he was never so happy in his life as he was on that day'. By 1840 he had a wife and five children to support, but he felt that he had helped to achieve something in 1830. The grievances of labouring people were at least being heard.[15]

The events clustered around November 1830 (Table 8.3). But on 22 November a new Whig administration, pledged to parliamentary reform, took office, and a more determined stance against the protests was deemed politically expedient. Troops moved into Sussex in late November to suppress the disturbances, which had mostly emanated from around Battle, to the east. For Ashdown troops were stationed at East Grinstead, Uckfield, Rotherfield and Maresfield to deter rioters and to guard the larger settlements. But feelings still ran high: Tom Brown, working on the Sheffield estate, was charged in December with inciting a riot, urging his fellow workers by letter to set fire to some of Lord Sheffield's buildings. Found guilty, he was sentenced to transportation for life.[16]

On Ashdown the outbreak seems to have petered out during 1831, with one final case of arson at West Hoathly in November. In Sussex more generally 52 men and women were tried; there was one execution (for arson) and 16 gaol sentences; 17 were sentenced to transportation and 18 acquitted.[17]

[15] Griffin, *Rural War*, 322. This updates and amplifies Hobsbawm and Rudé's publication, which includes an account of the incident (p. 82). Howis's widow Charlotte sold the Crowborough Warren estate in 1844 (Combes, 'The Crowborough Warren Estate'.

[16] *The Times*, 20 December 1830.

[17] Hobsbawm and Rudé, *Captain Swing*, 224.

Table 8.3. 'Captain Swing' and ongoing disturbances on Ashdown 1830–33.

Date	Parish	Incident	Note
17 Oct. 1830	Hartfield	Arson	
25 Oct.	Fletching	Threatening letter	Sheffield Park
1–6 Nov.	East Grinstead area	Threatening letters	
9 Nov.	Hartfield	Arson	Holtye Common
12 Nov.	Rotherfield	Assemblage	
12–16 Nov.	East Grinstead	Assemblage	Churchyard
13 Nov.	Maresfield	Attempt to excite a mob	
13 Nov.	Withyham	Assemblage	
13 Nov.	Rotherfield	Assemblage	
13 Nov.	Fletching & Newick – Maresfield	Threatening letter	
14–20 Nov.	East Grinstead	Arson (multiple cases)	
14–20 Nov.	East Grinstead	Threshing machines	
15 Nov.	Crowborough	Assemblage	
15 Nov.	Withyham	Assemblage	
15 Nov.	Crowborough	Threshing machine & other farm machinery	
15 Nov.	Rotherfield	Assemblage	
15 Nov.	Buxted	Assemblage	
16 Nov.	W Hoathly	Arson	
16 Nov.	Withyham	Assemblage	
16 Nov.	Rotherfield	Assemblage	
19 Nov.	Rotherfield	Assemblage	
27 Nov.	Fletching	Arson	Danehill
29 Nov.	W Hoathly	Assemblage	
'Post Swing'			
Mid–late Jan. 1831	E. Grinstead	Threatening letter	Overseer threatened with assassination
Mid–late Jan. 1831	E. Grinstead	Threatening letter	Magistrate threatened with assassination
6 Nov. 1831	W. Hoathly	Arson	Wheat and oat stacks

Source: Griffin, *The Rural War*, and Griffin, unpublished database.

There was undoubted concern among governing circles at the affront to authority, but this extended also to the condition of the agricultural workforce that it should feel moved to act in this way. Many blamed the beer-shops around Ashdown as 'most mischievous ... being in obscure situations, kept by the lowest class of persons; they are resorted to by the most abandoned characters – poachers, smugglers, and night depredators'.[18] Smuggling was also thought to be a factor, as 'labourers have acquired the habit of acting in large gangs by night, and of systematic resistance to authority'.

There were other sources of friction. In Fletching, for example, the second earl of Sheffield was caught up in a dispute between the curate, Charles Gaunt, and the Sheffield estate steward, William Cooke, during the tense times of the late 1820s and early 1830s. There were disagreements over arrears and leases between the steward and the tenants, men referred to by Cooke at one point as 'the Natives'. The earl and Cooke had a deteriorating relationship anyway, and now Gaunt and Cooke disagreed about how best to prepare for unrest in Fletching as reports of 'Swing' incidents came daily. There was also disagreement in 1829 about repairs required to estate fences against the vicarial glebeland, resulting in Cooke being barred from the tithe dinner that year. Later he wrote of the 'base and false aspersions' made against him from the pulpit, an allegation reported to the earl, who attempted arbitration while sympathising with Gaunt. Cooke was becoming extremely unpopular, not helped by another of his actions, recounted in his letter of 18 May 1831: 'I have lately discharged some labourers and employed their children on the same work who have done daily more than their parents ... '. Unwise, given the unrest of just a few months earlier.

By October, following further accusations from parishioners about Cooke's language and growing differences with the earl, Cooke, now aged 75, was dismissed, the earl regretting 'the warmth of your temper [and] language'. By 1840 Cooke was in dire financial straits, pressing for arrears and help from the earl.[19] Undoubtedly a competent steward, his inability to interact with tenants and parishioners and his 'warm' character had resulted in one of many such local conflicts in and around the Forest, and one which the second earl ultimately handled so as not to alienate the church. In this sense the case was not dissimilar to that of the Rev. Bingham 20 years earlier in Maresfield.

ONGOING UNREST AND THE POOR LAW ON ASHDOWN AFTER 1830

But when these events died down it did not end the disquiet around Ashdown, and the aftermath of the violent suppression of 'Swing' remained in peoples' minds. And now another form of protest saw the formation of embryo unions to support the poorest. In November 1832 'some of the paupers' of 'very remote agricultural' Horsted Keynes

[18] Great Britain Poor Law Commissioners, *Extracts from the Information received by His Majesty's Commissioners as to the Administration and Operation of the Poor Laws* (1837), 24–6, 'No. 1 Communication to the Poor Law Commissioners, from Ashurst Majendie Esq., Assistant Commissioner, on Kent, Sussex and Essex' (Report from the Disturbed Districts of Sussex). Majendie (1784–1867) was appointed as an Assistant Poor Law Commissioner in 1832.

[19] M. Lill, 'William Cooke, Steward of the Sheffield Estate, 1828–1832', *SAC*, 144 (2006), 177–90.

approached Brighton activists asking for a union branch to be formed. And, with pamphlets proclaiming 'Union is strength and knowledge is power', by February 1833 perhaps a dozen agricultural and political unions were said to be functioning in the vicinity of Horsham alone.[20] Radical political ideas were clearly being debated around the Forest, to the alarm of more substantial landholders.

Emboldened by a growing sense of common cause, there were also protests against the New Poor Law during the 1830s and 1840s in all the Ashdown parishes with the seeming exceptions of Hartfield and Withyham, perhaps more firmly under the eye of the Buckhurst estate.[21] When a national enquiry was launched into the state of the Poor Law in 1834, prior to the final sweeping legislative changes, the responses to queries from Ashdown indicated that considerable numbers were in the parish workhouses or receiving relief (Table 8.4). The numbers receiving some relief outside the workhouse at Rotherfield, given for September 1832, is alarming, being about one-third of the total parish population (assuming no clerical error), and was by some way the largest of all responses in Sussex to the queries. This did, of course, echo Young's thoughts about Rotherfield's poverty cited previously. The county's reliance on outdoor relief governed by a scaled allowance system certainly drew criticism. Assistant Poor Law Commissioner W.H.T. Hawley wrote of the Weald in 1834 that 'every abuse which has accompanied the career of Pauperism ... may be traced through the parochial annals of this district [including] that pernicious system ... giving Relief in aid of Wages'.[22]

Table 8.4. Indoor and outdoor Poor Law relief in Ashdown, 1834.

Parish	Population in 1831	Q.22	Q.23
East Grinstead	3364		
Fletching	1870	42	68
Hartfield	1445	33	103
Rotherfield	3085	32	1250
Withyham	1610	46	76

Q.22. 'have you a workhouse? State the number ... of its inmates.'
Q.23. 'What number of individuals received relief last week not being in the workhouse?'
Source: Report from HM Commissioners on the administration and practical operation of the Poor Laws: Appendix (B1), Part 2. Sussex. Answers to rural queries.... 1834.

20 R. Wells, 'Rural Rebels in Southern England in the 1830s', in C. Emsley and J. Walvin (eds), *Artisans, Peasants and Proletarians 1760–1860* (London: Croom Helm, 1985), 124–65.

21 A. Charlesworth, B. Short and R. Wells, 'Riots and unrest', in K. Leslie and B. Short (eds), *Historical Atlas of Sussex* (Chichester: Phillimore, 1999), 74–5.

22 TNA, MH 12/12854, 13138. Hawley to the Poor Law Commission 9 February and 15 August 1835. And see S. Shave, 'Poor Law Reform and Policy Innovation in Rural Southern England, c.1780–1850' (Univ. of Southampton unpub. PhD, 2010), 35–6.

To give a flavour of the replies to the queries, those from Hartfield were given by Henry Payne and William Morphew, churchwardens and men of local substance. They calculated that the parish needed at least 200 labourers for adequate cultivation but at present had 186; they were doubtful if that was sufficient. Many found summer work on the South Downs and hop districts but in winter, of those unemployed, about 30 were set to dig gravel and work on the roads. Of the 33 in the workhouse 16 were children under 14 years, four were under 21 and 13 over 21. This was in fact rather lower than the norm, for in 1827 and 1828 the average number was 55 per month. Children were put out to work, and indeed 40–50 were degradingly auctioned off to local farmers each year. Payne and Morphew also reported that agricultural capital in the parish was lower than hitherto owing to the numbers of illegitimate children and subsequent pauperism, and that the quality of labourers' work had considerably deteriorated. The parish authorities of West Hoathly, at the yearly hiring meeting in 1825, had been equally keen to set women and children to work, offering premiums and clothing for the children to farmers taking them on.[23] The final question put to the churchwardens was:

> Q.53. 'Can you give the Commissioners any information respecting the causes and consequences of the agricultural riots and burnings of 1830 and 1831?'
>
> Hartfield: 'In some instances from want of employment, general misconduct, and frequenting the beer shops, and being distressed (not wearied by working). The labourer is prompt to the act of poaching, the certain forerunner of other diabolical depredations.'
>
> Fletching, from Charles Gaunt, vicar: 'A determination on the part of the farmers to lower the rent and tithes by exciting or encouraging the men to riot, and the labourers feeling the pressure of the times, which feeling was fermented by newspapers and beer-shops into private malice and revenge.'

Other parishes told a similar story. At Fletching there were 40 men working on the roads or digging stone in summer and 80 in winter; in Rotherfield 120 on the roads in winter. While acknowledging the poverty that most labourers felt, such that Withyham thought 'they could hardly subsist without some parochial assistance', there was much concern over the numbers of illegitimate children to be cared for by the parish. At East Grinstead there was mention of 'improvident marriages' enabling men to receive a higher allowance, in turn leading to more children. Furthermore, 'within the past 5 years', it was claimed, 'we have had 50 bastards; the total expense, £1,104; recovered from putative fathers £373'.[24]

23 ESBHRO, PAR 360/12/12 Workhouse papers; E.M. Ainsworth, 'Changing Attitudes to the Employment of Women and Children on the Land between the 1830s and 1870s, with Particular Reference to the County of Sussex' (Univ. of Sussex unpub. MA, 1969), 9–13.

24 BPP, Report from HM Commissioners on the Administration and Practical Operation of the Poor Laws: Appendix (B1), Parts 1–4. Sussex. Answers to rural queries ... 1834. Many questions

Abuses and an apparent lack of 'social discipline' were reportedly prevalent around Ashdown. Accordingly, the assistant commissioners pressed ahead to implement the new arrangements stemming from the 1834 Poor Law amendment Act, shifting the onus for dealing with the poor from the parishes onto a new tier of government, the Poor Law Union. Now the East Grinstead Union, formed in September 1835, took over from the parishes of East Grinstead, Hartfield, Withyham and West Hoathly (and Forest Row from 1894); and the new Uckfield Union took over from Fletching, Maresfield, Buxted and Rotherfield (and Crowborough and Danehill later). Horsted Keynes became a constituent parish of the Cuckfield Union. All three Unions also incorporated other parishes away from the Forest. These Unions, with their workhouses and elected Boards of Guardians, were universally and enduringly unpopular. However, many administrators continued to offer outdoor relief even after 1834, as the continuation of outdoor relief was linked to 'common sense, morality, financial prudence and the enduring sense of parochial feeling'.[25]

But for the most part paupers would henceforth be maintained in the workhouse. Families were divided by gender; discipline was strict and the necessities of life parsimonious. Unfortunate circumstances, especially hard winters, forced some into the workhouse temporarily; others used it almost as a hostel, where a night's board might be had. In practice, as the 1840s progressed the workhouses became homes for the elderly and infirm and for abandoned or orphaned children. In 1834, prior to the construction of the new East Grinstead workhouse, a return 'of all lunatics and dangerous idiots in the parish', with a report on their health, was made to Edward Cranston.[26] Workhouses from the pre-1834 period might still be used to maintain the divisions imposed: initially East Grinstead workhouse received the able-bodied, with any overflow going to Withyham; the Hartfield workhouse was retained for women and girls only; Worth took the elderly and infirm; and Lingfield (Surrey) the boys.

Later, purpose-built central workhouses were used, as at East Grinstead from 1859 in Glen Vue Road (now demolished and converted to housing). By 1881 there were eight staff and 149 residents, 67 having been born in East Grinstead itself, eight each in Withyham and Hartfield and four in West Hoathly. The Uckfield Union began in March 1835 with no fewer than eight houses, with children, wives and husbands all separated. By 1838–9 a new workhouse had been built to the south of Uckfield using the approved cruciform layout (again, now occupied by housing). By 1881 there were six staff and 140 residents, with four from Fletching, 15 each from Buxted and Maresfield and 14 from Rotherfield. Altogether, then, the Ashdown parishes contributed 137 inmates to the two workhouses by 1881, 47 per cent of the total.[27] Horsted Keynes contributed ten inmates

put to the respondents were not answered, particularly from East Grinstead and Rotherfield.

25 K.D.M. Snell, *Parish and Belonging: Community, Identity and Welfare in England and Wales 1750–1950* (Cambridge: Cambridge University Press, 2006), 329.

26 WSRO, PAR 348/37/7 Return of all lunatics and dangerous idiots in the parish of East Grinstead to Edward Cranston, esq., with a report on their health by John Collins.

27 <http://www.workhouses.org.uk/EastGrinstead/EastGrinstead1881.shtml#Inmates>; <http://www.workhouses.org.uk/Uckfield/Uckfield1881.shtml#Inmates> [both accessed 23 September 2021].

to the 214 within the large substantial purpose-built Cuckfield Union workhouse (later becoming Cuckfield Hospital and now also converted to residential use).

Feelings still ran high, and the levels of poverty were scarcely ameliorated by any concessions initially made by farmers after the Swing Riots. By late 1835 riots, mass meetings and arson began, but then petered out. A growing population and a labour surplus was only partially offset by migration to the coastal towns or abroad. And farming was now in a depressed state in the post-1815 decades, with implications for farmers' incomes, the wages and conditions of Ashdown's farmworkers and, by extension, the incomes of local artisans and tradespeople.

The gulf in understanding between Assistant Poor Law Commissioner Hawley and Sussex agricultural workers seemed unbridgeable. In April 1835 he wrote:

> In no region it has been our fortune to visit have we ever seen a peasantry so completely disorganised. In no enemy's country that we have seen have we ever encountered the churlish demeanour which these men ... now assume. Perfectly uneducated ... in point of intellect little better than the horses they drive ... .[28]

At Maresfield in May 1835 several men, ordered into the workhouse, turned up to eat their dinners and then promptly left. In November 50 labourers, armed with sticks, marched on a meeting of the Guardians at Cuckfield, but were persuaded to disperse; and there was unrest at Horsham necessitating the intervention of the military. If overt protest had failed, however, more covert forms were adopted. Crops were trashed, fences and gates damaged, sheep and cattle maimed and incendiarism threatened throughout the 1830s. A host of minor complaints, insubordination and disagreements accompanied such activities. Horses' tongues were cut out at Buxted, and at William Wood's successful Woodlands nursery at Tyes Gate, Maresfield, prize-winning dahlias were damaged and petty thefts were common.[29] This nursery, begun in 1826 and lasting until 1919, produced French roses, forest tree saplings and dahlias, as well as mangel wurzel and swede seeds and 'turnip seeds direct from the East Lothians'.[30]

ASHDOWN'S 'OPEN' PARISHES

There has been much scholarly interest in what became known after the 1840s as 'open' or 'close' rural communities. The terms referred to the parish vestry meetings at which decisions were made, particularly the setting of the poor rate for the

---

[28] W.H.T. Hawley, 'English Charity', *The Quarterly Review*, 53 (1835), cited in Lowerson, 'The Aftermath of Swing', 61.

[29] R. Wells, 'Resistance to the New Poor Law in the Rural South', in J. Rule and R. Wells, *Crime, Protest and Popular Politics in Southern England 1740–1850* (London: Hambledon, 1997), 91–126. For attacks upon trees and other plants in southern England see Griffin, 'Protest Practice and (tree) Cultures of Conflict'.

[30] B. Abbs, 'A Maresfield Entrepreneur: William Wood and Woodlands Nurseries', *SAC*, 132 (1994), 161–72; *Sussex Agricultural Express*, 12 April 1850. For William Wood's funeral procession see Betty Turner, *Maresfield* (privately printed, 1991), 80.

ensuing year. 'Open' parish vestries allowed any interested ratepayer to attend, although these might descend into disarray, especially when farmers disputed what they saw as demands for excessive payments. By contrast, 'close' parishes were controlled by a small group including the landowner(s) or their representatives in a 'select vestry'.[31]

The vestry meetings both reflected and influenced the social and economic characteristics of the parish. 'Close' communities were more likely to be dominated by one or very few landowners who might also be church patrons and magistrates. The position around Ashdown has already been set out as it stood in 1785 and that serves to remind us that such differences were of long standing, having evolved over many generations (Table 5.1). The parishes might have larger farms but less resident labour living in nucleated villages under the squire's eye. Numbers of cottages were limited and harvest labour was imported when required. Smaller populations kept poor rates lower. These would probably be politically conservative parishes with small numbers of nonconformists and little manufacturing. About 90 per cent of Sussex parishes around the South Downs, smaller and with a richer farming base, could be classed as essentially 'close'.

By contrast 'open' parishes had a more diffuse landownership, with many small farms, and many farmers having secondary occupations. Scattered cottages on poorer land, perhaps owned by tradesmen or craftsmen, might be more plentiful. Populations might be larger with less control over settlement, and poor rates correspondingly higher, thereby diverting capital away from farm improvements. There might be self-governing village institutions, such as savings clubs, and a tendency to be more radical politically or nonconformist in religion. Worship in 'open' parishes might be conducted in a chapel by a man who was from the same class as the congregation, rather than by a wealthy rector or vicar. The Weslyan meeting house in Rotherfield was led by a local preacher, John Manser, who was a brickmaker. The Two Chimneys chapel in Withyham had been converted in 1836 from a barn.[32] In contrast to the downland parishes, over 70 per cent of parishes on the High Weald were 'open'.[33]

In practice these two types of settlement are extreme ends of a continuum within which facets of the two types might be present or absent. But clearly the Ashdown parishes conformed more nearly to the 'open' type. In some respects they resembled

---

31 For the basic premises of the 'open/close' concept see D. Mills, *Lord and Peasant in Nineteenth Century Britain* (London: Croom Helm, 1980) and B. Holderness, '"Open" and "Closed" Parishes in England in the Eighteenth and Nineteenth Centuries', *Agricultural History Review*, 20 (1972), 126–39; and elaborated in B. Short, 'The Evolution of Contrasting Communities within Rural England', in Short (ed.), *The English Rural Community: Image and Analysis* (Cambridge: Cambridge University Press, 1992); A summary is in A.J.H. Jackson, 'The 'Open–Closed' Settlement Model and the Interdisciplinary Formulations of Dennis Mills: Conceptualising Local Rural Change', *Rural History*, 23 (2012), 121–36.

32 J.A. Vickers, *The Religious Census of Sussex 1851*, SRS 75 (Lewes, 1989), 52, 60.

33 B. Short, 'Landownership in Victorian Sussex', in K. Leslie and B. Short (eds), *Historical Atlas of Sussex* (Chichester: Phillimore, 1999), 98–9.

the folk of the 'open' parish of Tadley in north Hampshire, 'reckoned a race apart – secretive, clannish and inbred'.[34]

Table 8.5. Ashdown landownership in the early 1840s.

	Date of tithe apportionment	Owner-occupiers	Total holdings	Percentage owner-occupier
Buxted	1841	68	164	41.4
East Grinstead	1842	150	424	35.2
Fletching	1839	76	222	34.2
Hartfield	1842	43	198	21.7
Horsted Keynes	1843	25	129	19.4
Maresfield	1841	86	318	27.1
Rotherfield	1839	128	367	34.9
West Hoathly	1841	47	135	34.8
Withyham	1843	37	263	14.1
Total		660	2220	29.7

Source: ESBHRO, tithe apportionments TD/E.

The structures of landownership are revealed more fully in the tithe apportionments accompanying the large-scale maps for each parish drawn up following the Tithe Commutation Act 1835, which state whether holdings are tenanted or owner-occupied (Table 8.5). The percentage of owner-occupiers is relatively high, in keeping with much of the High Weald, but drops below the Ashdown average in Hartfield and Withyham, where many holdings were tenanted from the Buckhurst estate.[35] By contrast, an estate parish such as Ashburnham had just 6.1 per cent of its holdings owner-occupied, and Frant, dominated by the earl of Abergavenny, had 9.4 per cent in owner-occupation.

Around Ashdown there were no parishes completely dominated by a large estate, as on the Downs, although there were certainly several large landowners, such as the Sackvilles or De la Warrs, the earl of Sheffield or the Newnhams/Shelleys. But the large parish of Hartfield demonstrates the complexity of categorisation: north of the Medway most land belonged to the Buckhurst estate and was characterised by larger farms; the area edging onto Ashdown, around Marsh Green at the Forest edge, was one of smaller cottagers, an almost 'peasant' district; while the Ashdown portion to the south was open heathland or scattered cottages, with a more 'open' feel. In fact, the large size of the parish caused the census enumerators to divide it into north

[34] Cited in A. Howkins, 'Types of Rural Communities', in E.J.T. Collins (ed.), *The Cambridge Agrarian History of England and Wales Vol. VII 1850–1914* (Cambridge: Cambridge University Press, 2000), 1297–353.

[35] The average for owner-occupation *c.* 1840 across the entire Sussex High Weald was 27.2 per cent, so Ashdown conformed closely to this figure.

and south parts from 1811. The northern part had a smaller population, whereas the south had the village itself, Marsh Green and the Forest (Table 8.6). The first half of the century saw little change in the agricultural north, contrasting with growth through to 1841 in the more 'open' south. The average holding size at the Forest edge was just ten acres, compared with 35 acres north of Hartfield village. Overall, in the parish the De la Warr property in 1842 was worth £855 out of a total £1285.[36]

Table 8.6. Hartfield population 1801–51.

	North			South			Hartfield total
	male	female	total	male	female	total	
1801*							1050
1811	220	223	443	398	409	807	1250
1821	247	227	474	495	471	966	1440
1831	231	206	437	555	463	1018	1455
1841	241	231	472	558	573	1131	1603
1851	248	217	465	563	545	1108	1573

*1801: no north and south division. The total figure for males was 493, for females 557.
Source: 1851 census table: BPP, 6 (1851–3), 458–9.

The relative lack of manorial or estate control, the intransigent nature of its Commoners and foresters, the continuing build-up of scattered cottages and the resulting pressure on Poor Law resources all combined to make Ashdown a difficult area to administer. In the 1820s Sir John Shelley from Maresfield Park despaired of the growing numbers of squatters on Ashdown:

> The foresters hereabouts are building up huts on the Forest which they finish in the course of a day or two and they then place somebody in them thinking that when once inhabited no body has a right to pull them down. How is all this? If these [sic] goes on I shall have an Indian town close to me of some of the greatest rapscallions that ever existed ... I mean to call a meeting of the parishioners and pull down these huts ... the respectable part of the parish will do anything I recommend...[37]

Shelley received the reply that four men had been targeted for legal action as examples to the neighbourhood. However, Duddleswell manor clearly continued to allow grants to individuals. A rental notes a grant near High Gate: 'John Blackstone for a piece of

36 TNA, IR 18/10352. Hartfield Tithe File 1842; ESBHRO, TDE 149 Tithe map and apportionment for Hartfield 1842.

37 ESBHRO, U of Sx Spec Colls, AFC II, 126.

land containing by estm. 2 roods being part of the Waste of the said Manor in East Grinstead – granted in 1851 … 1s'.[38]

The lawlessness of such communities was doubtless exaggerated by external and unsympathetic commentators, as had happened elsewhere. Wherever communities could assert some independence through access to common rights there were those who bemoaned their 'means of subsisting in idleness' and saw an inferior and primitive lifestyle as a barbarous hangover from a remote age. In 1827 William Luther Sewell of Twyford Lodge saw the Forest as a 'dreary waste of beggary and desolution, a few miserable cattle starve on it'.[39] Those living among heathy 'wastes' were seen as 'more perverse and more wretched' than those living in villages. There was resentment at the lack of deference to the elite: 'those people could go out and earn money there because they would not come and work a fair day's work; they could simply go out there and earn double the money.' The wastes were depicted as 'fruitful seminaries of vice' where pilfering and horse and cattle stealing thrived.

The presence of open heathland was certainly an attraction for travelling groups. In the early 1880s Wolff noted of Ashdown:

> Here the cattle may still be seen browsing among the bracken, and the wild tribe of foresters, popularly misnamed 'gipsies,' may be watched in their encampment or in their cabins, idling away the precious time, or plying their vagrant trades … . They are a wild tribe, these foresters, and education and civilization are making but slow inroads upon their uncultured manners and love of independence. Time was, when they were a really dangerous crew to deal with, and the wide sweep of country covering Ashdown Forest was a domain in which the Queen's writs could scarcely be said to run … .

And, harking back to the eighteenth century:

> Even then, however, some sort of self-imposed order was maintained. The gipsy headquarters of old was Crowborough, a hamlet only recently created a parochial district, after Sir Henry Fermor in 1732 left a sum of money with which to build and endow a church … it was at Crowborough that these same heathens annually assembled, when the fair came round, to hold their 'settling day', adjusting the accounts of the year's quarrels by a general fight. Carrying over was not permitted and the balance had to be promptly paid [in] kicks and blows, for which a receipt was given in the shape of bruises, black eyes, and broken heads.[40]

38 ESBHRO, ADA 98 (Rental of the Manor of Duddleswell, n.d.). Blackstone was later one of Raper's witnesses.

39 ESBHRO, U of Sx Spec Colls, AFC II, 135. I am grateful to Jenny Whitmore for her insights into William Luther Sewell.

40 H.W. Wolff, *Sussex Industries: a Series of Articles on the Principal and Peculiar Industries practised in Sussex* (Lewes: *Sussex Advertiser*, 1882). I am grateful to Chris Hare for alerting me to this publication.

At Fairwarp and Horney Common the families of the Smiths and Vincents were well-known gypsies. The Smiths were hawkers. At Beggars Bush, Duddleswell, in 1871 the Winson family, basketmakers, were living in a tent, the Clarks in a 'house cart'. William Brazil (20), his companion Sarah Ann (20) and their children (aged two and six months) were 'tramps' living in tents and carts. Benjamin Brazil (41) and his family were also hawkers living in 'tents or carts' in Duddleswell, as was John Brazil and his family at Chelwood Gate.[41] The Brazil family were longstanding visitors to the Forest; however, since October 1848 the children of William and Sarah Ann, Sarah Anne Jr, Emily and James, were baptised at Maresfield. William was then described in the parish register as a 'tinman' from Penshurst, Kent and the family were 'encamped on the forest as tramps', a description also applying to William and Philly Veney, pedlars from Burwash. The fairs around the Forest had long attracted all manner of pedlars as well as being the occasions for the sale of cattle and horses. Some travelling folk might put down roots if an encroachment could be secured: the Wright family of travelling tinkers, originally from 'some shire' or other, squatted on the east side of Burnt Hill, Nutley *c.* 1870, and were later permitted to buy the property (Figure 8.1). Two sons proceeded to encroach other ground in the area about 1876.[42]

The seasonal movement of travelling families was caught in an entry in the Guide for Uckfield and District in 1888:

> On the Forest reside numbers of those strange people who come down among the townsfolk in houses on wheels … . Their stock consists of toys, brushes, and brooms of all kinds. From these secluded valleys issue, just before harvest, a motley group of men, women and children, all wending their way to the extensive corn tracts along the coast, and 'under the hills', intent on earning as much as they can at harvesting, for to this source they look for their annual supply of clothes.[43]

Again, we might also note the ever-present phenomenon of poaching, with newspaper reports of convictions around Ashdown now appearing regularly. Agricultural labourers between 1860 and 1880 constituted 76 per cent of the cases reported in the *Sussex Express*. Many carried wires or catapults on their journeys to work. James Gilbert (Rotherfield, 1863), William Thorpe and Stephen Page (Maresfield, 1872) and the three men Arnold, Gorringe and Martin who came before Earl De la Warr at the East Grinstead Petty Sessions were discharged after claiming that they were birdnesting!

---

41 'CMM', 'The Romany Grave', *Ashdown Forest News* (Spring 1987), 17; ESBHRO, PAR 420/1/2/2: TNA, RG 10/1054, Buxted census 1871; 1055, Fletching census 1871.

42 H. Walter, 'Some notable happenings through the years from 1804 to the present time on Ashdown Forest from old stories heard' (unpublished typescript, 1950), 2. For more on the Wright family see ESBHRO, AMS 5762 Reminiscences of Harry John Peckham (1841–1922) Vicar of Nutley 1882–1913.

43 *Brookers Guide and Directory for Uckfield and District* (1888), 104–5. And see D. Smith, 'Gypsies, Tinkers, Travellers and the Forest Economy', in J. Langton and G. Jones (eds), *Forests and Chases of England and Wales c.1500 to c.1850* (Oxford: St John's College Research Centre 2010), 61–6.

Figure 8.1. The cottage of James 'Tinker' Wright at Nutley 1888 (source: Nutley Historical Society).

In 1875 William Bennett, pleading guilty to a charge of trespass in search of game on Ashdown, 'said he was sorry, he wanted a rabbit for a person who was ill, and he thought he should not be interfered with for getting one in the forest'. In 1879 Thomas Weller and Robert Turner pleaded not guilty to the same offence: 'on being spoken to they said they should do as they liked in the forest'.

The mixture of (perhaps wilful) ignorance and the assertion of ancient Forest rights, together with sanction by the community, was a feature of attitudes more generally. Most poachers were in employment, albeit often of a casual nature, but hoping to supplement food supplies, normally in their leisure time. It is significant that the offences above were committed during the winter months, when times might be pinching (or when winter nights made poaching easier). Some were inveterate poachers: Henry Langridge (Fletching), born in 1847, was first convicted in 1859 at the age of 12. By the time he was 23 he had accrued 14 convictions for game offences and had spent several months in prison.[44]

Charles Weller, known as Darry, was an Ashdown poacher and well-known character around Forest Row. When younger he had lived in a hut built with benders and heather, but later he used an old stone-built and thatched shed belonging to Major Darwin. He

44 S. Allen, 'Poaching and the Game Laws in East Sussex, 1830–1880' (Univ. of Sussex unpub. MA, 1982), 73–7.

was six feet tall, of slight build, unkempt and somewhat smelly. He had lost an arm in an accident with a gun but, helped by his lurchers, he lived by poaching and selling his catch, especially rabbits. As winter approached he would commit some offence sufficient to bring a short prison sentence, enabling him to spend the winter months in the relative warmth of Lewes prison.[45] This lifestyle was similar to that of 'Little Tinker' moving between the estates of Maryon-Wilson, the earl of Sheffield and the Shelleys:

> Old Johnny Burgess, that were his real name. Well, he burnt charcoal, and he was allus down in the woods getting wood for charcoal and also keeping an eye out to see where the pheasants was roosting ... Tinker was allus down there poaching and getting caught ... I think he had twenty Christmas dinners in gaol. He said he didn't mind being inside because he did like the 'skilly' so.

Gilbert Sargeant, 'Gillie', also remembered the lurchers: 'I've seen them down in the pub there, the Shelley Arms – seen as many as fourteen dogs in there of a Sunday dinner-time, all lined up beside their masters and every one a rabbit catcher.'[46]

Gamekeepers were often ostracised by the local community due to the nature of their work, and the role might be passed on to relatives. David Taylor Dadswell (1825–1880), keeper to the London solicitor and clerk to the Goldsmith's Company Walter Prideaux (of the Lake House, Hamsell, Rotherfield), had a cousin of the same name who was a gamekeeper for another local landowner.[47] John Turner, brought before the magistrate for poaching on J.V. Shelley's estate by gamekeeper William Gurr, remarked to the magistrate: 'But I never done nothing, sir, don't you see; but to tell the truth I might have done if I hadn't seen Master Gurr.'[48]

Poaching was not normally violent, but there were those whose activities did strike fear into law-abiding people. A series of crimes was associated, for example, with 'The Guildford Connection', a gang including nomadic hawkers and middlemen who disposed of stolen goods in London. In 1850 gang members undertook burglary with violence at Kirdford and burglary at a farm in Cuckfield, at a pub in Haywards Heath, at Knight's workshop in Maresfield, at the Six Bells in Chailey, at the White Hart at Holtye Common and at the elderly Farncombe sisters' mansion in Uckfield, where seven raiders were involved, meeting up later on Crowborough Common. Three of the gang, William Hillyer, Joseph Carter and James 'Butcher' Hamilton, were originally ploughboys from Rotherfield. Hamilton dealt in horses, sold brooms and went hop-picking in the autumn. The other two were known to have huts on Woking Common, an equally

---

45 L. Phelan, 'Economy to Amenity: the Commons of the New Forest and Ashdown Forest, 1851–1939' (Univ. of Sussex unpub. DPhil, 2003), 45–6, based on *Forest Row Magazine*, 3 (October 1986), 24. For Major Darwin see p. 323.

46 D. Arthur (ed.), *A Sussex Life: the Memories of Gilbert Sargent Countryman* (Tunbridge Wells: Chapel Press, 1989), 73.

47 And see W. Prideaux (Sir Walter Sherborne), *Memorials of the Goldsmiths' Company, Being Gleanings from their Records Between the Years 1335 and 1815*, 2 vols (London: Eyre and Spottiswoode, 1896).

48 Allen, 'Poaching and the Game Laws', 51–73.

difficult area for law-enforcement. A cottage in an obscure part of Crowborough Common provided a depot for stolen goods, and Copthorne offered another usefully marginal area for rendezvous. Hamilton was later arrested at the Rose and Crown pub in Groombridge, and Hillyer at a beershop near Guildford. All seven men involved in the Uckfield robbery were found guilty in March 1851. Hamilton had confessed and informed on the other six men and all were sentenced to life transportation.[49]

Transportation was the sentence for lesser crimes too. In November 1835 Samuel Walters, aged 27, stole 28 bundles of holly sticks at Fishers Gate, belonging to Buckhurst, to sell for firewood. Evidence of the theft, the carriage along the turnpike roads towards London and the borrowed wagons, was uncovered by De la Warr's steward David Edwards, and Walters was sentenced to seven years' transportation. Leaving behind his wife and a son, he never returned but died in Sydney, a local dignitary and a wealthy man.[50]

CHURCH AND CHAPEL

If a key feature of many 'close' parishes was the influence of the church, frequently acting to reproduce the hegemony of the gentry, then among Ashdown's 'open' parishes such an influence was diluted, although still present to some extent. However, if the parish church was difficult to reach in these large parishes, especially during the winter, or if the vicar was seen as the mouthpiece of the landowner, their cultural capital might be usurped by nonconformist ministers in their chapels and meeting places.

Some commentators were distinctly unimpressed by Wealden Christian attitudes. Following a series of 'pedestrian tours' Richard Heath wrote a number of essays between 1870 and 1884, assembled later as *The English Peasant*. In 'Sussex commons and Sussex songs' he wrote of the area:

> Now sunk in an almost incredible ignorance and immorality … . Those who exercise the Christian ministry in the Weald complain of the apathy and stolid indifference of the people, of the rowdyism of the boys, and the immorality of both sexes … from all I heard it would be difficult to exaggerate the latter evil. Let no-one suppose, however, that they are peculiarly corrupt. It is the poverty, ignorance, and a low state of public opinion which is at the root of the evil … . Of the depth of superstition, ignorance, immorality and poverty which prevails in the Weald of Sussex, it is impossible for me to give any adequate idea.

He continued in similar vein: there was a 'crabbed independence' possibly leading to 'petty jealousies', and prejudice against clergy by those 'just beginning to realise their power to think', for 'the rural clergy, as a class, have so closely identified themselves

49 R. Wells, 'Popular Protest and Social Crime: the Evidence of Criminal Gangs in Rural Southern England 1790–1860', in B. Stapleton (ed.), *Conflict and Community in Southern England* (Stroud: Alan Sutton, 1992), 135–82.

50 M. Parcell, 'The Case of the Missing Holly Sticks', *Hartfield Times*, 8 (n.d.), 1–4. I am grateful to Mike Parcell of Hartfield History Group for this information.

with the gentry as to give rise to the impression that they regard themselves as a sort of spiritual squirarchy'. This was contrasted with a visit to Buxted, where 'the pastor of the chapel proved to be an ancient labourer in a dark smock, a truly simple-minded, good old man'.[51]

Although many of the Ashdown churches had been judged to be in decent repair during the early eighteenth century, by the early nineteenth century many were suffering from the poverty of their parishes and the use of inferior materials that were now crumbling. But later in the century a renewed enthusiasm for church building, often associated with the Oxford Movement and High Anglicanism, came into the Weald, frequently helped by the entry of wealthy newcomers.[52] Fermor's All Saints church at Crowborough, originally built in 1744, was enlarged in 1881–3 and again in 1895–8 in line with a growing, and quite affluent, population (Figure 8.2).

In Buxted Park stood St Margaret the Queen, which had originally been surrounded by the village. But a new Buxted Park mansion had been built nearby in the eighteenth century and, although a drawing in the Sharpe Collection shows buildings near the church *c.* 1802, on the death of his wife in 1814 Charles Jenkinson, earl of Liverpool, became a recluse and wanted the village moved from the vicinity of Buxted Place (Figure 8.2). The villagers refused to move so Jenkinson, as landlord, stopped maintaining the properties. Soon many houses had become uninhabitable and the villagers did agree to be relocated, leaving the church behind. Signs of the medieval village street, a surrounding bank and ditch, a well and house platforms on either side of the church have been noted, together with thirteenth-century pottery within the churchyard.[53] The medieval church was restored between 1872 and 1885. Also at Buxted, the chapel of St Mary the Virgin was built in 1885–6, financed by the wealthy Anglo-Catholic Father Arthur Wagner of Brighton. This building, central to the village, is now nearly two miles from the original medieval church. In 1873 Wagner had leased Totease House in Buxted as a retreat and also made it available to a community of Anglican sisters and orphan girls who were trained for service and, when older, referred to as 'Industrial girls'.[54]

The church of St John the Evangelist was built in 1839 as a chapel of ease to serve the small settlement of Crowborough. It was situated on 14 acres of Forest land 'on the slope

---

51 R. Heath, *The English Peasant: Studies: Historical, Local, and Biographic* (London: T. Fisher Unwin, 1893), 186–200.

52 Father Robert Hugh Benson, a Catholic convert, lived as a boy in Horsted Keynes with his widowed mother, the wife of E.W. Benson, archbishop of Canterbury. Benson's novel, *By What Authority?* (New York, 1904), was set in a fictionalised Horsted Keynes, and the foreword is written from 'Tremans' in that parish.

53 C.F. Tebbutt, 'Two Newly-discovered Medieval Sites', *SAC*, 110 (1972), 31–6; and see G.R. Burleigh, 'An Introduction to Deserted Medieval Villages in East Sussex', *SAC*, 111 (1973), 45–83 and the Felbridge and District Historical Society site 'Buxted Park and the Medley Evelyn Connection', <http://www.felbridge.org.uk/index.php/publications/buxted-park/> [accessed 10 August 2020].

54 The house closed in 1911 (ESBHRO, SAS ACC7740 – Fitzhugh Gates, solicitors: clients' papers including records of St Mary's Home, Brighton, Ovingdean and Buxted).

Figure 8.2. Crowborough chapel (All Saints Church) and Fermor school; and St Margaret the Queen, Buxted, *c.* 1802 (source: Sussex Archaeological Society Sharpe Collection 65 and 101. Reproduced by kind permission of the Sussex Archaeological Society).

of a hill in a most wild region' .[55] The building was modelled on John Henry Newman's church at Littlemore, Oxfordshire, and offered an early Anglo-Catholic service, in line with Newman's Tractarian Catholic revival. Endowed by Countess De la Warr, it was built from local stone with carved panels from the chapel of Buckhurst Park incorporated into the pulpit *c.* 1870, at about the same time as the countess funded a chancel. By the 1860s there was a schoolhouse, schoolroom, accommodation for an incumbent and almshouses. In 1873 the church became the focus for the independent ecclesiastical parish of Withyham St Johns. Another strong Tractarian presence came from R.C. Carpenter's design of the chapel at Nutley, later St James the Less, opened in 1845.

Restoration certainly left its mark on St Michaels, Withyham. Initiated after Reginald Sackville-West, later seventh Earl De la Warr, became rector in 1841, restoration continued until 1864, with the interior plaster scraped and replaced by pointing. The rector himself probably oversaw the work, applauded by Sir Stephen Glynne in 1853 as making the church 'a very good model of ecclesiastical propriety'. The earl also painted the Last Judgement over the altered chancel arch in 1856. The south aisle dates from 1844, with a porch and three-bay arcade. Changes were made to the roof and north aisle, high oak pews were cut down and a new font donated by Mrs Hale from Holly Hill. In the south porch are the remains of a figure of St Michael, painted by Reginald's sister, Lady Arabella (née Sackville-West), shortly before her death in 1869.[56]

In 1851 a national census of religious worship was organised by the government in the face of increasing concern about the state of religion, with information on all places of worship and their congregations to be returned. The results are not without controversy and the incumbents from Buxted and Fletching refused information, which then had to be estimated, but in general the Sussex Weald was shown as an area of weaker Anglican presence within an otherwise predominantly Anglican county. Attendances here were somewhat below the county average, although poor weather conditions on the census day and large parishes helped to explain this. As the curate at Maresfield wrote, 'There was less than average attendance on March 30, there were several heavy partial storms of rain which may have prevented the distant parishioners attending.'[57] A fuller explanation also came from Jonathan Moses, minister at the Two Chimneys Chapel, Withyham, a Particular and Strict Baptist congregation:

> Local circumstances beside a stormy day lessened the usual attendance. Our congregation is gathered from considerable distances, and is therefore contingent upon the weather. Besides once a month there are places of worship or preaching stations nearer their homes where they attend on those days … .[58]

Thirty-six churches and chapels around Ashdown were returned in 1851, including 22 nonconformist chapels, but no Catholic church (Table 8.7). To the 1851 list might

[55] A detailed history is given in <https://cee55446-cfb0-48da-bc19-83b209e46dba.filesusr.com/ugd/efe475_00cc66c99e444b6585478e98f1180d46.pdf> [accessed 27 September 2021].

[56] <https://sussexparishchurches.org/church/withyham-st-michael-and-all-angels/> [accessed 27 September 2021]; D. Parsons, *Sir Stephen Glynne's Sussex Church Notes*, SRS 101 (Lewes, 2021), 319.

[57] Vickers, *Religious Census*, 57.

[58] Vickers, *Religious Census*, 60.

be added a number of later Victorian churches. There was a chapel of ease at Fairwarp by 1881, which became a separate ecclesiastical parish from Maresfield in 1901. The restoration of St Mary and St Andrew at Fletching was also carried out in 1880–81, and there were additions to Holy Trinity, Forest Row, built in 1836, but with a short chancel of *c.* 1850 and south aisle of 1877–8.[59] At Crowborough, nonconformist buildings multiplied: Jarvis Brook chapel-of-ease was built with a schoolroom in 1883, serving the flourishing brickworks there. The Congregational Chapel (1864, originally in a barn from 1862) and Ebenezer Gospel Hall (*c.* 1898), both in Jarvis Brook, and the Sweethaws Chapel (1896) and Crowborough Methodist church (1897) were also late Victorian additions but have now been demolished, as has Motts Mill Baptist Chapel (opened 1868). St Richard de Wych (1886) was built in the Gothic Revival style after Ashdown Park House was rebuilt in 1867, also in French Gothic style, for the wealthy Durham businessman and MP Thomas Charles Thompson. There was also a farm and workers' housing in the grounds.[60]

This was the period of Anglo-Catholic developments and around Ashdown these were supported by the controversial John Mason Neale (1818–66), warden from May 1846 for 20 years at Sackville College, East Grinstead.[61] His acceptance of the post was helped by his admiration for the views across the Forest to the south. The position was in the gift of the fifth Earl De la Warr, who became Neale's loyal supporter. In 1848 Reginald, later seventh earl, became an assistant warden. But by 1847 Neale's adherence to the Anglo-Catholic revival led to trouble with his bishop for his 'Romanistic observances' and for decorating the college chapel in 'an imitation of the degrading superstitions of an erroneous church'. Among his many writings were hymns and carols such as 'Draw nigh, draw nigh [O come, O come], Emmanuel' and 'Good King Wenceslas' – the latter written in 1853, when living near Ashdown, and containing the lines 'when a poor man came in sight, gathering winter fuel ... [who lived] right against the forest fence'.

Neale wrote of Ashdown, looking from his study window, 'now a wild waste of heath and down, where the trees have been recklessly felled for timber ... and one or two chapelries abut upon it, but can hardly be said to penetrate it'. He feared for both the spiritual and bodily health of the Foresters, 'heathen to all intents and purposes'. Sister Miriam, his biographer, wrote in 1907:

> The cottages were badly built and unsanitary; when sickness and fever laid its hand upon the peasantry they had no defence. In those picturesque, innocent-looking, whitewashed hamlets the morality was on a lower level than in a city slum, and deadly evil lay hidden from sight in the lovely sequestered woods and lanes, as a poison may be concealed beneath the fair semblance of a flower.[62]

---

59 Antram and Pevsner, *Sussex: East*, 315–17 *et infra*.

60 By 1920 St Richard had become a convent for Belgian nuns with its own chapel. The church itself gradually sank into disrepair and has now all but vanished. For more on churches at this date see D. R Elleray, *The Victorian Churches of Sussex: with Illustrations and a Check-list of Churches and Chapels erected during the Years 1810–1914* (Chichester: Phillimore, 1981).

61 Susan Drain, 'Neale, John Mason (1818–1866)', *ODNB* (2004).

62 E.A. Towle, *John Mason Neale D.D. A memoir* (London: Longmans, Green, 1906), 231–2, 334.

Table 8.7. Churches and chapels around Ashdown 1851.

Parish (poptn 1851)	No. of churches or chapels	Churches and chapels (congregation number)						
Buxted (1,694)	3	St Margarets PC, Buxted Pk (130)	St Marks, Hadlow Down (320 – 2 services)	Five Ash Down (Indep) (300)				
E Grin (3,820)	7	St Swithun PC (334)	Forest Row chapel (300)	Congreg Chapel, Below the Rocks, Indep (110)	Bethesda Chapel, Forest Row (Baptist) (70)	School Rm, EG Common (Weslyan) (20)	Zion Chapel (Countess of Huntingdon Connection) (109)	Sabbath school room, Ashurst wood (C of Hunt Connection) (35)
Fletch (2,007)	4	St Andrew PC (181)	Holy Trinity chapel (Danehill) (not given)	Short Bridge Weslyan chapel, Piltdown (40)	Advent Tabernacle, Piltdown (95)			
Hartfd (1,573)	3	St Marys PC (210)	St Peters chapel, Holtye (150)	Weslyan Chapel (46)				
HK (847)	3	St Giles PC (90)	Horsted Green chapel (Indep) (35)	Dane Hill Baptist (Particular Baptist) (47)				

Maresfd (1,805)	4	St Bart PC (154)	St James, Nutley (-)	School room Twyford Lodge (160)	Weslyan chapel, Duddleswell (32)			
Rother (3,531)	7	St Denis PC (170)	Crowboro Chapel (Anglican) (144)	Upper Chapel (Baptist) (263)*	Lower Chapel (Particular Baptist) (286)	Weslyan Chapel (95)	Weslyan meeting house (?Mark Cross) (40)	Boars Head St chapel (Weslyan) (136)
W Hoath (1,068)	1	St Margaret 's PC (152)						
Withy (1,692)	4	St Michaels PC (160)	St Johns chapel (100)	Two Chimneys chapel (Particular + Strict Baptist) (192)	Weslyan chapel (39)			

*services only twice a month

PC = parish church

Note: congregation numbers based on Q4. 'Numbers attending each service' – the table uses the figure for mornings, but does not include childrens' attendance, for which see Vickers.

Source: Vickers, *Religious census*.

Furthermore:

> The whole countryside was mutely crying out ... The parish of East Grinstead and those adjoining it were of great extent, with scattered population; the clergy were very few, and insufficient. The dread of infection was so great among the poor that it was sometimes impossible to get anyone to nurse a fever case, and at hopping time no one was at leisure for any nursing at all. Poor creatures were dying in lonely cottages without help for body and soul; in utter discomfort, and in the grossest ignorance. [Then he] conceived the idea of forming a community of women who should go and nurse those sick poor in their own hovels and care not only for their bodies but for their souls likewise. So St Margaret's Sisterhood began.[63]

Neale initiated the Sisterhood in 1854 for nursing sisters to minister to the poor and sick in their cottages, risking death from scarlet fever, among other dangers. At Woodside, Saxonbury Hill, to the east of Crowborough, a poor woman lay dying of consumption in July 1855. The sisters took turns to nurse her until she died in March 1856. There was a large family and the sisters slept in the loft of a nearby cottage, with a roof of dilapidated thatch. On returning at night to these lodgings the sister would be sometimes greeted with 'Come in Miss; you can have your tea; pig has had his supper.' Neale wrote: 'Such an extremity of wretchedness I never saw but in Portugal.' The sisters worked all day with the ill woman's family, but in the autumn any neighbourly help ended when all able-bodied persons went hop-picking. Over the winter a three-year-old daughter died just before her mother, and the two coffins were placed in one grave.[64]

The Sisters then returned to East Grinstead, starting an orphanage nearby, and eventually on to a new, Gothic Revival convent building that was not quite finished by the time of Neale's death. His time was certainly controversial: there had been 'No Popery' gatherings outside the college in 1851 and a riot at All Saints church Lewes in 1857 at the funeral of one of the sisters of his order. Sometimes, it was later recalled, they were welcomed in the Forest cottages, sometimes not.[65]

THE ARRIVAL OF THE RAILWAY

In the early years of Victoria's reign one of the greatest facilitators of change was the railway. Now even easier access to and from urban centres was made possible. The pre-modern world was now to give rise to a modernity that affected even Ashdown. Amidst great enthusiasm two main lines were developed early: Rennie's 'direct route' opened from London to Brighton in 1841, operated by the London, Brighton & South

63 See M. Leppard, 'The unregarded Sussex of J.M. Neale', *SAC*, 155 (2017), 165–79. He notes Neales's story in the *Penny Post* (1858) entitled 'Gills Lap', set around the Forest with accurate topographical detail. Neale was also keenly interested in the local Sussex dialect.

64 The late Michael Leppard kindly provided me with this information from his copy of Sister Miriam's *Memoir*, pp. 318–20.

65 Now the Old Convent, Moat Road and much redeveloped as housing. A.G. Lough, *John Mason Neale – Priest Extraordinary* (Newton Abbott: privately printed, 1976), 76–103.

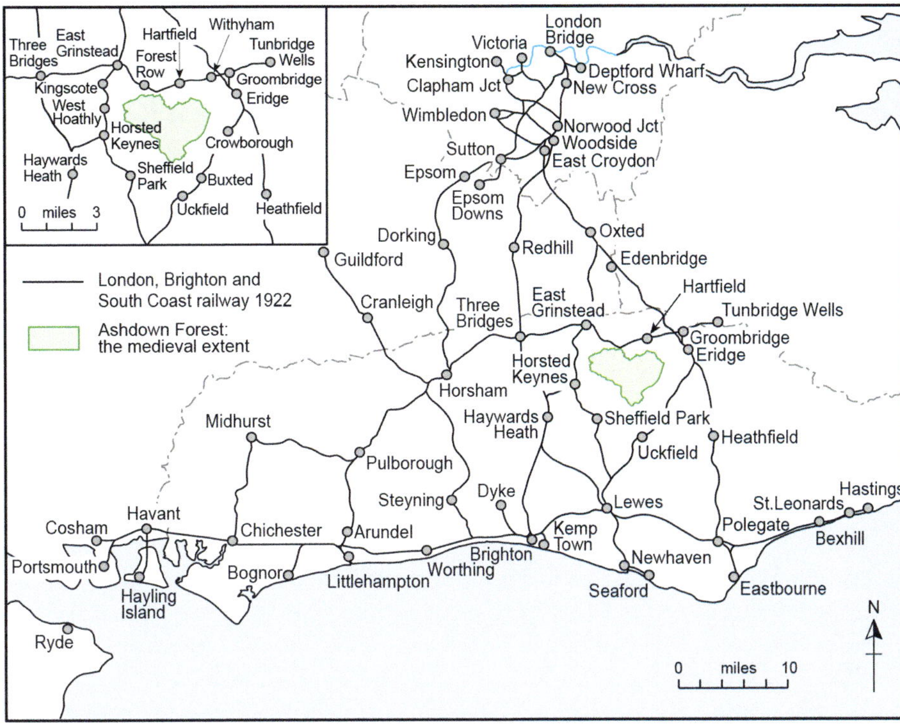

Figure 8.3. The LBSCR rail network and the lines around Ashdown (source: author).

Coast Railway Company (becoming the LBSCR after a merger of five companies from 1846). The other was the South Eastern Railway (SER), reaching Dover in 1844. The companies were great rivals but by an agreement of 1848 the area south of the Redhill–Dover main line was recognised as the territory of LBSCR, while the area to the north looked to the SER (Figure 8.3).

Many other possible lines initially looked to be doubtful investments, especially those serving rural areas. The lines linking Ashdown's villages therefore arrived late, mostly between the 1850s and 1880s, following the peak of national railway building, and were built primarily to define and defend the two companies' territories. The first branch line to approach Ashdown came in 1845, with an SER line to Tunbridge Wells from its main line at Tonbridge, thereby threatening the north-eastern flank of LBSCR territory. Rival proposals by the LBSCR and the SER for branches to East Grinstead from their respective main lines were then presented but shelved when the Railway Mania collapsed in 1847. However, the town was now losing trade to places that had railways; its cattle market was suffering from competition and passengers had first to reach Godstone or Three Bridges to catch London trains. In the absence of any other interest, the town's residents therefore ventured to build a line from Three Bridges themselves.

Parliamentary approval was given and an East Grinstead single-track branch, run by the LBSCR, opened in 1855 with local festivities and a public holiday. The era of the branch line had now begun. In 1866 the line was extended to Groombridge and Tunbridge Wells, being backed by the Sackville-Wests and with Earl De la Warr as a major shareholder. The line remained single-track, but stations were now built at Forest Row, Hartfield and Withyham. The latter two were built less for commercial reasons than because the railway's act of parliament stipulated them, as it did also a private siding near Withyham for the use of Earl De la Warr.

To the south-east of Ashdown, a line built in 1858 linked Uckfield and Lewes and thence on to Brighton. It was extended northwards ten years later as a single-track to Tunbridge Wells, merging with the line from East Grinstead for the final section beyond Groombridge. But around Ashdown construction proved difficult, necessitating a tunnel near Crowborough. Nevertheless, stations now also opened at Buxted and Rotherfield, with the latter changing its name to Crowborough in 1880 in acknowledgement of the increasing interest in that area. Crowborough was therefore now linked to Brighton, Lewes, Tunbridge Wells and London, and the whole line was made double-track in 1894. Crowborough's station was redeveloped in 1905–6, reflecting the increasing traffic.[66]

There was another surge in railway development during the 1880s. In 1882 the Lewes & East Grinstead Railway was opened, with stations at West Hoathly, Horsted Keynes and Fletching. This line largely resulted from the initiative of the third earl of Sheffield; as a cricket enthusiast, he wished to allow large crowds to access his Sheffield Park ground to see, for example, the visiting Australians and W.G. Grace.[67] But, despite support from gentry in Lewes and East Grinstead, this was always going to be only marginally economic and it was taken over by the LBSCR.[68] The line was extended from 1883 at Horsted Keynes to the Brighton main line at the rapidly growing town of Haywards Heath. It was also originally envisaged by the earl of Sheffield that the line would continue beyond East Grinstead to London, and this was duly achieved through a joint venture with the SER. This double-track line, opening in 1884, ran from the London–Brighton main line at South Croydon through Oxted to East Grinstead, meeting the Lewes & East Grinstead line at a low-level station at East Grinstead. But the west–east Three Bridges–Tunbridge Wells railway crossed at a high-level station, giving the town an unusual double-decker station and its third in less than 30 years. East Grinstead had become a railway hub.[69]

[66] The line south of Uckfield closed in the late 1960s, although proposals emerge periodically to reopen the track to Lewes, thereby providing additional capacity between London and the south coast, relieving the crowded Brighton line.

[67] The earl's cricketing interests are displayed at Sheffield Park today, where his cricket ground is still in use.

[68] Closing in 1958, even before Dr Beeching's axe fell, the Sheffield Park to Horsted Keynes section was taken over by the Bluebell Railway Preservation Society. From 2013 the line was again extended to East Grinstead. The line between Haywards Heath and Horsted Keynes was retained until the late 1960s.

[69] The line between Three Bridges, East Grinstead and Tunbridge Wells fell victim to Beeching in 1967 and is now partly a cycle path and partly the single-track Spa Valley heritage railway between Eridge, Groombridge, High Rocks and Tunbridge Wells West.

Ashdown had benefited from the desire of the two competing railway companies to defend their territories. The lines here were rarely economic in themselves and effectively cross-subsidised, but they transformed access to and from the Forest to the wider world. The turnpike roads now lost their commercial edge; the last Sussex turnpike scheme was authorised in 1841, the year of the LBSCR opening, and all the county's trusts were wound up between 1864 and 1885.[70]

AGRICULTURAL CHANGE

One of the most important facilitating effects of the newly acquired rail access was the impact on farming. Now London and the Sussex coast could be supplied more easily, and between about 1840 and 1880 the Weald as a whole showed some agricultural progress. Wealthy new arrivals required fresh food, and urban growth brought a more proletarian consumer population to East Grinstead, Uckfield or Tunbridge Wells.

But this remained an unpromising farming environment. Of the parish of Hartfield it was said in 1842: 'the soil is principally of cold clay; subsoils are on sand rock, gravely sand and clay. Timber (oak principally) and the underwoods are both of good average growth.'[71] Within the five major parishes within which Ashdown lay, 35 per cent was either common or woodland when the tithe surveys were taken *c.* 1840, and Ashdown itself took up 26 per cent of their area.

Arable farming remained precarious on these acidic soils and higher altitudes but some farmers did begin to use guano and 'artificials' such as superphosphates on their soils. Robert Crawfurd had, by 1851, begun to substitute guano for lime, calculating that it was overall more economical per acre.[72] Nevertheless, many continued to use farmyard manure, supplemented where possible by Forest litter, the latter proving a useful manure when trodden down in cattle yards over the winter. And although John Villiers Shelley spoke at agricultural dinners to advocate steam threshing, which, together with hops, made profits on his estate, most continued to use old turnwrist ploughs, thrash with flails and mow with scythes.[73] Even in the so-called 'Golden Age' of farming *c.* 1840–70 tenants here were normally constrained in leases from ploughing grassland for arable crops, such was the landlords' reliance on pastoral farming for their rents. For an incentive to market wheat from the High Weald in general prices needed to be above 56s–60s per qr, a level reached only briefly in the 1850s and between 1866 and 1874. The result was a very small increase in arable between 1840 and 1870 for some, but the necessary drainage, clearance of shaws and mechanised

70 BPP, Turnpike Trusts County Report no. 3; Sussex (1852), 26–55. And see J. Farrant, 'Growth of communications 1720–1840', in K. Leslie and B. Short (eds), *Historical Atlas of Sussex* (Chichester: Phillimore, 1999), 78–9. I am indebted to Martin Berry for this outline of railway development around Ashdown.

71 TNA, IR 18/10352 Hartfield Tithe File 1842.

72 R. Crawfurd, 'Farming economics series: lime and guano', *Sussex Agricultural Express*, 12 April 1851, 8.

73 Walter, 'Some notable happenings', relating to Nutley and Maresfield.

Figure 8.4. Hop picking in the Nutley area in the late nineteenth century. Note the older training of hops around individual poles, which have been pulled and stacked for the hops to be picked (source: Nutley Historical Society).

harvesting were simply uneconomic on most Ashdown holdings.[74] Additionally, with over two-thirds of holdings rented and generally held on Michaelmas annual tenancies investment was consequently less likely.

Hops remained a gamble that might sometimes pay off handsomely, but which consumed rather too much of the farmyard manure. Ashdown was at the economic margin of profitability at a distance from the mid-Kent heartland of production, with small hop gardens (few were as much as three acres) and hardier and coarser varieties compared with those in Kent. A maximum hop acreage was attained in the mid-1870s, after which they declined in profitability.[75] However, the crop was important socially, with many families helping to prepare and set the hop gardens, and joining autumn picking (Figure 8.4).

While hardly amounting to an agricultural revolution, some changes were perceptible. There was increased interest in breeding the native Sussex cattle from the 1840s onwards, compared with the older practice of fattening northern, 'Scotch' or Welsh runts, although these beasts still featured at the October Rotherfield fair in 1851. The older, rich mahogany red Sussex breed had been prominent here prior to

74 B. Short, 'Agriculture in the High Weald of Kent and Sussex 1850 to 1953 (a case study in the application of multivariate techniques in the field of Historical Geography)' (Univ. of London unpub. PhD, 1973), 67–73.

75 Short, 'Agriculture in the High Weald', 75–82. By 1908 contingents from Crowborough and Rotherfield joined the 'London and Provincial Hop Growers and Pickers Defence League' in a rally at Southwark and Trafalgar Square to call for duties of 40s per cwt against foreign imports (BL, General Reference Collection 08226. bb. 53).

the Napoleonic Wars, being championed by the earl of Sheffield, but thereafter its popularity had diminished, although herds of Sussex and mixed-bred cattle were still to be found. By the 1850s Shelley, for example, was fattening shorthorns, although still employing the strong Sussex as working oxen.[76] But in 1878 the Sussex Cattle Society was formed, followed by the first Sussex Herd Book in 1879; of its 234 entries, over half came from the Weald, and many from around Horsham. Careful breeding, feeding with oil cake and relinquishing common rights were seen as recipes for herd improvement and profitability. But turning these hardy cattle onto the Forest was still practised, although the scant pasture and the expense of a 'looker' (if a young family member was not available) might not repay the effort.[77]

With the expansion of the rail network, many farmers turned instead to the commercial production of milk, butter and even cheese. The Sussex breed, a poorer milker, could not easily compete in the subsequent enthusiasm for dairy farming. Robert Crawfurd at Saint Hill was purchasing polled Suffolks by 1851 as 'the dairy as well as the shambles is to be considered'.[78] In part this development was a local reaction to changing consumer tastes, but it was hastened by the widespread agricultural depression that gripped British farming from the 1870s for perhaps 30 years. Now cereal growers turned their land more to grass as urban consumers sought more dairy produce and from cleaner sources than the old town dairies. By the 1880s small Ashdown farms further increased their small milking herds, although often enduring poor accommodation. The trend was enhanced by a creamery at Sheffield Park and a cheese factory at East Grinstead.[79]

The effect of the late Victorian depression on Ashdown can be gauged by a vastly reduced area of cereals and decreases in hop production, but conversely a great increase in permanent pasture, particularly grazed by the increased dairy herds (Table 8.8). By the late 1880s the farmland around Ashdown had between 65 and 73 per cent of its area under permanent grass.[80] But the depression was not perhaps felt to the same extent here as elsewhere. Many relied already on grass and dairying, and to some extent on poultry rearing or fattening (see below). Families subsisted or earned modest incomes from relatively small acreages. Numbers of sheep fell away, as they were anyway supposedly precluded from Ashdown itself, and a decline in demand for mutton and for fat cattle meant that what pasture was available tended to be reserved for dairy cattle.

76 *Sussex Agricultural Express*, 13 December 1851, 3.

77 The Sussex Cattle Society remains operational. Additionally, Earl De la Warr is attempting to initiate a separate club to save the breed from extinction (*Daily Telegraph*, 25 January 2020).

78 R. Crawfurd, 'Farming economics series – the cow yard', *Sussex Agricultural Express*, 3 May 1851, 5. The 'shambles' was a butcher's slaughterhouse.

79 W.F. Ingram, 'Agriculture', *VCHSx* 2 (1907), 277; R.H.B. Jesse, 'Co-operative Cheesemaking Centres in Sussex', *Journal of the Ministry of Agriculture*, 30 (1923), 48.

80 Short, 'Agriculture in the High Weald', 133. The government began collecting agricultural data in 1866 but before 1917 the returns were voluntary, with any defects added by estimates.

Table 8.8. Changes in acreages and stock numbers around Ashdown 1870–1913.

	Wheat			**Permanent pasture**		
	1870	1900	1913	1870	1900	1913
Buxted	730	254	124	1740	3552	2953
East Grinstead	871	364	256	5282	6761	7373
Fletching	583	371	405	2282	2150	2383
Hartfield	599	280	219	1835	3848	4483
Horsted Keynes	403	121	128	909	1387	1891
Maresfield	364	165	107	1187	1987	2271
Rotherfield	1016	369	323	3947	6960	7428
West Hoathly	444	149	113	1385	2339	2621
Withyham	664	238	163	2182	3780	3594
Danehill		87	84		1085	1192
Total	5674	2398	1922	20749	33849	36189

	Cows			**Sheep**		
	1870	1900	1913	1870	1900	1913
Buxted	372	504	391	1188	685	467
East Grinstead	457	741	668	2999	2182	1260
Fletching	397	415	375	1456	876	889
Hartfield	299	373	516	1583	938	1887
Horsted Keynes	152	154	280	94	177	238
Maresfield	237	303	372	660	697	639
Rotherfield	522	870	736	1297	2070	995
West Hoathly	182	303	309	1450	720	164
Withyham	284	331	313	2256	1020	1379
Danehill		195	214		101	49
Total	2902	4189	4174	12983	9466	7967

Source: TNA, MAF 68/262; 1858; 2599 parish summaries.

The railways had also revolutionised the older practice of chicken rearing and 'cramming' (force feeding). Local higglers in their light carts had previously conveyed poultry to market but their role now shrank to conveying chicken crates to the railway stations for dispatch to London. The practice of 'cramming' had originated by the 1830s, and by the 1860s had centred around Heathfield, whence chickens were sent via Leadenhall or Smithfield to London's tables. By the 1870s, if not earlier, small-scale producers around Ashdown, especially on the eastern fringes nearer Heathfield, might therefore extend their holdings, with or without permission, for a chicken run, or perhaps keep a few coops on roadside verges.[81] When Henry Rider Haggard visited the Crowborough area in the final months of Victoria's reign he was impressed by the extent of the chicken-fattening industry, but his overall conclusions were significant: 'Were it not for the fowl industry, and for the fact that rich men from London occupy large houses which absorb much produce at a good price, it would go very hardly both with tenant and with landlords.'[82]

In general though, agricultural changes would be considered only if they could be accommodated within Ashdown's social and economic system and were applicable to small family farms, primarily under grass. Dairying, poultry and perhaps fruit cultivation might fit but any changes here were 'folk-led' and many nineteenth-century writers missed such key pointers, instead dismissing the backwardness of Wealden farming. James Caird wrote in 1850 of the small Wealden farms: ill-drained land, half-cultivated, inadequately stocked, with a great abundance of wood and hedgerows and with farmers 'unskilful and prejudiced in their methods of cultivation'.[83] The French visitor Léonce de Lavergne saw the Weald in the 1850s as 'nothing beyond the average', where farmers were 'men without capital and as ignorant as they are poor'. Indeed, 'in passing through the Weald one might fancy they were in one of our second-rate provinces'. Siday Hawes had concluded in 1848 that 'Nothing except the force of habit would make him [the Wealden farmer] endure such nuisances.'[84] Certainly farmers around Ashdown were overwhelmingly local and it seems that few incoming farmers coveted the poorer soils. Of 57 census entries for 'farmers' in the parishes of Maresfield, Buxted, Withyham and East Grinstead in 1851, for example, 31 were farming in their parish of birth, 12 were from adjoining parishes and four more were from the next-but-one away. Only six were from outside the High Weald.[85]

CHANGING VICTORIAN VALUES

The impact of the railway was also felt in other directions. Cultural appreciation grew and there was a notable change of tone: the combination of wild heathland and patchy woodland on undulating topography was now an object of aesthetic interest, and attractive for middle-class incomers. Not everyone was impressed. Mark Anthony Lower, for example, wrote in 1870, with disregard for the actual topographical history of Ashdown, that 'It has, like the other five Sussex forests, succumbed to the hewer of wood and the tiller of the soil, so that with the exception of a few patches of forest land here

81 B. Short, '"The Art and Craft of Chicken Cramming": Poultry in the Weald of Sussex 1850–1950', *Agricultural History Review*, 30 (1982), 17–30.

82 H. Rider Haggard, *Rural England: Being an Account of Agricultural and Social Researches carried out in the Years 1901 and 1902*, 2nd edn (London: Longmans, 1906), vol. 1, 134–5.

83 J. Caird, *English Agriculture in 1850–51* (London : Longman, Brown, Green, and Longmans, 1852), 126–7.

84 Lavergne, *Rural Economy*, 203; S. Hawes, 'Notes on the Wealden Clay of Sussex and on its Cultivation', *Journal of the Royal Agricultural Society of England*, 19 (1858), 194.

85 Short, 'Agriculture in the High Weald', 110–18.

and there, it, with its neighbours, has lost its distinctive character of a forest altogether.' However, he did admit that the Forest 'possesses much to gratify the geologist, the botanist and the lover of the picturesque', and that the view from Pippingford commands 'wild and romantic scenery'.[86] This was exactly the point: the Sussex countryside now came to be seen as a pastoral foil to the bustle and smoke of London. For artists the Weald represented a vast open-air studio, and for geologists and botanists an open-air laboratory. Landscape gardeners and architects, such as Gertrude Jekyll or William Robinson, also became receptive to this Wealden atmosphere, creating woodland 'wild gardens' or 'cottage' gardens; Robinson's garden at Gravetye, where he made his home from 1884, used vernacular architecture as an inspiration.

Towards the end of the nineteenth century these artistic trends began to encompass not only wild heathlands, woodland and isolation but also a yearning for a lost Arcadian world wrapped in late Victorian sentimentality, as illustrated, for example, in the cosy cottage world of Helen Allingham.[87] Lt Col. (later General) Francis Davies, an army officer living at Danehurst, Fletching, compiled a notebook *c.* 1845, listing examples of Sussex dialect.[88] The collectors of Sussexiana multiplied: in 1907 Anne Gilchrist noted down songs from Withyham while staying with her brother, the Presbyterian minister at Blackham, collecting songs, for example, from William 'Billy' Wickham.[89] And at this time Vaughan Williams, back from a visit to the Weald, could say to an audience: 'I am like a psychical researcher who has actually seen a ghost, for I have been among the more primitive people of England and have noted down their songs ... '.[90]

Although primarily referring to early modern Europe, Peter Burke's description is apposite: 'Craftsmen and peasants were no doubt surprised to find their homes invaded by men and women in middle-class clothes and accents who insisted that they sing traditional songs or tell traditional stories.'[91] A visitor to Rotherfield and Crowborough extolled the area's beauty and history in an article in *The Gentleman's Magazine* in February 1904 entitled 'Gossip in the Sussex Oberland', where the 'lower orders' might still exhibit some of their 'primitive gaucherie'.[92] There remained, wrote the author, a 'fidelity to environment' and a 'rusticity' that the Fermor school had failed to altogether eliminate because of the mud, hills, bogs and forest.

86 M.A. Lower, *A Compendious History of Sussex* (Lewes: Geo. P. Bacon, 1870), Vol. 1, 21; Vol. 2, 42.

87 P. Brandon, 'Wealden Nature and the Role of London in the Nineteenth-century Artistic Imagination', *Journal of Historical Geography*, 10 (1984), 53–74.

88 ESBHRO, ACC 10360/2.

89 See L.A. Wolz, 'Resources in the Vaughan Williams Memorial Library: the Anne Geddes Gilchrist Memorial Collection', *Folk Music Journal*, 8 (2005), 619–39.

90 V. Gammon, 'Folk Song Collecting in Sussex and Surrey, 1843–1914', *History Workshop Journal*, 10 (1980), 61–89; and V. Gammon, 'English Folk Song Collectors and the Idea of the Peasant', *Folk Music Journal*, 11 (2016), 44–66.

91 P. Burke, *Popular Culture in Early-modern Europe*, 3rd edn (London–New York: Routledge, 2016), 23.

92 W. Connor Sydney, 'Gossip in the Sussex Oberland', *The Gentleman's Magazine*, 296 (February 1904), 114–31.

This 'discovery' of an older lifestyle set among imposing 'natural' scenery brought newcomers, searching for 'Old England', which offered landscape beauty but also remnants of older rural cultures, a 'cultural primitivism'. Spurred on by the likes of Tennyson's *English Idyls* and other pastoral writings, they sought rest, comfort and spiritual enrichment while remaining close to the business and artistic ferment of London. Ashdown's byways were also visited by hardy country walkers, such as Sir Leslie Stephens, author and mountaineer, with his Sunday Tramps walking group of intellectuals.

A fair scattering of artists was to be found around Ashdown, although few at this time could be placed in the first rank, and no artists' colony arose here.[93] The interaction of Forest society and environment was instead caught in November 1852 by Thomas Pentecost (1792–1869), 'Uckfield's postman-poet'. Only coming to poetry in his sixties, much of his work, published as a collection in 1856, is too sentimental for modern tastes, but his 'Ashdown Forest in Sussex' is of interest as it does show something of the attitude of an Uckfield resident, which differs from that of many gentry:

> Seen is many a clay-built cabin,
> Thatch'd with brake or heathy stub,
> For chimney-top, half-anker tub;
> One of those kegs that find their way
> By stealth across the bounding spray;
> Nor rare to see a score of such
> In many a turf-built, lonely hutch …
>
> A heathy waste of huts and dens,
> Where human nature seldom mends;
> No polish'd schools, but turfy fires,
> Afar away from village spires;
> And should the chime of sabbath bells
> E'er reach these dark, benighted hills,
> Few heed the call – here all is wild …

However, he goes on to hope that Ashdown will:

> Be turn'd to cultivated soil:
> Where science holds, and order rules …
>
> No more a rude and darken'd scene,
> But turn'd to waving corn-fields green …
>
> And cottage homes and villas, wide
> Seen scatter'd up the mountain's side!

93 Brandon and Short, *The South East*, 336–59.

> Then, O my harp o' the southern winds,
> Breathe thanks to earls and noble friends,
> And to all who forward reformation
> O'er this dark spot of desolation.[94]

He also wrote poetry concerning Crowborough Beacon and Buxted Park, 'seat of the late Earl of Liverpool', and remarked on Queen Victoria's last visit in 1845. Despite his views on the future of Ashdown he also penned laments for the enclosures of Horley and Newick commons as well as commenting on topical issues such as the Crimean War and the Charge of the Light Brigade.

There was certainly a transformation at Crowborough. By 1870 Lower thought that 'Probably no spot within the county was, a century or two ago, in so uncivilised a condition', but, thanks to the good offices of the Fermor charity and school and the local chapel, 'this wild district' had been rescued 'from Heathendom to a more Christian condition of things ... but for wild and legendary interest it cannot be excelled in this part of England'.[95] The writer Richard Jefferies (1848–1887) came to Jarvis Brook and the 'The Downs' in Crowborough in 1885 for its 'fine invigorating air' during his final years before moving to Goring-by-Sea, where he died of TB. In an essay published posthumously, 'Hours of Spring' (1889), he described Crowborough's countryside, its wildlife and the bitter winter of 1885–6, where 'at a height of seven hundred feet the air was sharp as a scythe ... yet the forest was still beautiful'. People came to his door during that cold winter: a cottager with his spade asking if he could dig the garden to earn a little money; an old labourer trying to avoid the workhouse given bread and cheese; a gypsy woman selling heather brooms whose daughter had recently given birth to twins in a tent on the Forest.[96] So, a fleeting glimpse of Ashdown:

> a thousand acres of purple heath sloping southwards to the sun, deep valleys of dark heather ... you may walk, or you may ride, for hours and meet no-one; and if black game were to start up it would not surprise you in the least ... . The Forest is still in everyone's mouth 'on the Forest', 'by the Forest' in it or 'over' it. Here and there a house ... the deep Weald underneath ... .[97]

Crowborough was now expanding on the edges of Rotherfield and Withyham. The opening of the railway station here in 1868 reflected its growing attractiveness, a trend further stimulated by Dr Leeson Prince's book (1885) praising its fresh air and

---

94 Thomas Pentecost, 'Ashdown Forest, Sussex', in *Harp of Aeolus: Fugitive Poems* (London: Edward West, 1856), 274–8.

95 Lower, *A Compendious History*, Vol. 2, 125.

96 Typescript in author's possession: W. Salkeld, 'Richard Jefferies in Crowborough', *Journal of the Crowborough Field Society*, 1 (1973). And see S.J. Looker (ed.), *Field and Hedgerow: Being the Last Essays of Richard Jefferies* (London–Redhill: Lutterworth Press, 1948), 19–37. 'Hours of Spring', first printed in *Longmans' Magazine* (May 1886).

97 Cited in P. Freeland, 'People who knew Ashdown Forest: (John) Richard Jefferies (1848–1887)', *Ashdown Forest News* (Spring/Summer 2017), 4–6.

views. Leeson Prince (1822–1899) had moved from Uckfield to Crowborough High Street in 1872 and later to the Observatory, where he undertook meticulous weather observations. In the second edition of his book in 1898 he remarked that 'a great transformation has come over the scene ... of this favoured health resort'.[98] He remarked that 'preparations for gas illumination are progressing' and now houses, hotels and services multiplied to accommodate new residents and visitors, some enjoying shooting weekends and many arriving for healthy holidays. The High Street and Cross areas were now developing, and Lucas commented:

> In the spring of this year (1903) the walls and fences of Crowborough were covered with the placards of a firm of estate agents describing the neighbourhood ... as 'Scotland in Sussex'. The simile may be true of the Ashdown Forest side of the Beacon ... but 'Hampstead in Sussex' would be a more accurate description of Crowborough proper. Never was a fine remote hill so be-villa'd. The east slope is all scaffold poles and heaps of bricks, new churches and chapels are sprouting, and the many hoardings announce that Follies, Pierrots, or conjurers are continually imminent. Crowborough itself has shops that would not disgrace Croydon, and a hotel where a Lord Mayor might feel at home. Houses in their own grounds are commoner than cottages, and near the summit the pegs of surveyors and the name-boards of avenues yet to be built testify to the charms which our Saxon Caledonia has already exerted.[99]

The Goldsmiths Company purchased farmland from Lord De la Warr on which to build select housing, based partly on the premise that this was a healthy environment. This became the Warren Estate, initially owned by James Ramsbotham (an enthusiast for an abortive light railway scheme to Nutley) but thereafter passing to the Goldsmiths Company to join that company's Beacon Hill Estate and several other properties, including the site of the imposing Beacon Hotel.[100] Such development was on Crowborough Common, actually part of the waste of Framfield Manor, which had been enclosed under an award of 1862. At that time just under 25 acres had been awarded to Walter Prideaux, acting as trustee for the Goldsmiths Company, and duly conveyed to the company itself in 1905, the year that Crowborough became a separate civil parish.[101] An issue arose in 1909: what rights on the Forest should be enjoyed by the Goldsmiths Company, which had held land here – the Hamsell and Birchden estate in Rotherfield and Withyham and land near Groombridge Station – since the 1860s?[102] It was agreed that the properties entitled to rights totalled 117¾ acres, which would henceforth be rated accordingly. By 1910 there were large houses on the higher ground of the Warren and Beacon Hill estates on roads with Scottish names, such as

98 C.L. Prince, *Observations upon the Topography and Climate of Crowborough Hill, Sussex* (Lewes: H. Wolfe, 1885), 2–3.

99 E.V. Lucas, *Highways and Byways in Sussex* (London: Macmillan, 1904), 301.

100 Opened in 1900, demolished 1956.

101 I am grateful to Geoffrey Mead for sight of a copy of the 1862 award map and title deeds.

102 London Metropolitan Archives, MS 10274–97, Calendar of the Records of the Hamsell and Birchden Estate, Sussex.

Glenmore, Aviemore and Rannoch. Smaller service cottages, terraces and businesses occupied the area to the south and east of the main road and towards the station and brick works of Jarvis Brook.[103]

In 1890 Boys Firmin had produced his *Guide to Crowborough*, heralding subsequent guides issued by the parish council. With a population in 1911 of over 5,000, industrial, mercantile and finance capital now arrived, together with that derived from colonial service. Estate agents and builders such as Connor Bros thrived. Of the 13 parish councillors at their first meeting in 1905, five were involved in the building industry and four were local tradesmen.[104]

But there were those who were far from enchanted: in 1911 Arthur Beckett agreed with Lucas:

> I had come across Crowborough many years ago, and this place had so charmed me that now, full of memory … its fine air, its beauteous landscape, the sense of peace pervading the village itself … in so doing I had not taken into account the tergivisation of modern man, who, turning his back upon things simple and beautiful, will beat his brains to find out how he may mar and undo what has been well done by nature … I now find upon reaching it [Crowborough is] undone, violated, raped and soiled by the makers of red bricks and mortar, and the builders of modern villas.[105]

Arthur Stanley Cooke, in his Edwardian walking tours: wrote that 'Crowborough will not detain you long in passing … it is merely suburban … . The surroundings, of course, are charming, and the air superb.'[106]

East Grinstead, rather more plebian, did not share in the visitor boom to the same extent as Crowborough. But during the early 1880s 90 acres of the adjacent Sackville Estate were sold, offering 'pure bracing air, beautiful scenery, pleasant drives and walks, near proximity and good access to London, also to the favourable watering places of Brighton, Tunbridge Wells and Eastbourne'. Speculative development produced rows of terraces and artisans' homes, together with commuter houses. There were also scattered larger houses, but little Sussex vernacular was now to be found.[107]

The twin themes of accessibility and health were certainly emphasised. When the Glen Andred estate, comprising the house (originally called Rockhurst) and 37 acres of Old Birchden Farm (Withyham), was sold following the death of the marine artist

---

[103] See R.B. Harris, *Crowborough: Historic Character Assessment Report* (Sussex Extensive Urban Survey, 2008), <www.westsussex.gov.uk/media/2105/crowborough_eus_report_maps.pdf> [accessed 17 July 2019].

[104] A.J. Fielding, *Place, Work and Folk: a Study of the Development of Crowborough, Sussex*, Univ. of Sussex, Research Papers in Geography 5 (Brighton, 1981), 17. The photographer Ambrose Stickells captured much of the life of Crowborough before the First World War: see Payne and Batchelor, *Bygone Crowborough*.

[105] A. Beckett, *The Wonderful Weald and the Quest of the Crock of Gold* (London: Mills & Boon, 1911), 289.

[106] A.S. Cooke, *Off the Beaten Track in Sussex: Sketches, Literary and Artistic* (Hove: Combridge's, 1911), 346.

[107] J. Lowerson, 'Building on East Grinstead Common', *Bulletin East Grinstead Society*, 27 (1979), 5–8.

and geologist E.W. Cooke in January 1880, the sale catalogue emphasised two things: that it was just one mile from Groombridge Station and that 'the soil is sandy and dry, rendering the property extremely healthy and invigorating'.

The house had been designed for Cooke in 1866–7 by Norman Shaw in his domestic 'Old English' style, and the garden is now seen as an important late Victorian example. Cooke, moving from Kensington, had been keen to seek 'a pleasant spot where I can have a garden and grow conifers and roses to my heart's content without London smoke'. He described his house to William Gladstone, its name Glen Andred echoing an earlier Weald, as 'a thoro' Sussex or rustic Elizabethan structure surrounded by wonderful masses of rock and woods of oak, fir, holly and yew [...] a rapid brilliant trout stream running through'.[108] Cooke fashioned gorges, boulders, a rhododendron dell, pools, native and exotic trees – including bamboos, weeping beech (*Fagus sylvatica pendula*) and Wellingtonia (*Sequoiadendron giganteum*) – accompanied by glasshouses, a fernery and a pineapple garden. The conifers, especially Scots pine, were by this time becoming an iconic element of heathland landscapes, planted in clumps and rapidly self-seeding across the commonland.[109]

We have entered the era of the *Ferme Ornée*: small farms were now thought ideal for purchase by retired men or those whose occupations were outside farming and might take them to London for business. Cottages, now often referred to as 'pleasure farms', were sold and many were subsequently redesigned and enlarged. When the Saint Hill estate came up for sale in 1872 Holly Bush Farm was described as 'a beautiful pleasure farm' with 'fertile park-like meadows'; and, similarly, when the Sackville House estate (Buxted) was up for sale in 1880 the Rock Farm, of 45 acres, was advertised as a 'freehold pleasure farm' with a trout stream.[110] Finally, in 1884 the particulars for The Hall (Rotherfield) described it as '[comprising] every convenience for modern and profitable farming, and is admirably adapted for any gentleman wishing to combine business with pleasure, as the estate offers excellent shooting and other sporting facilities'.

Many further examples of the parvenu, City financiers, merchants and literary men arriving on the borders of Ashdown could be given, and some have been set out by Brandon.[111] The poet Coventry Patmore (1823–1896) married Marianne Byles in 1865 and converted to Catholicism. With his new wife's money he purchased two farms, Puxtys and Oldlands, on the south-western edge of Ashdown from the earl of Abergavenny for £14,000 in 1866, renaming the estate Heron's Ghyll in 1869. He turned the land into gardens, terraces, woodland, ponds and a small park. Peat was carted from Ashdown for the rhododendrons and stone for the rebuilding and even for

108 J. Munday, *Edward William Cooke 1811–80: a Man of his Time* (Woodbridge: Boydell, 1996), 253–366; and J. Munday, *E.W. Cooke, R.A., F.R.S.: His Life and Work* (Woodbridge: Antique Collectors' Club, *c.* 1996); <https://historicengland.org.uk/listing/thelist/listentry/1001709> [accessed 19 July 2020]; the sale catalogue is at BL, Maps 137.

109 For the 'rush to the pine woods' see P. Brandon, 'The Diffusion of Designed Landscapes in South-east England', in H.S.A. Fox and R.A. Butlin (eds), *Change in the Countryside: Essays on Rural England, 1500–1900*, Institute of British Geographers Special Publication 10 (London, 1979), 165–87.

110 Brandon, 'Diffusion', 179; BL, Maps 137 b.10 (11).

111 Brandon, *Kent and Sussex Weald*, 217–52.

sale to neighbours. Here he lived with his family from 1868 to 1874, opening a private Catholic chapel before retiring to Hastings.[112] Following his departure, the Puxtys part was finally sold in 1879 to the duke of Norfolk as a dower estate, and the 1881 census finds Augusta Marie *Minna* (née Lyons), the widow of Henry Howard, fourteenth duke of Norfolk, living here at Heron's Ghyll, aged 59, with three unmarried daughters, four grandchildren, a priest and 16 servants, as well as three more living adjacently by the stables or porter's lodge. She had a school and chapel built in 1880 and established a Catholic mission and a temporary iron church before her death in 1886.[113]

The newcomers did not always find the local people easy to understand. Some indication of Patmore's strained relations with his tenant John Minns, who he does not name, is given in his own account:

> I made acquaintance with my principal tenant one Sunday morning. I found him mending a wheel in his cart-shed. He had inherited the tenancy of the farm from his father, and held it at a rent of considerably less than half its value, as I found from offers which were made to me by persons in the neighbourhood, who rightly guessed that I should soon have to turn him out of his tenancy. It was the old story. Having to pay a ridiculously low rent, he had found himself unable to pay any, and had for years satisfied his late invisible landlord by sending in annually a bill for fencing and ditching the woods which amounted to or exceeded his nominal rent.[114]

The relations between Minns and Patmore were clearly less than cordial. Minns left the property on Patmore's arrival, being returned in the 1871 census as an agricultural labourer rather than a farmer. Patmore was also displeased with other 'inconvenient neighbours', the squatters on long-established encroachments who raided his woodland for 'dogwood' and the poachers and wood stealers who paid their fines or went to prison, although he admitted that in their dealings with the gentry there was nothing unconscientious or underhand.[115]

Minns returned to work for Patmore's successor, the scholarly Alexander Nesbitt (1817–1886), from a family of wealthy merchants and bankers. Moving initially to Kidbrooke, and then to other properties in Forest Row, he finally arrived at Oldlands,

---

112 J. Wright, 'Coventry Patmore in Sussex', *SCM*, 21 (1947), 24–7.

113 The present Catholic church of St John the Evangelist dates from 1896, after Minna's grandson, James Hope, bought the site from the Duchy of Norfolk. It should not be confused with the 1839 Anglo-Catholic church with the same dedication at Crowborough (see pp. 226–8). Coventry Patmore, *How I managed and improved my estate* (London: George Bell and Sons, 1886); R. Penn, *Portrait of Ashdown Forest* (London: Robert Hale, 1984), 48, 136; Arundel Castle Archives, Post-medieval deeds Buxted, D3127 (1880–92).

114 Patmore, *My estate*, 7. John Minns' father was Benjamin, former tenant.

115 Dogwood: underwood cut as faggots and used for making gunpowder. Harry Walter remembered the collection of dogwood by women, children and sometimes men when out of other work; it was taken to the dogwood yard at Fairwarp, where the bark was stripped from small sticks and sent to Faversham for the gunpowder factories there. The scraping of the bundles was remembered as hard, dirty and poorly paid work for women (Walter, 'Some notable happenings', 4).

purchased from his acquaintance Patmore in 1869. The new Oldlands Hall, designed by Matthew Digby Wyatt, was built by 1870 with sweeping views of the distant South Downs. Part of the original Oldlands Farm was converted into a 30-acre park, complete with imitation Highland glen. A paper that he wrote for publication in the *Sussex Archaeological Collections* 1883 was actually a critique of the so-called 'restoration' of churches ongoing at that time.[116] He died at Oldlands in June 1886.

VICTORIAN ASHDOWN

The period after 1830 saw several themes emerging and coalescing on and around Ashdown. Class resentment had emerged overtly after the Napoleonic Wars, most notably during 'Swing' and prolonged by enmity towards the New Poor Law. Furthermore, agriculture was hardly profitable and subject to marked fluctuations, with employment on any continuous basis consequently difficult for many labourers and cottagers.

But the area was now 'discovered' and gentrified by wealthy incomers looking to escape London. The *nouveaux riches* sought style and status through country living and new money easily found its way to the fringes of Ashdown, just as it did to the large houses of Belgravia and the West End.[117] Here was an excellent example of the declining interest towards British manufacturing by entrepreneurs who had made their fortunes, but who now, rather than reinvesting their money in industry, instead sought country living, large country houses and cheap farmland. For example, by 1849 the Fielden family, having made a fortune through their Todmorden cotton spinning empire, had purchased Crowborough Warren. This contrast between older attitudes and ways of life and the wealthy newcomers, 'the collision between old and new' is brilliantly portrayed in John Halsham's *Idlehurst* (1898), located in the 'deep lane country' near Ashdown around Lindfield and Haywards Heath.[118] Putting these trends together gives ample evidence that Ashdown's character was changing.

Arch Sherlock, who worked at Birchgrove, later drew a contrast:

> Some of them were gentlemen. Some were real people: others were nothing than bloody gasbags. And that's what the trouble with it was. Those that came up from nothing and made theirself are the worst people there are. The people that are landed gentry are the best people there are ... Not hardly any of 'em left.[119]

116 J.H. Stevenson, 'Alexander Nesbitt, a Sussex Antiquary, and the Oldlands Estate', *SAC*, 137 (1999), 161–73; A. Nesbitt, 'Thoughts on Church "Restoration"', *SAC*, 33 (1883), 225–36. No Ashdown churches were mentioned.

117 For the 'discovery' of the land see J. Marsh, *Back to the Land: the Pastoral Impulse in Victorian England from 1880 to 1914* (London: Quartet Books, 1982); M.J. Wiener, *English Culture and the Decline of the Industrial Spirit 1850–1980* (Cambridge: Cambridge University Press, 1981) and J. Mordaunt Crook, *The Rise of the Nouveaux Riches* (London: John Murray, 1999), 12–13.

118 P. Brandon, '"John Halsham": the Perfect Countryman', *SAC*, 148 (2010), 213–24. Halsham was the pseudonym of George Forrester Scott (1863–1937). See his *Idlehurst: a Journal kept in the Country* (London: Smith, Elder & Co., 1898), quotation from p. 4.

119 R. Hall, *I Never Played to many Posh Dances: Scan Tester, Sussex Musician, 1887–1972* (Rochford: Musical Traditions, 1990), 127.

CHAPTER 9

THE ASHDOWN FOREST DISPUTE

Custom is the most powerful protector of the weak against the strong … .[1]

THE PRECEDING CHAPTERS HAVE outlined how Ashdown's Commoners had on many occasions asserted their customary rights against those who threatened to enclose land, transgressed local customs regarding cattle numbers or cut and carried away the growth of the Forest to which they were not entitled. These sanctions were supposedly applied through the manorial courts, but, failing this, the Commoners might take matters into their own hands. This chapter relates how local custom encountered attempted legal suppression and how alliances and conflicts within the several strands of Ashdown's agrarian society were enacted.

ENVIRONMENTAL POLITICS

By the late nineteenth century the preservation of open spaces was becoming of greater concern among educated Victorians. It is often difficult to map the impact of such concerns onto particular places and across the ambiguities of rural class relations, and it is perhaps particularly difficult where forest and wood-pasture environments are concerned. Here semi-independent families, moving between subsistence and dependence on common rights, and with a marginal hold on capitalist relations through their paid labour, could be in more complex relationships with their elite neighbours. '[A]ntagonisms between labourers and landholders [were] … nowhere adequately explicable in terms of a simple capitalist–proletarian opposition',[2] so it is important to move beyond the elite perceptions of ruralist commentators in order to reveal the complexities and inter-relationships of culture, work and environment.

Ashdown was the scene of a struggle between intersecting concerns. It was the local stage for the promotion of environmental advocacy, expressed as the burgeoning conservation and 'outdoors' movements. But it was also the stage for local conflicts over the preservation of customary common rights in the face of private property

1 J.S. Mill, *Principles of Political Economy* (London: J.W. Parker, 1848; reprinted University of Toronto Press, Routledge and Kegan Paul, 1965), 240.

2 P. Joyce, 'Work', in F.M.L. Thompson (ed.), *The Cambridge Social History of Britain 1750–1950*, Vol. 2 *People and their Environment* (Cambridge: Cambridge University Press, 1990), 152–3.

interests. We can see the conflicts from several different perspectives, and now particularly from the autobiographical accounts of working people whose lives and collective practice were woven together across the Forest. These stories are drawn from the notebooks in which they were collated by William Augustus Raper, a young solicitor who set out to reveal the longevity of the exercising of common rights on Ashdown, and about whom more detail will be given below.

The Ashdown Forest conflict was a particular instance of a wider movement from the 1860s.[3] In 1851 the national population census recorded that, for the first time, a slight majority of the British population lived in towns. Such news was greeted by some with alarm, as many urban areas were already seen by public and philanthropic groups as unhealthy environments for mind, body and morality. Fresh air and space for recreation was needed to provide what was hoped to be civilising influences.[4]

Thus in 1865 the Committee on London Commons was formed to enquire into the preservation for the public of the forests, commons and open spaces within 15 miles of London. In the same year the Commons Preservation Society (CPS) was formed to protect the London commons in their 'wild, free and open state', and in the following year the Metropolitan Commons Act was passed.[5] The CPS aimed to enhance appreciation of the commons and wood pasture spaces for public leisure, rather than for their economic value to Commoners or to lords of the manor, who might seek enclosure and/or building development. To some extent this aim was helped by the lessened agricultural interest in reclaiming such marginal soils, as farming by the 1870s was entering a period of depressed prices for grain, followed in the 1890s for cattle, when costs of reclamation might be uneconomic.

During the 1860s, in the first case involving the CPS, there were struggles over open land on Hampstead Heath, where threats to build on the heath by Sir Thomas Maryon Wilson, lord of the manor of Hampstead, were interrupted by his death in 1868 and the transfer by his successor and nephew Sir Spencer Maryon Wilson (1829–1897) of his manorial rights to the Metropolitan Board (predecessor of the London County Council). He later sold a smaller area of land (60 acres) adjoining the heath in 1889. Interestingly, Sir Spencer was also a Commoner on Ashdown, becoming a legal adversary of the seventh Earl De la Warr concerning his own rights on the Forest in

3 The following section owes much to the revised edn of Lord Eversley's *Commons, Forests and Footpaths: the Story of the Battle during the last Forty-five years for Public rights over the Commons, Forests and Footpaths of England and Wales* (London: Cassell & Co., 1910). The first edition, *English Commons and Forests*, was published in 1894 under the author's original name of G.J. Shaw-Lefevre.

4 For environmental issues surrounding the growth of London at this time see B. Luckin and P. Thorsheim (eds), *A Mighty Capital Under Threat: The Environmental History of London, 1800–2000* (Pittsburgh, PA: University of Pittsburgh Press, 2020).

5 The CPS, the earliest national amenity body, now the Open Spaces Society, began from a suggestion by Philip Henry Lawrence (later to work on the Ashdown Forest case) with founders including John Stuart Mill, Shaw-Lefevre, Sir Robert Hunter (hon. secretary and solicitor for the society) and Octavia Hill. And see M. Roberts, 'Gladstonian Liberalism and Environment Protection, 1865–76', *English Historical Review*, 128 (2013), 292–322.

the 1870s. The solicitor Henry Freshfield, then living in Hampstead but later to live at Kidbrooke, and first chairman of the Ashdown Forest conservators, was also active in the preservation of Hampstead Heath.

Meanwhile Commoners' rights were being reinforced legally against attempts to enclose the commons at Berkhamsted, Plumstead and Tooting Graveney.[6] Similar proceedings upheld Commoners' rights at Wimbledon, Wandsworth, Coulsdon, Dartford and Wigley (New Forest). In 1870 a motion was introduced to the House of Commons to preserve Epping Forest for public use, followed by the 1871 Epping Forest Act. Also in that year a motion to the Commons proposed that in the New Forest there should be no new enclosures or felling of timber, and Commoners gave evidence before a Select Committee about their way of life, much as Ashdown deponents were soon to do. A Royal Commission first reported on the matter in 1875, with a final report in 1877 that subsequently offered more protection to the New Forest's amenity value. Meanwhile, in 1876 the Commons Act was passed, allowing for the better regulation of forests and commons through representative local bodies of conservators.

This is the important context to what happened on Ashdown Forest, as was the 1880s' dispute over rights of access across Knole Park, the home of Mortimer Sackville-West (created Lord Sackville in 1876), younger brother of the seventh Earl De la Warr. Here was another example of the complexities of 'day-tripper' access from London, local usage and a search for privacy by Sackville. The latter had already been financially weakened by a costly lawsuit with his litigious older brother about the title to Knole.[7]

It was against this legal background that the Ashdown Forest case began. It ran more or less concurrently with similar cases at Banstead, which was finally settled in favour of the Commoners in 1889, and the Malvern Hills, where common rights, enclosure and encroachment were regulated under a special Act of 1884 giving control to a body of conservators.

Catching the mood of this national discourse on open spaces, Ashdown's scenic qualities were now described for wider audiences. In 1879 it was noted that:

> it is to be hoped that by aid of the facilities offered by recent legislation, and under the guidance of that love of open spaces as such, which has arisen of late years, as one of the results of our crowding in towns, and the hurry of our lives, some mode of protecting and enhancing the beauty of the district, with due regard to all legal rights, may be discovered. In the meantime, if a tired and smoke-dried citizen wishes for a few days' ramble in a wild, hilly, and open country, accompanied by copious draughts of the freshest possible breezes, he cannot do better than pay a visit to Ashdown Forest.[8]

---

6 B. Cowell, 'The Commons Preservation Society and the Campaign for Berkhamsted Common, 1866–70', *Rural History*, 13 (2002), 145–61.

7 D. Killingray, 'Rights, "Riot" and Ritual: The Knole Park Access Dispute, Sevenoaks, Kent, 1883–5', *Rural History*, 5 (1994), 63–79.

8 'A Fragment of the Great Forest', *Spectator*, 15 February 1879, 207–8.

The important Decree and Award of 1693 had attempted to define the exact pattern of common land and common rights, but the precise nature of these rights remained an ongoing source of conflict between the lords of the manor of Duddleswell and the large body of Commoners, including in their number many representatives of the gentry, whose properties had acquired rights over several generations.

But many of these properties were also the small subsistence holdings that were, with the isolated cottages or hamlets, scattered within and around the Forest pale, and in which kinship, friendship and working links were maintained. They were the 'human fauna' of their land: '[They] lived with its seasons, they knew its history and its geography, they felt a sense of belonging in the routines of every day spent on it, not as an individual, but one of a tribe. The myriad small exchanges created "bonds of obligation"'.[9] The attitudes of different manorial lords to such small enclosures varied. They had been created at a fast pace in the 1820s and 1830s and again in the 1870s and 1880s, as the sales of recent encroachments after Ashdown's regulation in 1885 were later to testify. Capt. William Noble of Forest Lodge complained of a six-acre enclosure by William Turner adjacent to Turner's nursery at Nutley: 'as barefaced and impudent a piece of thieving as ever I saw'. In the past, the lord's men had levelled attempted fencing around the same spot.[10]

Many such encroachments were for gardens, chicken runs or dwellings, and the presence of squatters near Forest Row was blamed for the pollution of streams and an outbreak of fever in 1883.[11] A ranger noted in 1879 that 'Alfred Chapman is rebuilding a shed which was thrown down last year, and is also enclosing land. John Hobbs who lives in the Forge is adding to his garden, and Tidy is still enclosing land, though that is only what I have heard.'[12]

LITTER AND THE FOREST RESOURCES

Although championed by the growing open space movement, the Ashdown Forest Case essentially revolved around the intermingling of manorial rights, custom and common rights, which had been the basis, as we have seen, for ongoing disputes within the pale.

Much was focused around rights to litter. This, as we have seen, was the term used to encompass a number of species, but *Pteridium* – referred to as brake, fern or bracken, was dominant, together with heather and common and dwarf gorse (referred to as gorse or furze), plus grasses, herbaceous plants and tree seedlings.

9 George Bourne [George Sturt], *Change in the Village* (New York: G.H. Doran Company, 1912), 1–18.

10 Conservators' minute books 1887–1993 available as 12 separate transcriptions at: <www.ashdownforest.org/enjoy/history/docs/Ashdown_Forest_Historical_Minute_Book_1887_1897.pdf> [accessed 12 June 2018]. Thanks to Chris Marrable and Sharon Shipp for this work.

11 *Sussex Agricultural Express*, 24 November 1883.

12 Christian, *Ashdown Forest*, 3. John Hobbs accidentally drowned near Newick in June 1886, after leaving the Sheffield Arms late one night. I am grateful to Colin Hobbs for information on the Hobbs family.

Figure 9.1. Litter collecting on Ashdown (source: Nutley Historical Society).

Litter taken from the open Forest tended by the nineteenth century to consist more predominantly of furze and to be shorter than that taken from private land. It was cut, using a shorter and broader scythe than normal, during the months from October to January, following the corn harvest and hop picking. If the litter was mainly bracken (referred to as fine litter, as opposed to coarse litter, which had a high heather content) it might be cut with a hook, knife or sickle from August onwards, often by women and children. Benjamin Bridger, when aged five or six, 'used to go out on the Forest with a hook to cut fern as children do now being too young to cut litter'. Women and children raked and stacked litter and turned over turves to allow them to dry. One man would normally take between two and four days to cut a load, which would be raked and dried and then taken to his employer or sold to the highest bidder, at a price dependant on the ease with which the litter could be transported (Figure 9.1). A load for a two-bullock cart has been estimated at 11cwt 64lb, with 100 such loads being a typical amount cut for each farm – or about 56 tons of air-dried material.[13]

Most of the farms used litter extensively in their open cattle yards in winter, but it was also in demand for sheep, pigstyes and stables. Bracken was possibly even seen as better than straw, because it held less moisture, and straw was anyway far more expensive by the late 1870s because the agricultural depression saw much arable put down to grass, thereby rendering straw scarcer.[14] Some farmers might even sell straw to London buyers at good prices and use litter themselves instead. Bracken was preferred for yards under cover, but it might also be mixed with dung and road scrapings as a dressing for grassland. It was also used for covering charcoal during burning, heating ovens, firing bricks, burning limestone and a wide variety of commonplace uses as a

[13] Irons, 'Aspects of the Impact of Man', 101–9, 134–5; ESBHRO, ACC 3715/1–2.

[14] ESBHRO, ACC 3715/1, 3–4; Webb, *Heathlands*, 167; 'Ashdown Forest and Earl De La Warr', *The Echo*, 9 September 1878, 1, collected as a cutting in ESBHRO, AMS 4097.

quick burning fuel. Heather was used for thatching, covering drainage pipes, broom making or as the base of haystacks.

Gorse was used in covering stone-filled drains or for the filling of field drains subsequently covered over with clay or inverted turf which then consolidated, leaving a channel once the gorse decayed. The presence of a furze mill on the Forest also suggests that gorse was used as fodder, being crushed and mixed with turnips or cereals for cattle, a common practice in the New Forest, in Wales and on the Surrey heathlands.[15] It also served for stack and yard bottoms and for fuel, as in the case of the Chuck Hatch brickyard, or for domestic heating and cooking, where it gave out a fierce flame. It was also used to stop up hedges and on the bottoms and sides of stacks, clamps and turf piles to prevent cattle rubbing, as well as on the tops and sides of ricks as a bird and rat deterrent.

Turf, together with close-cut heather and ling, was dug one spit deep and dried in blocks before being carted away, mainly for burning, for potato clamps or in some cases even for building poor huts on encroachments. As supplies of wood became scarcer, so more turned to turf for fuel. But the stripping of turf was highly deleterious, as it removed soil nutrients, destroyed root systems, seed banks and soil fauna, and allowed the spread of weeds. Fears were expressed by the early nineteenth century that the Forest might indeed be 'skinned' by Commoners and non-Commoners alike.[16] As we have seen, many non-Commoners took turf: a list dated about 1830 named 83 who did so, and the practice seems to have continued unabated through to the 1870s. It was sold to nurserymen and for lawn making, as at Kidbrooke or Searles, but was mostly used for domestic fuel until lower coal prices rendered it less desirable.

John Gilham noted that 'We had turf until the last 6 or 7 years but now coal is so much cheaper we have not cared to trouble about it', a sentiment echoed too by Michael Maynard, who had not used turf for some years 'because it was so dirty and coal was better'.[17] Robert Turner of Tablehurst Farm, aged 70, concurred: 'I have always taken turf from the Forest for fuel till the last 3 or 4 years. I have not burnt it myself but my cottagers on the farm have had it for fuel. Lately the cottagers have not cared to use it as coal is so cheap and wages high.' By 1866 the railway could move coal between Tunbridge Wells and East Grinstead. The LBSCR was distributing coal from its Deptford wharf and by the mid-1860s was bringing Midlands coal around London to the Sussex area.[18]

Peat was cut and dried in blocks, again for domestic fuel. Martha Baker recalled that Colonel Young had peat off the Forest for his greenhouse fire at Holly Hill, and Bernard Hale, his successor, took 1,000 loads in the 1833–4 season. Again, the poor

---

15 Bourne, *Change in the Village*, 144.

16 Irons, 'Aspects of the Impact of Man', 110–14, 367. For the Wilson case see TNA, LRRO 67/1–3 Hewlett Record Agency: Ashdown Forest.

17 ESBHRO, ACC 3715/1; P. Warde, 'Woodland Fuel, Demand and Supply', in J. Langton and G. Jones (eds), *Forests and Chases of England and Wales c.1500 to c.1850* (Oxford: St John's College Research Centre 2010), 80–86. The former name of Clockhouse Lane, Nutley was Turfstack Lane, where there is a Turfstack Cottage at Forest Bank (Penn, *Portrait*, 134); Irons, 'Aspects of the Impact of Man', 292–9.

18 Farrant, 'Growth of communications', 81.

and small farmers were at various times allowed to purchase it, and some was being sold under licence in the 1830s. Peat was also used in estate gardens, especially for the growing of the popular rhododendrons. Peat digging appears not to have been extensive by the 1870s, although there were still cases of illegal peat cutting in the later 1880s. On 23 June 1888 it was reported that Joseph Ridley of Millbrook had been cutting peat for fuel. Leaf mould was also taken off the Forest for manure, and rushes might also be taken as a straw substitute in thatching or for baskets and mats.[19]

Other resources, such as clay used in brickmaking or sand and gravel used in road making, were of local importance. De la Warr owned a brickyard with two kilns near Chuck Hatch which was rented out, with furze and faggots for fuel carted from the area known as the Hundred Rough. The bricks were used locally or on the estate, but also went via Hartfield station for De la Warr's new development at Bexhill (see below). Brickyards also operated at Cackle Street and Horney Common in the nineteenth century, and a 25-ft clay seam was exploited at Nutley for a brickyard that was well-established by the 1870s, managed by William Turner and Sons the builders. Loam was also still used, as for centuries past, for covering the inside of walls and possibly for crowning ovens. With quarries for stone, such as that south of Marlpits, at Jumper's Town or near Gills Lap, and pits for marl, sand and clay, there were opportunities for associated employment and, indeed, exploitation appears to have been largely uncontrolled by the later Victorian period, producing an uneven Forest topography with old pits left unfilled.[20]

Finally, we might note that moss was collected for bedding or for sale to florists. Dung was spread around gardens or fields and the diet could be supplemented by nettles, berries, herbs, mushrooms, eggs and nuts. Wild plants might be gathered by gypsies for medicinal purposes. Flowers, such as marsh gentians, might be gathered and sold. Adders were caught and 'some people made a few bob... pin 'em down with a forked stick and kill 'em; split 'em open and hang 'em up, and catch the adder oil and sell it as a cure for deafness ... the chemist in Uckfield would buy the oil.'[21]

There was also the woodland. Undoubtedly much was surreptitiously cut but one addition to the stock of trees came in the form of conifer plantations. William Hobbs (1817–95) testified to Raper that:

> I heard 3 or 4 years ago of the present Lord D. planting fir trees. I never heard of any of his predecessors planting any. The clumps were planted before my time. When I was a boy there were no fir trees on the Forest except in the clumps, now there are a great many seedlings springing up.

In fact, De la Warr had planted 18 knolls and the spread of larch and pine became very obvious by 1879.[22]

---

19 Irons, 'Aspects of the Impact of Man', 115–6.

20 M. Beswick, *Brickmaking in Sussex: a History and Gazetteer* (Midhurst: Middleton, 1993).

21 Arthur, *A Sussex Life*, 89.

22 ESBHRO, ACC 3715/3; William Hobbs also declared that 'The Forest has been dug about a great deal more lately than it used to be for stone for building and for roads'. Short, *The Ashdown Forest Dispute*, 191.

All these resources were exploited to maximum advantage. Different groups still enjoyed different rights: the tenants of Maresfield and Duddleswell manors, for example, were permitted two loads of wood annually, fallen timber, brushwood, roofing ferns and mud for walls. In line with 'Levant et Couchant', pasture was also still normally allocated at rates commensurate with the stock overwintered on the holding.[23] Foreign tenants were still allowed fewer privileges, but meanwhile some landowners encroached onto the common land to protect their game and to recoup financial outlays by selling timber and underwood, house and kiln faggots, and hop poles.

Clearly, access to these Forest resources provoked internal tensions. Within the ranks of the Commoners themselves, landowners of some social standing might hold views quite different from those of their tenant Commoners, whose views in turn varied from those of the humbler cottagers and squatters. The protection of pastures by farmers did not sit well with turf and litter cutting, for example, and nor did the extension of quarrying or the quiet acceptance of encroachment in return for acknowledgement rents to the lord.

But all Commoners complained of the cutting and carting away of materials by those without common rights. There were certainly still many squatters and cottagers around the Forest edges illegally using the Forest. Captain George Greenland lived at Hartfield Grove from 1828 to 1841, and soon after arriving wrote:

> We the undersigned being entitled to rights of Common ... do hereby give you notice that if at any time hereafter you break up any part of the said forest or wastelands or otherwise injure the pasturage thereof or cut or carry away any turf soil mould heath litter or other produce thereof you will be prosecuted for so doing. Witness our hands the sixth day of May 1829.[24]

John Wheatley, interviewed by Raper, remembered:

> Captain Greenland living at Hartfield Grove. I used to work for him. I helped dig stone for him to build 2 cottages, with one by the stable and another which has since been pulled down. We dug some of it outside the Furnace and some up by the Four Acres. I paid nothing for digging it to the Lord. Captain Greenland used to have litter.[25]

With this level of use, tree regeneration and cover for game remained a matter of concern to manorial interests. The fifth Earl De la Warr, through an appointee, joined with others in drawing up a management plan for the Forest in 1830, whereby a committee of four Commoners and a nominee of the lord would control litter cutting, with four lookers enforcing the regulations. The committee was designated to guard

---

[23] Levancy and couchancy was the widely accepted procedure in England, although complicated by agistment. Instead stinting – a numerical agreed allowance of animals to be grazed – came to be used more widely and certainly on Ashdown by the nineteenth century.

[24] ESBHRO, U of Sx Spec Colls, AFC II, 140a.

[25] All the material and quotations used herein are taken from the transcription in Short, *Ashdown Forest Dispute*, unless otherwise noted.

Commoners' rights against trespassers and, perhaps significantly given the 'Swing' events of that year, to issue tickets for the poor to use the common. A full list of those with properties having common rights according to the 1693 decree was also drawn up. At the meeting in 1830 the Commoners agreed to a restricted cutting time between September and November, and that the committee should designate the areas to be cut. Small farmers situated on or within half a mile of the Forest pale without rights to cut litter were to be sold litter at 2s 6d per load.

The fifth earl seems to have taken a rather different view of litter cutting from his predecessors in recognising the customary rights of Commoners to cut litter to be used on their own holdings. The brief tenure of the sixth earl similarly saw no interference with litter cutting by those who were so entitled, but 'vagabonds, gypsies and tramps' were believed to carry out much of the litter cutting, for sale to any who would buy from them.[26] Every winter litter cutting, legal or not, continued. These shifts in the attitudes from Whitworth to the fifth and sixth earls and then again to the seventh earl may have created some confusion as to litter cutting entitlements. Changing ideas about custom served to show that it was not an ahistorical structure rigidly adhered to by all, but a changing and constantly challenged set of ideas.

George Edwards, the reeve, wrote at one point in the early 1870s:

> considering the great extent of the waste lands it is exceedingly difficult for me to check the depredations of unauthorised persons and to prevent tenants, whose right to cut for their own use is not disputed from cutting under colour of such right and selling to strangers ... the tenants of the manor are more interested than the Lord in preventing unauthorised and undue cutting of litter, and I hope to have their assistance in my endeavours to protect their rights.

So, by the early 1870s the right to cut litter by *bona fide* Commoners was seemingly not in dispute. And in December 1872 Messrs Hunt, Currey and Co., the manorial stewards, even stated that 'the persons entitled to cut litter were the same as are entitled to the rights of pasture; which it is in fact dried'.[27] Nevertheless, the restrictions against non-Commoners do appear to have been enforced: at the East Grinstead Quarter Sessions in 1872 there was a conviction for 'larceny of litter', and two offenders were sent to Horsham gaol.

However, the seventh earl, on his accession in 1873, took a different view. Henceforth the removal of litter by anyone, Commoner or otherwise, other than by 'bite of mouth' of cattle, would be prosecuted, and litter deemed to have been cut illegally was seized by the reeve and police and sent to Buckhurst Park. One apprehended cutter was fined £2 for damage done while cutting and carting heather. Levels of cutting seem to have fallen, but hostilities correspondingly increased and the loss of litter brought inevitable damage to farming operations.[28]

---

26 The name of James ('Ginger') Wheatley occurs in this instance. Raper said that 'This man is a regular litter cutter for any who will buy' (Short, *Ashdown Forest Dispute*, 76).

27 Phelan, 'Economy to Amenity', 94.

28 Irons, 'Aspects of the Impact of Man', 108–9.

This new regime was swiftly enforced against Sir Spencer Maryon Wilson from Searles, Fletching, who was taken to court in 1876 by the seventh earl for removing turf and hollies from a site near Nutley in 1874 and 1875, together with stone from Dodds Bank for renovation work in 1872. No fewer than 700 wagonloads of stone had been taken between 1870 and 1872. The action was settled out of court in 1877.[29] A description of the Forest in the early 1870s is contained in the papers for the lawsuit, and makes grim reading:

> The timber has almost entirely disappeared. The birch, willow and alder have entirely disappeared – large tracts of the surface have been stripped of the turf. The heath has been pared off with the soil for fuel. The ferns and young furze are constantly cut and carried away for litter and very little pasture is to be found. The marl pits in consequence no doubt of the revolution in agriculture have long been nearly if not quite disused and a very rough class of people have taken up their abode upon the forest and built hovels and enclosed small parcels of the ground and use the forest as their own – grazing their cattle and cutting and selling the turf heath ferns furze and stone etc.[30]

THE ASHDOWN FOREST CASE[31]

These conflicts over custom and common right were now brought to a head by Reginald Windsor Sackville, seventh Earl De la Warr (1817–96), who had inherited the title from his brother in 1873 and who, as we have seen, as lord of the manor of Duddleswell, began soon afterwards to intensify legal action against the Commoners over their rights to collect litter.

George, the fifth earl, had died in 1869. He had been Lord Chamberlain of the Royal Household 1841–6 and Queen Victoria had become a regular visitor to Buckhurst in the 1830s and 1840s. He and Elizabeth Sackville left six sons and three daughters, although the eldest, George, died in 1850. The oldest of the surviving sons, Charles Richard, became sixth earl and was a major-general, serving in the Crimean War, but, feeling himself unsuited to the duties of his new earldom, drowned himself in the River Granta at Cambridge in 1873.

Reginald, his younger brother (Figure 9.2), then assumed the title on 23 April 1873. Cultured and somewhat ascetic, he had chosen the church as his career, becoming

[29] ESBHRO, AMS 4005–6; High Court Common Pleas Division (1877 D no. 134).

[30] TNA, LRRO 67/1, p. 8.

[31] The background documents for the case have been referred to previously as ESBHRO, U of Sx Spec Colls, AFC I and II. Their title in ESBHRO is *Ashdown Forest. Earl de La Warr v. Miles and Hale. Book of Documents* I and II. Two lithograph volumes were produced of the documents for the case: Volume I, covering 1297–1693, and Volume II, 1693–1878. The copy in ESBHRO was that used by R.E. Webster. The volumes are virtually duplicated from the De la Warr side in the De la Warr papers (ESBHRO, DLW 466–8), and in the House of Lords Library: FCP/1/61a (formerly Box 25 Number 56).

Figure 9.2. Reginald Windsor Sackville, seventh Earl De la Warr, 1876, by Lazzaro Pisani (1854–1932) (source: National Trust, Sizergh Castle, Cumbria, collection number NT 998466).

the 'pronounced ritualistic' rector of Withyham 1841–65 and a chaplain to the queen 1846–65. He wrote a book on the Sackville Chapel (1857) and edited the works of Thomas Sackville, Lord Treasurer (1859).[32] As noted previously, he was appointed assistant warden of Sackville College in 1848, two years after John Mason Neale's arrival. But although the two men had similar Tractarian views, they quarrelled by the 1850s about internal matters at the college and about Neale's offering of confession to the inmates. At one point Neale even unsuccessfully sued Reginald for libel, while on another occasion the inhabitants of East Grinstead petitioned the earl to intervene to eject Neale, something he refused.[33]

By his aunt Mary's initial will Knole and the estate in Kent would pass from Charles to the fourth son, Mortimer, bypassing Reginald. A subsequent codicil reinstating Reginald was left without legal status at Mary's death in 1864 and legal cases over the inheritance of Knole and the title of Baron Buckhurst rumbled on within the family for many years, eventually moving to the House of Lords by 1876.[34]

Reginald had, in fact, succeeded to the tenure of Knole as Baron Buckhurst, in line with Mary's intended codicil, where he resided for three years, changing his name in 1871 to Sackville.[35] But on Charles' death in 1873 he had to return to Buckhurst, with Mortimer now inheriting Knole. Between 1876 and 1879 he was in dispute in the Chancery court with Mortimer in an attempt to bring the divided estates into one hand under the Barony of Buckhurst.[36] The Buckhurst estate was in fact far larger than Knole. Bateman's 1883 publication showed Knole consisting of 1,960 acres yielding £3,450 a year, compared with Buckhurst with 17,185 acres yielding £10,827 a year.[37]

Reginald was clearly very active and litigious. He was prominent in debates in the House of Lords on a wide variety of subjects, with recorded contributions on no fewer than 280 occasions between 1871 and 1892, although nothing touched on Sussex. He was also contemplating the development of land at Bexhill in the early 1880s, with a sea wall, parade, drainage works and roads that he hoped would emulate the developments at Eastbourne. Initial success had not, however, fully materialised by his death in 1896, and he had anyway handed over much of the responsibility for further development to his son Gilbert.[38]

---

[32] Reginald Windsor Sackville, Earl De la Warr, *Historical Notices of the Parish of Withyham in the County of Sussex, With a Description of the Church and Sackville Chapel* (London: Chiswick Press, 1857); and *The Works of Thomas Sackville, Lord Buckhurst, Afterwards Lord Treasurer to Queen Elizabeth and Earl Of Dorset* (publisher unknown, 1859).

[33] Lough, *John Mason Neale*, 97–9.

[34] Sackville-West, *Inheritance*, 155–69. And see Kent Archives and Local History Centre, U269/L19/4–6 relating to family disputes on Reginald's accession to the earldom.

[35] Sackville-West, *Inheritance*, 160–64.

[36] Kent Archives and Local History Centre, U269/L19/7 and L92.

[37] Bateman, *Great Landowners*, 125, 392.

[38] The manor of Bexhill had been granted to the Sackvilles by Elizabeth I in 1560. Gilbert George Reginald Sackville (1869–1915), eighth earl and mayor of Bexhill-on-Sea in 1903 and 1904, died on active service on his way to the Dardenelles.

Such were the varied interests of the seventh earl, but his energies were also turned upon Ashdown after 1873. In some measure his actions were precipitated by the longstanding issue of litter cutting by poorer families who lacked Commoner status and thus any entitlement. In February 1867, for example, Daniel Heasman and Thomas Comber, 'both of them being persons of no ostensible occupation and not even being tenants were discovered cutting underwood'. They were duly convicted, although it was noted that the effect was to 'create a defiant feeling in the foresters' rather than setting an example.[39] Similar cases were brought throughout the late 1860s and through the 1870s. In March 1868 George Edwards, the reeve, had discovered six men cutting and tying up oak and beech underwood, another three with handbills but who were not actually cutting at the time, and Abraham Card, 'a woodbuyer, etc.', loading the wood onto his wagon. Edwards had cautioned the men against cutting but 'they laughed and went on cutting. I then gave them all into custody.' Apparently, the men were handcuffed and led away. Daniel Heasman, who was again involved, was imprisoned for 21 days, the others fined. But all ten men then signed a paper promising not to cut any more wood:

> I, Daniel Heasman do hereby express my regret for having trespassed upon the Estate of Earl De La Warr by cutting underwood on the Forest in the Manor of Duddleswell on the 25th day of March 1868 and I pledge myself never to cut any underwood on the said forest in future without the leave of the Steward or Reeve of the manor for that purpose first obtained ... The mark of X Danl Heasman.[40]

The magistrates thereupon remitted their fines, overlooked previous poaching offences and reduced Heasman's prison sentence to ten days.

Overt confrontation with authority was rare and when men were brought before the magistrates for wood cutting or other misdemeanours they invariably apologised and hoped to escape large fines. The chairman of the bench stated that 'he hoped everyone in the court would understand that whatever might have been done before, no person had a right to go to the forest and cut wood, and he hoped the bench would have no more of these cases before them'.[41] But unauthorised cutting continued unabated. In 1865 at the East Grinstead Petty sessions City Bannister (1828–1872), a customary tenant of a cottage and five acres in East Grinstead, had been sentenced to three months' imprisonment in Lewes Prison for cutting oak sticks for walking sticks, and then in 1869 was again sentenced to four months' imprisonment with hard labour.

The conflict now escalated with the seventh earl's accession, with repeat offenders incurring increasingly severe penalties. Accusations were also made that there were many willing purchasers of litter who themselves were not entitled to it but who bought from Commoners. Thus Bernard Husey-Hunt, De la Warr's steward, wrote to Bernard Hale, a leading Commoner, in December 1873 that:

[39] ESBHRO, U of Sx Spec Colls, AFC II 255–82.

[40] ESBHRO, U of Sx Spec Colls, AFC II, 262.

[41] *Sussex Agricultural Express*, 4 April 1868.

> I shall be very glad if the tenants of Duddleswell Manor will bestir themselves and do something for the protection of their own interests. To talk of dispossessing squatters is idle the flight must be at higher game. I fear there are many willing buyers of litter who don't trouble themselves to ask the seller where it was cut.[42]

Bernard Hale (1824–1889), a Duddleswell copyholder, was also a barrister in the Middle Temple, deputy lieutenant of Sussex and chairman of the East Grinstead Bench and Board of Guardians. He was then living at Holly Hill, described at its sale in 1818 as consisting of a house, a 65-acre park and 220 acres of farmland. After the sale part of the holding subsequently became the basis for the Hartfield Grove Estate, which passed to Robert Melville in 1879. The other part of the original Holly Hill Estate went in 1818 to H.H. Jackson, whose widow Charlotte subsequently married Hale in 1845, who also himself brought Cotchford, Posingford, Whitehouse and Claypits Farms into the complex, together with Forest House. The Return of Owners of Land 1873 assessed this land at 218 acres worth £360. In 1884, following Charlotte's death, the estate, including Cotchford and Posingford, was put up for sale, with common rights over Ashdown proudly advertised.[43]

In 1865 Hale had been a trustee for Mary, dowager countess of Amherst, whose will was to split the Sackville family, as we have seen. Hale, as a trustee, was left £500 in her will. It can be appreciated, therefore, that he had already encountered Reginald as a contender in the will.[44] Furthermore, there had already been correspondence around the time of the fifth earl's death about common rights. Hale had received a letter on 14 December 1869:

> Those who are now actual dwellers on the Waste have no rights at all. It is only those holding properties outside the Forest pale, who have commonable rights on the waste within the pale, all these as you know are named in the Decree, one object of the Decree was to prevent the Lord of the Manor enclosing for his own benefit, therefore it was decreed that he must first obtain the consent of the Homage. But it is quite impossible to prevent Copyholders on the Waste from cutting litter, they fancy they have rights and they cannot be persuaded they have not … .[45]

But now, by December 1873, strongly worded letters were being exchanged between the earl's solicitor, Husey-Hunt, and Hale.[46] The earl had almost immediately begun to act against Commoners for trespass and to buy up land on the Forest with Commoners' rights attached. He was reputed to hold about 60 acres of such land by 1874, something which aroused suspicion among other Commoners. In that year a self-appointed Commoners' Association was thus formed, based largely on the 1816 precedent and

[42] ESBHRO, U of Sx Spec Colls, AFC, II, 300.

[43] Charlotte left a personal estate of over £18,000, and Bernard Hale then moved to Forest House, where he died in 1889 (England and Wales National Probate Calendar 1889).

[44] *Illustrated London News*, 20 May 1865, 491.

[45] ESBHRO, U of Sx Spec Colls, AFC II.

[46] Bernard Husey-Hunt (*c.* 1812–94) was a solicitor, and friend and executor of Edward Lear.

known as 'The Conservators of the Commonable Rights of Ashdown Forest'.[47] The small committee consisted of Hale, Alexander Nesbitt and Freshfield, together with Raper, who suggested that the new earl might join the Association. But, although moral approval was given, no financial support was forthcoming – indeed, the earl had not technically been a Commoner and thus was not entitled to the herbage.

By July 1875, however, Husey-Hunt was increasing the pressure by writing to Raper to ask that the Association stop turf being dug, as this would be an appropriate action for the Association to take. Notices duly went out in August from Raper on behalf of the Association warning those without rights against depasturing animals, cutting turf and litter or doing other damage, or making enclosures. Liberal rewards were offered for information resulting in the conviction of trespassers.

During 1875 De la Warr had clearly changed his mind on the interpretation of Commoners' rights to cut litter, and tension increased when, on 8 October, his solicitors wrote to Raper:

> On looking into the matter more closely, lately we find that no custom and common of turbary exists, but the practice of cutting turves like that of cutting litter is a trespass … . The right of pasture we thought at one time might carry that to cut litter but recently we have altered our opinion. The tenants have no right to touch the soil or anything growing on it in any way, their only right is to have estovers (if there are any) assigned to such as have the right to take their common of pasture by the bite of cattle. In fact litter cutting etc. may be said to impoverish the land as the manure resulting from it is not brought back, whereas if cattle are depastured they manure the land.

Raper replied, asking upon what grounds this view was reached, but letters were already being sent to leading Commoners such as Hale warning against employing men for litter cutting. In return, Hale informed the solicitors that he believed the owners and occupiers of Holly Hill had always enjoyed such rights and 'I do not like to abandon a right so enjoyed'. Instead, the lord should have been proceeding, in his view, in a 'raid upon marauding trespassers outside whose name is legion and who are now cutting and carrying away Brakes and Litter to the great detriment of the tenants of the Manor'. He also asked that De la Warr stop the quarrying activities, which he regarded as a more serious incursion on the Forest than litter cutting.[48]

The sense of grievance was compounded by the suspicion that De la Warr might be aiming to profit by the planting of fir trees or, more probably, through the enclosure of the Forest and the substitution of allotments in lieu of common rights. De la Warr has been referred to as 'a greedy protagonist, who was motivated by a desire for personal profit'.[49] This may be true, but in the absence of hard evidence as to his motives it is difficult to sustain.

47 ESBHRO, ACC 5640/6/10. Also referred to as the Association for the Preservation of Forest Rights.

48 ESBHRO, U of Sx Spec Colls, AFC II., 329–36.

49 Phelan, 'Economy to Amenity', 106.

Figure 9.3. The enclosure landscape of Chelwood Common (source: Ordnance Survey Explorer Map 135: Ashdown Forest © Crown Copyright and database rights 2022. OS 100065054).

The possibility of enclosure, after all, had been raised by Charles Abbott in 1808, and was the fate of Chelwood Common at Danehill in the manor of Horsted Keynes Broadhurst, where the Brands from Glynde were manorial lords.[50] By 1859 a provisional enclosure award had been drawn up, which was finally approved in August 1864. The rectangular plots at the northern end of the common resulting from the surveyed land are still visible in the present landscape, with 176 acres enclosed into 115 small plots, mostly given in lieu of former common rights. The straight lines of the roads can still be seen, the layout contrasting vividly with that of the irregular former assarts elsewhere in the vicinity (Figure 9.3).[51]

In this respect it is noteworthy that when the Freshfields purchased Kidbrooke Park from the Abbott family (including the game preserve of Hindleap Warren) the 1874 sale catalogue contained the following rather uncertain phrase:

[50] The common would have been part of the original Ashdown, but one of those excluded by the building of the pale in the thirteenth century.

[51] D. Rawlings, 'Farming and land-owning in Victorian Danehill', *Danehill Parish Historical Society Magazine*, 3 (1987). For the enclosure award and map see ESBHRO, QDD 6/E13.

> These estates are understood to have attached to them certain rights of common etc over the extensive tract known as Ashdown Forest. The vendor does not undertake to define or prove such rights, but such as there are will be included in the sale of the respective lots, and it is supposed that in the probable event of an enclosure of the Forest, they will entitle the owner to a considerable allotment.[52]

The 'probable event of an enclosure' was unsettling, and now Husey-Hunt wrote again to Hale on 20 September 1876 stating that De la Warr's reeve was reporting litter being cut on Hale's express wishes and asking him to acknowledge his mistake in issuing the instructions. Furthermore, 'his lordship complains so much of the damage done that we are compelled to take some steps in the matter. It is of course useless now writing to the persons who actually cut the brakes etc as they do not understand anything but prompt and decisive action.'[53] At this stage De la Warr's wishes may not have been fully supported by his stewards, and his actions might yet be stopped as being based on little more than a whim. But Husey-Hunt wrote to Hale again in more challenging terms on 27 July 1877:

> In determining as far as possible to put a stop to what he believes to be unlawful acts Lord De la Warr has decided on proceeding by Action at Law against any person who is found cutting Fern or Litter of any kind. He does not thereby wish to assume a hostile attitude towards the tenants of the Manor but rather to preserve their and his legal rights. He hopes for their assistance, but if there should be among them any who really entertain a belief in their right to cut Fern and Litter, his lordship will be ready in the best spirit and for the sole purpose of setting at rest a vexed question to try the right in an action at law.[54]

Bewildered at the turn of events, others in the neighbourhood expressed their anger. Elphinstone Barchard (1827–1893), whose wealthy merchant father had built Horsted Place (Little Horsted), and who had been called to the bar in 1853, wrote 'I was much astonished by the receipt of your letter which endeavours to make out that I have been guilty of a serious crime.' He had been a customary tenant for about 15 years at that time. He continued: 'If any man has committed a crime and ought to be proceeded against it is Geo. Jenner and he is answerable to the Lord.' Barchard had purchased litter from Jenner, who had approached Barchard's gardener with litter to sell. In later years Barchard became a prolific writer of letters concerning Ashdown to local newspapers. The west window at Fairwarp church is dedicated to his memory.[55]

Luke Tester, who had lived in Duddleswell all his life, deposed to Raper that:

---

52 BL, Maps 137 b 9. Plan of Kidbrooke Park, For sale 1874.

53 ESBHRO, U of Sx Spec Colls, AFC II, 334.

54 ESBHRO, U of Sx Spec Colls, AFC II, 340.

55 Christ Church Fairwarp, 'the church in the Forest', consecrated in 1881, was later enlarged through the influence of the Eckstein family in the 1930s. *Kelly's Directory* of 1882 notes that 'about half-a-mile from the church, is "The Fox," a temperance coffee-tavern, formerly an inn, established by the rector, T. Brown esq. of Old Lodge and Elphinstone Barchard esq.'

> I know the property belonging to Mr. Barchard now called Duddleswell House ... Captain Barwell purchased [it] and when Mr. Barchard married his daughter they rebuilt the house and have lived there ever since. Mr. Barchard has had brakes and litter. I have cut them at times for him.[56]

A similar letter came from Dr William Wallis (1814–1893), 'surgeon and apothecary', who wrote saying that he and his father had habitually purchased litter cut on the Forest for 60 years. But he would not contest any legal proceedings and would pay any required compensation. His daughter Katherine remembered her father as 'devoted to all the pursuits of the country: farming, hunting, shooting, and fishing and his long drives and rides were made interesting to him and his children by his botanical knowledge ...'. He was an orchid collector, and 'always liked to remember his walks on Ashdown Forest with Charles Darwin [who] was glad of a guide to the habitat of sundew and *Malaxis* [dragon's mouth] when writing the *Fertilisation of Orchids*'.[57] Dr Wallis supplied orchid specimens to Darwin, who was also interested in studying the wood ant colonies near the bottom of Kidd's Hill. They had met when the Darwins visited Charles' sister-in-law Elizabeth Wedgewood at 'The Ridge' in Colemans Hatch in July 1860. The Ridge site, near Newbridge, had been 'a little group of fields, formerly filched from the open heath and bordered with hollies, beeches and firs'.[58]

Finally, we might note Robert Melville (1841–1891), another barrister, who responded to the earl. Born in Sierra Leone, where his father had been a judge and Lt-Governor, he had moved his family to Hartfield Grove in 1870. He noted that litter had been cut 'from time immemorial and has never been questioned until now'. He also stated that he would 'far rather see the Forest whilst uninclosed in all its pristine wildness, unmarred by shovel or scythe'. However, he moved to London soon after the conclusion to the Ashdown Forest Case.

Here, then, was a flavour of the standing of those Commoners aggrieved by the earl's actions. And this was, of course, on top of the case he had brought in February 1876 against Sir Spencer Maryon Wilson. In September that year Raper again suggested a liaison with the earl in applying to the Inclosure Commissioners for the regulation of Ashdown under the recent 1876 Commons Act, to establish a body of elected Conservators to more fully protect the Forest. But this was declined, De la Warr apparently not wishing to relinquish his mineral rights.

Barchard had played first-class cricket for Cambridge University. His son Edmund was killed in the *Lusitania* disaster in 1915.

56 ESBHRO, U of Sx Spec Colls, AFC II, 345, 345b.

57 Katherine Wallis, 'Village Vignettes of late Victorian Days', *SCM*, 13 (1939), 468–71. In the *Post Office Directory* (1862) for Hartfield he was described as a surgeon and registrar of births and deaths, living at Oaklea, which he had built that year and where he died aged 79. His only son Percy, also a doctor, practised at East Grinstead. William's father, also William, had been 'surgeon, apothecary and man-midwife' to the Hartfield workhouse and was overseer to Hartfield parish in 1818 (ESBHRO, PAR 360/10/2/1).

58 http://darwin-online.org.uk/content/search-results?freetext

Typical of the increasingly febrile situation at this time was an incident on Church Hill narrated by Samuel Hinckley of Martins Hill, Crowborough and Philip Heasman of Friar's Gate:

> On 29th October had been cutting nearly a month. The litter was just ready for carting. There were 6 loads cut. We had delivered 2 loads to Mr Read, Rector of Withyham and he had ordered four more. On the 29th Geo. Edwards the Reeve of Duddleswell came up to us with two teams and said he had warned us not to cut often enough and he was now going to cart the litter away. He did not ask us for whom we were cutting it. He said he was not going to do anything with it, but to cart it into the Park, and let it rot. He had seven or eight men with him. We shook the first two loads of litter about so as to prevent its being taken. Edwards then said he would fetch a policeman. A policeman from Hartfield came up. We then left off shaking the litter about & Edwards carted it all away, about 6 loads in all. He carted it into Buckhurst Park where it is now lying. Edwards said we could send in our bill for cutting it to Mr Gregory and perhaps it would be paid. If not, we could sue him in the County Court. We have been in the habit of cutting litter in the Forest for many years. We cut for anyone who would buy of us. Mr Hoath of Fryars Gate has had a great deal of us. Mr Reed [sic] of Withyham has had litter for several years past.[59]

Clearly, litter cutting was continuing. But finally came the earl's opportunity to proceed. On 13 October 1877 John Miles (1823–1896) of Red House, Harts Lane, was cutting litter with others, including his wife Ann, near New Lodge Park for Bernard Hale. He had loaded one wagon when William Pilbeam, one of De la Warr's keepers from Wren's Warren, appeared and instructed him to stop cutting (Figure 9.4). Pilbeam was particularly busy at this time warning off other litter cutters, some of whom merely laughed at him. When later giving evidence in court John Miles explained:

> I went out and cut 1½ loads of litter and my wife cut ½ a load of brakes. It took me about 2 or 2½ days to cut the litter. While I was about it William Pilbeam … came to me and told me I must not cut it. I told him I was employed by Mr. Hale and should not leave off unless Mr. Hale said so. I told him he had better go and see Mr. Hale about it but he told me I must go myself. I replied that if he would pay for my time I would go but I should not go without as I could not expect Mr. Hale to pay me.
>
> I continued to cut. I told Mr. Hale what has passed when I saw him 2 days after. The litter I cut consisted of coarse grass, heath and furze – it grew 4 or 5 inches above the ground not higher (having been cut down close the year before) and I had a job to cut it. There were some seedling firs on the spot, some as high as my boot and some

---

59 ESBHRO, AMS 4071 Notes and memoranda as to carting of litter by Earl De la Warr 6 November 1878. The reference is to the Rev. T.F. Rudston Read (1811–92), who succeeded Reginald as rector of Withyham in 1865. He was energetic in further developing Withyham parish church and in creating St Thomas's at Groombridge and a chapel (originally an 'iron church') at Blackham.

as high as my knee but I cut round them and did not damage them and I am positive from the lowness of the litter that I cut no seedling firs.[60]

One load of litter went to Suntings. I helped load and unload it. The other half load of litter and half load of brakes went to Holly Hill. About a month after the same Pilbeam helped me to a paper [the Writ in this Action]. I cannot read and gave it to Mr. Hale. That is the only occasion I ever cut litter in my life. I have been in Mr. Hale's employment ever since.[61]

Husey-Hunt wrote to Hale two days after the incident:

My dear Sir,

It has been represented to me that you have quite lately given directions to one of your men to cut litter on the waste of Duddlewell ... and that in accordance with such directions he has so cut. Lord De la Warr disputes the right of tenants to cut litter and is anxious to have a judicial decision on the point; in no spirit of hostility therefore but solely with the view of getting this vexed question disposed of, he is prepared to institute proceedings. May I assume that what you have done has been purposely intended to raise the issue and that the facts are as they have been represented to me. If so, I will at once put the matter forward yours is the first case which has been brought under his notice where a tenant has cut.[62]

The earl initially proposed that the action should be tried at Lewes or another assize town, but it was actually begun in the Common Pleas Division of the High Court in 1877 as Earl De la Warr v. Miles. The papers for this complex and expensive case may easily constitute the most extensive set of material relating to a Victorian dispute over common rights. During the early months of the case there was still some attempted cooperation between Raper as secretary of the Commoners' Association and Husey-Hunt over dismantling illegal enclosures and stopping illegal litter cutting. Repeatedly, witnesses told Raper that everyone who wanted litter could take it without repercussions. Samuel Heasman told him '[I] took to litter cutting myself as soon as I was big enough and have cut every year since for anyone who wanted it ... everybody now takes what he wants'. This despite strongly worded warnings about such activities (Figure 9.5). But De la Warr's Statement of Claim was delivered on 3 January and any thoughts of cooperation were abandoned. In July 1878 the case, 'enormously full

[60] The reference to seedling firs here contradicts De la Warr's claim that the cutting of litter by Miles had 'destroyed thereby the cover for game and young seeding firs' (ESBHRO, AMS 3780/1).

[61] ESBHRO, AMS 3904 Evidence of Hale and John Miles, given two years after the incident. In 1871 John and Ann Miles, both aged 47, were living at Posingford, John being returned as a farmer. In 1861 he was returned as a woodreeve. He appears to have had a varied career. According to Hale's list of employees, given as part of his evidence, Miles had been working for him since 1853.

[62] ESBHRO, U of Sx Spec Colls, AFC II, 350.

of matter', was transferred to the Chancery Division because of its complexity.[63] Delays were incurred because of the extensive historical researches now undertaken by Raper in the Duchy of Lancaster records, Duddleswell court rolls and elsewhere in his capacity now as solicitor for the defendants, and the Statement of Defence was eventually produced by mid-November 1878. An amended Statement of Claim was then produced by De la Warr and an amended Statement of Defence was delivered in March 1879.[64]

Figure 9.4. William Pilbeam, gamekeeper for Earl De la Warr and apprehender of John Miles, 1877 (source unknown, possibly late Peter Kirby private collection).

Earl De la Warr was represented by Sir Henry Jackson and G.I. Elton, and the Commoners by Joshua Williams QC, Sir William Harcourt and R.E. Webster (later Lord Alverstone).[65] The trial began in Lincoln's Inn on 18 February 1880 before Vice-Chancellor James Bacon and lasted in total for 11 non-consecutive days. The heavy costs incurred by the hiring of three such expensive counsel indicates the extent to which the defendants were being backed by wealthy Commoners. The Junior Counsel for the defendants, P.H. Lawrence, had acquired considerable experience in defending Commoners' rights in other such cases around London.

The elderly Vice-Chancellor Bacon then delivered judgement against the Commoners, emphasising the wording of the 1693 decree:

> at no period of that history nor in any of the documents, is to be found any trace of a claim of right by the commoners to cut and carry away pasture and herbage, or brakes, heather, or litter. On the contrary, there is more than negative evidence that no such right was ever claimed or lawfully exercised … . There is no ground

63 Phrase used by Sir Henry Jackson QC at the opening of the case.

64 For a fuller account of the proceedings see Short, *Ashdown Forest Dispute*; for the legal aspects of the case, including biographical material on the counsel, see R. Cocks, 'The Great Ashdown Forest Case', in T.G. Watkin (ed.), *The Legal Record and Historical Reality* (London: Hambledon Press, 1989), 175–97.

65 Jackson and Williams were in poor health, both dying in early 1881.

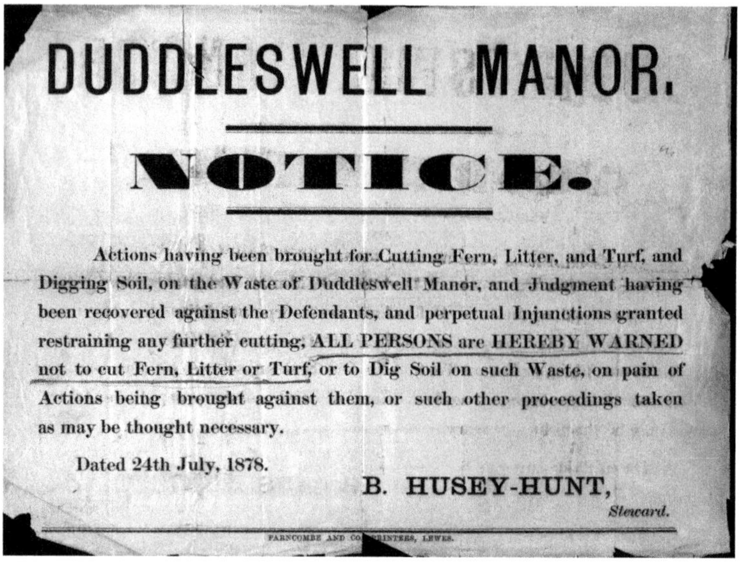

DUDDLESWELL MANOR.

NOTICE.

Actions having been brought for Cutting Fern, Litter, and Turf, and Digging Soil, on the Waste of Duddleswell Manor, and Judgment having been recovered against the Defendants, and perpetual Injunctions granted restraining any further cutting, ALL PERSONS are HEREBY WARNED not to cut Fern, Litter or Turf, or to Dig Soil on such Waste, on pain of Actions being brought against them, or such other proceedings taken as may be thought necessary.

Dated 24th July, 1878.

B. HUSEY-HUNT,
Steward.

FARNCOMBE AND CO., PRINTERS, LEWES.

Figure 9.5. Warning notice against litter cutting (source: ESBHRO, AMS 4070).

> on which I can hold that at any time there existed within the Forest of Ashdown a special custom conferring a right on the commoners to cut and carry away any part of the growth of the soil.[66]

The longstanding custom of litter cutting was thereby dismissed as not constituting a right on the part of the Commoners. The undisputed right to common pasture did not, he felt, also include the right to cut and take away plants of any kind. But the judgement was curiously light on the issue of prescription – the long-accepted local access for litter cutting. He had concentrated overmuch on the 1693 decision rather than the issue whereby a Commoner who had exercised common rights for 60 years would also have a valid claim to continue to enjoy that right.

The Commoners appealed from early February 1881, and the decision was reversed. Much of the appeal was based around the issue of prescription: user by the Commoners since 1693, resisting interference by the lord.[67] Although in failing health, Robert Edwards ('a little slow in manner but memory remains good. He has a great store of information') of Lines Farm, Hartfield, was recalled to give evidence again, now emphasising that Hale's Suntings Farm had used litter since 1816, a crucial date for 60 years' user. The cutting, it was established, was undertaken not by way of concession or permission, but as of right.[68]

---

66 *The Law Times* XLII, 5 June 1880, 474, 479; ESBHRO, AMS 3859 The order from Bacon, V.C. lists the witnesses and evidence supplied; see also Eversley, *Commons, Forests and Footpaths*, 116; and see *Law Journal Reports*, n.s. 49 (8 August 1880), 476–90.

67 The Prescription Act 1832 had dealt with claims to right of common and other profits, such that after 60 years' enjoyment the right was to be absolute, unless had by consent or agreement.

68 ESBHRO, AMS 3904.

However, the Court of Appeal also agreed that the Commoners had failed to establish their claim to their right to cut and carry away litter and similar materials, based on the 1693 Decree, thereby agreeing with the previous judgement. But, importantly, it was accepted that the only defence left to Hale was under the terms of the 1832 Prescription Act. And under this heading Hale and the previous owners of his property had established a right by user during a period exceeding 60 years. Appeal judge Lord Justice James observed that Hale had:

> proved at first instance that for a period of upwards of sixty years he and his predecessors in title had claimed to take, and had taken, not by way of permission, not by way of concession on the part of the Lord, but has taken as of right ... the litter from the Forest to his farm for the use of that farm.[69]

To protect the rights of all Commoners, a cross-action that had originally been brought in December 1878, but postponed pending the result of the appeal, was undertaken by Bernard and Charlotte Hale, Robert Melville QC and De la Warr's old foe Sir Spencer Maryon Wilson. Raper and Webster acted for the Commoners against De la Warr.[70] They claimed an injunction restraining the earl from enclosing any part of the Forest or from interfering with their rights by encouraging the impairment of their common pastures by stone quarries and gravel pits, by the growth of firs and larches, and by the settlement of squatters on the Forest to create a labour force. A settlement was subsequently made by consent, finally establishing the right to cut litter by all those entitled to rights of common under the 1693 decree. A monument sits in Hartfield church to Robert Melville, with somewhat exaggerated wording, as 'primarily responsible for securing in 1882, the Commoners' rights in Ashdown Forest'.

The earl had received little popular support. The press had reported his 'feudal bluster' and during September 1878 had debated his motives, which appeared to be a desire to turn Ashdown into 'mere sporting ground, to stock it with black game and deer, cover it with fern, abolish all such nuisances as the cow of the poor man at once and for ever, and in short, treat the place as his own freehold'. More darkly, 'A Dweller on the Forest' in August 1878 noted that he had brought actions against 'poor people of the labouring class which mostly occupy small plots in or near the common'. Having no money for lawyers and being ignorant of legal procedures, judgement went against them by default, with copies of the judgements served on 'ignorant men', to be followed by executions put into their houses to recover the nominal damages (ranging from 10s to £6) as well as costs ranging from £30 to £150 or more. As such, the cottagers were threatened with destitution and enforced sales. De la Warr was put

---

69 Cocks, 'The Great Ashdown Forest Case', 192; ESBHRO, AMS 3977–8. See the *Solicitor's Journal and Reporter*, 26 March 1881, 391–2.

70 A copy of an early draft of an agreement to compromise in Hale and others v. De La Warr dated 9 December 1881 is in the De la Warr papers (ESBHRO, DLW 459–60). And see ESBHRO, AMS 4014.

into a very bad light.[71] However, there was some support for the earl: a reply to the 'Dweller on the Forest', for example, from 'Salva quoe amissa' read:

> as a resident interested in the picturesqueness of the neighbourhood, that what was once a beautiful forest has been converted into a barren moor, owing to these oppressed litter cutters having in their depredations cut off every young tree as soon as it appeared above the ground, and that whatever his lordship's legal rights may be, I earnestly hope that he may be successful in the actions yet unfought, and so be enabled to restore to its former grandeur and beauty what is now but a monument of your correspondent's friends the 'ignorant men'.[72]

The antagonism between De la Warr and the gentry Commoners was all too apparent, despite the superficial civility of the letters passing between them. This was a high-profile case, widely reported nationally, but there were many local repercussions. One vivid illustration is provided by the case of John and James Brazil, brought before the magistrates for unlawful cutting of underwood to the value of 3d in January 1879 – therefore, while the case was ongoing. The Brazils had lived at the margins of the law for some time, as we have seen. John was a hawker aged 54, reported as living at Chelwood Gate in 1871 with his wife Priscilla and two children. The 1851 census recorded that two of his children had been born in the Ashdown area, although the enumerator could not ascertain where two others were born. Others of the Brazil family were to be found at Beggars Bush near Duddleswell by 1871, living in 'tents or carts'. William Brazil's family members were described as tramps and those of Benjamin Brazil as hawkers, as were other household members in that district.[73] Three of the magistrates were Commoners: Bernard Hale, the chairman of the bench, Colonel Moor and Henry Freshfield. Having heard the case put by the reeve, Hale then stood up and declared that three of the Justices present were tenants of the manor and that the property (i.e. the underwood) that it was claimed had been maliciously injured did in fact belong to the Commoners rather than the earl, and furthermore the earl had no right whatever to summons the defendants. The summons should have been taken out by those with common rights, he claimed, and at that point the three Commoner magistrates left the court and another magistrate took the chair. The remaining justices decided that 'it mattered not who were the owners of the property, it was quite clear that it did not belong to the defendants'. They had claimed that they 'thought it belonged to anybody' but James was fined 13s 6d and John 15s 6d, with a fortnight to pay.[74] After the decision was made the three magistrates returned and Hale explained that there were two ongoing lawsuits surrounding these very issues and involving him personally so that it was thought

[71] *Daily Echo*, 9 September 1878; *Daily News*, 29 August 1878, and reprinted in the *East Sussex Advertiser*, 3 September 1878, the cutting together with others collected in ESBHRO, AMS 4096–9.

[72] *Daily News*, 7 September 1878.

[73] Other members of this extended traveller family lived as pedlars elsewhere in Sussex.

[74] ESBHRO, AMS 4098 *North Sussex Gazette and East Grinstead General Advertiser*, 18 January 1879, 1.

right for the three not to take part in the proceedings. The magistrate's court had been the setting for another rehearsal of the dispute, with the hapless Brazils looking on.[75]

The Commoners had financed the defence and subsequent actions, sharing all their costs between them. Mutual agreement was reached on the proportions to be paid by 'an influential body ... not disposed to submit quietly to such treatment'. The total cost to the Commoners was £6,194 11s 7d, or about £1 per acre saved.[76] But they were quite able to afford this, containing among their number nearly 70 people owning about 200 properties to which rights were attached, such as the third earl of Sheffield, the third Baron Colchester, Louisa, widow of Sir John Villiers Shelley, Sir Spencer Maryon Wilson, and even Henry Fitzalan Howard, fifteenth duke of Norfolk. Attribution of the costs between the litigants dragged on during late 1881.

However, the earl now dropped his opposition to the future regulation of the Forest and the Committee of the Commoners' Association then obtained a Provisional Order for the regulation of the Forest under the Commons Act 1876, although the earl refused to associate himself with their application. Given that normally the support of the lord of the manor, as well as one-third of the Commoners, was required to proceed to regulation, it is unclear how the Ashdown case could technically proceed. Nevertheless, this landmark was indeed confirmed by act of parliament in 1885.[77]

Significantly, it was now necessary for Ashdown to be seen to 'benefit the neighbourhood', implying access and even recreation. Numbers of cattle to be commoned were now to be agreed, together with the procedures for appointing Conservators, who were to formulate byelaws and oversee Ashdown's management. The Committee of the Commoners' Association thereafter became the nucleus of the new Board of Conservators, with power to regulate the use of the common land of the Forest. Their first meeting was convened at the Nutley Inn on 5 May 1887. But the Commons Act 1876, under whose aegis the Conservators met, had lacked clarity as to its aims: on the one hand it was based on a centrally driven concept, with ideals of public access and preservation, but on the other hand by an older culture of enclosure and agricultural improvement.[78] And the Act could not account for local requirements, so that the Ashdown Forest Conservators found themselves having to invent their roles and powers as they proceeded.

Ashdown became the twelfth such area of common land in England to be dealt with under the terms of the 1876 Act. An enclosure award with a large-scale map in

---

75 The withdrawal of magistrates who were manorial tenants was also reported for the 30 March 1868 East Grinstead Petty Sessions case involving the ten underwood cutters noted above. Bernard Hale and Major Moor withdrew, but there was no reported statement of defiance against the then fifth earl on that earlier occasion (ESBHRO, ADA 100).

76 ESBHRO, AMS 3862, 3979–4004.

77 Commons Land Regulation (Ashdown Forest) Provisional Order Confirmation Act 1885 (48–9 Vict.c.56); Eversley, *Commons, Forests and Footpaths*, Appendix II, 333–4; Irons, 'Aspects of the Impact of Man', 60–61.

78 E. Straughton, *Common Grazing in the Northern English Uplands, 1800–1965* (Lampeter: Edwin Mellen, 2008), 204–17.

two parts, signed off by Raper, was confirmed and dated 28 March 1887, and further endorsed by the Land Commissioners for England on 31 March.[79] The Ashdown award was one of a number of such later awards, which have not received the same degree of attention as the earlier body of enclosures.[80] Importantly, this was not an enclosure award in the sense that it divided up the Forest, but merely that it delimited the area to be governed by the Board of Conservators.

Under the terms of the award and registration, various arrangements had to be made and one of the first concerned land ownership on the Forest. The Land Commissioners agreed that those who owned Forest land before December 1869 be allowed to retain it without further expense, while under the terms of the Award those who had enclosed their land since that date were allowed to purchase it at current market value. However, the sale of Ashdown lands did arouse some anxiety, even resulting in a question in the House of Commons in April 1886. Dr R. McDonald, MP, asked the Secretary of State for the Home Department whether, since the Land Commission

> are now attempting to sell to about 120 squatters on Ashdown Forest, Sussex, the small bits of land that they have reclaimed from the forest during the last 12 to 18 years, [...] whether compensation is to be given to those squatters for their houses and reclamation of land if they are not able to buy their plots and are consequently evicted?

He continued by pointing out that the price being demanded was £50 per acre, thought locally to be far too high for Forest land, and also that the Land Commissioners were forbidding the cutting of peat, 'a right unquestioned till lately and rather beneficial than otherwise to the land'. One man, he continued, was fined for cutting peat by the Uckfield Bench, with one of the magistrates sitting thereon being a Commissioner (Conservation Board member) for the Forest.

The Secretary of State, Mr Childers, replied that the Land Commissioners held that:

> the small bits of land referred to are recent encroachments, illegally made; and they are now being sold under the Provisional Order, confirmed by Parliament, for the regulation of Ashdown Forest as an open space. The encroachers comprise all classes, from landowners to labourers. In the event of their not being able, or willing, to buy their plots, they are entitled to such compensation as is provided by the Enclosure Acts. The price asked is the fair market price [which] ... varies with the quality and convenience of the land sold. Most of the encroachers have agreed to the price asked, and are willing to buy more land at the same rate. The

79 A copy of the subsequent inclosure award (1887) with two maps of Ashdown annexed is at ESBHRO, QDD 6/E15. 'Inclosure' was the version on the map itself, rather than the modern preferred variant 'enclosure'. See also ESBHRO, ACC 5786/55 for a draft version on 25-inch sheets. Also TNA, MAF 1/377 Enclosure awards and awards of regulation of commons under the Enclosure Acts 1845 to 1899: 1887 Ashdown Forest (Regulation).

80 But see A. Howkins, 'The Commons, Enclosure and Radical Histories', in D. Feldman and J. Lawrence (eds), *Structures and Transformations in Modern British History* (Cambridge: Cambridge University Press, 2011), 118–41.

> Commissioners have not the power to interdict people from cutting peats; but the Provisional Order, before quoted, prohibits this practice, which is damaging to the surface of the forest.[81]

No fewer than 158 encroachments were rapidly sold to their occupiers, 123 of them consisting of between half and three-quarters of an acre, and only four exceeding one acre. By October 1886 all the plots had been sold.[82]

WILLIAM RAPER AND HIS NOTEBOOKS

It is appropriate to give a little more detail about the man to whom so much is owed for the collection and submission of the copious material relating to the Ashdown case. William Augustus Raper (1845–1940), born in Portsmouth, qualified as a solicitor before he was 21. He practised in Liverpool for three years but refused a partnership and instead arrived in Battle in 1870, beginning by practising alone but soon entering into partnership with Mr Ellman in a business that later became Raper and Son and into which R.W. Fovargue entered as a partner in 1922. The partnership flourished until Raper, aged 90, finally retired in January 1935 with failing sight (Figure 9.6).

In 1871 he married Mary Anna Ellis, his wife of 50 years until her death in 1922, with whom he had four sons (one of whom, Robert, was killed at the First Battle of the Somme) and a daughter, Millicent, with whom he lived in St Leonards for the final ten years of his life, dying on the last day of December 1940.[83]

During his long years in Battle he was extraordinarily active. He was a long-time chairman of the Local Board of Health, predecessor of the Battle Urban District Council, which he also chaired 1894–1918, and was registrar of the Deanery of Battle. He belonged to the Battle Abbey Masonic Lodge and was a churchwarden, clerk to the commissioners of the Pett Level and Pevensey, a member of the Sussex Archaeological Society from 1872 and the Sussex Record Society and, nationally, a president of the Law Society. He was a county councillor from 1907 and steward of up to 20 local manors as well as estates such as those of the Fullers at Ashdown Park, the latter familiarising him with the developing issues on Ashdown. Following involvement with a case of trespass to cut litter on the Forest in 1872–3 soon after arriving in Sussex, he became clerk to the Conservators of the Common Rights of Ashdown Forest in 1874.

He possessed a powerful personality, and something of his character was caught in an obituary:

---

81 BPP, 'The Land Commission (England and Wales) – Ashdown Forest', 1 April 1886, 433–4. McDonald was MP for Ross and Cromarty 1885–92, with a particular interest in crofters' welfare.

82 For the sale of plots see ESBHRO, ACC 6815, Minutes of the Ashdown Foresters Protection Association, further discussed below.

83 B. Short, 'Raper (William) Augustus (1845–1940)', *ODNB* (2004); Anon., 'Modern South Saxons: No. 88 – Mr William Augustus Raper', *SCM*, 10.4 (April 1936), 224.

Figure 9.6. William Augustus Raper (1845–1940) (source: *SCM*, 10.4 [1936], front cover).

> He was a solicitor of the old school, his 'professional manner' being heightened by a touch – characteristic of that full-flavoured period – of what can only be called courtly arrogance. In politics he was, naturally, a staunch, one might almost say a militant, Conservative.[84]

For our purposes, however, his most significant work was that on behalf of the Commoners in the Ashdown Forest case. He assiduously collated material relating to the case, which was transferred from the London solicitors who had dealt with the Chancery proceedings to the East Sussex Record Office in 1959. These papers are extremely wide-ranging, covering the history of common rights on Ashdown from the late thirteenth century onwards. Unsurprisingly, when the Conservation Board was established from 1885 Raper was installed as clerk, a position he continued to fill until December 1929.

Raper himself also deposited seven volumes of material with the Sussex Archaeological Society in 1934, including five small notebooks containing depositions, extracts from which he published in 1935 and which we examine more fully below.[85] Other relevant material is extant at the National Archives relating to the earlier case of De la Warr v. Sir Spencer Maryon Wilson as private papers of the Hewlett record agency, researchers into land disputes. There are also private papers in the De la Warr archives and material resting with the Ashdown Conservators, such as minutes of the committee from 1887, accounts and other miscellaneous papers.[86]

Following the lodging of De la Warr's claim in the Chancery Court in January 1878, Raper set about collecting his evidence on rights of user: 'I interviewed all the old men living round the Forest and noted in these books their life histories and all they could tell me.' These five books contain the testimonies of Ashdown's small farmers, tradespeople, farmworkers and domestic servants, who had lived most or all their lives

[84] *Sussex Daily News* (Battle edn), 1 January 1941, 8.

[85] W.A. Raper, 'Ashdown Forest', *SNQ*, 5 (1934–5), 146–7. The depositions by James Baker, Caesar Gurr and John Brooker were singled out in this short paper.

[86] TNA, LRRO 67/1–3.

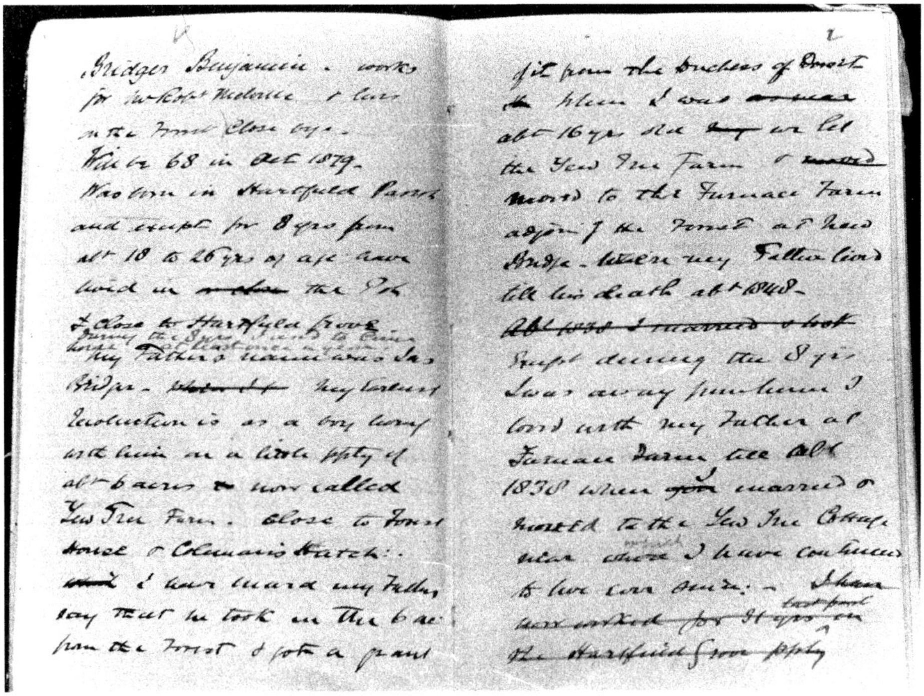

Figure 9.7. Raper's notebook 1: entry for Benjamin Bridger of Hartfield, 1879 (source: ESBHRO, ACC 3715/1–6).

on the Forest. There is evidence from 119 Ashdown residents, written down by Raper as they spoke to him in their cottages or farmhouses. The large number of deletions, corrections and insertions in the text indicate that we have the original notes taken at the time, rather than a later, corrected and tidied version.[87] They were mostly obtained from elderly men, frequently poor, with only five being from women – Martha Baker, Sophia Constable, Elizabeth Fenner, Sarah Page and Susannah Rice.[88]

The earliest date given for an interview was 26 April 1878, three months after the lodging of the claim, when he met Benjamin Richardson at Mudbrooke Hill (Figure 9.7). All the interviews thereafter in Notebooks 1 and 2 were carried out during late 1878, mostly September, around the northern margins of the Forest. The first date entered in Notebook 3 is for a visit on 9 October 1879 to Abraham Edwards at Suntings Farm,

87 For analysis of the notebooks see B. Short, 'Conservation, Class and Custom: Lifespace and Conflict in a Nineteenth-century Forest Environment', *Rural History*, 10 (1999), 127–54.

88 The Raper notebooks were transcribed in 1976 by Mrs V.B. Lamb, Honorary Librarian to the Sussex Archaeological Society. This faded typescript, illegible in places and containing some inaccuracies, partly due to the difficulties of Raper's handwriting, is in the Library of the Sussex Archaeological Society (Accession No. 20412). The Notebooks themselves were later deposited with ESBHRO and listed as ACC 3715. The depositions have now been transcribed afresh and published, with supplementary material, as Short, *Ashdown Forest Dispute*.

Hartfield, suggesting a year's gap in collecting depositions, and we now find Raper moving primarily along the northern and eastern boundaries of the Forest from East Grinstead and Forest Row through Hartfield, Duddleswell and Buxted. Notebook 4, dated 21 November 1879, is concerned primarily with the Nutley, Fletching and south-western edges of the Forest, except for one or two entries relating to the Duddleswell area. Overall, then, evidence was collected between April 1878 and November 1879.

Notebook 5 differs from the other four. It contains entries relating to the Maresfield/Fletching area, the only date entered – 31 December 1878 – belongs to some notes at the rear of the volume, and the entry is sufficiently obscure as to render its attachment to the depositions in the volume somewhat problematic. Most of the first-person singular statements in this notebook have been overwritten at a later stage by a third-person statement ('I went' became 'He went').

In fact, the evidence in Notebook 5 exactly duplicates that given by witnesses interviewed for the De la Warr v. Wilson lawsuit that had begun in February 1876. The witnesses are identical and it becomes clear on closer examination that Notebook 5 merely copies material already available from the earlier action. Four points are revealing. One is that in the notes to Notebook 5 Raper lists the witnesses in the volume, as usual, and includes Jacob Tyler, although there is a later reference to his being dead. Secondly, there is one man, Henry/Harry Carr/Card, who was interviewed on both occasions, but in Notebook 5 he is referred to as Henry Card, aged 81, and in Notebook 4 he is referred to as Harry Carr, aged 84. His evidence makes it indisputable that this is the same man, but the three-year difference in his age between the two notebooks is now explicable by the fact that he was interviewed for the Wilson case perhaps three years earlier. Thirdly, we find Thomas Osborn also interviewed on both occasions: in the earlier instance he was 53 and living at Fairwarp Farm, but in Notebook 4 we find him aged 56 and stating that he left Fairwarp Farm at the previous Michaelmas. Finally, it should be noted that all the deponents in Notebook 5 are from the southern end of the Forest and nearer to Searles, the home of the Wilsons.[89] It is not known whether the fact that this latter evidence was originally collected three years earlier made any difference to the case.

The extent to which the witnesses were able or willing to expound varied. Some were very expansive and must have occupied a great deal of Raper's time; others gave only scanty evidence; some were thought by Raper to be unsound and therefore not to be used at the hearing itself. There is a discernible structure within the testimonies, pointing to some prompting and ordering of questions by Raper: 'One almost hears the helpful, persuasive voice of the "lawyer gentleman" as he sits by smoky fires in Forest dwellings and with the greatest patience and humour wheedles their stories from one after another.'[90]

At the beginning of each deposition he included a frank, unpunctuated comment of his own about the witness: John Heaver was 'rather deaf and decrepit. Health uncertain', Isaac Slater was an 'intelligent man but health uncertain recently had a

[89] For the earlier witnesses and their evidence see TNA, LRRO 67/1.

[90] Barbara Willard, *The Forest: Ashdown in East Sussex* (Poundgate: Sweethaws Press, 1989), 49.

fit' and Sophia Constable 'has had a paralytic stroke but has got over it. Has a little difficulty in pronunciation. She says she could travel very well.' But John Walter was 'Too ill to be any good', while 'Wheatley, John. Suffers dreadfully from neuralgia in the head and will be impossible as a witness' and 'Winn, John. 61. Has been in Lewes Gaol for cutting litter on the Forest and for poaching.'

The depositions retained much of the witnesses' own phrasing and words. They spoke firstly about their early life and work, recounting their successive workplaces. Each then spoke specifically about the removal of litter from the Forest, recounting their own role or describing the actions of named others. The very mention of so many names from 119 depositions, many being repeated by several individuals, allows us to build an overall impression of the primary activists upon this contested rural stage.[91] They normally listed farms, cutters and the tenants and landowners who received the materials. To finish, they often said a few words about other activities: the hunting carried out by the gentry was frequently described at this point. Scattered throughout the deposition would be insights into their daily lives or the recounting of events that had stuck in their memories.

We never 'hear' Raper's questions, except through the implicit structuring of the interviews, and we need to remember that he was reducing a deposition that was probably rich with long pauses, interminable sentences, digressions, gestures and intonation into a hastily written note. Of Benjamin Richardson he noted 'a good witness as far as his evidence goes but views everything from the squatters' point of view'. But Richardson's evidence in Raper's notebook contains no such bias, and one wonders what was omitted as irrelevant to the case in hand.

We might note that the depositions were mostly taken from people who were not themselves Commoners, but who worked for Commoners or cut and sold litter to farmers and other purchasers, legitimate or not. Many of the more substantial landowner–Commoners might privately condone this, while outwardly defending Commoners' sole rights to litter.

The Edwards surname features largely, with three deponents, Abraham (*c.* 1813–1899), from Toll Bar Cottage, Hartfield (later Galleypot Street), George (b. 1815), and Robert (b. 1805). Given that Abraham, who worked for Bernard Hale, alone fathered 12 children in two marriages, the presence of such a prolific family can be understood.[92] Abraham received a subpoena to give evidence in person in London, although there is no mention of him receiving 'conduct money'. Apparently unrelated to Abraham was Robert Edwards Jr, a 73-year-old farmer when he gave evidence to Raper. His father, Robert Sr (d. 1831), had combined the offices of manorial wood reeve and estate bailiff for Buckhurst and the manor of Duddleswell until his death, and hence was remembered by many for his institutional policing role.[93] Robert Sr had lived at

---

91 B. Short, 'Environmental Politics, Custom and Personal Testimony: Memory and Lifespace on the late Victorian Ashdown Forest, Sussex', *Journal of Historical Geography*, 30 (2004), 470–95.

92 I am grateful to Stephen Bullar for information on Abraham Edwards and Robert Edwards, father and son.

93 See p. 202.

Andrews Bank, which he owned, and his son Robert was also living there in 1842 and at Lynes Farm by the 1870s, when he indicted James Wheatley for 'larceny of litter', the case being dismissed through Wheatley showing he had a right to cut.[94] Robert Jr was called upon to appear in person as a witness for the defendants. Confusingly, Robert Sr had been succeeded as reeve by George Edwards, uncle of the deponent George, and a very unpopular figure, who appeared at the High Court in 1880 to give evidence against the defendants and who had been actively sending men to the East Grinstead Petty Sessions for cutting underwood for the best part of 30 years. The Edwards confusion is also magnified by the presence of David Edwards, Lord De la Warr's steward in 1835, who was Welsh and unrelated to the Hartfield/Withyham families.

The wider Tester family provided six separate deponents to Raper and no fewer than 45 separate individuals were referred to in the depositions. George Tester, aged 93, had been born on the edge of the Forest *c.* 1789 and Raper noted 'his faculties appear very clear and he is vigorous for his age'. George recalled 'This is the 8th Lord of the manor I have known. They all stirred up about the Forest and wanted to claim it as their own but never could yet.' Of Job Tester (born *c.* 1834) Raper noted, 'an intelligent man – one of a family of litter and turf cutters'.[95]

Ashdown was a space for family work, including that of the children, who also cut litter or wood, tended cattle or minded babies. The deposition of Martha Baker (née Langridge, 1810–1894, and who died in East Grinstead workhouse), a blacksmith's daughter and 'capital witness', typically intertwines children's work and play, family relationships, memories of landowners and the use of the Forest resources. Birch, willow and alder were used for the family's fuel, together with peat and turf. She remembered cutting litter with her mother and other children for her family's landlord, Col. Young, for their own pigs and for other purchasers:

> I had a new frock out of the money. I remember the occasion because we took the baby in its cradle out on the Forest and my sister Martha and I upset it between us. My Father was one who bit before he barked and he gave me and my sister a good hiding when we got home.

She spoke of seeing litter in Col. Young's yards at Holly Hill, and of peat, which was regularly cut from Newbridge and used in the greenhouse. 'I used to help him lay out the pieces and I remember on one occasion … falling into the water there and getting into a dreadful mess. I was frightened by the "efts" swimming over my arm and when I got out my Father boxed my ears.'[96]

---

94 See ESBHRO, TDE 149/2/591 for residence at Andrews Bank; also AMS 3891/52.

95 Short, *Ashdown Forest Dispute*, 66, 172. For a portrait *c.* 1905, reputedly that of 'Granny Tester', a redoubtable Nutley midwife, see P. Kirby, *Forest Camera: A Portrait of Ashdown* (Poundgate: Sweethaws Press, 1998), 25. This work contains many vivid Ashdown photographs from about 1890 by Arthur 'Daddy' Francis (1855–1934).

96 Short, *Ashdown Forest Dispute*, 80–82. Martha's excellent account is made even more interesting because Raper also interviewed her husband, James Baker, a farmer of 63 acres, whose lengthy testimony complements Martha's. Raper commented: 'a most intelligent old

So everyone helped: and through work and parental observation children were inducted into the traditional use of Forest resources. And in seeing places, people and events from a variety of locations on the Forest, the notebooks offer a triangulation of memory and a confirmation of the events.[97] The view from a southern Forest dweller might differ in detail from that of a northern resident, but the congruencies and differences can be examined for their significance.

A good example relates to an event remembered by many, the driving of the Forest, by which cattle were rounded up to establish their ownership and impounded if illegally depastured by non-Commoners. There are variations in the telling from person to person, and there are enclosed narratives within the overall story, such as the activities of the poor and the involvement of the larger farmers and those with authority. Some recalled the wrecking of the manorial pound at Nutley, which, when triangulated, indicates that the event seemingly happened in the unsettled year of 1830, and was possibly undertaken by women and/or the poor:

> I remember some of the farmers used to drive the Forest occasionally. I do not know whether the Reeve went with them. My uncle Edward Heaver used to go ... some on horse back and some on foot. They drove all the cows, sheep and horses they could find to Nutley pound. I have seen them at it. I always understood they got the fine imposed and spent it. I remember hearing a great row on one occasion when they drove the cows of a poor old worn out man named Chedle. Some of his neighbours went and got them back again by force. (Edward Heaver, Leggs Heath)

> The Forest used to be regularly driven. I often saw it driven. The tenants drove it and used to take horses and sheep and impound them at Nutley. They once tried to pound our horse but we scared it away from them... I remember the Association drove the cows once and pounded them at Nutley and the women broke the pound down and got their cows out. (Henry Carr, Nutley)[98]

Of those giving depositions to Raper, several were subsequently called upon to give evidence of user for the defence at the London trial, and accordingly paid conduct money.[99] The clarity of evidence by Martha Baker or Edward Heaver ensured their inclusion, together with those who had also testified at length and more substantial members of the community, such as Robert Turner, the miller and farmer.[100] Eighteen

lady and will make a good witness. Bring family bible' (ESBHRO, AMS 3715/1). James accidentally drowned in 1880, aged 75, and Martha's nephew Thomas Langridge, a farm bailiff from New Lodge, committed suicide in 1886. I am grateful to Dean Langridge for this information.

97 For the importance of internal consistency, confirmation of accounts, bias and early memory see P. Thompson, *The Voice of the Past*, 3rd edn (Oxford: Oxford University Press, 2000).

98 For a full analysis of driving the Forest see Short, 'Environmental Politics, Custom and Personal Testimony'.

99 Conduct money: payment (here mostly 10s) to witnesses who had been subpoenaed, sufficient to defray reasonable expenses of travel and attendance. Edward Heaver and William Young were each paid £1 10s and Robert Turner 21s.

100 ESBHRO, AMS 3904–5.

deponents received subpoenas for the trial, of whom 14 were paid conduct money. Another 16 were said by Raper to have given good evidence, such as Martha's husband James, but did not receive subpoenas. And there were also those who were called as witnesses, such as Hale himself, his wife Charlotte, John Miles and Robert Melville.

Of course, documentary sources are rarely neutral. Raper structured his deponents' responses in ways appropriate to the case and most favourable to the Commoners. But importantly, there appears to be no material setting De la Warr's case in the same detail. Indeed, what were his real reasons for acting against litter collecting by the Commoners? It is possible that he had a disinterested concern to establish once and for all time the limits of Commoners' rights, and perhaps he was genuinely concerned at the loss of ground cover and hence the impoverishment of his shooting. Concern at the cutting of bracken seems justified if the disappearance of the formerly ubiquitous black cock during the nineteenth century is remembered. Certainly, Edward Turner, vicar of Maresfield, recalled in 1862 that 'so numerous were these birds at the beginning of the present century that it was hardly possible to walk or ride across it in any direction without disturbing some of them ... [but] in this locality [they] are now very rarely to be met with'. Petley, writing in 1896, thought the last one recorded had been shot about 30 years before.[101]

De la Warr's action against Maryon Wilson over the taking of turf and stone might also offer evidence that he was genuinely concerned at the environmental deterioration that he witnessed. James Walter (b. 1817) recalled that 'The Forest in my young days was not dug about nearly so much as it is now. They dig gravel and stone a deal more ...'.[102] Long before, Husey-Hunt had written to George Wheatley on 29 June 1855 that:

> A complaint has been made to me on behalf of several tenants of Duddleswell Manor that you persist in scamping off the surface of the Waste land of the Manor. Notice has already been given to you that this is an illegal act, and I now again require you to discontinue it ... .[103]

The Forest appearance by the late nineteenth century, as testified by contemporary photographs, was indeed rather forlorn and the long years of extracting resources had taken their toll (Figure 9.8). Most of the great trees were gone, there was little underwood, there were patches of bare ground where turf had been taken, unfilled pits, stone quarries and even previously grazed land was returning to heath or to bracken infestation. De la Warr asserted that in the act of litter cutting seedling trees

[101] Turner, 'Ashdown Forest', 35–64. Edward Turner (1794–1872) was rector of Maresfield 1837–72. His tombstone is on the south side of Maresfield church, and a memorial window to him and his wife sits over the choir stalls. See also Petley, *Maresfield Old and New*, 60.

[102] Short, *Ashdown Forest Dispute*, 192–4.

[103] ESBHRO, ACC 3715/3, Evidence of Samuel Heasman, 68, of Pages Gill, Forest Row; letter Husey-Hunt to George Wheatley in ESBHRO, ADA 99 Rental and Custumal of the Manor of Duddleswell (envelope enclosed within the volume of miscellaneous letters concerning enclosures and Court Barons of the Manor).

Figure 9.8. Ashdown at the end of the nineteenth century. Aaron Scott, turf cutter 1890, skimming the turf from Ashdown. Much of his life was spent in a turf hut near Fairwarp. All turf and peat cutting was made illegal in 1885 (source: Nutley Historical Society).

were destroyed. Cutting holly, oak and beech was well known as prohibited, but it is likely that many took what they could get. Certainly, the Forest trackways were now wide and susceptible to gullying. It appears that a 'robber economy' was now in full swing, with all parties covertly participating. Even the earl, for example, had developed his stone quarries prior to the lawsuit.

TELLING ASHDOWN STORIES

The material gathered by Raper allows glimpses into lives of the less privileged within Victorian rural society. We see that rural custom was contested and that rural working communities, caught between manorial, estate and parvenue interests, sought actively to turn the inevitable overlapping interests to their own advantage rather than merely succumb to those seeking to control them and their environments.

In 1847 Rev. John Cox, in discussing the social segregation affected by the decline in living-in service with the Select Committee on Settlement and Poor Removal, very candidly and revealingly admitted that 'I do not know how they reason at all.' George Coode, in his 1851 Report on Settlement and Removal of the Poor to the Poor Law Board, saw 'a wide distinction ... between the two classes [masters and servants]'.[104]

Raper was a Victorian 'collector' (and amateur archaeologist and historian), and he has bequeathed a rich source from which we can discern a discourse on resource utilisation, custom and local political power in ways that complement the educated perceptions of the elites.[105] There is certainly no veneer of rosy retrospection in these accounts, but rather a radical conservatism through the remembrance and cultural construction of litter-cutting activities within a tradition of resistance. The evidence was assembled for a specific legal purpose rather than a 'passion for enquiry' into the lives of working people, and therefore we need to be aware of the context of the depositions and indeed of issues relating to the very act of collecting the accounts.

With prosecutions of litter cutters and others mounting, many elderly people had long been engaged in providing evidence on one side or the other, and sometimes with more than a hint of rebellion. At the East Grinstead Petty Session in March 1868, at which Daniel Heasman was sentenced to imprisonment for cutting underwood, John Heaver, who ten years later was to give evidence to Raper, was called to give witness. He was then aged 68 and living on the edge of the Forest, and 'had cut underwood 42 years last month, when he had an opportunity'. He had seen 'other persons cutting, but never saw them interfered with until last Wednesday, and never saw that (laughter).' In cross-examination he stated that he had cut wood in the previous week and 'took whatever he could get'. James Todman (aged 77) from near Forest Row deposed that he had cut underwood on the Forest and seen boys and girls cutting it and not being interfered

[104] P. Joyce, 'The Sorrows of Edwin Waugh: a Study in "Working-class" Identity', in P. Joyce (ed.), *Democratic Subjects: the Self and the Social in Nineteenth-Century England* (Cambridge: Cambridge University Press, 1994), 23–82.

[105] For problems associated with the evidence see Short, 'Environmental Politics, Custom and Personal Testimony'.

with. He used to cut any kind of wood and 'the reeve never caught him but might have done if he had looked after him (laughter)'.[106]

Raper's notebooks do contain some material of an interpretive and reflective, rather than a factual nature, though not a great quantity. We have, for example, the deposition of the somewhat radical Henry Scott (1806–1898) from Duddleswell, 'intelligent but rather decrepid [sic]', who remembered a previous earl sending teams at night to carry away turf cut by squatters, who resisted and took the turf back off the wagons. Unusually, Raper reverts to a third-person note within his transcription of Henry Scott's evidence, writing that 'he also speaks of the litter cutters breaking into the bounds or reservations of 1816 and cutting the litter but he does not give any intelligible reason'. And he earnestly wrote down, presumably verbatim, Scott's feeling towards the seventh earl that 'I don't see why one fat hogg that has got plenty should swallow up the lean ones.'[107]

The depositions also reveal the continuing existence of living-in farm service in this part of Sussex, with successive places of service being outlined.[108] Much was very short-distance, but a constant ebb and flow of younger men across and off the Forest is a feature of the accounts. A rather sad example was recounted by Michael Maynard (1809–1885), 'born twentieth May 1809 expects he was baptised in East Grinstead'. He was living at Coldharbour, near Plawhatch Gate:

> I was born on the property and lived there till about 14 when I went to service on Cansiron Farm for about 1 year, then to Holtye with a Mr. Gasson who had Courtlands Farm for 1 year. Then went brickmaking in the summer and hoopshaving in the winter for 3 or 4 years. During this time I went backwards and forwards to home. Then I had my leg off at Brighton hospital from a cold caught brickmaking. Then went turnpike gate keeping and other occupations till I came home about 30 years ago.[109]

He was indeed returned in the 1841 census as a turnpike gatekeeper – in Aldrington, near Brighton. Interestingly he was returned as the only inhabitant, and as a consequence of his wooden leg the parish population was reckoned as ¾![110] By 1851 he was described as a 'proprietor of houses' in West Hoathly with his widower brother George. By 1871 he was farming 16 acres at Coldharbour Farm, East Grinstead, aged 62, so his disability seemingly did not prevent him from leading an active working life. When he died he left estate worth just over £1,000, a not insubstantial sum.

---

106 ESBHRO, ADA 100 (press cutting 30 March 1868).

107 Short, *Ashdown Forest Dispute*, 179.

108 The living-in arrangements of Abraham Edwards and James Everest from when starting out as teenagers within the parish of Hartfield are shown to have lasted between 6 months and 25 years for James Everest at the time of the interviews. See Short, 'Environmental Politics, Custom and Personal Testimony', 470–95.

109 Short, *The Ashdown Forest Dispute*, 112–13.

110 *Illustrated London News*, 11 November 1843, 317. The remains of Aldrington were absorbed into Hove in the later nineteenth century.

Some, like Michael Maynard, might find wage labour when available, but cash might be scarce and barter was also common: families in debt to miller Edward Heaver at Tablehurst repaid him in litter. His nephew recalled: 'The poor people used to get into his debt for flour and he used to work out doubtful debts by putting them to cut litter. I often saw his team carrying the litter.'[111] Henry Scott mentioned that his father had cut litter as payment for a farmer who had sold him some pig pens.

We might also notice intersections of the deposed evidence with a world beyond Ashdown. Walter Weeding (1829–1879) of Hartfield, meeting Raper shortly before Walter's death, was a Crimean War veteran, having served with the 3rd Battalion Grenadier Guards 1854–5 and received the Alma Inkerman Crimean War campaign medal. He had served for 25 years and, although in 1861 he was still recorded as a soldier, he was lodging at a beer shop in Lewes. He must have returned to Hartfield soon after his discharge, where his employer, Bernard Hale, referred to him not only as a Crimean veteran but as a hero. Weeding had been employed by him, both before and after the war, up until 1875.[112]

One national personality was Professor St George Jackson Mivart FRS (1827–1900), the eminent but controversial barrister-turned-biologist who had converted to Catholicism in 1844. As author of *On the Genesis of Species* (1871) he was a contemporary and adversary of Huxley and Darwin, and attempted to reconcile evolutionary theory with Catholicism, falling foul of both camps and being excommunicated shortly before his death. He owned 96 acres and lived at Wilmshurst, Fletching, according to the 1874 *Post Office Directory*. By 1875 he also had an address at Great Russell Street, London, when he purchased Lighthouses (or Light Rowe) of 25 acres near Mark Street, Fletching, for £200. In 1876 Mivart's property, as with many similar ones, was enfranchised, then leased to Jasper Botting among others and eventually sold to Sir Spencer Maryon Wilson in December 1882.[113] Botting told Raper that about five years previously:

> Professor Mivart who had bought it [Light Rowe] enlarged the house and laid out grounds and let me part of the farm. I have no buildings there but have carried manure from Woolpack Farm made from the litter on the land. Professor Mivart has used the rest of Light Row and I have carried litter there for him from the Forest.[114]

Mivert's connection with Ashdown is previously little known, and his living on the Forest at the same time that Darwin was visiting Dr Wallis and walking the Forest is an interesting illustration of the burgeoning importance of the Weald as an open-air laboratory at this time for natural science.

---

[111] Short, *The Ashdown Forest Dispute*, 99, 178–9.

[112] ESBHRO, AMS 3904; TNA, WO 100/25. I am extremely grateful to those who have researched Walter Weeding's life and put so much of this information on Ancestry.co.uk.

[113] ESBHRO, SAS G/ACC/904, 154–5, 192; SRL 2/16/18a and 18b. Mivart's evangelical father was the wealthy owner of Mivart's Hotel (now Claridge's), selling it in 1854 to the neighbouring Mr and Mrs Claridge.

[114] Short, *Ashdown Forest Dispute*, 231.

One word of warning about the depositions comes from a statement made relating to the witnesses' proofs for the De la Warr vs. Maryon Wilson lawsuit:

> Nearly all the witnesses whose evidence was not taken are cottagers or in a small way and the whole of the witnesses either treated the matter as a joke and implied that those who were clever enough might help themselves to what they wanted, or else were rather shy of admitting what they had done, thus showing a doubt as to their rights.[115]

Coventry Patmore, who as we have seen was rarely at ease with his Ashdown neighbours, noted 'another moral peculiarity'. 'They were', he writes, 'for the most part always ready to sell "affidavies" made to the fashion desired by the purchaser, at extremely low prices.' All this was done with 'Arcadian simplicity'. He attributed this tendency to:

> the almost total absence of a soul-disturbing faith. Baptism was not commonly practiced even among the more settled of the country-folk ... When I asked my ploughman if he had had his baby baptized, his answer was, 'No, I don't hold with baptism; it never did me no good.'[116]

So, were Raper's deponents shy of admitting to litter-cutting, for fear of legal repercussions? Seemingly not. To Raper at least, the depositions were given with no anxieties about their actions, with few traces of self-justification and even with a sense of composure; what they were doing was customary practice constituting part of belonging to Ashdown's culture. At this stage in his career Raper remained something of an outsider, so the levels of confidence among his deponents were possibly less, at least initially, than if he were an insider. But, on the other hand, they might not be coy about expressing opinions to him that they might not articulate to a member of the local gentry.

THE STRUGGLE FOR ASHDOWN

The interactions between common rights, class fractions and local interpretations of national ideologies of conservation assume different forms within different environments. Ashdown offers a detailed study of the conflict between custom, as perceived by the communities living around the Forest, and nineteenth-century reform and legal 'rights', as expressed by the De la Warrs as owners of the soil. Under threat were the older understandings of those Forest customs to which both landed and landless clung. Whether legally correct or not, the collecting of litter was seen as a tolerated custom, indeed a right into which many Ashdown residents were born. Here was a constituent part of the nexus that bound families rich and poor together on

[115] TNA, LRRO 67/1 pp. 48–9.

[116] Patmore, *My estate*, 11–13.

common ground, resulting in a greater sense of belonging, being a part of a community, however scattered across the Forest.

The Forest permeated daily speech and activity, and was a taken-for-granted element in everybody's lives. Access to the Forest bound landowners, tenant farmers, cottagers and landless labourers in a mutuality of exchange and a network of employment contracts dependent upon a local understanding of what could and could not be undertaken on the Forest, although, as we have seen, the interpretations of such rights varied from group to group and indeed from one time to another. Here were place-specific customs, with common rights fluid, negotiable, not immutable, defined by practice and oral tradition rather than written code, reaching back beyond memory and whose continuation was recalled in oral tradition. Each side in the dispute could claim historical precedent. Forest rights could be interpreted to suit the case being made by all concerned.

This struggle, fractured as it was along class and cultural lines, illustrated well the use of what Scott has referred to as 'the weapons of the weak'.[117] There were no large-scale acts of collective resistance by the poor, and the legal fights were left to the wealthy. Instead we can discern innumerable acts of defiance – encroachment, depasturing, raids by women on the manorial cattle pound, litter gathering, wood and turf cutting, stone quarrying: all low-profile attrition sanctioned within local working culture and over generations becoming 'rights' through daily practice. Some acts were clandestine, some showed a makeshift creativity, but here were the tactics of survival against manorial, estate and parish officers, and perhaps also increasingly against those bringing with them metropolitan capital.

In these ways the legal and local cultural boundaries were constantly tested within what was a complex socio-cultural environment composed of elite landowners (some with common rights, some without), larger tenant Commoners and their labourers who cut litter for them, poorer squatters who helped themselves, illegal litter cutters who sold the litter on, and non-resident itinerant families. The struggle for making a living on Ashdown was thus constantly played out between and within fragmented social groupings.

In a very real sense, when De la Warr mounted his challenge to Hale and Miles, he also challenged fundamental working practices among the poor. And it is significant perhaps that, although he won the first action on the technicalities of Commoners' rights as set out in 1693, the subsequent appeal succeeded because it was based precisely on the everyday understandings of the inhabitants as to their normally accepted rights of user over at least 60 years.

The dispute set out here was social, cultural and economic, but the legal outcome also ensured that the Forest provided the landscape for a notable piece of environmental legislation, culminating in the establishment of a Board of Conservators. But, as we shall see, this did not end the social and environmental conflict. By the 1880s Ashdown was at a low ecological ebb. The depressed rural economy during the 1870s and 1880s

[117] J.C. Scott, *Weapons of the Weak: Everyday forms of Peasant Resistance* (New Haven, CT: Yale University Press, 1986).

was felt, and the Forest was economically marginal and threatened, its communities rent by underlying factions and feuds. But help for the landscape, if not for the cottage economy, had arrived in the shape of the emergent conservation interests, which, reinforced by a growing concern for the scientific analysis of habitat types, continued to attempt to protect the Forest into the twentieth century.

CHAPTER 10

THE EARLY YEARS OF FORMAL CONSERVATION, 1885–1914

> If the judgement of the Court of Appeal [1881] had been in favour of Lord de la Warr, there can be little doubt that ultimately he would have been able to compel the Commoners to submit to inclosure; as it was, the Commoners' rights saved the Forest, which is an exceedingly beautiful and valuable Open Space.[1]

IT MIGHT BE THOUGHT that the conclusion of the Ashdown Forest Case, the resulting registration of the Forest and the establishment of Conservators would have subdued conflict and ushered in some tranquillity; and that the gradual erosion of the Forest's rural economy as structural, economic and social changes in the twentieth century brought more leisure time, more commuting and a greater significance for conservation would have remodelled Ashdown into the apparently peaceful open space that the casual visitor sees today.

Not a bit of it. This chapter instead relates a familiar theme as conflict continued through to the First World War. The Conservators faced ongoing challenges of many kinds, some of which are indeed still faced at the present time. As noted in previous chapters, the Forest was now being actively sought as a fashionable address, transport links were easier to London, and Ashdown's heathland was seen as beautiful and wild, rather than an ugly, unproductive space.

The chapter therefore further depicts the impact, both positive and negative, of those wealthy families. For example, their arrivals also ushered in feelings among cottagers and smallholders, many of them Raper's deponents, that their own space and customary uses of the Forest were under threat. After all, 'Regulation of access to common resources necessarily involves some degree of exclusion which may well be socially differentiated.'[2] The very fact that an area was demarcated for conservation could itself provoke social unrest, as activities and groups were now

1 Eversley, *Commons, Forests and Footpaths*, 117.

2 L. Shaw-Taylor, 'The Management of Common Land in the Lowlands of Southern England circa 1500 to circa 1850', in M. de Moor, L. Shaw-Taylor and P. Warde (eds), *The Management of Common Land in North-west Europe, c.1500–1850* (Turnhout: Brepols, 2002), 59–81.

to be measured against a new set of criteria.[3] And, to add further context, another theme addresses the impacts of modernity, such as motor cars (with attendant need for 'motor houses' or warning signs to slow down near schools), mains gas and drainage, and telegraph poles.[4]

THE CONSERVATION BOARD

Ever since the park pale was erected over a prior area of common land in the thirteenth century there had been issues over commoning and many attempts to find compromises. The most recent had involved a Committee of the Conservators of Ashdown Forest since at least 1875 and there was also a somewhat shadowy Committee of the Associated Commoners in existence by 1882, but neither had any legal heft.[5]

But now all that was to change with the appointment of a Conservation Board. Formal and legally binding arrangements were to be enacted to ensure Ashdown's future protection. The task would be 'Janus-faced', looking to the interests of the Commoners but also to the protection of the landscape. Furthermore, the interests of the wider general public were also to be included, with provision for recreation grounds and access to sites of interest. Could all three be accommodated? Rev. George Maryon Maryon-Wilson, longstanding chairman of the Conservators, later set out their task in a letter to the Commons Preservation Society in 1919:

> the protection of the Forest from encroachments and trespasses and its preservation in its natural condition ... for the benefit of the Commoners, setting out specific areas within the Forest, as recreation grounds for the use of the inhabitants, and securing for the public the enjoyment of the woodland scenery and extensive views over the surrounding country.[6]

To finance the Forest management, rates would be levied on those benefiting from the regulation and any improvements, and income might also come from the sale of any outlying or small areas of Forest, not exceeding one-fortieth part of the whole. In 1889 Raper calculated that the rate, at 6d per acre and levied every 15 months, would yield about £250. It was, however, a highly unpopular rate and seen by many as a tax on ancient commoning rights. There was also a cap, so that

3 For examples from the early years of conservation in the USA see Karl Jacoby, *Crimes against Nature: Squatters, Poachers, Thieves and the Hidden History of American Conservation* (Berkeley: University of California Press, 2001) and also C. Griffin and I. Robertson, 'Moral Ecologies: Conservation in Conflict in Rural England', *History Workshop Journal*, 82 (2016), 24–49.

4 The East Grinstead Gas & Water Company was actively spreading its networks around the Forest.

5 ESBHRO, ACC 5640/6/8 The Report of the Committee of the Associated Commoners (1882).

6 <www.ashdownforest.org/enjoy/history/HistoricalMinuteBooks.php. Transcript Historical minute book 1913-23> [accessed 9 June 2019]. Unless otherwise stated, the material for this chapter is derived from the minutes of the Board of Conservators.

the larger Commoners were not charged overmuch. Refusal to pay could incur legal action: one Commoner whose rate had gone unpaid on two occasions had his pony cart impounded at Maresfield.

The membership of the new Conservator body initially reflected, to some extent, those who had been involved in the Ashdown Forest lawsuit and the Committee of the Conservators of Ashdown Forest. They had to be Commoners and they were elected on the basis of one acre/one vote, thereby prioritising those with larger holdings and giving rise to a longstanding grievance on the part of the smaller foresters.

Their first minuted meeting was held at the Nutley Inn on 5 May 1887, which was to become the normal venue for their future meetings.[7] The minutes of these meetings are relatively full and allow a reconstruction of the early years of Ashdown's new management structure. Sometimes we hear subaltern voices, as minutes recorded cottagers' activities, their perceived misdemeanours, anxieties over gypsies and sometimes hints of class consciousness as the attitudes of the wealthy were revealed. There were 12 members plus the representative of the lord of the manor. Four members, chosen by lot, would retire in the January at the end of each of three years, although they would be eligible to stand for re-election, which in practice they generally did, and members might retain their place for many years. The Conservators themselves were unpaid, and on occasion had to miss meetings because of other business commitments or summer sojurns.

William Augustus Raper was appointed clerk; interestingly, he was nominated by Earl De la Warr, his old legal adversary. The initial Board, perhaps inevitably, showed a wealth bias (Table 10.1). The Searles estate was represented by Sir Spencer Maryon-Wilson (a member 1887–92) and later by the Rev. George Maryon Maryon-Wilson (1860–1941) (a member 1897–1929 and chairman 1900–29). By the early twentieth century others included the earl of Sheffield from 1887 to 1903 and the leading Catholic layman James Fitzalan Hope (1870–1949), grandson of the fourteenth duke of Norfolk, a member from 1892 to 1903. Until his death in 1903 Guy Hardy, Maryon-Wilson's brother-in-law, also sat as a Conservator. And in 1905 they were briefly joined on the Board by the Rt Hon. Thomas Baron Brassey (1836–1918), the first Earl Brassey from 1911, former governor of Victoria 1895–1900 and an inlaw of the De la Warrs. He had Chelwood (or Chapelwood) Manor designed in 1904, but resigned in late 1906, citing difficulties in attending the Board meetings.[8]

7 The Nutley Inn later became the Shelley Arms Hotel, which became a private house in 1994.

8 The son of Thomas Brassey, the railway contractor. His third daughter, Lady Muriel Agnes, married Gilbert Sackville, eighth Earl De La Warr, and was the mother of Herbrand, ninth earl. Hope was Conservative MP for Sheffield Brightside 1900–06 and Sheffield Central 1908–29, serving under Asquith and in Lloyd George's wartime cabinet. In 1932 he was elevated as Baron Rankeillour of Buxted. He had a strong regard for his grandmother's Heron's Ghyll, and was buried there [Geoffrey Throckmorton, 'Hope, James Fitzalan, first Baron Rankeillour (1870–1949)', rev. M. Brodie, *ODNB* (2004).

Table 10.1. The Ashdown Forest Conservators at their first meeting 5 May 1887.

Name	Address	Note/occupation
B. Husey-Hunt (representing Earl De la Warr)	Offices in Lewes	Earl's steward. Replaced quickly by E.A. Nicolson
Henry R. Freshfield (chair)	Kidbrooke Park since 1874	Senior partner in solicitors
Bernard Hale	Holly Hill	Co-defendant in Ashdown Forest Case
Albert H. Hills	Perryman's Grove, Buxted	Landowner
Sir Spencer Maryon-Wilson	Searles, Fletching	Landowner
Capt. William Noble	Forest Lodge, Maresfield	JP chair of Uckfield Bench etc
Joseph Ridley	Misbourne Farm, Horney Common	14 acres
Earl of Sheffield	Sheffield Park	Landowner
Mark Sandford (representing Lady Louisa Shelley)	Dairy House, Maresfield Park	Agent for Landowner
Albert Turner	Nutley	Property owner, timber and coal merchant
Joseph Wallis	Coopers' Arms, Crowborough	Publican
William Carr	Nutley	8 acres
Rev. J.M.B. Butler	Maresfield	Vicar of Maresfield

Source: Historical Minute Book 1887–97.

Representing De la Warr, Husey-Hunt soon gave way to E.A. Nicholson; and Bernard Hale, such a leading figure in the Ashdown Forest Case, was forced to step down owing to ill health by 1889. There was also a constant, though minority, presence of working foresters – Joseph Ridley of Misbourne Farm, Horney Common and William Carr, often supported by Mark Sandford, representing Lady Shelley. Sandford was the most consistent of all the Conservators, present from the first meeting in 1887 through to May 1913, when illness caused him to miss a meeting. Although meetings of the Board were less well attended during the summer months, only on one occasion was a meeting adjourned because it was inquorate.

Importantly, alongside these meetings there was an annual meeting of the Commoners each December, although this was rarely well attended except by the Board members themselves and might last for only 30 minutes, primarily being concerned with the election of Conservators for the following year. It became normal for the latter to reassemble for a Board meeting directly afterwards.

The first Board meeting on 5 May 1887 was concerned almost entirely with the formulation of byelaws. There were 23, which were submitted to the Land

Commissioners for approval by the Home Secretary. The byelaws incorporated many of the controls that had been exercised historically by manorial courts or by the short-lived associations of Commoners. The Conservators reserved to themselves the power to control numbers of authorised grazing animals, with any transgressor liable to a fine of 40s. Legal action to recover the fine could thereafter be instigated. These initial byelaws now forbade vagrancy and camping; the taking of birds and game; lighting fires; turning out unauthorised animals; the taking of turf, peat, gravel and stone; damaging gorse, furze, broom, brakes, fern, heather, rushes, grass, litter or herbage; cutting trees and underwood; defacing authorised noticeboards; dumping timber, manure and refuse; fighting and obscene language; the use of firearms and unlawful assemblage; and unauthorised enclosures.

It is as well at this point to confirm that there were now three groups of interested parties closely involved with the Forest: the Conservators from 1887, the Commoners, and those who lived around the Forest, who might enjoy its benefits but without necessarily having any common rights. Members of this latter group were generally referred to as foresters, and they soon formed a Foresters' Protection Association. At their second meeting on 11 June 1887 the Conservators were faced with the latter group when a deputation from the Ashdown Foresters' Protection Association, consisting of T.C. Thompson (Ashdown Park), Mr Slack (Forest Row) and Mr Bunce (Brighton), requested some amendments to these byelaws. After careful consideration, some amendments were made that were duly agreed by the Home Secretary. In fact, the byelaws were amended again in 1905 when the Commoners again forced some changes.

In September Harry Hickmott was appointed the first ranger. At his initial attendance in October 1887 he had to report that already ten of the new noticeboards erected to display the byelaws had been variously defaced with paint, tar or cow dung, or removed altogether. Such actions were reported again in December and remained a problem thereafter, an early sign of the poor local regard in which the Conservators were held, although some damage was undoubtedly attributable to casual vandalism. Hickmott resigned after just three months, to be replaced by the unpopular and aggressive John Robinson of Duddleswell. However, by August 1888 Robinson had absented himself from his duties and resigned. Hickmott deputised pending a replacement and in November James McQuillan was appointed, but in turn he lasted only until the following August.

Clearly this was a difficult and unpopular post within the local community. An advertisement in the *East Grinstead Observer* of 27 July 1889 stated that 'A policeman or non-commissioned officer of Engineers or Artillery, about retiring on a pension preferred … '. Four men applied but all withdrew after visiting the Forest and meeting the Conservators, and three others then followed suit. Eventually John Taylor, Inspector of Nuisances for the Uckfield Union, was appointed temporarily. This too was unsatisfactory, since by February 1891 Taylor was unable to produce a report owing to demands upon his time from the Uckfield Union. The Conservators reprimanded him, expressing dissatisfaction with his work over the previous three months, and in August he too resigned, to be replaced by William Faulkner, an ex-Sergeant of the

Metropolitan Police, who lasted until November. Then came Samuel Soper from Danehill, who held office rather longer – until September 1904, when family concerns prompted his resignation. He was replaced by Alfred Pilbeam for two years until he left to join the police and was replaced by William Manners of Chelwood, until he too resigned at the end of 1911. Thus, there had so far been eight appointments in the 24 years since 1887. But now at last came the long-serving Herbert Kirby of Maresfield.[9] His duties were set out:

> to include particular attention to the removal of gypsies from the Forest, to superintend and assist in making and draining roads and tracks on the Forest and clearing out ditches, to keep a sharp outlook for the outbreak of fires and be prepared with assistance to extinguish and to do any other work connected with the Forest which the Conservators may require … salary to be 20/- per week with house rent 5/- per week, his engagement to be weekly … and to be terminable by either party four weeks previous notice as at present.

In 1907 the Board had decided that a second Ranger was needed, to be based near Forest Row and funded by donations from wealthier Commoners, with a rent-free cottage in Broadstone Warren provided by Douglas Freshfield, only son of the late former chairman of the Board. William Brown, a retired police constable, was duly appointed and proved zealous, working until his retirement in 1926. Many of the wealthier Commoners duly subscribed, but by 1910 there were those who objected:

> Lord Robert Cecil has replied that owing to the many calls upon him it is impossible for him to continue his subscription; Mr. Misa has declined to renew his subscription on the grounds that he never sees either ranger, and that whenever he sees gypsies, dead donkeys or other things on the Forest he cannot get hold of a ranger, and that they do him no good.

Another rather obvious issue was addressed in 1899. Who now exactly controlled Ashdown? Ever since the Conservators were established in 1887 there had been difficulties in what was in effect a system of dual control over the Forest. The lord of the manor of Duddleswell had retained many powers as owner all the while that the Conservators were attempting to enforce their duty as managers under the Commons Act 1876 and Ashdown Forest Act 1885. In an effort to end a frustrating situation the Board now sought a lease that would effectively hand over four important powers to the Board: the power to cut underwood, in part to raise money to pay the rent (£75 annually) to De la Warr; the power to authorise drainage where necessary, although not to extend this to non-Commoners; the power to authorise the digging of stone – an ongoing issue between the two parties, as De la Warr retained the rights to minerals over Duddleswell manor; and, finally, sporting rights which the Board wished to control. But these powers

9 Herbert was the grandfather of the late Peter Kirby, whose photographic collection provided the basis for his *Forest Camera*. Herbert had previously worked on the Maresfield Park estate for 19 years. On taking up his ranger's post he moved to Fords Green, where Peter also later lived. Herbert retired in 1948, becoming manor reeve to the De la Warrs until his final retirement in 1961, to be succeeded by Fred, Peter's father.

were rejected by De la Warr. The Board expressed its thanks to the sub-committee negotiating with him and regret at 'the futility of their efforts'.

Nevertheless, the issue was not allowed to drop, and in 1902 it was raised again. This time the Board suggested that 'three substantial persons' might take on the lease, enabling 'certain gentlemen [who] were interested in the Forest ... [to assist] the Conservators in the protection of the Forest from injury and the Commoners in the protection of their rights'. In fact, four stepped up: newcomer Count Alexander Munster (Maresfield Park), Douglas Freshfield (Kidbrooke), George M. Maryon-Wilson (Searles) and Charles H. Corbett (Dane Hill Park). The rent for such a lease was now proposed at £150. But again the proposal foundered and little more could be done. Following 'discrete enquiries' in October 1910, the executive committee of the National Trust also suggested that they might pay up to £6,500 to acquire the rights from De la Warr, with the Conservators retaining their management oversight. Discussions proceeded between January and May 1911, but, once again, came to nothing.[10]

THE COMMONERS 1887–1914

In 1887 the names and addresses of the Commoners were reported, together with the acreage of their holdings, to facilitate the levying of 6d per acre on land with common rights. Initially there were 66 properties identified, ranging in size from just ¼ acre up to the 1,706 acres of Lady Shelley. By 1902 the number of registrations had risen to 156, although subsequent subdivisions of properties left the Conservators somewhat puzzled as to the exact number of those entitled to common rights. By 1916 the number had risen to 219.

The largest registered properties are listed in Table 10.2. Most of the names are familiar from earlier chapters, but we might note that Miss Clara Tapps-Gervis (1831–1910) held Ashdown House (and the manor of Lavortye) through descent from Clara Meyrick Fuller, her mother, who had married into the Tapps-Gervis family. Although she spent much of her time on Anglesey, Clara was residing at Ashdown House in 1881 with her five household servants. Mrs Nesbitt was Alexander Nesbitt's widow. On her death Oldlands was put up for auction with its 11 tenanted farms, leading to problems for Raper, when these were sold in June 1914, in trying to assess which of the properties were entitled to common rights.[11]

Finally, the Cranston Trustees were listed. A house at East Court, East Grinstead, had been built in 1769 by John Cranston, a London lawyer who had been Assistant Warden of Sackville College. He married Catherine Green, whose family owned the nearby farm known as Estcot's, and it was here that East Court was constructed and

[10] National Trust Executive Committee minutes October 1910–May 1911. The Trust had been founded in January 1895. Sir Robert Hunter on the executive committee of the Trust provided a link between the Trust and Ashdown from his involvement with the dispute 30 years before. I am grateful to Daz Beatson, archivist for the National Trust, for help with tracing the NT minutes.

[11] ESBHRO, ACC 448/2/14 Printed particulars of sale of the Ashdown House Estate 1910; SAS PS/334 outlying portions of the Ashdown House Estate.

the estate developed. Edward, his only son, inherited East Court in 1781 and died in early 1841, upon which his five daughters jointly inherited the estate, together with his widow Harriet. By 1887 the property remained in the hands of trustees, who could both sell and purchase assets.

Table 10.2. Landowners with the largest commonable acreages, 1887.

Name	Address	Commonable acreage
Lady Shelley	Maresfield Park	1706
Sir S. Maryon-Wilson	Searles, Fletching	1150+
Earl of Sheffield	Sheffield Park	950
Miss Tapps-Gervis	c/o Hastie, E. Grinstead	925
D. Larnach	Brambletye, E. Grinstead	500
M. Snooke Grosvenor Woods QC	9 Old Square, Lincolns Inn	498
Cranston Trustees	Bedford Rd WC, London	323
Mrs Nesbitt	Oldlands, Buxted	300*
R. Streatfeild	The Rocks, Uckfield	240
Lord Colchester	Carlton Club, Pall Mall	213

*285 acres in 1889; + 1265 by 1893
Source: Historical Minute Book 1887–97.

The expansion in Commoners' numbers was partly due to improved registration, but there was also now a conscious and proactive desire to incorporate more Ashdown residents into Forest benefits. By this time Elphinstone Barchard and the Foresters' Protection Association were arguing that the Conservators had little interest in the smaller cottagers, and so this represented an attempt to head off such criticism. A lively meeting of the Association had upheld the idea that the Conservators had no powers to prevent men from exercising rights which had been theirs for many generations. The Board was clearly stung by such comments. Raper assiduously drew up a list of those small occupiers who could claim user over 60 years, and in October 1889 he issued a lengthy general notice to the Commoners:

> None of these small holders originally had any rights on the open Forest. It is suggested that some of them, by 60 years user, have acquired rights, and that nearly all have used the Forest without legal warrant or authority. In some cases offenders have been prosecuted, but detection is difficult, and legal proceedings are costly and undesirable.
>
> Gross misrepresentation has been made as to the action of the Conservators in putting a check upon these practices, and ill-feeling has in consequence arisen. It has been repeatedly argued that, as the boon would be great to these small occupiers, it might be expedient that they should be allowed to take brakes, litter, and peat-fuel for use

> on their holdings, and to turn out in the summer such stock as their land will maintain in the winter. No doubt the acreage of these holdings being comparatively small, the brakes and litter which those occupiers can consume on them is also small. It is not from such user that the Forest suffers, but from thefts by persons who cut and subsequently sell the litter to people who have no common rights, and many of whom reside at a long distance from the Forest.
>
> After careful deliberation, the Conservators, undeterred by the unworthy and groundless imputations which have been made upon their proceedings, have come to the conclusion that it is desirable to admit these small occupiers within the pale to the enjoyment of Forest rights, on payment of rates similar to the Commoners; and they have reason to believe that in the event of this scheme being carried out, these small occupiers will see the importance of co-operating with the Conservators in preventing the illegal cutting above mentioned, and generally in protecting and preserving the Forest.
>
> The soil of the Forest being vested in Lord De la Warr, he could prohibit any person, not a Commoner, from cutting or turning out, and the Commoners could do the same. The Conservators have no legal power in themselves to confer these privileges; they, therefore, deemed it right to consult his Lordship before moving further in the matter, and I have the satisfaction of stating that he has expressed his willingness to consent to the proposal.

De la Warr did consent, although expressing concern at the large numbers involved. The scheme was also approved by the Commoners, who even agreed to peat being taken for fuel in 1891. Licences would be issued for such purposes but some smaller occupiers were reluctant to pay the rates. On one occasion ranger Brown called upon George Philcox Sr at Newbridge to collect his rate, only to be severely assaulted with a pitchfork and run off the premises by George Philcox Jr. in 'a violent and unprovoked assault'. The two men were summoned to the East Grinstead bench, Raper attending to support the prosecution, and they were convicted and fined.

The Board deliberated at length over the growing number of claimants for Forest rights. To give one example, Elphinstone Barchard attended the meeting in July 1893 in support of his claim and, echoing Raper's interviews for the Ashdown Forest case, he was accompanied by Mrs Sophia Inns, aged 75, resident near Duddleswell nearly all her life. Based on her evidence Barchard established his claim and his name was duly added to the Schedule of Commoners. Similar evidence was brought forward on behalf of Samuel Heasman (Hoadly Farm, Withyham) and William Heasman (Grubbs Farm, Withyham), this time based largely on the evidence of George Taylor, charcoal burner from Crowborough. Mary Diplock, owner of the Crow and Gate pub, also brought witnesses to successfully demonstrate user, as did Sylvanus Ridley, now of Forest Farm, Danehill, who claimed Forest rights in respect of a house and about 3½ acres of land near the Old Duddleswell Turnpike Gate. But at the same time the Board investigated an alleged encroachment on the frontage of his property, which was duly thrown out. At a meeting on 19 August 1893 the chairman, Albert Turner, stood down, allowing claims in respect of eight of his own properties in and around Nutley to be accepted, although some others were rejected for lack of evidence for 60 years' user. Lady Shelley had five properties accepted.

The depositions in support of common rights continued to provide insights into life around the Forest. For example, in February 1894 Charles Corbett claimed Forest rights in respect of Forest Farm (142 acres in Fletching, of which 108 acres were wood), Danehill or Botches Farm (Fletching and Horsted Keynes, 168 acres, of which 80 acres were wood) and Woodgate Home Farm (Fletching and Horsted Keynes, 140 acres, of which 29 acres were wood).

> George Baker of Chelwood Common aged 83 stated he was born at Dane Hill and with the exception of 4 or 5 years had lived there all his life. When he was a youth Forest Farm belonged to Squire Mabbott who lived at Uckfield ... and employed a Bailiff named Tyman who hanged himself and the next Bailiff named Whistler shot himself. ... Witness worked on the Farm when a boy. He knew that the successive occupiers of the property had litter, they used it while he was working there, a great part of it came from the Forest round the Vetchery. Witness cut litter for them in 'Passion Bottom' 50 years ago and knew they had continued to use it since. He knew the Farmhouse ... near Wheelers Wood. John Day lived there when witness was a boy and kept 3 horses and 3 cows. Witness knew Wood Gates House which belonged when he was a boy to Squire Hutchinson who occupied it and farmed the farm. ... the successive tenants of both properties grazed their cattle on the Forest and not in Chelwood Commons.

It was resolved that Corbett had established his claim and his name was duly added to the Schedule of Commoners.

THORNS IN THE CONSERVATORS' SIDES

Clearly the beginning of the Conservators' oversight of the Forest did not run entirely smoothly. It took time to settle into an agreed form of meeting, with such standard items as turf cutting, encroachments and applications to dig for gravel and stone. But it soon became apparent that there were now four main intertwined issues of concern: the extended Ridley family, gypsies, the Ashdown Foresters' Protection Association, and fires on the Forest.

In an early example, Joseph Ridley of Millbrook and Samuel Ridley of Fairwarp were reported almost immediately the Conservators took office for cutting turf, but before the byelaws were in force. But Joseph was also reported later for cutting peat for fuel as well as attempting an encroachment. Raper was directed in June 1888 to take legal action against him. One must be careful with the name of Joseph Ridley, because there were two men with this name living on the Forest at this time.

One Joseph (1842–1922) was elected as a Conservator and lived at Misbourne Farm, Horney Common, where he had a 14-acre holding. He was also a carrier and was referred to as a carpenter in 1881. He was a constant presence at Board meetings until his resignation at the end of 1900, and was then again elected from 1906. He and wife Abigail (1845–1894) had at least ten children, two of whom died young. The other Joseph (1829–1912) was farming 12½ acres at Millbrook, north of Nutley (later 19½, and then 12 after being ordered to abate his encroachment), and his was a constant name before the Conservators with turf cutting, encroachments and other minor misdemeanours. For example, his son, Spencer Ridley, was seen cutting peat at Millbrook but on being approached by the ranger he claimed the right to cut peat. The Board resolved to prosecute him for a breach of the byelaws. Father and son attended court at Uckfield and were duly fined.

The Ridley name was certainly pervasive: William Ridley owned 3½ acres of copyhold land at Marlpits, Nutley, and was claiming Forest rights in 1892. One of eight children, he was the nephew of Timothy John Ridley (or Gurr) (b. 1842), from Cripps Corner near the Goat Farm, and Sylvanus (1850–1929), tenant of Forest Farm near Chelwood Pale.[12] By 1895 there were no fewer than five Ridleys recorded as having common rights on the Forest: the two Josephs, William, Sylvanus and Timothy John. Their names feature regularly in the Conservators' minutes for minor breaches of the byelaws.

Another Ridley, Samuel (b. 1841), who had come to the attention of the Conservators for cutting turf in 1887, exemplified the spread of the name by fathering ten children by 1881. In 1888 we also hear that Charlotte Josephine Ridley of the windmill, Nutley, had been cutting brakes, and James Ridley at Marlpits was reported for encroaching with a potato plot there in June 1889. And by 1903 we also hear of Mrs Lucy Rose Ridley (b. 1861) at Whitehouse Farm, Horney Common, with 19½ acres of land entitled to common, a larger acreage than any of the other Ridleys. She had married Henry Jesse Ridley (known as Jesse) in 1895, the son of Joseph Ridley, the Conservator. We therefore have the situation where Joseph, a member of the Conservation Board, had a son, Jesse, who became a leading opponent of the Conservators, and another son, Albert, who, as we shall see, was involved in a scuffle during attempts to put out a fire.

It would seem that Joseph Ridley of Misbourne had divided loyalties: again and again his was a dissenting voice on the Board, and he had, after all, been put forward by the Protection Association in 1887. In 1907, for example, when the Board was considering prosecuting Alfred Coleman of Fairwarp, apprehended in the act of setting fire to furze, his was the lone voice against prosecuting. And, even more seriously, at the December 1908 annual meeting of the Commoners he put forward a motion to enclose Ashdown and divide it into freehold properties. He felt that this was necessary because the Conservators seemed unable to grant Commoner status to others around the Forest and were unenthusiastic about the creation of smallholdings. He was supported in his motion by his namesake of Nutley and Sylvanus Ridley. A vote was taken and, based on the acreages held, the Ridleys were outvoted by just under 55 acres to over 800,

12 Timothy John Ridley or Gurr was referred to in this way because he was born out of wedlock to John Ridley and Jane Gurr in 1842 and baptised as Gurr. The 1871 census uses the term 'alias Gurr'. John and Jane subsequently married in 1844, with Sylvanus Ridley being born in 1849. Timothy married Martha Chatfield in 1865, when he was referred to as Gurr, and they went on to have ten children. He seems to have used the names Ridley and Gurr interchangeably, dying in 1924 and being recorded as Gurr.

with other Conservators such as Douglas Freshfield and seven other members of the Board voting against. Others unable to attend also protested against the motion, such as Count Alexander Munster and Countess De la Warr.

Joseph Ridley did not attend the Conservators' meeting that followed on from the Commoners' meeting that day. It is perhaps also revealing that he was never selected to serve on any of the standing sub-committees of the Board, such as those dealing with the Maresfield and Forest Row areas, or the audit committee, or those later dealing with the golf clubs. And in December 1909 he was effectively silenced by chairman Maryon-Wilson at the Commoner's annual meeting while making a speech about the benefits to the Forest of the Protection Association.

But in May 1897 Timothy John Ridley/Gurr at Cripps Corner really overstepped the mark. He was found to have diverted a stream around his property further out into the Forest and was enclosing the resultant Forest land between the two watercourses when his actions were seen by Raper, accompanied by the ranger and two Board members. Other members of the Board were outraged at such a blatant transgression. The Board itself lacked the power to initiate proceedings under the 1876 Commons Act and it was agreed that County Court proceedings should be initiated by three members with the backing of Lord Sheffield and Mrs Pechell.[13] An injunction with costs was found against Ridley, but on 6 November the Board learned that, although Ridley was infilling the ditch, he was also encroaching onto the Forest again. The ranger was asked to monitor Ridley with a view to taking further proceedings against him, and it was reported that Ridley had attempted to hide his encroachment behind young trees, underwood and gorse around his property. It was agreed that all this growth should be cut away to reveal the true situation. Four years later the Forest Row sub-committee reported that, following an inspection of Ridley's premises, including a sand pit near the Goat:

> [we] found that a considerable area of Forest ground had been destroyed by it, that there had been no attempt to level the debris or to preserve or re-spread the top spit and that a larger area of surface of the Forest round the pit had been seriously injured by the making of a number of cart or wagon tracks from the pit, persons drawing sand from the pit appear to have taken any route convenient to them over the Forest regardless of whether there was a track way or not and where the ruts in any tracks had become too deep to have made fresh tracks in the vicinity.

Wider matters also impinged. In June 1891

> Joseph Ridley gave notice of his intention to move at the next Meeting that this Board make application to the Board of Agriculture to take a Bill in the House of Commons to finally abolish the copyhold system now existing in Duddleswell Manor and in all other Manors where the burdens are grievous to be borne throughout the United Kingdom.

13 Blanche Pechell (1836–1898). As she was in poor health the Conservators requested the Board of Agriculture to grant her a delegate who could vote on her behalf at Commoners' meetings, but this was refused. The estate was sold by a devisee under her will on 14 December 1898 to Count Alexander Munster.

Unfortunately for him the chairman at the next meeting ruled his motion out of order, precluding any discussion. In fact, attention was being given nationally to just this question, as archaic copyholds were being converted to freeholds or leaseholds, and under various Acts, including that of 1894, the enfranchisement of copyholds had actually become compulsory. Rights of common were unaffected.[14]

And so the seemingly continuous battle against the Ridleys continued. In 1899 Sylvanus Ridley had by his holding a chicken coop, a large mound of turf pared off one of his fields, battens, old tree roots, remains of a collapsed cart lodge, a harrow, an old cart, two old wheels, some bricks, an old tub, a heap of sand, a mixen (near a pond, so a health hazard), a faggot and a peat stack, two stacks of litter and some old clothes posts. Typically, when upbraided by the Conservators he apologised, cleared away the debris and promised not to do it again. But right through to 1914 his name appeared in the Conservators' minutes, sometimes for committing an action to be investigated, sometimes as a complainant and, very occasionally, as a forester congratulating the Board on dealing with a particular issue. He was certainly a regular attendee of the annual December Commoners' meetings and frequently spoke out, agitating for some minor or, in some cases, more important change. He was, for example, a vocal opponent of the voting system for appointing Board members, arguing that the system of voting by acreage of commonable land should be replaced by a more democratic one-man-one-vote system.[15]

A second area of concern for the Board was the ongoing prevalence of gypsies on the Forest (Figure 10.1). In 1902 Timothy Ridley/Gurr again came to the Board's attention, this time for allowing gypsies to camp on his land at Cripps Corner. But the Board was powerless to act, as this was on private land. A similar issue concerned George Inns at the same time, who was also allowing large numbers of gypsies onto his land. At one point Inns stated that the ranger entered his land and one of the caravans, a claim that was denied. Where court attendances were required the travelling groups had often disappeared, and recurrent fines did little to deter them from reappearing. Janetta Smith, daughter of a travelling hawker family, seems to have been a regular visitor. In 1871 the census finds her, aged six, among five 'travelling house carts' at the Sussex Oak Inn, Withyham, another family there being that of John Weston, 'Ginger bread seller'.[16] One of the Ridley extended family also seems to have been among them: Alfred was convicted at Petty Sessions of having a tent near Plaw Hatch in breach of the byelaws and fined, but did not appear and went to prison in default of payment.[17]

In December 1887 Board chairman Henry Freshfield reported the case of 'a man named Jones who, with his family, has been living in tents on the Forest near Tompset's Bank'. The ranger had served a notice to remove, but to no avail. Bernard Hale then

---

14 G.E. Mingay, 'The Farmer', in E.J.T. Collins (ed.), *The Cambridge Agrarian History of England and Wales Vol. VII 1850–1914* (Cambridge: Cambridge University Press, 2000), 800; Copyhold Act 1894 [57 & 58 Vict. Ch 46].

15 A system not changed until the 1970s (Phelan, 'Economy to Amenity', 170).

16 TNA, RG 10/1056 Withyham census 1871. In 1881 she was living in a 'house cart' in Westerham, Kent, where she had been born.

17 In 1891 Alfred Ridley, aged 25, a farm labourer, was one of six young men living in farm outbuildings at Chiddingstone, Kent (<https://www.ancestry.co.uk/>).

Figure 10.1. Gypsies arriving at Millbrook, 1895 (source: Nutley Historical Society).

reported three similar cases near Kidd's Hill and Raper was instructed to take action to remove the families. On 11 August 1888 Raper reported that proceedings at Uckfield and East Grinstead Petty Sessions had resulted in fines for illegal peat cutting by Messrs Hazelden, Izzard, Tester and Jenner; that Absolom Jones was ordered to remove his van from the Forest; and James Brazil was fined for camping there, the alternative to fines being 14 days in prison.[18] Large numbers of gypsies were fined in December 1893, and in 1898 very many were reported as camped between the Crow and Gate pub and the National School on the Uckfield Road, to the annoyance of the school authorities.[19]

By 1906 gypsies were still being prosecuted for pitching tents or having vans on the Forest, and also for leaving a dead horse and later a dead donkey on the Forest, the latter dangerously near a source of drinking water. W.H. Cowan MP (1862–1932), a new arrival at Crows Nest, wrote that gypsies collecting near Poundgate were encouraged by Mr Morphew, the broom maker, and the Board wrote to Morphew's landlord R.J. Streatfeild to ask him to intervene. But, more positively, the travelling families were also thought to be:

> pretty good step-dancers ... they could rattle out a tune on the old mouth-organ and tambourine. They used to get in the pubs over Jarvis Brook ... and when they wasn't fighting, they was pretty musical ... you got all the gypsy families staying there: Smiths, Lees, Fullers, Willetts, Brazils.[20]

Two area sub-committees were formed by 1906 to report on activities in the Forest Row and Maresfield areas respectively, covering the Forest between them. But the Forest Row report to the Board noted that the Jones family of travellers were once again seen: 'John Jones had 3 carts lying on the Forest which must be removed ... [also that] he had horses turned out on the Forest. The ranger is to enquire as to this.' Alfred Inns at Mount Pleasant, Fairwarp, was reported to the Board in 1912 as particularly culpable in

18 James Brazil was in the magistrates' court in 1879 when the magistrates withdrew in connection with the Ashdown Forest Case (see pp. 272–3).

19 Plans for the Church of England National school are dated 1857, the date given in the 1895 directory, and confirmed in ESBHRO, PAR 286/25/1–19.

20 Arthur, *A Sussex Life*, 103.

providing a rendezvous for gypsies, who turned their horses onto the Forest. He denied responsibility but the ranger reported seeing gypsies' carts and horses there on seven separate occasions during January and February 1912 (Figures 10.1 and 10.2).

A third area of concern was the emergence in 1886, as noted above, of the Ashdown Foresters' Protection Association, aimed at protecting the rights of smaller Ashdown cottagers against the new Board of Conservators, who were seen to be more preoccupied with the landscape than with the well-being of the working people. There were some who held that the 1885 Ashdown Forest Act had been 'smuggled through Parliament'.[21] It was certainly the case that much of the Conservators' legislation hit the poor hardest: the cutting of brakes and collecting of litter were restricted to certain times and places; the removal of turf and peat became illegal; gypsies were harried and taken to court. The byelaws specifically excluded what was thought to be rough language or fighting, lighting fires or holding gatherings on the Forest. What has been called 'a patriarchal system of middle-class control was often conducted under the guise of protecting the Forest'.[22]

As we have seen, acts of defiance saw the noticeboards advertising the byelaws damaged. More serious was the ongoing issue of fire-setting (see below), which in some cases was begun to provide fresh grass for cattle, but in many cases was viewed as incendiarism in response to fears about the curtailment of stinting rights. The Association claimed that there were no such malicious fires before the Conservators began. It was when they tried to 'deprive inhabitants of privileges, then the storm begins'.[23] An intimation from the Conservators that the fires stemmed from the promptings of the Association was vigorously denied. So here was the latest manifestation on Ashdown of resistance to elite control, joined frequently by dissimulation, feigned ignorance of the byelaws and many similar proceedings.

Raper, as clerk to the Conservators, now had to face this more rural working-class criticism. He was, however, himself seen by the Association as a great drawback; indeed, his character was perhaps not well suited to dealing with the foresters, and, coming from Battle, he knew little about 'the men of the Forest and their wants'. Raper's careful accumulation of depositions from working people in connection with the Ashdown Forest case 15 years or so previously seems to have been forgotten.

Although resistance to local power had always exerted itself around Ashdown, as in many other areas of common land, the place-specific coalition that bound the Commoners together, whether rich or poor, now broke down along class lines. Smaller Commoners had benefited during the Ashdown Forest Case from the resources of the wealthier Commoners, while the latter's common rights had been safeguarded by enlisting the depositions of many poorer residents. The Protection Association was

[21] ESBHRO, ACC 6815/1/1. I am grateful to Christopher Whittick for drawing my attention to the archive of the Ridley family of Horney Common (ESBHRO, ACC 6815) containing minutes and correspondence of the Ashdown Foresters' Protection Association 1886–1890 and various other letters, lists etc. For ongoing abuses of common rights see ESBHRO, ADA 98, Duddleswell Rental and Custumal *c.* 1883 to 1915.

[22] Phelan, 'Economy to Amenity', 175.

[23] *Sussex Advertiser*, 7 June 1905, and ESBHRO, ACC 6815/1.

Figure 10.2. Thorns in the Conservators' sides: above, gypsies on the Forest and Jesse Ridley canvassing for Liberal Charles Corbett, 1906; opposite, 'Tinker Wright' (left) and Sylvanus Ridley (right) (source: Nutley Historical Society).

formed largely because of a view held by poorer cottagers that the new Board was too elitist: all those appointed were either large landowners or their representatives.[24] The smaller cottagers and working people thus aligned themselves with the Association, which closely monitored the actions of the Conservators and prompted debates over the byelaws and forest fires. The Association also listed those with common rights in 1896 and the properties entitled to estovers.[25]

---

[24] ESBHRO, ACC 6815/1/1 Association meeting held at Nutley school 21 April 1886.

[25] ESBHRO, ACC 6815/1.

By 1887 it was claimed that there were as many as 183 members of the Association. Listed occupations give an indication of class difference compared with the Conservators: included were farm labourers, farmers, carpenters, publicans, well-sinkers, sweeps, brickmakers, squatters, sawyers, broom-makers, bricklayers, blacksmiths, coal merchants, plumbers, shoemakers, grocers, tinkers, corn dealers and roadmen. Many worked at T.C. Thompson's Ashdown Park.

However, the Association's committee actually also came to include wealthy residents: Thompson was joined by Elphinstone Barchard, who became chairman from about 1890; and the Liberal politician Charles Corbett of Woodgate, Danehill, who in fact also became a Conservator 1898–1905. Barchard explicitly stated in a newspaper report that the Conservators supported the rich while the Association supported the poor. Jesse Ridley toured the area in a donkey cart canvassing for the Liberal Corbett in the 1906 general election (Figure 10.2).[26] Thompson also advocated at one point that Ashdown 'should be parcelled out in small pieces to be cultivated', with more cottages to be built near schools and residents' work, since 'in its present state it was a disgrace to the county'.[27] One inevitable response came from the Conservator Rev. Butler, who criticised the 'radical politicians' striving to set 'the masses against the classes'.

The Association did need to be taken seriously. In 1887 it was reported that De la Warr had suffered the destruction of 150 of his small fir trees. Four men were taken to court but found not guilty, with bail provided by C.J. Heald, president of the Association. Three were the sons of copyholders and one was himself a copyholder.

26 ESBHRO, ACC 6815/1 Postcard of H.J. Ridley in donkey-cart.

27 ESBHRO, ACC 6815/1.

The men asserted their rights to estovers and claimed that there would be no grass for the Commoner if the firs were allowed to grow – 'then the whole Forest would become the Earls'. In return, it was further reported that the earl 'feared the strong feeling amongst [the lower] classes'.[28]

In December 1888 the Conservators met a deputation from the Association, a group they now referred to as the foresters, and attempted to deal with no fewer than 47 questions that had been put to them. The deputation comprised Barchard, James Heasman, who had worked at Tablehurst, and Edward Taylor, a tenant farmer from Stroods and Spring Garden. The foresters claimed entitlement to cut litter, bracken, turf and peat for their own use and for sale. The Conservators replied at some length in a written letter to each member of the deputation, setting out the existing situation, which enabled those who could legitimately claim rights of user for 60 years to file a claim to be a Commoner, but that no cutting for onward sale was permissible.

It took some time to organise the Association, but in November 1906 the first AGM was held. Thereafter, and in the first decade of the new century, the Association became the Ashdown Forest Commoners' and Foresters' Protection Association, signalling the inclusion of Commoners as well as foresters, and thus the wider Ashdown community. Whatever the reasons, relations between the Board and the Protection Association were now strained. It was reported in April 1905 that there were still several foresters who felt that they had been deprived of rights because their 60 years' user had not been recognised. Raper was therefore to contact Jesse Ridley, secretary of the Association, to request the particulars of the holdings involved.

Another deputation from the Association was received at the June 1905 Board meeting. This deputation included the then currently ex-Board member, Joseph Ridley, as well as William Carr (Jr), Jesse Ridley and 14 committee members, including Sylvanus Ridley, together with their solicitor. Various representations and proposals were put forward, including potential changes to the byelaws. They also asked that the claims of many more Forest people to be included as Commoners be reconsidered. Many previous applications had been poorly framed; many believed that they should be entitled to common rights with fewer than 60 years' user, or that their claims should be based on their personal commoning practice rather than on their property. They also asked that some of the poorer Forest roads be repaired, including those tracks used by Commoners collecting litter. But they also requested a larger say in the running of the Forest, and that the voting system be changed to allow smaller Commoners a more prominent role on the Conservation Board, rather than being outvoted by those with larger acreages. Finally, they asked that the close time when carting of litter was prohibited be changed to give a further 21 days after 31 March to carry away the cut litter.

The Board appears to have considered these points carefully, and agreed that the Commoners might assist in making road improvements and that the further 21 days be allowed, but that no changes to the voting system or any other fundamental alterations

[28] Phelan, 'Economy to Amenity', 178, citing 'The Ashdown Forest crofter case', *Sussex Advertiser* (People's edition), 6 April 1887; and ESBHRO, ACC 6815/1.

be conceded. A new set of byelaws was entered into the minutes of the August 1905 Board meeting. Among them remained the prohibition on cutting and carting of brakes from February to July, or any litter between 1 April and 30 September.

As in 1889, Raper also wrote to Jesse Ridley asking him to provide names of those who felt aggrieved at being deprived of common rights. The Association quickly returned with the names of no fewer than 100 people who now claimed common rights. Furthermore, in August 1906 Joseph Ridley, in a statement agreed by Jesse Ridley, returned to the Conservators and again asked that:

> all occupiers of land within the Forest pale (excluding land enclosed under the Decree of 1693, and land sold under the award of 1887), whether held as freehold, copyhold or on tenancy at will, who and whose predecessors in occupation can show that they have turned out stock on the Forest and taken litter and brakes therefore continuously for 60 years last past, in respect of such land should be admitted as legal Commoners and pay the Forest Rate.

In October 1906 a specially convened sub-committee of the Conservators produced a report on the situation. They had sifted through nearly 100 applications for individuals and/or properties seeking common rights, dividing them into those for whom rights could be granted and those to whom such rights could not, under the terms of the 1885 Act. The minutes of the meeting include a list of the applicants and their properties and allow a glimpse into the landholding arrangements of smaller cottagers. One thing is immediately clear from the list: that some applicants were small rentiers with several properties claiming such rights, such as Mrs Elizabeth Bellingham, who claimed on behalf of more than ten dwellings, with land, gardens and orchards in Hartfield and Forest Row, but totalling less than five acres altogether. These were allowed as properties entitled to common rights, but a row of cottages and another cottage and garden was disallowed. At this stage, eight smallholders living within the Forest and judged to have met the criteria of user were now admitted as Commoners. By 1908 Commoner status ran to 187 people, although the fact of such status being inherited by successors in the properties meant that some were actually living at considerable distances from the Forest, many in London, but also in Ringwood, Yeovil, East Sheen, Addlestone and Blackheath.

Finally, as a further problem for the Conservators there was another ongoing issue: fires on the Forest (Figure 10.3). The controlled burning of the heathland had, of course, been a management tool for centuries. The 1273 survey recorded that: 'if it be necessary for the improvement of their common pasture they may burn the aforesaid, so that they burn no wood'.[29] But by 9 April 1892 we hear of incendiarism and also of friction within the Conservators. Over two days and nights the Forest was fired, and five Conservators wished for an emergency meeting. They wrote to the chairman, Capt. Noble:

[29] Irons, 'Aspects on the Impact of Man', 299. Modern conservation techniques do now include controlled burning of small areas on the Forest (see chapter 12).

Figure 10.3. Fire at Chelwood Vachery, March 2011 (source: courtesy of Rich Allum).

> In consequence of the serious incendiary fires on the Forest which have taken place the last two days and nights we think it necessary to request you to summon a Meeting of Conservators for Saturday next. We would too suggest that before the Meeting you should see for yourself what mischief has been done & possibly it may be advisable to ask some Commoners dwelling near to attend that Meeting.
>
> Thomas Colgate
> Albert Turner
> Mark Sandford
> Joseph Ridley
> William Carr

However, Capt. Noble protested against the meeting, stating that he had received the notice but would not attend. Another refused to attend because the meeting had not been called by the chairman. Others, including Thomas Colgate, said they were away on business, but called for measures to protect the Forest. The meeting nevertheless went ahead, with E.A. Nicholson joining the other four who had called it, and Albert Turner being elected to the chair.

A minute, as recorded by Raper, noted Capt. Noble's response. If true, it was somewhat extraordinary:

> Mr Turner stated that after the serious fires that had taken place he wrote on Sunday to Captain Noble urging him to prepare & sign a requisition calling a Special Meeting of the Board to consider the matter of the fires [and] that he had received a letter in

> reply stating that while one of the Conservators (whom he named) remained on the Board he declined to identify himself with the Conservancy & that in the meantime the Forest must take care of itself.

From the testimonies of various witnesses attending the meeting, it would seem that a series of fires was started on the night of Thursday 31 March and on the following days, resulting in nearly 2,000 acres of Forest being burnt, and that some private property had also been burnt. But no one could give information as to who was responsible, although all agreed that it was the work of incendiaries. A substantial reward for information leading to prosecution was offered.

The rift in the Conservators' ranks worsened. Capt. Noble did not appear at the May 1892 meeting and Albert Turner continued in the chair, as indeed he was to do for seven more years.[30] And in July Raper read out:

> Dear Sir,
>
> The Meeting of the Conservators appointed for the 9th inst: having fallen through, we hereby request you to summon a Meeting of Conservators for Saturday the 23rd July instant to receive the resignations of six of the Conservators & letters thereon from certain Commoners, and for the discharge of general business.

The somewhat enigmatic letter of resignation had in fact been sent on 13 June: 'We regret having to take this course, but we feel that we cannot continue to act with Mr Sandford – the nominee of Lady Shelley – with due regard to our own position and to your interests.' The resignations came from Maryon-Wilson, W.R. Arbuthnot, Rev. Butler, Henry Freshfield, John McAndrew and William Noble. Despite attempts by Mr Streatfeild and Mrs Nesbitt to smooth over the differences, the six would countenance returning only if Mark Sandford was removed, something the Conservators had no power to do. The departing members were replaced by the earl of Sheffield, James Hope, Frederick Augustus Du Croz, Thomas Hyde, Llewellyn Midgley and Charles William Bell. The minutes frustratingly reveal little more about this disagreement, and in the minutes of the previous meeting there is a large deleted section that might have shed more light on the controversy. The problem may have originated from an unauthorised meeting between Mark Sandford and Elphinstone Barchard of the Commoners group. But it may be noted that those calling the original meeting in April were from lower echelons of Forest society – Ridley, Carr, Colgate and Turner – and those resigning were from the gentry class. Social friction seems likely.

The fact that the Board received reports of these burnings in late March and April probably indicates spring burning to improve pastures, rather than incendiarism or even accidental burning. This was accepted by the Board but, nevertheless, they were unsympathetic and determined to stop the practice:

[30] The chairman was initially elected by the Board into the chair at each meeting of the Conservators, although by 1900 the position was held on an annual basis.

> the Conservators are of opinion that the fires are deliberately caused for the purpose of improving the feed, that they are not likely to occur again this year but that it is desirable before Easter next year to request the Chief Constable to draft plain clothes detectives into the district with a view to discovering the offenders if fires should again be started.

But the fires did occur again: in April the ranger reported numerous fires on the Forest, and Raper:

> at once saw the Chief Constable and arranged for two plain clothes constables to be employed for four weeks to try to detect the incendiaries, although the later bill for so doing was thought, at £81, to be excessive. The Chief Constable also specially instructed the police around the Forest to use every effort to detect the perpetrators.

By 1904 two men were appointed as plain-clothes constables. They clearly found their task difficult, but on 10 April one of the plain-clothes men apprehended Joseph Welfare of Nutley in the act of firing the gorse near the Fox Coffee Tavern, Duddleswell. Significantly, he was rescued by a crowd, but was rearrested on the following evening and committed for trial at the ensuing assizes, bail being allowed. At the same court Jesse Ridley was convicted of obstructing the police while they were apprehending Welfare and was fined £3 8s costs, which he paid; in addition, his brother Albert, of Horney Common, was charged with refusing to assist the police when called upon at the time of Welfare's rescue and was committed for trial at the next assizes. Both were the sons of Joseph, the Conservator. At a later meeting Raper reported that Welfare had pleaded guilty, the judge administering a serious caution and releasing him on his entering into a recognisance of £5 to appear when called upon for judgement. The prosecution of Albert Ridley for refusing to assist the police was then withdrawn. The two plain-clothes men were employed for longer than originally planned, with the Board resolving that they be retained for as long as the chairman considered expedient and also that the chief constable be requested to provide two more plain-clothes men who were strangers to the district.

The incident illustrates the gulf in understanding between Conservators and working families and constituted another attack on accepted local practices. While the Conservators were attempting to care for the vegetation of the Forest, as they saw their remit, the foresters were continuing a tradition of burning, which in this case resulted in a fracas as they attempted to defend that perceived right. There were also fires of a different kind: Raper reported that two boys aged about ten, boarded out by Dr Barnardo at Crowborough, had set fire to the gorse when playing with matches. No prosecutions ensued, since the fire was accidental and the boys so young. Similarly, in 1914 two boys, Frank Moon and Roy Ridley (the son of Albert), both aged nine, were caught setting fire to the Forest. Their fathers were contacted and subsequently a letter was received from the Moons saying 'I am sorry that I set the Forest alight, and I won't play with matches anymore.'[31]

But by early 1905 the fires were again serious and the Conservators now considered cutting furze themselves around properties liable to be damaged, and burning it in controlled heaps. Quite clearly they blamed the Commoners for the fires, and so they wrote to the Protection Association asking for their cooperation:

> The Conservators however trust that the Commoners will by every means in their power, take steps to prevent a repetition of these fires this year, and ... thus render it unnecessary for the Conservators to put in force the scheme which has been described, with the consequent expense which it will entail.

In 1892, as we have seen, the fires had been the catalyst for divisions to appear within the Board and the issue was to provoke dissent again in 1905. John Edward Eggleton had been elected as a Conservator in December 1904 but in June 1905 he sent a letter of resignation:

> as Lord Sheffield has accepted the position of President of the Ashdown Forest Commoners and Foresters' Protection Association he [Eggleton] feels as his Lordship's Steward his remaining on the Board would be inconsistent with the views taken by his Lordship and he therefore resigns his seat as a Conservator.

The third earl of Sheffield had been a member of the Conservation Board until 1903, albeit rarely attending, but he now switched his loyalties to the Protection Association, which also prompted his representative Eggleton's resignation. The reasons for the earl's switch are not recorded. Perhaps he was attempting to assist his tenants in whatever way possible, as this was not a prosperous time for his estate.

At the June 1905 meeting a telegram from the earl of Sheffield himself was read out, asserting that the Conservators' letter to the Commoners contained 'a grave imputation of organised incendiarism against the Association'.[32] This was repudiated by Raper, but at the same time the Board received a letter from Muriel Countess De la Warr with reference to fires that had occurred close to Old Lodge that she felt placed her property in danger. Again, Raper was busy, this time detailing the action the Board had taken in the matter.

---

31 In 1911 Albert Ridley was an 'apiarian' living in a four-roomed cottage at Horney Common, with his wife Annie and three children, of whom Roy was the youngest (TNA, RG 14/4962, Maresfield census 1911). Roy had a chequered life: invalided from the navy in the Second World War, he served time in prison and was a writer of letters from Buena Vista Farm, Heron's Ghyll, to local newspapers criticising, among other things, the 1947 Agriculture Act for its continuing oversight of farms by local committees (https://www.ancestry.co.uk/, family tree).

32 Henry North Holroyd, third earl, died without issue. The Sheffield Park estate passed into the hands of a creditor, Arthur Gilstrap Soames (1854–1934), brewer and maltster, who was in fact entered as a Commoner in a 1908 list for 946 acres with common rights, a year before he actually purchased the estate following the earl's death.

Figure 10.4. Arrivals: W.H. Cowan MP (knighted 1917), by Walter Stoneman, 1917 (source: © National Portrait Gallery x39414).

The hostility of much of the local population was obvious. Rewards for information leading to convictions for arson were being offered by 1907, and in April that year Harry Miles, a bricklayer of Horsted Keynes employed at the Vetchery, was taken to Uckfield Assizes charged with arson, although he was discharged for lack of evidence. The fires were certainly serious now: Henry (W.H.) Cowan at the Crows Nest was in regular correspondence with the Board about access to his property, which he wished to improve and for which he did receive a licence (Figure 10.4).[33] But he complained that a fire had reached his premises and nearly affected his horses. He actually raised

[33] Sir William Henry Cowan was a Liberal candidate in Guildford in 1904 and elected to a Scottish constituency in 1910 and 1918, before switching to the Conservatives and being elected in Islington North in 1923. He was an active member of the Select Committee on Commons, 1911. On his death his widow, Lady Cowan, in selling the property, was instructed by the Board in December 1932 to take down the 'For sale' sign on the Forest.

Figure 10.5. Mr Sparks driving the Chelwood Vachery private fire engine, 1910 (source: Nutley Historical Society).

the issue of the Ashdown fires with the Board of Agriculture, only to be told that the powers to act were vested in the Conservators. He later suggested the instigation of a fire brigade to help tackle the blazes, as was the practice at Hindhead. The Board now decided that existing fire brigades around the Forest should be rewarded for attending to Forest blazes, and more informal groups of men be deployed wherever necessary. Sir Stuart Samuel had his own horse-drawn fire engine at Chelwood Vachery by 1910, with the appropriately-named Mr Sparks the driver (Figure 10.5)![34]

Another proposal was for the coarse heather and furze to be cut into sections separated by rides across the Forest to act as fire breaks. Douglas Freshfield stated that he had also organised his own employees to deal with any fires in his immediate neighbourhood, a tactic he thought might be adopted more widely. The gravity of the fires and their social context was made clear in a report by Douglas Freshfield in October 1906. He related that he had personally attended four fires and had deployed his own men in helping to extinguish them. However:

> The last and most extensive fire was on the night of Sunday September 30th. It broke out between the Isle of Thorns and Vetchery Wood among some high gorse covering disused gravel pits. It was clearly the act of incendiaries and had been lighted in several places. The flames advanced in a line in 3 directions and many acres were burnt. It was present for about 2 hours from 9 to 11 p.m. Two if not three constables were on the spot, and a large crowd including three notorious poachers with their guns and dogs. The crowd cheered and hooted, jeering the police and anyone they

[34] Kirby, *Forest Camera*, 48.

> saw working to limit the conflagration. One of my men was told anyone putting out the fire ought to be thrown into it, and some lads I induced to work were soon made to desist by their fellows' ridicule and abuse. During this time the constables remained spectators and inactive except one who used a small branch in beating down the fire for a few minutes.

Cutting back the furze around the boundaries of properties was also now ongoing, and newcomer Buenaventura Pablo Misa (1868–1939), a senior partner in a firm of sherry importers of Jerez de la Frontera, Spain, who arrived in 1905, soon cut a protective belt 30 feet wide at Oldlands, where he lived with his family and a large retinue of servants.[35]

At least the Ranger now had a new bicycle to help him in his rounds, the previous one donated by the Hon. Portman having collapsed. By 1904 there had been a request for money for a new machine, and thereafter the repair costs for these bicycles became a regular feature of the Conservators' meetings. He was now also given a deer-stalking telescope to assist him in his work. On many nights he was out on the Forest until midnight or even 4 am to detect fires being started, for which he received an honorarium. Even so, the fires continued, and in early 1909 the Board formally minuted, for the first time, how they saw the causes of the fires:

> the Conservators are of opinion that they are mainly due to an impatience of any control on the part of the Foresters, and secondly to a desire on the part of some of them who are annoyed at the inability of the Conservators to admit them as Commoners to retaliate on those who are responsible for the administration and preservation of the Forest, and probably in a minor degree to a belief on the part of some of the smaller Commoners that the fires will by destroying all the coarse growth and underwood on the Forest improve the quality of the grazing in the immediate vicinity of their homesteads.

This was provocative. The Association and Lord Sheffield had already bridled at any implication that the residents of smaller Forest properties were culpable, and if the list of causes was comprehensive it could actually be argued that the order of motives should be reversed. The list also omitted casual incendiarism, sometimes by children, and accidental burning, such as by sparks from steam engines. Douglas Freshfield, by now a very active Board member, pointed out in 1909 that the younger members among the foresters should try to stop the fires, as 'it is to the interest of the whole community that these fires should cease as they deter persons with means from settling in the neighbourhood that would spend large sums in the locality and would benefit all classes of the community' (Figure 10.6).[36]

---

35 Stevenson, 'Alexander Nesbitt, a Sussex Antiquary'. Buenaventura Pablo Misa y Busheroy, second marquis of Misa, was master of the Southdown Hunt. Oldlands was sold to Frederick Eckstein in 1920.

36 Douglas Freshfield became a renowned mountaineer and travel writer (T.G. Longstaff, 'Douglas Freshfield 1845–1934', *The Geographical Journal*, 83 (1934), 257–62). His Wych Cross

Figure 10.6. Douglas Freshfield (1845–1934) (source: T.G. Longstaff, 'Douglas Freshfield 1845–1934', *Geographical Journal*, 83 (1934), 258 © Royal Geographical Society (with Institute of British Geographers)).

His remarks made no difference: more than 30 fires were reported in early 1910 and another 15 in April and May, although Board members thought this actually an improvement on previous years, so clearly many had previously gone unreported. Freshfield thought that the very wet weather had helped and that the fires were 'principally due to ne'er-do-wells'. But the sparks from steam traction engines seem to have been a growing problem: by 1911 at least three fires were reported as having been started in this way, because not all were fitted with efficient spark catchers. Picnickers coming by car, sometimes from south London, were also blamed, and letters to the

Place was the subject of a prominent article in *Country Life*, 24 December 1910.

Automobile Association and 'the motoring papers' were dispatched, warning of the hazards of fire.[37]

In 1912 there were reports of more than 60 fires between late March and May. One incident reported to the Board, again by Douglas Freshfield, was particularly disturbing:

> Mr. Spear is Bailiff to Mr John McAndrew, and mentioned that when putting out a fire he was threatened with a stick by a certain Commoner, who said if the fires were extinguished again Mr McAndrew's ricks would be fired. He (Mr. Freshfield) had expressed his view very fully in a letter he had written to the *East Grinstead Gazette*. It was perfectly clear many of the fires were deliberately kindled, one had been started in Lady De la Warr's private property the evening she returned home, and fires had also occurred in his (Mr Freshfield's) Prest Ridge Warren.

THE ONGOING WORK OF THE BOARD

Apart from the Ridleys, the gypsies, the Association and the fires, there were, of course, other regular issues to be faced. By March 1889, for example, the Conservators were busy dealing with many minor encroachments for cart lodges, pigsties, potato plots, hen houses and sheds. On 1 June Raper's report, following an exhaustive survey of the encroachments, included:

> 'Alfred Hoath, Friar's Gate. 3 hoop shaving lodges on the green in front of his house and shavings and poles strewn about … .'

> The Ranger found a man taking moss, who gave a false name, he could not at the time identify him but with the aid of Brown P.C. we are clear he is a man named Brooks of Wandsworth who is a notorious moss stealer and sends large quantities to London by rail. Moss is not mentioned in byelaws but I propose that the Ranger should summon him under Byelaw 8.

> Grunter Stevenson has placed a pug-mill on the Forest near his encroachment …. He is well aware it is a breach of the byelaw. There is some brick earth dug on the Forest close by.

Later we hear that William 'Grunter' Stevenson was convicted at the Uckfield Bench for breaching a byelaw in having a pug-mill on the Forest and was fined 1s and 13s costs 'to be paid within 14 days or in default of distress 7 days imprisonment – the amount has not yet been paid'. In default of payment Grunter's pig was destrained, causing much ill-feeling against the Conservators, especially as Stevenson's father was dying inside the cottage. A critical doggerel poem was published in the local newspaper. That 'Grunter' should have a pig taken was not seen as funny in the neighbourhood! When it was taken for sale at Nutley Inn no-one would buy it and it was then taken off to the Chequers at Maresfield and to an indeterminate fate.

---

[37] The Automobile Association, founded in 1905, provided bicycle patrols by 1914, especially on the London to Brighton routes. They warned unwary motorists of police ambushes awaiting those who broke the 20-mph speed limit set down in 1903.

At their July 1893 meeting the Conservators were informed that sheep were being depastured on the Forest. Raper was directed to prepare a notice to be served on anyone who took in sheep over winter – the so-called 'keep sheep system', which allowed their normal pastures to rest and which remained prevalent in the High Weald – warning that sheep were not allowed on the Forest. The following year he reported that 21 residents had agreed to remove their sheep but that a further 19 had failed to comply, and that it was difficult to ascertain to whom the remaining sheep belonged. But in August 1894 it was reported that Alfred Tester at Forest Row was charged with keeping sheep on the Forest. He was convicted but Raper stated that the Conservators had no wish to press for a heavy penalty but to make it clear to all that sheep were not commonable. Tester was fined 2s 6d with 9s costs which he paid. Following this, the Ranger reported that nearly all the sheep had been cleared off, except for 30 near Plaw Hatch belonging to William Heasman of Page's Gill, who was duly convicted and fined. In 1895 the matter went for legal opinion, as Lord Cantelupe, with some legal opinion on his side, was claiming that his tenants could graze sheep on the common by prescription and within the constraints of levancy and couchancy, whereas Lord Sheffield urged the Commoners to stand by their older prohibition. But thereafter the matter was allowed to rest somewhat ambiguously, so presumably sheep were henceforth allowed on the Forest with impunity, as indeed is the practice today.

Another problem was the placing of faggot stacks on the Forest, which was also now held to be illegal. An 1896 survey revealed no fewer than 118 stacks all around the Forest, often accompanied by piles of litter, old tree trunks, poles, bricks and manure belonging to small cottagers as well as non-Commoners, including the inevitable Ridleys. As a partial solution it was agreed that licences should be issued to all such stack owners unless they could demonstrate a right of 60 years user. Turf was also to be included in this concession, as were bricks, stacked where a house was being built. There were discussions about deposits of hop poles that were being stripped, although no decision was recorded.

As one example, Henry Harding, tenant at Little Claypits, Hartfield, who attended with witnesses in September 1897 to unsuccessfully claim rights of common, 'had taken brakes and litter from the Forest ... [stating] that he did so under the impression that his holding had a right but that in future he would not take any without obtaining a licence from the Conservators.' At that time he had nine stacks or heaps on the Forest: a faggot stack, a brake or litter stack fenced round and seven stacks or heaps of wood, besides various appliances used in his trade of a trug and scythe-handle maker. The chairman stated that it contravened the byelaws to lay property on the Forest and that the Conservators had no power to make exceptions for tradesmen, but that that they would grant him a licence for one faggot stack and one brake or litter stack. But the rest would have to be removed.

Various other matters were tackled: licences were issued for 5 November bonfires on the Forest, six being granted in 1898 on condition that no damage ensued to the Forest itself. In 1902 permission was also given for bonfires to celebrate the coronation of Edward VII. But the Conservators clamped down on Rev. H.P. Dunlop's taking of turf near St Johns Crowborough for use on graves, and also on charcoal burning by Henry Norman of Fairwarp, despite his protests that the practice had been carried on for 20 years. And by 1898 the old issue of unauthorised bracken cutting was also clearly continuing:

> The Ranger reported that on the 1st August he found Henry Tester of Tompset's Bank cutting for his landlord Martin, who on enquiry he found was dead. Tester then stated he was cutting for Martin's daughter Mrs. Wells who denied employing him. Tester then said he was cutting for Mr. Histed. The Ranger stated that Tester is an old man and appears to be of weak intellect.[38]

Another ongoing issue involved the taking of stone, sand and gravel from pits and quarries around the Forest. Although it was of an inferior quality, authorised highways officials from the parishes, rural districts and East Sussex County Council applied for annual licences to remove material for road improvements, and these were normally granted. But many helped themselves without any licence, often leaving the quarries dangerously unfenced. The mineral rights technically still belonged to Earl De la Warr under the terms of the 1885 Regulation, and on several occasions the Conservators attempted to persuade him to surrender such rights, but he declined. It was at least agreed that no new pits were to be created, that licences for taking stone and gravel were to be granted only for existing pits, and that adequate fencing of the pits was to be undertaken by the highways authorities.

Such hopeful agreements were not always successful: the Board continually heard of new pits, extensions of existing pits and of pits not being adequately backfilled. Up to 1914 Job Luxford, a Forest Row builder and the earl's tenant, was particularly busy in this regard, taking as much stone, sand and gravel as he could obtain from both private and common pits. Luxford was also chairman of the East Grinstead Rural District Council. His building work, for which he was in great demand at this time, provoked several collisions with the Board. And as we have seen, T.J. Ridley/Gurr was equally cavalier with his quarries, leaving deep cart tracks across common land. In fact, so great was the damage being done that the Conservators declined to grant any licences for digging for road mending from 1899. East Grinstead Rural District Council was accused of extending their search for gravel across large parts of the Forest by 1909 and were warned that their licence to dig gravel would be in serious doubt if such diggings continued. The rural district council denied the accusation.

Another alarm sounded in 1907 when W.H. Cowan asked 'what steps the Board proposes to take in connection with Lord De la Warr's recent experimental works on the Forest with the view of finding and working coal and iron'. He was informed that the situation would be carefully watched. But in the following year it was indeed reported that De la Warr had authorised further exploration and that there were no fewer than 150 pits dug; some had been infilled, but others were left open and endangered Commoners' cattle. And again in 1913 the Board heard that the earl had

38 Henry Tester (b. 1824, if the same man) had attempted suicide in 1866 and was twice convicted of larceny and sentenced to two months' hard labour as a second offence (England and Wales Criminal Registers, via https://www.ancestry.co.uk/).

granted a concession to the Perkins Macintosh Petroleum Tool and Boring Company to drill at Gills Lap for coal or oil, with a rig erected in an old stone quarry.

The affair brought the eighth earl, the Hon. Gilbert Sackville (1869–1915), his solicitors and steward into conflict with the Conservators. The earl believed that under the 1885 Act his predecessor had reserved the right to minerals on the Forest but that without going through the Forest surface he could not enjoy this right. He authorised a consortium to start proceedings, but during a visit by the Board chairman workers at the site were cautioned not to proceed. The Board's stance was firmly of the opinion that the lord's rights were subject to the Board's agreement. A special Board meeting was called on 22 November 1913, attended by representatives from the concessionaires. An expert informed the Board that he believed oil was present below Ashdown and that, given the national need, they should acquiesce to a trial boring down to 3,000 feet; if nothing was found, then the surface would be reinstated. Nevertheless, the Board refused, reinforcing this view at a meeting on 19 December 1913. Once again, Joseph Ridley voted contrarily, in favour of allowing the trial, as 'the minerals were placed there for the good of man and that the Conservators ought not to prevent their being worked'. Messrs Turner, Card and Kenward abstained. In February 1914 the matter escalated further when the Conservators received a letter from De la Warr's solicitors intimating that the earl would now take matters into his own hands and conduct boring for oil. The Board threatened legal action if the earl proceeded, but little further seems to have been heard of the proposal and by late 1915, with the death of the eighth earl at sea on active service, all signs of exploration near Gills Lap had gone.

Other events impinged on the Conservators: on 8 November 1890 they asked Nicholson to convey their sympathy to the seventh earl on the loss of his young son and heir, Viscount Cantelupe, in a boating accident on Belfast Lough. The Board was in discussion with the new Lord Cantelupe concerning his wish to take over some of the Forest land near the Five Hundred Acre Wood, while themselves requesting that he might consider relinquishing the lordship of the Forest to them. There was a stalemate, during which time he became the eighth earl in succession to his father, as his second but only surviving son, at the age of 25.[39]

On 30 March 1895 the death of Louisa, Lady Shelley was reported and members paid fulsome tribute to her for her support and the manner in which she had hastened to defend common rights against Earl De la Warr. In 1898 Blanche, her only daughter and successor at Maresfield Park, also died, followed soon afterwards by her husband Hervey Pechell. Messages of commiseration from the Board were now received with thanks by Count Munster. The estate (now 2,657 acres in Maresfield, Danehill and Fletching and 2,400 acres of commonable acreage by 1909) had been left to Hervey's friend, Count Alexander Otto Munster (1858–1922) of Derneberg, Hanover,

[39] The Hon. Gilbert Sackville spent much time in the 1890s at Bexhill, managing his father's development, moving into the manor house with his wife Muriel Brassey, becoming mayor and generating a glittering social circle. A Kursaal and hotel were built, and cycle racing, later motor car racing, were initiated. But further speculative ventures stalled, and the family's fortunes became precarious, not helped by the divorce of the young couple in 1902. Muriel (1873–1930) became an active supporter of women's suffrage.

Figure 10.7. Count Munster at Maresfield recreation ground 20 March 1897. Empress Frederick is about to finish the tree planting ceremony, aided by Count Munster (right) and Mark Sandford, Conservator and agent for the estate (holding a trug with soil) (source: ESBHRO, P/420/2/1).

Germany. Before his death Hervey Pechell had funded land for a recreation ground near Maresfield church, and a tree was now planted by Victoria's oldest daughter, the Empress Frederick of Germany, accompanied by the count, standing in for the Pechells, who were living in Italy. As such, this underlined the close elite Victorian association between the United Kingdom and Germany (Figure 10.7). In 1899 Count Munster made the transfer of the land to Maresfield parish council official.[40]

WEALTHY ARRIVALS

The advent of still more wealthy residents around the Forest brought its own challenges to the Board. A 1904 letter from Julius Lawrence Hamilton of Forest Row complained that children playing and shouting on the Forest outside his house seriously interfered with his literary work and that they should play on the recreation ground.[41] The ranger

[40] I am grateful to Charlie Stevenson for information concerning the transfer of land.

[41] This was a second home for Hamilton (1845–1913) a retired physician, with his main home in Sussex Square, Brighton.

saw their parents, who all stated that it would be useless to send their children to the recreation ground, as the man employed there always ordered them off.

When wealthy newcomers arrived they frequently required the rebuilding of older cottages or the building of new houses on enclosures hitherto used only for pastoral or agricultural purposes. This now generated the carriage of heavy building materials and increased traffic along access routes. Cart tracks across the Forest were a problem: extensive damage was now being done to the surface of the Forest, often by near-parallel tracks carved out to avoid deep ruts or quagmires in bad weather. Some tracks were also being infilled with stone and gravel and thus converted from their original character of green ways. By the beginning of the twentieth century the Board needed legal opinion on their powers in such cases.

There was certainly an ongoing need to monitor the laying of materials to make metalled access roads onto these properties, such as in 1898 for the access road to Kidbrooke for Douglas Freshfield, and again in 1902 at the same property as the Board gave permission for Freshfield's lessee, Major Darwin, to harden a road onto Hospital Farm, where he was having a house built.[42] Similarly, in March 1903 Countess De la Warr gained permission to widen the entrance into New Lodge so that her motor car could be manoeuvred more safely. At that time nothing more was done, but she reapplied in 1913 for a modified entrance and permission was again granted. Yet another application to metal a trackway came from the fourth earl and countess of Norbury for access to their house to be built near Duddleswell.[43] Such problems with development multiplied. John Bashford at Quabrook and the ubiquitous builder Job Luxford were instructed to restore surfaces or face fines. Luxford was fined although acting on the instructions of his client the Rev. Canon Maberly-Smith (who was also fined) at Little Shepherds, on Shepherds Hill, where he was building a house.[44]

In 1902 came the delicate case of Lord Robert Cecil, who had purchased Rocks Plat and was needing bricks delivered across Forest land. Lord Cecil had complained in February 1899 of being harassed by the Conservators, who were monitoring the work being undertaken on the access route. In addition to their succession of London residences at increasingly fashionable locations, the Cecils built a relatively modest country house, 'Gale' (in an exposed location) at Chelwood Gate.[45] A sub-committee was formed especially to deal with Lord Cecil's case, and on reporting back with a

---

42 Major Leonard Darwin (1850–1943), youngest son of Charles Darwin. Following an army career he was a Liberal Unionist MP in the 1890s, President of the Royal Geographical Society (1908–11) and Chairman of the British Eugenics Society 1911–28.

43 ESBHRO, DW/A/15/286 (November 1912) William Brabazon Lindsay Graham-Toler, fourth earl of Norbury (1862–1943), a title in the Irish peerage, and Countess Lucy (1876–1966).

44 George Maberly-Smith (1831–1917), rector of Penshurst, Kent, from 1878 and an honorary canon at Canterbury from 1897. His link to Ashdown is otherwise unknown.

45 Lord Robert Cecil (1864–1958), lawyer, and from 1906 a highly influential politician, diplomat and an architect of the League of Nations, for which he was awarded the Nobel Peace Prize in 1937. Later Viscount Cecil of Chelwood. The house was a favourite and survived a Luftwaffe bomb falling just outside its front door in November 1940 (M. Ceadel, 'Cecil (Edgar Algernon) Robert Gascoyne [Lord Robert Cecil], Viscount Cecil of Chelwood (1864–1958)', *ODNB* [2004]).

recommendation that a metalled road be approved the Conservators voted to accept, with the exception of Llewellyn Midgley and with Thomas Colgate abstaining. But in 1903 the Board complained that the grass verges that should have been laid were still missing from the new drive. Lord Cecil's peevish reply in complying included:

> I confess that the exaggerated importance that has been attached to this question of turfing, in itself a perfectly unimportant matter to the Conservators and to everyone else, makes me very doubtful whether the business capacity of the Board is sufficient to make it worth while to assist them any further in any of their undertakings.

Another sub-committee was hastily formed to consider this new situation!

Applications to metal more tracks multiplied. By 1907 Sir Spencer Walpole was granted permission to improve the track to Noah's Ark Cottage to allow building materials for the enlargement of his dwelling, although he was shortly to die.[46] He had purchased Hartfield Grove in 1900 on his retirement from public office, although in 1902 he became a director of the LBSCR, in 1904 chancellor of Oxford University, and in Sussex a magistrate and golfer. After his death his widow, Lady Walpole, assumed his common rights. A similar access request came from another recent arrival, Sir Stuart Samuel (1856–1926), the Liberal MP for Whitechapel 1900–16 and a wealthy banker. He desired a new access onto the Wych Cross road from his property, the Vetchery, purchased from De la Warr. By 1906 he had employed William Flockhart to build a neo-Tudor, part-crenellated mansion and Leonard Rome Guthrie (from 1912 Flockhart's business partner) to landscape the grounds with a series of terraces with sandstone walls, rose garden, lily ponds and water channels, bosquet, tennis and croquet lawns and kitchen garden. Samuel later added a rockery, an arboretum and an extensive collection of rhododendrons.[47] He also acquired two farms at Millbrook with attached common rights. It was not long, however, before he took issue with the Conservators. A letter from him on 30 July 1908 stated:

> I regret to be unable to subscribe to the salary of the [second] Ranger, as I do not agree with the policy of the Conservators towards the poorer Commoners. If they were to make arrangements to themselves [to] burn undergrowth so that the Commoners might have grass for their cattle I am sure they would have no trouble. If this entailed a payment to the Lord of the Manor I should be pleased to subscribe.

[46] Sir Spencer Walpole (1839–1908), KCB 1898, former Secretary to the Post Office, an efficient administrator and historian. On his father's side descended from the famous Walpole family, on his mother's from the murdered prime minister Spencer Perceval, whose biography he wrote.

[47] Bosquet: a French term for a formal and strictly regular plantation of trees. Samuel was created First Baronet Samuel of Chelwood Vetchery, Maresfield, in 1912, at a time when he was fighting antisemitism within ruling British circles. The property is now known as Chelwood Vachery.

He was also given leave by De la Warr in 1911 to enjoy sporting and fishing rights over a considerable area adjacent to the Vetchery, including fishing on both banks of the Mill Brook.

As the older cottages were now being extended there were now also the regular visits of tradesmen who were less careful about Forest trackways. Problems of sewage also became more obvious, as older methods of waste disposal were now inadequate for the larger houses, with 'offensive liquid' issuing onto the Forest. In 1903 the Forest Row sub-committee reported:

> that foul liquid in considerable quantities is flowing from a tank at the northeast corner of Mr. Lucas' property into a grip cut on the Forest for the purpose of conducting it away. Also that there are signs of foul liquid overflowing down the bank on to the Forest from a tank or cesspool near the middle of the north side of Mr. Lucas' property and that broken glass and old tin cans are lying on the Forest at the back of Mr. Lucas' coach house which appear to have been thrown from his premises. The Committee recommended that steps be taken at once to compel the abatement of these nuisances.

Additionally, in 1903 the wheels of the 'Sanitary van' were cutting up the Forest, although the Board felt that this was a lesser problem than forbidding the van access. Sewage pipes were also a nuisance, but the Board gladly consented to them being dug into the subsoil rather than tolerating the noxious alternative ways of sewage disposal. The growth of Crowborough later entailed such underground pipes being laid with the agreement of the Conservators.

Domestic and other rubbish also accumulated. Near Forest Row there was much refuse thrown onto the Forest by late 1913, causing an ongoing disagreement between the Board and East Grinstead Rural District Council as to where responsibility lay for its removal. In the same year a motorcycle rally involving about 300 participants *en route* to the Beacon Hotel, Crowborough, left large amounts of refuse near Duddleswell for ranger Kirby to collect. And in 1909 a potential danger to cattle had come with the proposal by Mr Mansell to have an access road lined with laurel while developing a new cottage on the edge of the Five Hundred Acre Wood. Laurel being poisonous to cattle, the Board rejected his proposal.

Clearly Ashdown was, in detail at least, not yet an idyllic landscape. But the wealthy invasion from London continued. Meanwhile, we find in early 1900 the Conservators faced with the somewhat mischievous making of a metalled road by Joseph Ridley of Millbrook, who stated that 'he was making it like [Lord Cecil's] road at Rocks Plat'. He was warned by the Ranger not to continue and to appear at the next meeting of the Board, which he failed to do. Later, however, on proof of user, he was allowed to continue improving the trackway.

A different kind of application came from Dr Louis Waldstein (1853–1915), a Commoner at Posingford, who applied successfully to take peat in 1913 as part of a scientific experiment. He had built his own house at Posingford, where he died in 1915, and his brother later wrote:

> Posingford seemed to combine everything for which he had been looking, with its old Sussex cottage, which he retained for his bailiff, and its miniature park, glorious trees and a rippling brooklet running among them a short distance from the house. He added a new house and stables, to carry out his own ideas, and especially arranged the upper passage-ways so that his mother, in case she could not leave the house, should have a sunny walk indoors. During the short period in which he resided there before his death, he loved to take distant walks through the fields and woods of Ashdown Forest, and even on his last sick-bed, would drink in the beauty of the view through his window, gaining for the time being peaceful respite from his suffering.

But 'the failure of his great life-work, the death of his mother, and the outbreak of the Great War, banefully affected his physical and moral health'.[48]

Charles Corbett (1853–1935), as we have seen, was certainly a prime example of a wealthy person with connections to a wider world arriving on Ashdown. A Liberal politician and barrister living at Dane Hill Park, he was on the Conservation Board 1898–1905 before retiring because of the pressure of political work, winning the seat at East Grinstead in the Liberal landslide of 1906, but then losing it in 1910. He had married Marie Gray (1859–1932) and they had two talented daughters and a son. Their daughters were Margery (later Dame) Corbett Ashby, an international feminist campaigner and Liberal candidate, and Cicely Corbett Fisher, a suffragist and workers' rights activist.

Corbett was himself a passionate campaigner for female suffrage, although his East Grinstead constituents hardly shared such radical views. Marie also campaigned for women's rights and the Liberals in East Grinstead High Street, once enduring egg throwing and stoning in 1913. In 1911 she joined with Muriel, Countess De la Warr, to form the East Grinstead Suffrage Society. Also passionate about poor relief, she was a member of the Uckfield Board of Guardians for 36 years, becoming one of the first women Poor Law guardians and the first woman to serve as a rural district councillor in Uckfield. She oversaw the removal of all children from the Uckfield workhouse and placement with foster parents through the Ashdown Forest Boarding-Out Committee for Poor Law Children, and was also a founder member and secretary of the East Grinstead Women's Soroptimist Society.[49]

Another Liberal politician, living at Hindleap from 1898 (as well as at Buckingham Gate, London), was Lord James Bryce (1838–1922). A respected Oxford historian, advisor to Queen Victoria and kings Edward VII and George V, he had been an MP 1880–1907 and in 1913 had returned from his post as British Ambassador in Washington when he

48 L. Waldstein, *The Subconscious Self and Its Relation to Education and Health* (London: G. Richards, 1897; 2nd edn 1926), xvii–xlviii, with a biographical sketch by his brother Sir Charles Walston (Walstein). Waldstein had been born in New York to German emigrant parents and had been German-educated, but briefly met Henry Freshfield in London. He had a thriving New York medical practice but later joined his widowed mother in London, continuing his 'great life-work', scientific studies especially related to blood and infectious diseases.

49 Dame M. Corbett Ashby, *From the Feudal to the Jet Age* (Danehill Parish Historical Society, 1982). She was a president of the society.

was created Viscount Bryce of Dechmont. He had also been president of the Alpine Club 1899–1901, bemoaning the impact of railways upon the scenery of the Swiss Alps. In this context it is clear from his remaining letters that he was in contact with Douglas Freshfield, another Alpine enthusiast and near neighbour; both were also members of Sir Leslie Stephens' Sunday Tramps group, mentioned above. According to his wife, Marion, 'life there was very quiet for he desired above all things an undisturbed time for meditation'.[50]

We also have Maurice MacMillan (1853–1936) living at Birch Grove Lodge from 1906 in somewhat austere fashion with his American wife, Helen (Nellie) Belles, who had purchased the property on their marriage.[51] His social circle was small but did include Lord Cecil, at nearby Chelwood Gate. And by 1913 we find L.F. Salzmann at Wroth Tyes, Hartfield, with common rights relating to just under four acres which he still held by 1920. A renowned historian, both in Sussex and more broadly as General Editor of the *Victoria County Histories* from 1934, he was returned as living at Hope Park, Bromley in 1911 and again in 1915.[52] His acquisition of common rights on Ashdown is hitherto unknown.

The Conservation Board itself was still dominated, if not by the nobility, certainly by wealthier men and larger landowners and their representatives (Sandford, Eggleton until his resignation in 1906 and Nicholson until his death in 1913). As an example, we might note the presence of the wealthy Hon. Henry Berkeley Portman (1860–1923), later third Viscount Portman of Bryanston, a Sussex JP who inherited the Buxted Estate through his ancestral mother's line from Cecil Cope Jenkinson, third earl of Liverpool, himself the half-brother of Arabella Whitworth. There was also the charming gambling addict Col. Charles Needham (1844–1934), 1st Life Guards, who had purchased Tylehurst, Forest Row, in 1902 for a summer residence, spending the winters in London. He had been military attaché in Rome 1895–1901 and had married the Dutch heiress Baroness van Serooskerken (1853–1936) in 1874; despite a chaotic marriage, they lived at Forest Row until their deaths.[53] Their second daughter, Amy Violet Needham (1876–1967), was later the author of popular children's books and a close friend of Douglas Freshfield, her near neighbour.

50 Marion (1853–1939) was also a keen Liberal. They shared anti-suffragist views. For letters between Bryce and Freshfield see the online Bodleian catalogue: <https://archives.bodleian.ox.ac.uk/repositories/2/resources/3201> [accessed 26 September 2021].

51 Maurice, a senior partner in the family publishing business, was the father of Harold, later prime minister, who died there in 1986. The house itself was built in 1926 by Maurice and Helen, the latter holding a separate 44 acres at Birch Grove with common rights in her own name by 1913.

52 Salzmann (1878–1971) was returned in the 1911 census as 'historian', working for the VCH (TNA, RG 14/4977 Hartfield census 1911). I am grateful to Colin and Judy Brent for help on Salzmann.

53 Col. Needhams' father was the second earl of Kilmorey, but Charles was illegitimate and passed over for the earldom. He is depicted, aged ten, in bas-relief with his parents on the impressive Kilmorey Mausoleum in Richmond upon Thames. Violet left Forest Row following the deaths of her parents and Freshfield in the mid-1930s <http://www.violetneedhamsociety.org.uk/> [accessed 27 July 2020].

THE ARRIVAL OF GOLF

There was a huge national boom in golf at this time, and many exponents looked to marginal land, and especially open heathland with scattered trees, as being appropriate for their needs. Given the numbers of wealthy arrivals, Ashdown looked ideal. The Ashdown Forest and Tunbridge Wells Golf Club near Forest Row and the separate Ladies Club were established by 1890.[54] The earl had given his permission for the latter, but without sanction from the Conservators.

In fact, the Conservators did express concern at changing the nature of the northern part of the Forest through earth moving for the greens, although bunkers were not allowed, and also through the employment of litter cutters who had no such rights. But an agreement was reached that allowed the clubs to continue. Nevertheless, further expansion was difficult, and Commoners complained in 1902 that their access was inhibited and that there were dangers from stray golf balls. It must have helped that the first club secretary was Robert Birch, a Conservator from 1891 until his death in 1897. In that year the 'Cantelupe' was established as an artisan section, intended to persuade the foresters to join the club and to maintain the course, a cheaper option than paying for staff.

Another request to the Conservators came from Crowborough for permission to establish a club there. This was more controversial, as many Commoners in that area felt that there was already insufficient litter or pasture and this shortage would be exacerbated by the construction of a golf course. Nevertheless, the course was opened in 1895, with Lady Cantelupe driving off from the first tee. Sir Arthur Conan Doyle was club captain in 1910.

The new courses certainly raised class issues. It was claimed that the Conservators were a self-serving clique, working for the rich against the poor.[55] Why was a poor man penalised for cutting heather while the golf club was allowed to do so, even without 60 years' user? In April 1904 Raper read a letter from Henry Harding of Tompset's Bank and commented upon it:

> denying that he had deposited the building material on the Forest in front of his premises. Complaining that not long ago the Ladies Golf Club were allowed to cut turf before his house to put on their greens and that the Gentleman's Golf Club were allowed to put any filth on their greens and stating that there is one law for the rich and another for the poor. The ranger subsequently ascertained that the building material was deposited on the Forest by Mr. Luxford. The Clerk in reply pointed out that if the building was done for him the person who carried out the work was his servant. Expressing surprise that turf had been cut in front of his house and that

[54] Later the word 'Royal' was added from 1893, following a review of troops on the Forest by the duke of Cambridge, commander in chief of the British army. The Ladies' Golf Club was entirely independent, with their own clubhouse and nine-hole course, gaining royal status in 1932. After the Second World War the two clubs merged (J. Penninck, 'Royal Ashdown Forest golf clubs 1888–1953', *SCM*, 26 (1953), 366–72). The Cantelupe was named after Gilbert Sackville.

[55] Phelan, 'Economy to Amenity', 179–82.

> the matter would be enquired into. That the Board endeavour to treat rich and poor exactly alike and that the Golf Club were permitted to use the Forest because it is understood that the money they bring to Forest Row tends to the general prosperity especially of the working classes. And that if he would at any time point out any abuse it should at once be enquired into.

The fact that income from the golf courses tended to remain in the northern parts of the Forest also caused resentment. Thus, in April 1904 Joseph Ridley of Horney Common, at that time not a Conservator, attended the Board to explain that the roads in Maresfield parish, which were not the responsibility of the local authority, were in a poor state and received no help, unlike those in the north around the golf courses. However, this failed to convince the Board, who felt that they had no powers to take the matter any further. There was even an approach from the Association to construct a golf course between Nutley and Fairwarp to redress the northern bias, but this was refused.

The golf clubs certainly required ongoing attention, and the local working population was not above damaging the courses. In 1907 James Reed of Bank Cottage, near Broadstone:

> having on the 2nd May last seriously damaged the fifteenth green of the Royal Ashdown Forest Golf Course by driving his milk cart round and round the flag on the green, was summoned before the East Grinstead Bench on the 21st May last, for breach of Byelaw 15 ... was convicted and fined £1 and 14s costs.

This did not stop Reed from being a problem to the golf club in driving his horse and trap across the course right through to 1914. And now it was reported that members were parking motor cars on the Forest, contravening the byelaws, so in 1913 the chauffeurs of such cars were to be warned not to leave their cars on the Forest roads near the Royal Ashdown club house.

MODERNITY AND THE RAILWAY THREAT

In February 1898 a different threat arose that revealed social conflicts once again. A light railway was proposed from Sheffield Park station to Groombridge, crossing through the Forest itself, and through Minepits at Nutley to Duddleswell, emerging near the Crow and Gate (Figure 10.8). Raper was instructed to take counsel's opinion, and in December the proposal was put before a well-attended Commoners annual meeting. It was there agreed, thanks largely to the acreage voting power of George Maryon-Wilson, that a committee of Commoners be appointed to examine the likely impact of the proposal, although 33 Commoners had already signed a memorial assenting to the scheme. Charles Corbett and Albert Turner were particularly enthusiastic. In January 1899 a further meeting heard that the committee of Commoners backed the 10-mile-long scheme subject to several provisos, among them being aesthetic safeguards, such that a ridgeline route should be avoided, that no station be built within the Forest or passengers set down. But it was felt that such a railway, linking Nutley with the main line at Groombridge, would particularly benefit Nutley residents, and allow local produce to reach markets more quickly.

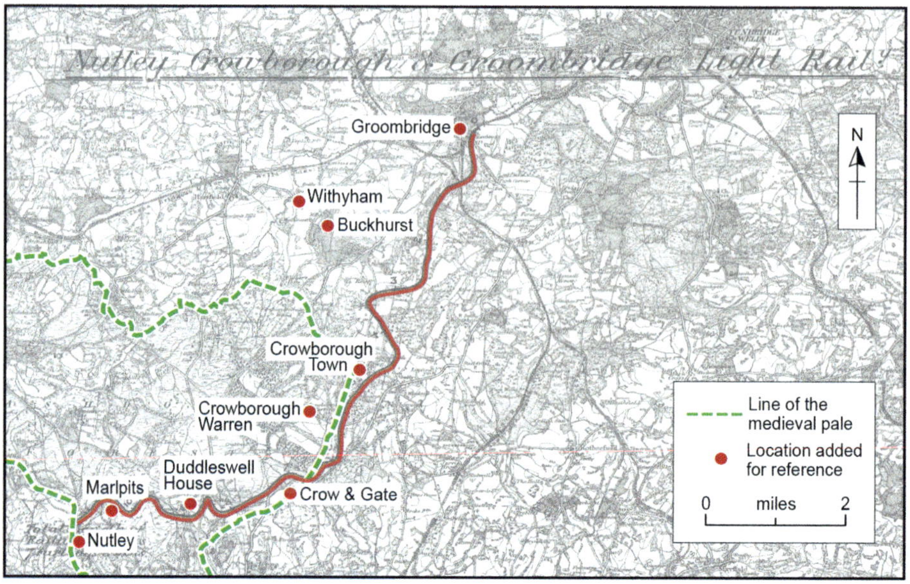

Figure 10.8. Plan for the proposed Nutley, Crowborough and Groombridge Light Railway, 1898 (source: ESBHRO, Q/D/P/603A). Locations superimposed for clarity and the proposed line emphasised. The line of the pale is also added for context.

On this basis the Commoners' meeting agreed to the scheme, with only Douglas Freshfield in opposition, and with the chairman, Mark Sandford, and messrs Hyde and Midgley abstaining. Freshfield's 200 acres was far outweighed by the other 21 voters' 573 acres. Five of the Conservators also voted for the scheme, thereby splitting the opinions of the Board. At the next meeting of the Conservators the votes were reported and the wish of the Commoners that the scheme be approved was received. Now the Board of Agriculture became alarmed, wishing to be kept informed of the situation, as did the Light Railway Commissioners. The Commons Preservation Society and the National Trust agreed to jointly fund legal opposition.[56] At the Board meeting in March 1899 it was recorded that:

> while accepting the report and recommendation of the Commoners' meeting the Conservators feel themselves precluded by the provisions of the Commons Regulation (Ashdown Forest) Provisional Order Confirmation Act 1885 from taking any active course in the matter. And … it was resolved nem con (Mr. Midgley not voting) that the Clerk be directed to attend before the Light Railway Commissioners and hand them a copy of the preceding resolution and that he take such action as he may deem necessary for giving effect to the recommendation of the report.

Exactly what 'any active course' meant is unclear, but it was certainly obvious that the Board was divided on the issue. Capt William Noble, Nutley resident and

56 The National Trust executive committee minutes November 1898 to March 1899.

then chairman of Uckfield magistrates and the Board of Guardians, did not think 'objectionable excursionists' would be attracted to the district. He continued by saying that what that class of people wanted was 'a minstrel, a photographer, and a public house' – they did not admire beautiful scenery.

In April 1899 Raper duly attended a meeting at Groombridge of an adjourned enquiry into the scheme, where R.J. Streatfeild, the Goldsmiths Company and many Crowborough residents were opposing it. In July there was something of an anti-climax: 'The Clerk reported that since the last meeting the inquiry into the Nutley and Groombridge Light Railway Scheme was concluded and the Light Railway Commissioners declined to grant the order without stating any reasons.' But opposition from the Goldsmiths Company may have tipped the balance, together with the dubious profitability and difficult engineering required to deal, among other things, with the ghylls on Ashdown, requiring some 1:40 gradients and tight curves, together with several crossings, six bridges and halts. This threat to the beauty of the Forest had at least been overcome, but the issue had shown once again how the Commoners could be at odds with the Conservators, and indeed how the Conservators could be divided among themselves.

But if not the light railway, then other signs of modernity appeared: telegraph poles at Fairwarp and later both telegraph and telephone wires and posts across the Forest, and letter boxes (e.g., at Friar's Gate and Marlpits). Later the telephone wires were carried underground to the larger properties, but the Board was powerless to halt the erection of telegraph poles on private property, even though the Rev. H.H. Lowe of Forest Row described them in 1913 as 'an unwarrantable blemish of a beautiful neighbourhood'.

CONSERVATION SUCCESSES BEFORE THE FIRST WORLD WAR

The Board was conscious of its duty to environmental management and to the protection afforded to *bona fide* Commoners. Little further social policy was articulated, but in administering the Forest they tried hard to clear it of illegal squatters, encroachment, gypsies and the taking of raw materials – thus in a way favouring and even underpinning the wealthy as well as the 'deserving poor'.[57]

By the early twentieth Century the Board initiated a 'Protection and Improvement Fund' comprising income from licences, acknowledgements and other sundry monies. Another innovation was to produce an annual resumé of their work for the benefit of the Commoners, to demonstrate, in the words of chairman Maryon-Wilson, 'that they [were] not a mere paper body'. The Board also wrestled at this time with establishing precedents. Raper was assiduous in his perusal of older documents in this respect.[58] In 1908, for example, the issue arose as to whether someone could be classed as a Commoner who only took estovers and did not turn out cattle. Moreover, did that person have entitlement

[57] Phelan, 'Economy to Amenity', 167. She notes the similar situation arising at this time in the New Forest, where the Commoners' Committee and the Commoners' League were formed between 1877 and 1900.

[58] Raper clearly devoted a great deal of time to the interests of the Conservators. A 1909 estimate showed that during that year he had sent out no fewer than 1,000 letters and communications for the Board. He was being paid just over £100 pa.

to take estovers without the permission of Forest authorities, and were they to be directed only to certain areas? Raper combed through documents going back to the time of Queen Phillippa, consort of Edward III, but could find only conflicting evidence. Forest authorities claimed that assignment was necessary, whereas the Commoners thought otherwise. The Board made little progress on this but did agree that taking estovers over at least a 60-year user could constitute the acquisition of common rights.

In the years before the Great War the Forest had faced several existential threats: proposals to subdivide it into small plots, to plant fir trees (including a proposal from the Royal English Arboricultural Society to use 500 acres as a school for forestry), to build a railway across it, to bore for minerals, and always to allow more and more development – for housebuilding and roads, for a school near Forest Row, for golf courses or for the taking of stone, gravel and sand by local authorities. Encroachments and eyesores were contested, pollution tackled and urban pleasure-seekers warily watched. In reacting to threats, the preservationist ethic was obvious. But always in the background was the question of class – for whom was this work being done? For those visitors and weekenders who sought the Forest for its beauty and tranquillity, or for its residents? And, of the latter, for the wealthy or the poor?

CHAPTER 11

ASHDOWN IN WAR AND PEACE, 1914–1945

> The fighting services are bound to become serious enemies of what is left of England. Wherever they see a tract of wild, unspoiled country they naturally want it for camps … .[1]

THE FIRST AND SECOND World Wars impacted upon Ashdown as they did elsewhere, superimposed onto all the other issues facing the Forest. The lives of many of the Forest's families were to change for ever after 1914 as sons were lost or maimed in the fighting, and the military camps spread over much of the Forest as men were readied for advancing to the Front, whether to Flanders or the beaches of Normandy. But by 1914 there was some continuity: there was still a lord of Duddleswell manor, still a Board of Conservators, still an ongoing tussle between those upholding the aesthetics of landscape and those conscious of the needs of those who lived there and worked the unrewarding soils. The interwar period too brought its own challenges, particularly to the Forest landscape at a time of growing travel, tourism and housebuilding.

POPULATION CHANGE IN THE FOREST PARISHES 1911–1951

Population continued its growth through to the middle of the twentieth century, increasing by 52 per cent through to 1951, although there was little overall pattern (Table 11.1). For most parishes the highest totals were in 1951, although that for Maresfield came in 1921 because of its wartime army camp. The population of Buxted in 1921 was similarly boosted by the military, but some other parishes, such as Fletching, Rotherfield and West Hoathly, saw little change. East Grinstead continued to grow as the largest settlement around Ashdown, helped by the early electrification of the LBSCR from 1909 such that all their lines were electric from 1938.

Subdivisions of civil parishes occurring at the end of the nineteenth and in the early twentieth centuries brought Forest Row, Danehill, Crowborough and Hadlow Down to join the ancient parishes around Ashdown. The two largest parishes were thereby

[1] E.M. Forster, 'Havoc', in Clough Williams-Ellis (ed.), *Britain and the Beast* (London: J.M. Dent and Sons Ltd, 1937), 45.

affected: the carving of Crowborough from the old parish of Rotherfield meant that the latter's 1851 area of 14,733 acres had shrunk to 11,597 acres by 1961, and East Grinstead, losing Forest Row, shrank from 15,071 acres in 1851 to just 6,600 acres by 1961.

The Poor Law Unions had formed the basis for newer district authorities, first urban and rural sanitary authorities under the Public Health Act 1872, and then under the Local Government Act 1894 these were further rationalised into boroughs, urban and rural districts. The Ashdown area was under the aegis of East Grinstead and Uckfield rural districts, with East Grinstead itself constituting a separate urban district. East and West Sussex County Councils had also been established in 1889 following the Local Government Act 1888, and under another Local Government Act of 1929 county councils were urged to review the number and boundaries of district councils; by 1934 this led to the loss of East Grinstead rural district, partly absorbed into the expanded Uckfield and Cuckfield rural districts.[2] Subsequent changes entailed the 1930s Cuckfield rural district (and East Grinstead) becoming part of Mid Sussex District and West Sussex County Council in 1974, and Uckfield rural district was incorporated into the new Wealden District, East Sussex. East Grinstead, West Hoathly and Horsted Keynes are now therefore in West Sussex, the other main Ashdown parishes in East Sussex.

Table 11.1. Population changes in Ashdown 1901–51.

	1901	**1911**	**1921**	**1931**	**1951**
Buxted	2038	1665	2869	1913	**2956**
E. Grinstead	6094	7089	8106	8728	**10,942**
Fletching	1088	1116	**1151**	1014	1008
Hartfield	1501	1628	1861	2026	**2332**
Horsted Keynes	888	931	961	960	**1148**
Maresfield	2001	2278	**3610**	2309	3172
Rotherfield	2823	**2870**	2821	2821	2795
West Hoathly	1446	1522	1615	**1632**	1626
Withyham	2147	2479	2498	2724	**3041**
Crowborough	3639	5148	5846	6095	**7126**
Danehill	1140	1131	1087	1153	**1271**
Forest Row	2516	3035	2519	2989	**3258**
Hadlow Down		**1043**	933	912	823
Total	27,321	31,935	35,877	35,276	41,498

Note: there was no census in 1941 because of the war; population maxima indicated by bold type.

Source: General Register Office, *Census 1951 England and Wales* [London: HMSO, 1958].

2 J. Godfrey, 'Local Government in the 19th and 20th Centuries', in K. Leslie and B. Short (eds), *Historical Atlas of Sussex* (Chichester: Phillimore, 1999), 126–7.

The overall pattern of population growth, of course, concealed many local variations. To what extent, for example, was the area hit by the terrible Spanish influenza pandemic of 1918–19, which had reached Brighton in July 1918? Phoebe Eames (15), from King's Standing, reportedly died within 12 hours of developing symptoms and was buried in a lime pit. Rural Sussex was certainly spared the terrible loss of life experienced in northern and urban areas, but nevertheless in October and November 1918 there were also deaths in Uckfield, and the Medical Officer of Health reported in 1918 that East Grinstead Urban District had experienced 'the highest death rate ever recorded in East Grinstead since the Public Health Act was passed ... largely due to the heavy mortality in two epidemics of Influenza, which accounted for 17 deaths'.[3]

Cottage sanitation was often rudimentary, but Ashdown continued to be an attractive place for wealthy newcomers, a point noted in 1918:

> For agricultural purposes Sussex is, or has been, too much in possession of rich people, who do not depend upon the cultivation of land for their livelihood. This tendency has, of late years, been increased by a fashion with a group of literary men, who have come to live in Sussex, of extolling this county in verse and prose, thus attracting other invaders.[4]

Georges Lemaitre saw in 1931 that: 'Le Weald fournit aux citadins fatigués et déprimés par la vie modern un élément qui est devenu aussi indispensable que la nourriture: le repos et la détente nerveuse.' [The Weald offers town dwellers who are tired and depressed by modern life something which has also become as indispensable as food: rest and mental relaxation].[5] When the Conservators were attacked again for what was seen as their favourable attitude to the golf courses on Ashdown, their reply was that the Forest rates could be kept lower because of dues paid by the clubs, that employment was offered, and that the courses attracted still more wealthy residents who created a local market for farm produce.

The mix of wealthy incomers, retirees and agricultural workers around Ashdown is shown by the occupation tables in the 1921 census (Table 11.2).

---

3 J. Kaminski, 'A terrible toll of life: the impact of the "Spanish influenza" epidemic on Brighton 1918–19', *SAC*, 147 (2009), 193–210; Peter Freeland, personal communication October 2013; *Report of the Medical Officer of Health, East Grinstead UDC* 1918, 1; C. Bambra, P. Norman, N. Philip and A. Johnson, 'Visualising regional inequalities in the 1918 Spanish flu pandemic in England and Wales', *Environment and Planning A: Economy and Space*, 53.4 (2020), 607–11. This was the H1N1 influenza A virus. Lasting from February 1918 to April 1920 in four waves, the death toll was 228,000 in Britain alone.

4 BPP, Report of B.H. Holland to the Enquiry on Wages and Conditions of Employment in Agriculture (1919), 340 (written in May 1918). Rider Haggard had earlier expressed an opposite opinion, feeling that 'rich men in big houses' were providing a market for local jobs and produce (Rider Haggard, *Rural England*, vol. 1, 134–5.

5 G. Lemaitre, *Le Weald des Comtés de Kent, Surrey, Sussex, Hampshire* (Paris: Les Presses Universitaires de France, 1931), 251.

Table 11.2. Employment categories in the Ashdown Forest area, 1921.

Type of employment	East Grinstead rural district		Uckfield rural district	
	males	females	males	females
Farmers	186	20	588	57
Gardeners	558	18	799	22
Agricultural labourers	770	31	1309	43
Builders & bricklayers	308		463	1
Painters & decorators	103		160	
Wood workers (carpenters etc.)	234	3	325	3
Commerce & finance	286	105	601	212
Transport & communication	468	8	602	15
Personal service	186	1289	332	2062
Retired or unemployed	716	4236	1371	7831

Source: adapted from C. Lockwood, 'The Changing use of Land in the Weald Region of Kent, Surrey and Sussex, 1919–1939' (Univ. of Sussex unpub. DPhil, 1989), 55, and BPP, *Census of England and Wales 1921*, County Report of Sussex, Table 17: occupations.

The broad nature of work patterns was now well defined. There remained a considerable agricultural element, but this was now joined by large numbers in the building trades and by women in domestic service. Matching the number of agricultural workers, however, was a large number of retired men and retired or non-working women. Apart from the farmworkers, there is almost a suburban feel to these figures, and the much larger numbers of older women is also marked. Betty Turner, writing of her younger days living in Maresfield Park in the 1940s, remembered 'the retired "colonial" types who had settled in the developed part of the Park … ',[6] a trend echoed elsewhere around the Forest and an ongoing *motif.*

THE CONSERVATORS 1914–1945

There were few changes initially in the membership of the Conservation Board from the immediate pre-war years, and the Board continued to meet during the war, although founder member Albert Turner was ill for much of the time. Indeed, the February and May 1918 Board meetings were held at his house, in deference to his indisposition, rather than the normal Nutley Inn.[7] By early 1919 Turner had died. Llewellyn

6 Betty Turner, *Maresfield*, 1.

7 The Board was informed that their meetings at the Nutley Inn (Shelley Arms Hotel) were not always convenient, but they continued to meet there.

Midgley resigned in mid-1919, following his wife's death and having relinquished his commonable property.

Arriving late from a meeting of the ongoing East Sussex War Agricultural Committee (see below), Maryon-Wilson informed the Board that he was now chairing the Board for the twenty-first year in succession but felt that his time was now taken up with county business, especially as chairman of the County Council Small Holdings Committee, and that 1920 would have to be his last year as chairman. In fact, he continued until 1929, suggesting in 1927 that a vice-chair be appointed to assist with correspondence. Admiral Colville was approached, but declined owing to pressure of other work.George Maryon-Wilson's place as chairman was eventually taken by H. Ronald Hardy, stepping up from vice-chairman, with Admiral Colville now agreeing to be vice-chairman.

Admiral the Hon. Sir Stanley Colville (1861–1939) was one example of the distinguished servicemen joining the Board after the war. He joined in 1920, having been highly decorated for services in the Royal Navy.[8] He lived at Larchwood House, Crawley Down, an Arts-and-Crafts building west of East Grinstead, serving on the Board until leaving the area in 1933. There was also St John Douglas Townsend Loftus (1866–1950), a retired army officer with the King's Royal Rifles, now living at Crowborough. And elected for 1924 was Rear-Admiral Tufton Beamish (1874–1951) of Chelworth, Chelwood Gate, who joined briefly following a naval career but then promptly resigned to pursue politics, becoming Conservative MP for Lewes 1924–31 and 1936–45, when he was succeeded by his son Tufton.[9] The ex-service presence was further amplified by Major R.E.S. Barrington, who joined in 1931 having served at Gallipoli, Egypt, the Suez Canal and Salonika with the Black Watch, but who resigned in 1935 because he no longer held land with Forest rights.[10]

In a different field was Sir Francis Henry Champneys Bt (1848–1930), an eminent obstetrician, created baronet in 1910 and living at Littlemead, Nutley. He was a county councillor and composer of hymns. The barrister William W.S. Follett (d. 1934), from Crowborough, was a trustee for the minor Earl De la Warr in 1920.[11] Between 1925 and 1928 another distinguished man appeared on the Board, and one who illustrated very well the advent of new men and wealth. Freidrich (anglicised Frederick) Eckstein (1857–1930) was a German-born businessman, Rand gold mine developer and chairman of the Sudan Plantations Syndicate. He was created Baron Eckstein of Fairwarp and the Sudan in 1929 in recognition of his work benefiting

8 Colville was a senior officer at the vital naval base of Scapa Flow 1914–16 and commander-in-chief Portsmouth Harbour 1916–19. Retiring in 1922, he was appointed rear-admiral in 1927 and vice-admiral and lieutenant of the admiralty in 1929, ancient offices revived by Edward VII in 1901. He was awarded many honours, including a knighthood in 1921, together with several foreign decorations. He had strong royal connections.

9 ESBHRO, CLW 2/1–3 Papers of Baron Chelwood (formerly Sir Tufton Beamish MP).

10 Imperial War Museum, Documents 9714 Private papers of Major the Hon R.E.S. Barrington DSO.

11 ESBHRO, AMS 6627/1.

Figure 11.1. Fairwarp churchyard. Tomb of Frederick Epstein showing Sudanese plantation workers, with Hermione's statue (source: the author).

Lancashire cotton operators.[12] He had purchased Oldlands in 1920 from Ventura Misa, passing it on to his son Sir Bernard (1894–1948), who added Furnace Wood (Buxted) and undertook many alterations to the house and gardens. When Bernard died the estate was broken up, but Fairwarp churchyard has three impressive monuments to the Eckstein family, the tomb of Sir Frederick with sculpted Sudanese plantation workers and miners, and with a poignant bronze statue of Frederick's daughter Hermione (1893–1945) as a young girl (Figure 11.1).

In 1929 came the end of an era. On 29 October Raper wrote to ask the Board to 'release' him from his office after 42 years with the Conservators. He had been highly instrumental in defending the rights of Commoners against De la Warr, establishing the Board from 1885 and drawing up the constitution and its procedures and regulations, as well as in the Commoners' meetings. Raper was now nearly 85 and, although he continued to be active in Battle, his decision was probably influenced by the knowledge that both Maryon-Wilson and Douglas Freshfield, longstanding colleagues on the Board, were also shortly to retire. He was now to be replaced by his

[12] The Sudan Plantations Syndicate was formed in the early twentieth century to help create and then manage the huge Gezira irrigation complex, then jointly administered by the UK and Egypt, an imperialist project that changed traditional social and economic relations in that area.

partner Fovargue on the same terms. In fact, in 1950 another William Augustus Raper (1911–1987) was to be appointed as Fovargue's deputy. This was Raper's grandson, who became a junior partner in the Battle law firm after the war, taking on the role of clerk to the Conservators and replacing Fovargue in June 1951.[13]

When the Board first met in 1887 there were 67 registered commoners; now there were some 450, a number likely to increase further, making the collection of the rate and other work increasingly onerous. As the numbers of Commoners increased with the break-up of some estates (see below) there were added difficulties of assessing the liability to pay, actually collecting the money and dealing with defaulters. There were some who really wished to give up their common rights, thus avoiding payment, but they were informed that the rate was levied on the rights attached to their property, not to them as individuals.

In an unprecedented move at the December 1919 Commoners' meeting Sylvanus Ridley and Thomas Osborne proposed a woman, Mrs Lucy Rose Ridley, as a Conservation Board member for 1920. They put her forward because her husband, Jesse Ridley, had been proposed but refused because not then a Commoner. Perhaps unsurprisingly, she was defeated, by Sir Francis H. Champneys Bt, who had been nominated by Maryon-Wilson. Mrs Ridley had three supporters voting, including her two nominees; Sir Francis had seven, interestingly including Joseph Ridley. When the vote went against her Jesse Ridley exclaimed 'The smallholders are not doing their duty, they are under the rich man's thumb.'

This was a particularly troubled meeting. Even before it had properly begun Sylvanus Ridley had attempted to move that the Conservation Board be dissolved: 'we can get on much better without them than with them ... we want to get rid of you lot, we have had enough of you. You are a disgrace to the neighbourhood.' And, when W.W.S. Follett was then elected, Jesse Ridley intervened: 'this is another smuggling job ... this is another rotten lot, no doubt about it, it was all cut and dried before we came here.' The interventions by the Ridleys continued during the meeting and, following a final vote of thanks to the chairman, Jesse Ridley said 'I hope before we all meet again there will be a Labour Government in power and shoot the lot of you out of it.' In fact, the first Labour government was not to come until the minority government of Ramsay McDonald in 1924, although no radical legislation then ensued.

Then, according to their normal procedure, the Conservators moved to convene their own Board meeting without the Commoners. But Sylvanus Ridley said 'I don't believe you can compel us to leave the room. We shan't go and if you want us to do so you must fetch a policeman.' After fruitless argument the Board arranged to move to an adjoining room but Sylvanus also endeavoured to force his way in, but was prevented by the two rangers posted at the door. The 'warm' meeting then continued.

---

[13] Reginald West Fovargue (b. 1892) had married the widow of Raper's son Robert. The large archive of Raper and Fovargue papers are at ESBHRO, RAF and include papers concerning the Ashdown Forest dispute 1877–81. William Augustus Raper Jr was articled to a Brighton law firm before the war and had a distinguished war record as a major by 1942, being awarded MBE in 1945.

None of this, of course, was included in the Conservators' minutes, but rather in a newspaper report headed 'The fighting Ridleys' Battle of the Forest – trouble at the Conservators' meeting'. And the December 1920 meeting was another occasion for the fiery Jesse Ridley to argue for the rights of smaller foresters, finding himself in a dispute with Raper, who at one point exclaimed: 'I do not expect to be cross-examined.' The chairman had patiently pointed out (again) that Jesse still had no actual right to be at the meeting, but was being indulged.[14]

Such then, was the ill-temper of many of the meetings. In February 1920 William Osborne of the Foresters Arms, Fairwarp, who had been elected a Conservator in December 1919, declined to hold office. At the Commoners' meeting in December 1921 Sylvanus was himself put forward as a Board member for 1922 and declined to stand, but promptly brought five issues concerning the state of local roads and other matters to the attention of the meeting. Joseph Ridley, the oldest surviving member of the original Board, died in March 1922. He had frequently been at odds with other Board members, but his passing was greatly regretted and sympathies to his family were sent from the Board.

By 1926 Jesse Ridley had become a Commoner in his own right, having moved to the Old Forge, and at his first meeting of the Commoners, in December 1926, he called for the resignation of all the Board members, to be replaced by others 'with a view to introducing practical blood'. The proposal was ignored. At the meetings of December 1932, 1933 and 1934 he put forward Miss Frances Rose Ridley (later Penfold), but without success. But at least F.T. (Frank) Ridley, who had repaired rangers' cycles and provided petrol and technical assistance, was elected in 1933.[15] His shop in Nutley was converted to a garage and petrol station when motor cars arrived (Figure 11.2).

GOVERNANCE

The Forest was now governed through committees. The two area committees for Maresfield and Forest Row were joined by a Fire Committee, which in 1936 became a General Purposes Committee, together with various ad hoc committees established as necessary, such as one convened to discuss funding difficulties or another to deal with planning matters, especially road widening schemes.

---

14 *Sussex Agricultural Express*, 26 December 1919, 'The Fighting Ridleys'; 31 December 1920. I am grateful to David Ridley for drawing my attention to the reports.

15 Frank Ridley began with cycle repairs at Horney Common, then moved to Nutley high street and to another location with petrol pumps and car repairs. His garage was requisitioned by the army in early 1940. He was also a member of Maresfield parish council during the Second World War and involved in placing evacuees <http://www.nutleyhistorical.org.uk> [accessed 11 January 2019]. His brother Albert Joseph was the renowned beekeeper (see above) but also clock repairer and diarist whose diaries and letter books, chronicling much of village life from 1883 to 1932, are at ESBHRO, ACC 4669 (David Ridley personal communication 17 April 2000 and <http://www.bbc.co.uk/history/ww2peopleswar/stories/96/a6781296.shtml> [accessed 12 July 2018]. The first woman eventually to be elected as a Conservator was Mrs Munn, representing East Grinstead, who joined the Board in 1942 but died in late 1943.

Figure 11.2. The Ridley cycle shop at Nutley 1911 and the later garage (source: Nutley Historical Society).

One existential issue facing the Conservators was finance. There were two main sources of income: the Forest rate at 6d per eligible acre from the Commoners, and fee income from licences issued to residents or utility companies wishing to improve roads, erect noticeboards or, in the case of the golf clubs, to expand their courses further into the Forest. Some of the income from the latter source was in fact divided between the Board and De la Warr, as lord of the manor. The income was channelled into four accounts: two for payment for the rangers; one for the clerk; and the 'Protection and Improvement Fund', for work on the Forest. Any other money went into a deposit account.

But as the tasks became ever more complex during the 1930s the income was clearly insufficient. By 1935 the inadequacy of the 6d rate was clear to all. Approaches to the Ministry of Agriculture for financial help were rebuffed. Sir Lawrence Chubb, secretary of the Commons and Footpaths Preservation Society, advised that an approach to the appropriate local authorities might ensure income, in return for representation on the Board. It was now estimated that £1,200 was needed annually, but that at present only £480 was expected. Economy measures were deemed necessary, one being the termination of ranger Hatchett's position in April 1936. He had been with the Conservators for ten years but now he was given three months' notice and his motorcycle was sold back to Frank Ridley. The Board also attempted to recover outstanding rate contributions, or failing this, to distrain on property.

In order to sound out the local authorities on a proposal that they contribute to Forest expenses and have Board representation, an exploratory conference was convened at what had now become the Shelley Arms, Nutley, on 20 January 1936. Present were representatives of Uckfield and Cuckfield Rural District Councils, East Grinstead Urban District Council, East Sussex County Council and Board members. Legal positions were discussed, local authority sub-committees formed and they met again on 14 May, when it was now agreed to proceed, but in the knowledge that a Bill would be necessary to change the governance of Ashdown. This became the Ashdown Forest Act 1937 and Royal Assent was given on 1 July 1937, the whole event being overshadowed by the ongoing abdication crisis.

This new Act ushered in annual grants from East Sussex County Council and the three districts for the preservation of the Forest.[16] Increased income was now assured, and the Board could now afford to appoint a head ranger (who had to be Sussex-born and under 50); H.T. Nicholas from Heathfield was appointed, together with Gilbert Day of Cripps Corner as under ranger alongside Herbert Kirby.[17] Nicholas was to have a motorcycle and the two under-rangers pedal cycles, something of a comedown for ranger Kirby, who relinquished his position as the first ranger with a motorcycle, although apparently he had not applied for the Head Ranger position.

---

16 Ashdown Forest Act, 1937 [1 Edw 8 & 1 Geo. 6].

17 Gilbert Day, promoted to senior ranger in 1948, died of head injuries in 1963 when a tree collapsed on him during felling operations (Christian, *Ashdown Forest*, 43), the coroner recording death by misadventure.

Half of the new income was to come from East Sussex, allowing them to have five Board representatives, and the rest from Uckfield and Cuckfield Rural District Councils and East Grinstead Urban District Council, with one representative each. The Act also brought concessions about car parking for 'trippers', hopefully lessening random parking on the Forest. This was separate from the Board's decision to allow a car parking space at Fairwarp Church in 1936, since on Sundays cars had been blocking the road.

The Conservation Board was, for the first time in its history, no longer to consist solely of local men and Commoners. Nominations for the Board now came from the local authorities and a new Board was elected as from 1 January 1938, with seven of the previous Board retaining their seats, including messrs Hardy, Beamish and F.T. Ridley. The local authority representatives tended not to be as long-serving as the Forest men, since much obviously depended upon local election success. The new Board, given its increased size, would now meet in the village hall, Forest Row. Two East Grinstead MPs now joined the Board as East Sussex County Council representatives: the Conservative Henry Cautley (1863–1946), later Baron Cautley of Lindfield; and Lt-Col. Ralph S. Clarke (1892–1970), the son of Colonel Stephenson Clarke, who had purchased the Borde Hill estate near Haywards Heath in 1892, to which Ralph moved in 1949, following his father's death.

ONGOING ISSUES

Beneath the surface of these changes in governance, the Conservators continued to meet and deal with tasks some of which were already longstanding. Once again, much was heard about the presence of gypsies. The Commons Act 1876 had allowed local authorities to treat Gypsy intrusions as a 'nuisance' affecting 'the health, comfort, and convenience' of local people, who found their itinerant culture unsettling. Many agreed that the gypsies should be moved on, whether by the police, locals or Conservators. Here, as elsewhere, discrimination and eviction were their lot.

Familiar names recurred: Harry Walters was twice fined for not removing his caravan in 1920, and by 1924 was reportedly sleeping in an old stone quarry near Fairwarp vicarage, but without either tent or van. No action was taken, although he did threaten ranger Kirby with a pitchfork on one occasion.[18] Many frequented the Forest Row area, with some being removed by the rangers between 1927 and 1931; Henry Brazil was fined 10s.[19] But the numbers increased again throughout the 1930s, and in 1932 ranger Hatchett, second ranger, was attacked by Rose Jones while moving 20 caravans off the Forest.[20] Gilbert Sargent recalled the fights between the gypsies

---

18 F. Kirby, 'The Reeve's tale 1', *Ashdown Forest Newsletter*, 6 (1984), 14–16.

19 Members of the Brazil family continued as regular visitors. In February 1953, for example, Fred Brazil was taken to court by the Conservators but by December 1958 the Brazils were reported as having been on the Forest for some time.

20 Ranger Hatchett was somewhat accident-prone. Soon after taking office he sprained an ankle while putting out a fire, then contracted influenza, had an accident on his new motorcycle and took a chill that incapacitated him in February 1930. He was then reprimanded for

and the village lads, especially at the Shelley Arms, where a field was set aside for Nutley Fair or Feast Day. He told of the bare-fist fights among the foresters themselves to see who would be 'Don of the Forest' and how PC Dadsell could do little because of the closeness of the community. 'Oh, they was pretty rough.' Drunkenness and violence might characterise the later hours of these days, with fights between Nutley youths and those from Maresfield or Fairwarp.

But gypsy families had built up strong links with the Ashdown landscape over many generations, with kinship groups keeping place traditions alive. Ashdown as a marginal environment would be sought out and the making of birch brooms, pegs or wooden toys depended on opportunistic access to Forest resources. The rangers were urged to take out summonses against the gypsies, and with prosecutions and fines the issue abated somewhat by the late 1930s, but not before the Board heard in 1936 that increased numbers were accumulating because clothing was being distributed from the Convent at Ashdown Park. The Mother Superior was to be contacted to discuss this matter.

There was also the ongoing problem of one Weller, who had achieved local notoriety by shooting deer and then selling venison on the Forest, sheltering under bushes at night and roaming the area with a gun. A ranger

> reported that he had found on the Forest in a shack with the clothes of Weller, the homeless and half-witted vagabond who frequents the Forest, a loaded gun. He had after consulting the Clerk handed the gun over to the Police at East Grinstead, on whom Weller had called and claimed it.

A stern rebuke was given by the police. Nevertheless, Weller was still camping on the Forest in October 1934 and refusing to move.

There were also issues surrounding the commoning of cattle. One female Commoner complained that her heifers were being interfered with by a bull, but was informed that, rather alarmingly, there was nothing in the byelaws to prevent bulls being turned out on the Forest. And, in response to a complaint from the Ladies Golf Club that cows were wandering over their greens, the Conservators replied that 'the cows were on the Forest by right, while the ladies were there only by sufferance'.[21] But the passage of timber tugs across the fairways, reported in 1920, was halted.

The Automobile Association was similarly rebuffed after complaining about cattle on the roads, but in 1922 three gates were installed to stop cattle from reaching the busier main routes, although fencing along the roads was criticised as compromising the character of the Forest. And when a motorist killed a sheep in 1925 it raised the

---

carrying pillion passengers unconnected with Forest duties. On 1 July 1931 he reported another motorcycle accident, damaging the machine. In summer 1932 he underwent X-ray treatment at Brompton Hospital and had been absent for this purpose on various occasions. Ranger Kirby seems to have fared better, although in summer 1932 he had to report an accident on his motorcycle when he rode into a swarm of bees.

21 C. Lockwood, 'The Changing Use of Land in the Weald Region of Kent, Surrey and Sussex, 1919–1939' (Univ. of Sussex unpub. DPhil, 1989), 208.

question again of whether sheep were commonable: an 1893 decision had decided against, although they were in practice allowed to remain. In fact, a count taken in summer 1945 found 450 being grazed, mostly around Colemans Hatch. Other non-commonable animals, such as geese and goats, were also suffered to remain, especially in the food crisis years immediately after the war.

However, social friction continued. Relations between the smaller Commoners and larger landowners were not necessarily harmonious: many felt that the earl should devolve more of his manorial powers to the Conservators, especially over the cutting of gorse and rank vegetation, which were fire hazards. The Board also felt that the lord's representative should again attend meetings, a practice that had fallen into disuse. Mr Adams, representing the lord, duly attended the October 1929 Board meeting and continued to do so thereafter. In the interests of compromise, Lord Castle Stewart at Old Lodge stepped in. He offered to host the first of what might become annual lunches or dinners for Commoners and Board members to help liaison, an offer that was politely declined by the Board, although allowing members to attend in their private capacity. In fact, the dinner did take place in late April 1930, to general satisfaction.[22]

ASHDOWN AND THE FIRST WORLD WAR

The use of Ashdown during the Napoleonic Wars has already been noted, but the twentieth century brought still more demands on such large open spaces, including in the South East. In fact, as early as July 1892 a letter had foreshadowed events to come: an informal enquiry from the General commanding the South Eastern District enquiring whether the Conservators would allow a brigade to camp on the Forest during 1893. It was decided to recommend the request to the forthcoming Commoners' AGM 'not only on patriotic grounds but also on the ground that it would be a great pecuniary benefit to the neighbourhood'.

Following this came an application for a rifle range near King's Standing for the 1st Cinque Ports Rifle Volunteers Corps. This request was subsequently withdrawn, but it was later agreed that a rifle range at King's Standing be allowed for the Rotherfield & Crowborough Company of Volunteers, subject to safeguards, including public safety and compensation for damage to Commoners' cattle. A rifle range went ahead near Tompset's Bank, but it was not without incident. A genteel game of croquet on the lawn of Dr Walsham of Warren Side, Forest Row, was interrupted by two bullets passing close to Mrs and Miss Uhthoff before cutting into nearby bushes! The incident

---

22 Arthur Stuart, seventh earl Castle Stewart, MC (1889–1961), an Anglo-Irish peer and Unionist MP for Harborough 1929–33, was a major in the Machine Gun Corps in the First World War. He committed suicide at Old Lodge on 5 November 1961, leaving his widow Eleanor (née Guggenheim) to reside there until her death in 1992, at which time the property was sold to His Highness Torki M Bin Saud Al Kabeer, of the royal house of Saudi Arabia. Lady Castle Stuart had initiated Nutley's Women's Institute in 1924 and was a very active President throughout the Second World War.

Figure 11.3. Royal Engineers at Chelwood Gate 1912 (source: Danehill Historical Society).

was duly reported by the Conservators but an Inspector of Musketry later declared the range safe, and the Conservators were powerless to help Dr Walsham further.[23]

Amid the wave of patriotism engendered by the Boer War, in July 1900 the Conservators were informed that the 1st Sussex Volunteers Royal Engineers would be encamped at Sheffield Park from August to September and were asking permission for about 600 men to use the Forest Row Rifle Range during the encampment, although again the application was later withdrawn. Later another request came from the Royal Artillery in mid-1908 for a camp near Crowborough accompanied by manoeuvres on the Forest. The Commoners agreed that a force of about 1,200 Artillery and 600 horses might camp in July and August, subject to all necessary repairs and compensation. But a subsequent agreement that the horse manure arising from the camp was to be spread around the Forest was not enacted and instead the manure was removed by a contractor.

Whatever the impact on the Forest ecology, these camps were certainly welcomed by the tradespeople: in 1911 Jesse Ridley wrote to the Conservators expressing the inhabitants' satisfaction because of the sales of dairy and poultry produce as well as fruit, vegetables and general farm goods. There was also plenty of laundry work available, and permission for temporary refreshment bars and marquees was given. Now the number of camps multiplied for both enlisted men and territorials: in 1912 the Royal Engineers Signal Company was near Chelwood and the 1st and 2nd County of London Yeomanry, about 900 men and 800 horses, was on St John's Common (Figure 11.3). By mid-1914 military camps had become a regular feature, although a fire probably caused by visitors leaving the St John's camp late at night raised concerns. The fire was extinguished by about 60 soldiers.[24]

---

[23] In 1891 William Johnson Walsham was a surgeon living at Warren Side, Forest Row, with his wife Edith. *The Medical Directory* of 1895 listed him as a widely published FRCS at 77 Harley Street.

[24] An excellent account of military activities is M. Smith, 'The military on Ashdown Forest 1793–1925', <http://www.nutleyhistorical.org.uk/index_htm_files/THEMILITARYONASHDOWNFOREST1793-1924.pdf> [accessed 19 July 2019].

And then, at the August Board meeting:

> It is Resolved that, having regard to the war, in case any application is received from the War Office to use the Forest for camping or other military purposes, the Clerk be … authorised to place the Forest at the disposal of the War Office … and the District Committees be requested and empowered to make all necessary arrangements.

The Conservators were quick to open the Forest for any military use, although stipulating that when a camp ended all excavations were to be properly backfilled and the surface restored; the horse manure be spread; the surface be reseeded; and the utmost precautions taken to prevent fires, with the military responsible for their extinction. Raper now reported that regulations issued under the Defence of the Realm Acts 1914 and Military Manoeuvres Acts 1897 and 1911 empowered the use of land, including the temporary suspension of rights of way. By October there were temporary camps at St John's Common and Camp Hill, King's Standing and in Kidbrooke Park. Freshfield Hall (Forest Row) was also turned over to the military. The 'pecuniary benefit to the neighbourhood' was quickly seen again: almost immediately the Conservators were asked to consent to George Levick of Primrose Farm, Uckfield, erecting a tent on the Forest near Poundgate School, to supply the soldiers encamped on the Forest with tea, coffee and other light refreshments.

Support for the troops in Flanders was unrelenting, but on Ashdown the inconveniences were many. There were still more fires through soldiers' carelessness; footpaths and bridleways were closed; roads and drains were damaged by heavy military traction engines drawing hay. There was damage to trees by tethering of horses, polluted watercourses and injuries to cattle falling into trenches (Figure 11.4). There was also the destruction of gates, unauthorised cutting of litter for horses' bedding and shallow burying of cans and other rubbish. The soldiers' privies erected alongside the hedge of Ashdown Place were also a concern, but no-one seemed to have the power to move them again. The constant turnover of troops meant that the offences were repeated, to the Conservators' frustration. On the other hand, the military authorities at Crowborough complained that Commoners' cattle were straying into their camps and doing damage, asking that fences be erected or the cattle be impounded. The authorities were firmly reminded that there were about 250 Commoners and that fencing was illegal. Clearly there was some tension on both sides. The military were, however, encouraged to join the Commoners in cutting back coarse furze and heather, perceived as a fire hazard and a danger at bends on roads.

Somewhat typically, on being re-elected to the Board for 1916, Joseph Ridley was forthright: 'some people think the Conservators might bring more pressure to bear on our Military visitors, but that we must make up our minds to exercise forbearance'. And then, at the December Commoners' meeting, he proposed that Board meetings be postponed until the Forest was again clear of the military. This was seconded by Sylvanus Ridley, but all others present voted against, as it was felt that the Conservators were essential in dealing with Forest issues arising from the military presence at this time.

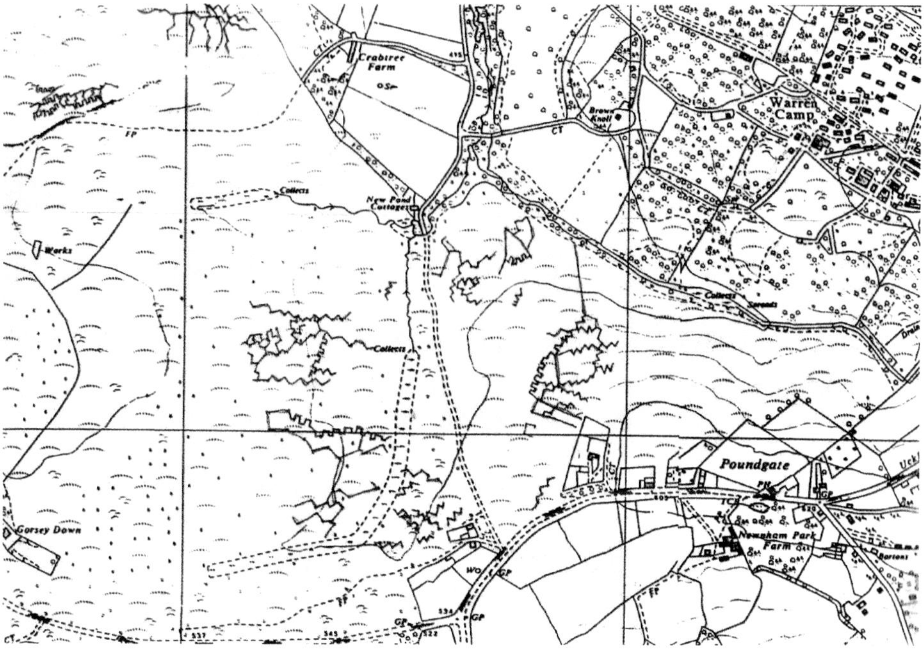

Figure 11.4. Practice trenches west of Poundgate (source: P. Chassaud, 'A mirror to Armageddon: the landscape of Sussex in the First World War: Trench systems, Defence plans and Military training in Sussex 1914–1918', *SAC*, 152 (2014), 153–87, map p. 160).

Military activity did entail the loss of some grazing land. As a result, there were some refusals by Commoners to pay their Forest rates, or to pay for the second ranger, as their commoning activity was curtailed. One who certainly suffered from the outbreak of war, although in a different way, was Count Alexander Munster at Maresfield Park. He was highly regarded locally, but declared an enemy combatant under the Trading with the Enemy Act 1914 and his property of 2,657 acres was confiscated in May 1915.[25] He and his family were apparently transported to Germany by a zeppelin which landed in his grounds. The park was then commandeered by the military and partly used for an army camp, from which the Royal Corps of Signals evolved. But the camp gave cause for concern: there were conflicting reports of litter-taking and stopping Commoners' wagons on the Forest, causing interruptions to their activity, and the estate threatened to stop its £60 rate payment.

The Royal Ashdown Golf Club also claimed that the war was affecting its finances, making it difficult to pay the rate. The club appeared to be facing ruin, informing the Conservators in December 1917 that 'The Club has lost this year six Members by resignation, and nine by death, and has about forty-one serving their country. A considerable number of resignations must be expected in December. Not a single new

[25] *Hansard*, House of Commons Debates 14 June 1915 vol 72 cc 481–3.

member has been elected.' The club hoped that if their rent was remitted then that due to the De la Warr trustees might also be remitted. They felt that they had provided work for the smaller Commoners, had attracted wealthier visitors and residents to the area and had guarded the Forest growth. In December 1916 the Board had agreed to a substantially reduced licence fee, as very little golf was apparently being played, but thereafter remained adamant that economies should be found from within the club.[26]

Many of the private enclosures within the Forest were certainly affected. By 1918 large numbers of pines on the Pippingford and Ashdown Park estates had been felled for pit-props. In 1918 the Oxcroft Colliery Company purchased Ashdown Park, felling much of the timber for pit-props.[27] Albert Turner's firm alone purchased 1,000 trees from Campfield Rough on the Streatfeild estate near Fairwarp. In all cases permission was sought from the Conservators to stack the poles on the Forest temporarily until they could be transported to Forest Row station by heavy traction engines, permission readily granted, although sparks from the engines were again a concern. In the north of the Forest permission was also granted at the end of the war for an oil-driven engine and tram line to carry away the 1,500 tons of timber cut from Tile Lodge Wood, and Connor Brothers were still deploying their horses to haul timber tugs from Crowborough Warren by 1920.[28]

The immediate post-war depletion of woodland was a concern. In 1921 the newly created Forestry Commission approached the Board asking for an enclosure in which to plant fir trees. They were flatly refused, as had been the proposal before 1914 to plant 500 acres with conifers. In 1924 the first census of woodland was undertaken. At the Conservators' meeting on 25 June that year Douglas Freshfield called attention to:

> the inaccurate and inconsistent manner in which the various parts of Ashdown Forest were coloured and distinguished in the maps in course of preparation by Messrs. Wood and Walford of East Grinstead acting as local agents of the Forest Survey of England undertaken by His Majesty's Government. He had recommended the Firm to communicate with Mr Raper.

In 1927 Lord Lovat, then Forestry Commission chairman, again approached De la Warr for permission to plant an area with trees, and although De la Warr was strongly in favour, once again the Board refused, as 'to approve such a scheme would be to betray the trust vested in the Board on behalf of the Commoners'.[29]

During the war the house at Ashdown Park was taken over by Lady Sybil Brassey for a hospital and convalescent home for Belgian army officers, together with her

---

26 See C.J.L. Strachan, *Fair Ways in Ashdown Forest: the History of an Elite Golf Club playing on Common Land for 125 years* (St Andrews: St Andrews Golf Press, 2013).

27 The Oxcroft Colliery Co. was formed in 1901 at Bolsover, Derbyshire. In April 1919 an explosion in the colliery claimed several lives, although whether there was a link with the sale of its Ashdown estate two months later is unknown.

28 See the photo of Connor Bros hauling timber in Payne and Batchelor, *Bygone Crowborough*, no. 33.

29 Lockwood, 'The Changing Use of Land in the Weald', 252.

Figure 11.5. Forest Row war memorial 1920 and Freshfield Hall (1892, rebuilt after fire 1895) (source: author).

own house at Chapelwood Manor, Chelwood Gate. But in June 1919 the Ashdown Park estate was again put up for sale by the Oxcroft company, and Sister Marie Des Saints Anges, head in England of the Order of Notre Dame, considered it ideal for convent purposes. In the spring of 1920 the first Sisters of Notre Dame, together with 33 novices, arrived. Building extensions, such as the beautiful Church of Our Lady and St Richard, were completed in the 1920s, and the convent's influence developed throughout the interwar period.[30]

30 <http://www.ashdownpark.com/historyashdownparkhotel> [accessed 21 June 2020]; the Convent remained here until the United States International University took over in 1971, to be followed by Barclays Bank (from 1974) and the present Hotel and Country Club (from 1993). By then T. C. Thompson's church of St Richard de Wych had already been allowed to fall into disrepair.

Women and children were involved in the war effort. A training school for women farmworkers was established at Woodgate Farm, Danehill, although the prime minister Lloyd George singled Sussex out as a county too conservative in its attitude to employing women on the land.[31] Schoolchildren such as those from Fairwarp school collected sphagnum moss, which was then sent away to be sterilised and packed into bandages to absorb blood.[32]

Inevitably there would be losses of life from the Ashdown area. All households either suffered directly from the death or wounding of a loved one or knew intimately of a neighbour experiencing loss (Figure 11.5). The war seemingly spared no-one, and there was no 'thankful village' around Ashdown.[33] At their October 1916 meeting the Conservators expressed their regret at the death of the clerk's son, Major Robert Raper, killed leading an action in France on 2 July, and following on from the death of the eighth earl in 1915.[34]

It is perhaps invidious to single out one Forest family, but the Kekewich family, with seven children, was very severely affected, with four sons fighting. In 1909 Lewis Pendarves Kekewich (1859–1947), a metal merchant, and Lilian his wife had purchased Kidbrooke, which they held until 1916, when the remains of the family moved to Devon. Of the four sons only Sidney survived the war, but was severely wounded. Capt. John Kekewich, with the East Kent Buffs, was killed at the Battle of Loos on 26 September 1915 and his body never found. George, with the City of London Yeomanry, fought in the Dardenelles but died of wounds on 28 October 1917; Hanbury died in November 1917 in Gaza fighting with the Royal Sussex Regiment. Kidbrooke was sold to Sir James Horlick, whose son Gerald was killed in 1918, and thence to the banker Olaf Hambro in 1921, in whose family it remained until 1938.

Many poorer families were also affected and such Forest surnames as Miles, Tester, Heasman, Wheatley, Divall and Weeding are to be found on war memorials in the north of Ashdown. Some had no known grave. Private George Heasman (1882–1917), from Hartfield, was killed near Ypres on 9 October 1917. His sister Ethel had married William Brown, of Fincham Farm, Hartfield, in summer 1914 but he too had been killed on 26 September 1917 in the third Battle of Ypres.[35] Edward Luxford, son of Job Luxford, the Forest Row builder, was killed on 8 May 1918, also at Ypres.

---

31 *Sussex Express*, 28 December 1917, 4.

32 Peter Freeland, personal communication 31 October 2013.

33 As at March 2020 East Wittering was the only known 'Thankful village' in Sussex. These were the villages recording no fatalities from the First World War and was a term coined by Arthur Mee in the 1930s (<www.hellfirecorner.co.uk/thankful.htm>).

34 Major Robert Raper (1877–1916), 8th Btn, North Staffordshire Regiment, was killed in Fricourt Wood during the Battle of the Somme. The main street in Fricourt is 'Rue du Major Raper' and the family restored the heavily damaged church in his memory (Jonathan Raper personal communication).

35 Ashdown Forest Research Group, *Men of Ashdown Forest who fell in the First World War and are commemorated at Forest Row, Hartfield and Coleman's Hatch: A Collection of Case Studies*, <www.ashdownforest.org/enjoy/history/docs/WW1CaseStudies-AshdownForest_000.pdf> [accessed 12 June 2016].

One shortly to join the military was ranger Kirby, who was liable for call-up. The Board felt that his services were indispensable and that a replacement would require long training, but an appeal served only to postpone Kirby's military service for about two months, and he was called up in 1916. The Board agreed that his rent should be paid and job kept open. Henry Walter, a carpenter of Dodds Bottom, was appointed temporary ranger in his place for two days a week and one evening a week when required. In December 1917 news came that Kirby had been badly gassed in France and was hospitalised. However, by July 1918 he was discharged from hospital and able to attend for work again.

Amidst the immediate post-war grief, on Christmas Eve 1918 Maryon-Wilson wrote to Lord Milner asking when he thought the Conservators might be able to take back control of the Forest. The Board was anxious to repair damage done by the trenches and camps at Forest Row and Crowborough and for the compensation to be agreed for the damage.[36] One new danger now came from live hand grenades and discarded ammunition, which could detonate during forest fires. One month later a non-committal reply from the War Office gave no date for leaving the Forest, although a later communication stated that the Forest Row camp should be vacated by the end of November 1919.

In fact, in May 1919 the Board heard that some Forest Row residents wished to retain the camp buildings after the military had gone to use as housing, many feeling that accommodation was scarce. The rural district council investigated, but the Board felt strongly that new housing on the Forest would be inimical to their responsibilities; that the shortages were temporary, being caused by military families; and that there was anyway sufficient space within the village if more housing were needed. 'The proposal to plant a permanent hamlet on the Forest is a preposterous one. It would be a disastrous precedent for further encroachments … .' Any such schemes must therefore be 'frustrated'. They were supported by Lawrence Chubb from the Commons Preservation Society, who noted, however, that under various pieces of legislation the huts could be retained for a further five years after the war. But shortly afterwards the scheme was dropped by the rural district council on cost grounds. A reminder of the camp is the war memorial unveiled in 1926 on the Royal Ashdown golf course.[37]

Nor was this the only cause for concern. By July 1919 there had been an enquiry from East Sussex County Council stating that the Sanatorium Sub-Committee of the Public Health Committee were seeking a suitable site for a sanatorium for tuberculosis and the Local Government Board were urging the provision of accommodation for ex-service men. They enquired whether the camp at Crowborough might be used for this purpose. Raper had replied that the camp was under the jurisdiction of the

---

36 For the Crowborough camp, including a plan of the layout, see L. Barber and J. Russell, 'Training for War: Plans of the three Great War Divisional Camps in Sussex', *SAC*, 153 (2015), 191–201.

37 'Forest Row camp War Memorial', <https://historicengland.org.uk/listing/the-list/list-entry/1441756> [accessed 23 June 2020].

Conservators, but referred to the Statutes under which no part of Ashdown could be enclosed without parliamentary sanction.

Finally, during 1919–20 a series of further letters passed between Raper, Freshfield and the War Department about the restoration of the Forest. The department argued that labour was now difficult to obtain and the Conservators were asked whether they would accept £1,000, later increased to £1,500, in lieu of any further obligation by the military. Board members inspected the damage and felt that the government should either bring in unemployed men to do the work, housing them at the Crowborough camp, or make other arrangements to bring the Forest back to its original state. Consequently by 1921 the army had still not repaired the Forest surface.

At this stage there were essentially three areas of concern: the Forest Row camp with its huts; the unfilled trenches; and the cultivated areas around Wych Cross (see below). For the latter the Board was willing to accept £300 compensation; as far as the trenches were concerned, they insisted that the government infill them; and, as regards the camp, discussions were hindered by the continuing existence of the huts. But none of this touched upon Maresfield Park, where a military presence continued throughout the period.

ASHDOWN AS RURAL IDYLL

Following the horrors of the First World War, Sussex seemed to offer the tranquillity to help mend the broken minds and bodies of those who returned from the Front. While many looked forward to a new life, others found solace in the past, either real or imagined; indeed, there was 'a swooning nostalgia for the rural past'.[38] When the *Poetry Review* in 1921 published an edition based on Sussex, 300 poems were submitted for inclusion.[39] The countryside was a haven for creativity and landscape writing, such that by the 1930s much of rural England had almost become a cultural artefact, a literary construction, and seemingly valued all the more as its economic contribution declined.

As ever, there were social distinctions: nostalgia for past countrysides was a more elite pursuit. 'Genteel' countryside was admired, but its real working inhabitants might provoke greater ambivalence. England was, to its writers, a *rural* England, but any poverty was seemingly invisible to many artists and writers. As in many other regions of rural interwar England, the Sussex countryside was that of the townsperson's, the outsider's, perception, with little awareness of agricultural decline, unemployment and the frequently abysmal living conditions. For there to be an Ashdown *landscape*,

---

38 P. Mandler, 'Against "Englishness": English Culture and the Limits to Rural Nostalgia, 1850–1940', *Transactions of the Royal Historical Society*, 7 (1998), 155–75.

39 I owe this reference to the late Peter Brandon, whose unfinished last work was *Sussex Writers in their Landscape: Self-fulfilment in the Age of the Machine 1850–1939*, ed. Brian Short (Chichester, forthcoming 2022).

there needed to be an outsider looking at it, because people *working* there did not enjoy the luxury of detachment.[40]

Among the many candidates for idyllic landscape in Sussex, Ashdown Forest, with its varied environments – heathland and farmland, villages and wooded countryside – should have ranked highly. The open space, the views to the South Downs, the traditional cottages and elite parkland could surely offer an impression of timelessness, stability and peace. Moreover, London was close. But Ashdown had no Kipling or Belloc to draw out its sense of place in literature, no equivalent to Charleston below the Downs with its Bloomsbury Group, no network of writers. Instead, Ford Madox Ford recovered from shellshock near Pulborough and craved ordinary commonplace fields, hedges and country lanes, but 'not heaths, moors, forests or famous views'.[41] Perhaps, then, Ashdown did not quite possess the appropriate cosiness, seclusion and tranquillity. It should be remembered, of course, that the Forest was much more open than it was later to become, and scarred by the remains of military occupation. Instead it was the South Downs that inspired and where the writing of 'Englishness' really blossomed. Much of the output was, of course, thin sentimentality, but hungrily devoured.[42]

Nevertheless, there certainly were artistically creative people in the area. In 1913 Sheila Kaye Smith had written her novel *Isle of Thorns*, her descriptions of Ashdown replete with literary allusions to Hardy's Egdon Heath in *The Return of the Native*. She also wrote *Sussex Gorse* in 1916, with new 'Sussex' editions of both novels coming in the 1920s amid a flurry of country writing. *Isle of Thorns* does illustrate exactly the kind of naturalistic hyperbole later to be satirised by Stella Gibbons in *Cold Comfort Farm* (1932):

> Mist like a thick blanket lay round Isle of Thorns. At sunset it put on a ruddy glow, which deepened uncannily to crimson. For a few minutes it flushed round the cottage, casting strange lights into it – then it sank to purple – then dropped into the blackness of night.[43]

The evocation of earlier times was also displayed through pageants invoking meaningful historical events, which were popular at this time. They usually consisted of short, chronologically arranged episodes with a large cast, including many local people from all sectors of society. They were an exercise in the 'invention of tradition', an appropriation of the past to cement community cohesion and to legitimate and

40 See R. Williams, 'Plaisantes perspectives: invention du paysage et abolition du paysan', *Actes de la Recherche en Sciences Socialiales*, 17 (1977), 29–36; and R. Williams, *The Country and the City* (New York: Oxford University Press, 1973).

41 P. Brandon, *The Discovery of Sussex* (Andover: Phillimore, 2010), 197.

42 See A. Howkins, 'The Discovery of Rural England', in R. Colls and P. Dodd (eds), *Englishness: Politics and Culture, 1880–1920*, 2nd edn (London–New York: Bloomsbury Academic, 2014), 62–88.

43 S. Kaye-Smith, *Isle of Thorns* (London: Cassell and Co., 1913; 'Sussex edition' 1924), 288. The Isle of Thorns, for the purpose of the novel, is a ruined cottage.

affirm the present.[44] In July 1929 Ashdown's cultural heritage was displayed in a pageant organised in the Hambro's Kidbrooke Park in the form of a romanticised history including Anderida 'the Spirit of the Forest', with a prologue written by Vita Sackville-West, eight episodes and a grand finale. On the second day of the pageant the duchess of York (later Queen Elizabeth, the Queen Mother) attended, as did Kipling and his wife Carrie, coming over from their home at Batemans, Burwash.[45] In 1937 Christopher Fry, who had directed a repertory theatre in Tunbridge Wells and was living at Colemans Hatch, was asked to write a pageant celebrating St Cuthman, who, according to local legend, pushed his mother from Cornwall to Steyning in a wheelbarrow. The resulting 'The Boy with a Cart' (1938) was performed in the bishop's grounds at Chichester.[46]

Arthur Conan Doyle (1859–1930) lived at Windlesham, Crowborough, from early 1908. In the 1901 census, aged 41, he was staying at the Ashdown Forest Hotel with his mother Mary and future second wife Jean, so he would have known Ashdown prior to moving to Crowborough.[47] His success with the Sherlock Holmes stories was already established, and *The return of Sherlock Holmes* (serialised 1903–4 and published as a book in 1905) contained a story, 'Black Peter', that concerns a murder in the grounds of 'Woodman's Lee', a stone cottage near Forest Row.[48] Conan Doyle died at Windlesham in 1930.

One location notable for its connection with literary development was an actual Stone Cottage, on the Forest south-west of Colemans Hatch, where for three years from the winter of 1913–14 Ezra Pound and the older W.B. Yeats rented rooms. The two poets sought the isolation and inspiration of living by the Forest, escaping from the 'bourgeois state of mind' and also, unsuccessfully, the jingoism and atrocities of the war.[49] Here they could discuss Japanese Noh plays, Irish folklore, the fashionable occultism and the emergence of what became literary modernism.[50] There could be no better example of contrast between the Forest residents of the 'Waste Moor' and their intellectual visitors. But they were not altogether isolated from the war, as military

44 P. Readman, 'The Place of the Past in English Culture c.1890–1914', *Past and Present*, 186 (2005), 147–95; and see Virginia Woolf, *Between the Acts* (London: The Hogarth Press, 1941) for a Sussex play/pageant based loosely on Alfriston. And E. Hobsbawm and T. Ranger (eds), *The Invention of Tradition* (Cambridge: Cambridge University Press, 1983). In 1986 the Friends of Ashdown Forest (see pp. 395–6) also chose to mark their twenty-fifth anniversary with a pageant.

45 M. Russell-Goggs, 'The Ashdown Forest pageant', *SCM*, 3 (1929), 410–11. *The Times* reported the pageant, noting that many were overcome by the heat on the opening day. In 2001 the discovery of a brief film of Christopher Robin Milne at the pageant, with friends dressed as characters from the Pooh stories, was revealed (BBC News 26 November 2001, <http://news.bbc.co.uk/1/hi/entertainment/1677945.stm> [accessed 27 November 2017].

46 M. Billington, 'Fry, Christopher (1907–2005)', *ODNB* (2009).

47 TNA, RG 13/905, East Grinstead (Forest Row) census 1901.

48 Sir Arthur Conan Doyle, *The Return of Sherlock Holmes* (London : George Newnes, Ltd., 1905, Harmondsworth: Penguin, 1981), 139–61.

49 J. Longenbach, *Stone Cottage: Pound, Yeats and Modernism* (Oxford: Oxford University Press, 1988), 63.

50 Conan Doyle was also interested in the occult and in spiritualism, used by him and many others as a way of making contact with those killed in the war.

manoeuvres were held on the Forest in front of the cottage, and both wrote war poems. From the cottage Pound wrote his '1915: February':

> ... This war is not our war,
> Neither side is on our side:
> A vicious mediaevalism,
> A belly-fat commerce, Neither is on our side:
> ...
> We have about us only the unseen country road,
> The unseen twigs, breaking their tips with blossom.[51]

Pound married Dorothy Shakespear in 1914, who thereafter became a third person in the cottage arrangement. Yeats married her friend, the wealthy Georgie Hyde-Lees, in 1917. The two women were friends, often staying at 'The Prelude' nearby. Yeats and Georgie seem to have honeymooned on the edge of Ashdown, presumably not too far from Stone Cottage. However, the poets were never to visit Stone Cottage again after 1916, although Yeats, just before his death in 1939, did visit Dorothy Wellesley's estate Penns in the Rocks (Withyham), where his name is carved on the floor of the 'Temple of Friendship' she erected in 1938 'to those poets who loved Penns'.[52]

Such esoteric worlds could not have been more different from that of perhaps the most famous resident of Ashdown. Alan Alexander Milne (1882–1956) occupied the old farmhouse at Cotchford as a country retreat from 1925. In the Second World War he was captain of the Home Guard in Hartfield and Forest Row. As is well known, he created for his son Christopher Robin the stories of Winnie-the-Pooh, which achieved long-lasting global appeal as a result of their invocation of a world that was part-real, part-fantasy. Again, as is well known, the Hundred Acre Wood, Galleon's Leap, Poohsticks Bridge and Pooh Corner are all based on nearby locations.[53] Of Gills Lap,

---

[51] In Longenbach, *Stone Cottage*, 119–20. Because Pound was not completely satisfied with the poem it remained unpublished until being included in Longenbach's book. Reproduced also at <https://www.poetryexplorer.net/poem.php?id=10013433>.

[52] Dorothy Wellesley (1889–1956), herself a poet, had met Yeats in 1935, and he spent time in later years at Penns. She was with him when he died at Roquebrune. She had separated from Gerald Wellesley (later seventh duke of Wellington) in 1922, although never divorced, and was for some years a lover of Vita Sackville-West, who had recommended the purchase of Penns in 1928 (Vita Sackville-West, 'Wellesley, Dorothy Violet, duchess of Wellington (1889–1956)', rev. C.L. Taylor, *ODNB* [2004]). Pound moved to Italy in 1924, celebrating Mussolini's achievements.

[53] Among myriad publications on Milne, Christopher Robin and Winnie-the-Pooh are the beautifully produced book relating the stories to actual Forest locations by K. Aalto, *The Natural World of Winnie the Pooh* (Portland, OR: Timber Press, 2015) and Christopher Milne, *The Enchanted Places: A Memoir of the real Christopher Robin and Winnie-the-Pooh* (New York: E.P. Dutton, 1975). The last contains a frontispiece map (the original auctioned at Sotheby's for £430,000 in July 2018, about three times the expected amount) credited with drawing hundreds of visitors to the spot. The film *Goodbye Christopher Robin* (2017) testified to the ongoing fascination, as did the December 2017 V&A exhibition 'Winnie-the-Pooh: exploring a classic'.

Milne wrote: 'In that enchanted place on the top of the forest a little boy and his bear will always be playing.' The books were illustrated by Ernest Shepherd (1879–1976).[54] In 1978 the Conservators approved a memorial to Milne and Shepherd on the north side of Gills Lap, overlooking Cotchford Farm.

Much more could be written about the bear, his friends and their Forest environment, as well as about the enormous impact that the ongoing industry has had on tourist visits to 'the Enchanted Places' and the translation of Milne's books into myriad foreign languages. Cotchford Farmhouse, also later the scene of Rolling Stones founder Brian Jones' death, remains a private residence (with common rights), advertised for sale in early 2017 with a guide price of £1.895 million, and last sold in June that year.

PRESERVATION AND FRAGMENTATION

Twin themes of the interwar period can also be illustrated on Ashdown. The break-up of many great estates during the first quarter of the twentieth century has received much attention. Increased taxation, or even the threat of taxation, together with heavier death duties, higher maintenance costs and the loss of heirs on the wartime battlefields, induced many to consider selling land. At the same time there was an increasing awareness of the need for countryside preservation in the face of agricultural depression, increased development and visitor numbers.

Around Ashdown four significant estates were sold or fragmented between 1914 and the mid-1950s. Following the departure of Count Munster and its use as an army camp, the Maresfield Park estate was put up for sale by the Public Trustee in 1924 in 90 lots. There were soon 24 different owners, but only Flitterbanks and Lampool farms exceeded 100 acres.[55] The Maresfield Park sale catalogue specifically referred to the 'valuable Forest rights and privileges' but also the 6d rate levied on Commoners (Figure 11.6). Some took the opportunity to purchase properties they had previously rented, such as the Cottingham family at Reedings Farm (75 acres, mainly dairying), but this was a difficult time to purchase when farming was in a depressed state. Raper now anticipated a great deal of work dealing with subsequent applications for common rights. Already in 1925 permission was being sought for access over the Forest to one of the newly purchased plots, where four cottages were planned. The purchasers of these small plots included Herbert Kirby, who thereby became both a Commoner and a ranger enforcing regulation of the Forest, although this did not apparently compromise his position. One growing problem as a result of speculators' empty building sites, and one certainly noted at Maresfield Park, was the proliferation of weeds.[56]

---

54 He drew the illustrations for *When we were very Young* (1924), *Winnie-the-Pooh* (1926), *Now we are Six* (1927) and *The House at Pooh Corner* (1928), as well as Kenneth Grahame's *Wind in the Willows* (1908) and Richard Jefferies' *Bevis* (1932 edn).

55 Maresfield Park Estate sale catalogue 1924. On 18 September 1924 the Public Trustee sold the estate to William Henry Abbey of Uckfield House for £61,714. One week later most of the estate was again sold off to Thomas Place of Northallerton, Yorkshire, a wealthy timber merchant who was purchasing land around England for development and timber.

56 Lockwood, 'The Changing Use of Land in the Weald', 105.

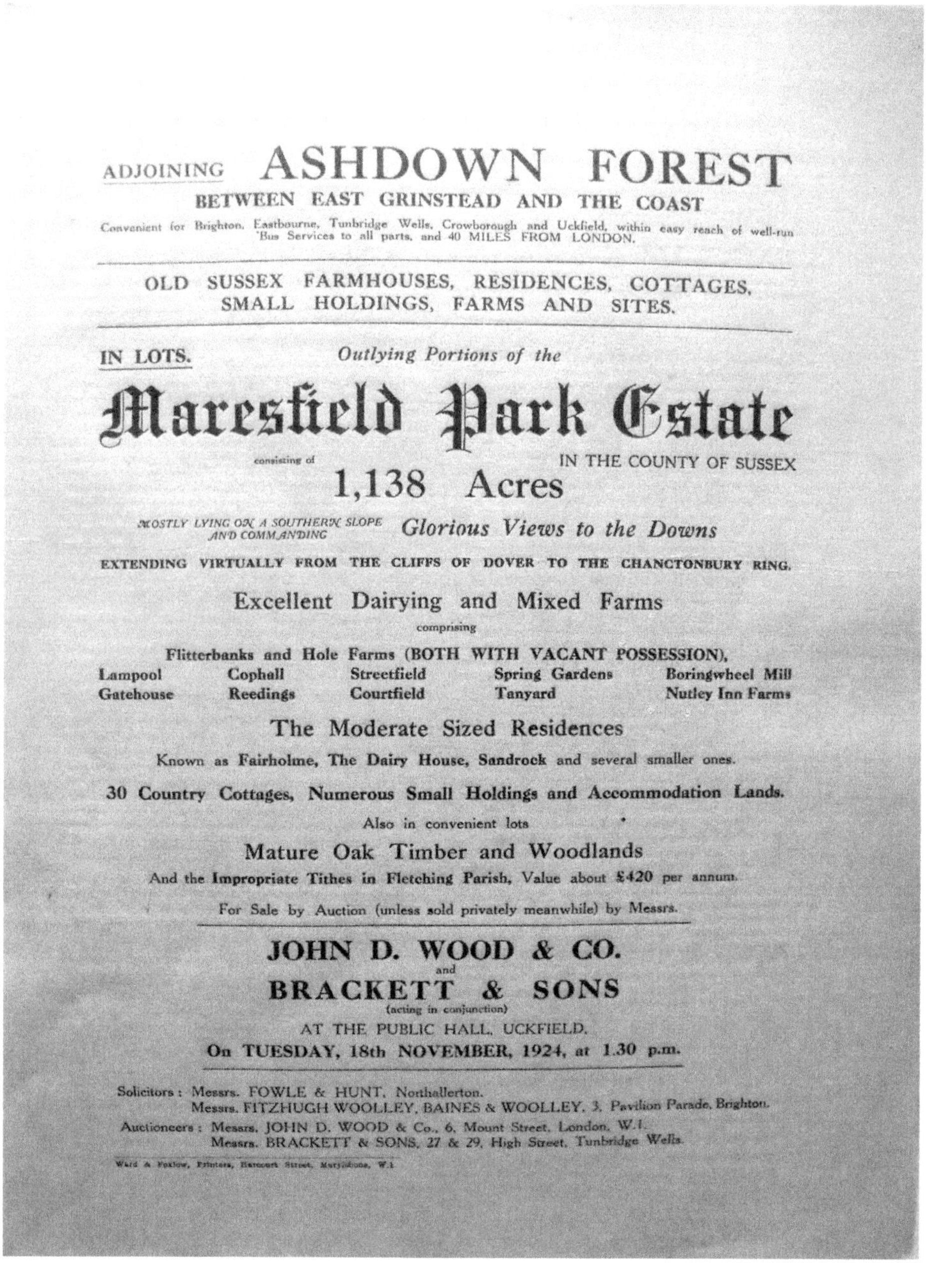
ADJOINING ASHDOWN FOREST

BETWEEN EAST GRINSTEAD AND THE COAST

Convenient for Brighton, Eastbourne, Tunbridge Wells, Crowborough and Uckfield, within easy reach of well-run 'Bus Services to all parts, and 40 MILES FROM LONDON.

OLD SUSSEX FARMHOUSES, RESIDENCES, COTTAGES, SMALL HOLDINGS, FARMS AND SITES.

IN LOTS. *Outlying Portions of the*

Maresfield Park Estate

consisting of 1,138 Acres IN THE COUNTY OF SUSSEX

MOSTLY LYING ON A SOUTHERN SLOPE AND COMMANDING *Glorious Views to the Downs*

EXTENDING VIRTUALLY FROM THE CLIFFS OF DOVER TO THE CHANCTONBURY RING.

Excellent Dairying and Mixed Farms

comprising

Flitterbanks and Hole Farms (BOTH WITH VACANT POSSESSION),

Lampool, Cophall, Streetfield, Spring Gardens, Boringwheel Mill, Gatehouse, Reedings, Courtfield, Tanyard, Nutley Inn Farms

The Moderate Sized Residences

Known as **Fairholme, The Dairy House, Sandrock** and several smaller ones.

30 Country Cottages, Numerous Small Holdings and Accommodation Lands.

Also in convenient lots

Mature Oak Timber and Woodlands

And the **Impropriate Tithes in Fletching Parish,** Value about **£420** per annum.

For Sale by Auction (unless sold privately meanwhile) by Messrs.

JOHN D. WOOD & CO.

and

BRACKETT & SONS

(acting in conjunction)

AT THE PUBLIC HALL, UCKFIELD.

On TUESDAY, 18th NOVEMBER, 1924, at 1.30 p.m.

Solicitors: Messrs. FOWLE & HUNT, Northallerton.
Messrs. FITZHUGH WOOLLEY, BAINES & WOOLLEY, 3, Pavilion Parade, Brighton.

Auctioneers: Messrs. JOHN D. WOOD & Co., 6, Mount Street, London, W.1.
Messrs. BRACKETT & SONS, 27 & 29, High Street, Tunbridge Wells.

Ward & Foxlow, Printers, Harcourt Street, Marylebone, W.1

Figure 11.6. Maresfield Park sale catalogue 1924 (source: original in possession of Dr Geoffrey Mead).

In 1930 the Georgian mansion at Buxted Park was sold, with 400 acres broken up into six separate lots as pleasure farms, building lots, smallholdings or 'cottages suitable for weekends or conversions'.[57] And, following the Second World War, large parts of the Eridge estate, to the north-east of Ashdown, were sold in 1947, and the Sheffield Park estate was sold in 1953. Rather different was the purchase of the Isle of Thorns by Alfred Wagg, who, through the Manor Charitable Trust, an institution particularly concerned with the welfare of boys from London's East End, developed the site as a camp for underprivileged children. At one point it was managed by Major the Hon. R. Barrington, himself briefly a Conservator. It too became an army camp during the Second World War, as the murals on the gymnasium testified until recently.

Such fragmentation helped to double the number of Commoners from 232 in 1919 to 446 by 1930, although commoning practices were now fast disappearing.[58] The Conservators recognised that Ashdown offered 'natural beauty', false as that concept actually was, since little was 'natural' about the landscape. One way to keep it 'natural' was to graze it, but cattle numbers were now declining. One estimate in 1933 put the number now being turned out as low as 800. Poorer grazing as oak and birch scrub encroached, the costs of cutting and collecting litter, dangers to stock grazing at roadsides from walkers' dogs or motorcars and the transfer of cottages from former Forest dwellers to wealthier incomers all pointed in this direction. However, the Commoners through the centuries had mostly been small cottagers, and in the interwar period this did not change: holdings below five acres remained very much in the majority (Table 11.3).

Table 11.3. Ashdown Commoners' holding sizes 1919–30.

	1919	%	1925	%	1930	%
100 acres +	25	10.8	22	5.7	25	5.6
50–100 acres	15	6.5	22	5.7	21	4.7
15–49 acres	24	10.3	49	12.6	55	12.3
5–14 acres	59	25.4	75	19.3	86	19.2
Under 5 acres	106	45.7	211	54.4	247	55.4
unknown	3	1.3	9	2.3	12	2.7
Total	232		388		446	

Source: based on Lockwood, 'The changing use of Land in the Weald, 53.

57 Lockwood, 'The Changing Use of Land in the Weald', 57.

58 Commoners' numbers have continued to rise. In 1965 there were 580 Commoners registered under the Registration of Common Land Act, although it was stated that there were actually over 1,000 at this time, so that half the number did not reregister. By 2017 there were around 730 (Phelan, 'Economy to Amenity', 231–2; <www.ashdownforest.org/living/past/decline.php> [accessed 27 September 2021].

Along with the estates, some of the old social and economic certainties were disappearing. To some extent their place was now to be filled by public and statutory bodies, such as the district and county councils. In 1932 a new national Town and Country Planning Act was passed 'to prevent the desecration of the countryside' and, although it was permissive rather than compulsory, it was nevertheless a step forward, and it was hoped that it would stimulate more planned intervention in the countryside.[59] These fledgling planning laws did affect Ashdown. Under section 34 of the Act it was to be preserved as an open space, a point emphasised in the 1937 Ashdown Forest Act, and furthermore in November 1940 it was agreed with East Sussex County Council and Uckfield Rural District Council that the preservation of Ashdown in its existing state as a private open space should be included under their town planning schemes. The agreement reserved to De la Warr his rights of sporting, minerals and quarrying. No-one was to use the Forest otherwise than as a private open space and there was to be no building or fencing except as required for the registered users of that space.

Preservation was now linked with planning, a combination lauded by many influential intellectuals and writers.[60] In 1926 the Council for the Preservation of Rural England (CPRE) was formed and the Board subscribed in 1929, together with the Commons Preservation Society in 1928. Elsewhere in Sussex, 1923 saw the founding of the Society of Sussex Downsmen, and in the later 1920s all attempts at countryside preservation in Sussex were supported by Douglas Freshfield.[61] A subscription was promised in 1930 to a new county Anti-litter League, but a request in 1933 from the Society for Checking the Abuses of Public Advertising (SCAPA) for a subscription was refused by the Conservators since it was felt that they already subscribed to similar organisations. However, in December 1934 the Board agreed a donation of 10s 6d to 'East Sussex Rural Community for the Preservation of the Countryside', this a reference to the Sussex Rural Community Council (SRCC), which had been formed in 1931.[62]

In this atmosphere of enhanced vigilance and intervention, the Board received a letter in April 1935 from the SRCC intimating that development was being planned for the Broadstone Warren area following the death of Douglas Freshfield. The Conservators agreed that concerted action was required to prevent any building in this area and in October the solicitors acting for the disposal of Broadstone asked whether the Conservators would purchase it for £12,000. This was impossible, but the Board determined to act together with the Uckfield Rural District Council planners to guard against any inappropriate development. Once again the Manor Charitable Trust stepped in and, from 1937, 400 acres of the Warren was used by the Boy Scouts Association, a facility continuing to the present day.[63]

---

59 J. Sheail, *Rural Conservation in Inter-war Britain* (Oxford: Clarendon Press, 1981), 63–93.

60 Clough Williams-Ellis, *England and the Octopus* (London: G. Bles, 1928) and Clough Williams-Ellis (ed.), *Britain and the Beast* (London: J.M. Dent and Sons Ltd, 1937).

61 Brandon, *Discovery of Sussex*, 224.

62 SRCC, now Action in Rural Sussex.

63 *The Times*, 3 November 1936.

There was now also an increased awareness of Ashdown's botanical resources: the Conservators were alerted in 1933 to the cutting and selling of marsh gentians (*Gentiana pneumonanthe*), and rangers were instructed to stop such acts of vandalism.[64] With the reduction in grazing, this heathland specialist plant was now in some danger of being shaded out and in 1934 it was suggested that the Board incorporate a new byelaw for the protection of fauna and flora along the lines of one recently adopted by the county council. In 1935 new byelaws were introduced to prohibit the uprooting of plants, heather, moss and rushes. This was needed: holly was being cut for resale in London, especially near Christmas; moss collecting was still an ongoing business, and one couple had been collecting moss between Crowborough and Withyham for 17 years by 1937, armed with a permit and using a pony and trap. The moss was sold to florists far beyond Sussex.[65]

The concept and practice of planning was now seen as necessary and on Ashdown this led to the formation of a special Board sub-committee from 1930 to consider plans being formulated by East Grinstead and Uckfield districts, mainly relating to road widening and improvements but also to the designation of areas of Ashdown as open spaces. The Conservators resisted county council road-widening schemes, fearing, as in 1930, that land might be taken for a new dual carriageway between Wych Cross and Nutley, cutting across the Forest. If this were to happen, a compensatory amount of land would have to be offered, but the road did not materialise. In 1932 the sub-committee commented on proposals from the North Sussex Regional Town Planning Committee, again declining to support any road widening and definitely opposing a proposed East Grinstead bypass scheme via Legsheath, but approving a scheme to bypass Nutley between Nutley and Shortbridge, Piltdown (neither of which materialised). And the Board was constantly pressurised by the county council, who wished to take increasing amounts of material for road repairs and improvements. In 1933, in response to a suggestion by the Conservators, De la Warr did give permission for stone to be quarried by unemployed men, a sign of the difficult times.

By now motor traffic was increasing rapidly, and London's commons, and indeed accessible parts of rural England more generally, were now seen by many as 'the people's playgrounds'. But such visits could be problematic. In a letter to the Commons Preservation Society in 1928 the Conservators stated: 'We are being troubled during the late spring and summer and early autumn by multitudes of motor cars, mainly from London, the occupiers of which draw off the roads onto the common, many of whom picnic, others only roam over the common.'[66] But discerning visitors were also being encouraged:

---

64 The plant has been recorded on Ashdown since at least 1835, https://diversionsinnaturalhistory.wordpress.com/2015/09/07/marsh-gentians-in-ashdown-forest/ [accessed 8 November 2020], M.E. Reader, 'Marsh Gentian on Ashdown Forest', *Sussex Botanical Recording Society Newsletter*, 51 (January 2001), https://www.sussexflora.org.uk/resources/articles/marsh-gentian-ashdown-forest/ [accessed 28 September 2021].

65 W.J. Boorman, 'The Sussex Moss Collectors', *SCM*, 11 (1937), 299–300.

66 ESBHRO, ACC 5640 CAF1/5 (22 March 1928).

> we who love the Ashdown Forest can still have it to ourselves ... go by foot ... and while you pity the 'road hog' who misses all your joy, you can have the gratitude in your heart because the litter of the picnic parties has not yet come to mar the wonders of Ashdown Forest ... Choose your day, travel on top of the bus and then, if you do not mind the keen air of the moorland and its fresh breezes, you will be amazed at the grandeur of the scenery, almost Caledonian in its wildness and spacious charm.[67]

Similarly, S.P.B. Mais wrote: 'as you go deeper and deeper [into the Forest] you will find yourself going deeper and deeper back into your own and countries past'. Guide books extolled the beauty and wildness of Ashdown, 'a place for a long day's dreaming'[68] – or, indeed, for a weekend, now an established part of middle-class London life.

Ramblers and cycling clubs also visited. The Lewes section of the Cyclists' Touring Club organised a ride, also attracting about 20 non-members, taking in Crowborough for lunch and then through the lanes around Ashdown to West Hoathly for tea.[69] No doubt such visits offered income for some residents, as did the golf courses, requiring ground staff and caddies. But the Royal Ashdown club had suffered from the effects of the First World War: membership had fallen and the Ladies' course, commandeered for the military camp, was unusable for several years. The Conservators agreed to cut the acknowledgement fee but refused permission to extend the course, although permitting a car park extension in 1928. The Crowborough golf club suffered similarly.

There were also the continuing fires, the ongoing scourge of the Conservators. In March–June 1920, for example, there were over 60 incidents covering over 50 acres and, in 1921, 69 incidents covering 90 acres. In 1922 extra unemployed men were hired to assist the rangers in fire watching, but the fires continued – 126 acres were burnt in March and April 1926. Now organised by a separate Fire sub-committee, the litter areas were to be cut into blocks separated by rides, and extra help for the rangers was supplied during the spring fire-starting season.

Some help came. In 1928 Major James Denning, a local resident, offered either his old Ford motor or a motorcycle to the Board so that rangers or a patrolman would be better able to watch for both fires and gypsies. Shortly afterwards new Board member G.D. Haslam also offered a motorcycle. The offers were gratefully accepted, as was the offer from Herbert Hardy, District Commissioner of Boy Scouts and another newcomer to the Board, for scouts to alert Board members or rangers when fires were spotted. In January 1929 the rangers' houses were also linked via new telephones to speed communication.

Nevertheless, the fires continued, with spring 1929 particularly acute; two young men were caught in the act of setting fire to the litter, now profuse because few Commoners could afford the labour to cut it, and were prosecuted, with one given three months in gaol with hard labour. But no fewer than 2,000 acres of land, both Forest and private, were burnt that year, and many private owners criticised the Board's lack of action. In 1933 the Board even tried again (unsuccessfully) using private detectives

67 J.B. Firth and J.A. Royle, 'Romantic Ashdown Forest', *SCM*, 1 (1927), 360 and 362.

68 S.P.B. Mais, *Round about England* (London: Richards, 1935), 90; J. Prioleau, *Car and Country* (London: J.M. Dent, 1929), 8.

69 Lockwood, 'The Changing Use of Land in the Weald', 161.

to watch for incendiaries and who 'arrayed in suitable disguise should frequent the Colemans Hatch Inn' to infiltrate the ranks of the suspects. Board member James Owen of Colemans Hatch also promised more precise information 'as to the hours and nature of employment of the suspects in his area'. But the culprits were difficult to catch. Fred Marshall remembered that the fires:

> were lit in such a way that ... the Commoner could be home and sitting by his fire (laughs) when the fire really got going ... and I know one Commoner who used to go to a house and knock at the door and come in and have a talk with them so that he had got an alibi when (laughs) there was an enquiry ... .[70]

Board members demonstrated a socially distant and elitist preservationist ethic typical of much interwar countryside thinking, hoping that Ashdown would retain its charm for the golf clubs, literary men and a diligent smallholding class. Compromises allowed horse riders to use specified gallops, but Commoners' wives who hung out their washing to dry on the Forest were now forced to buy a licence to erect poles and clothes posts, since it was felt that the sight of washing might detract from the Forest's beauty. Similar sentiments underpinned the Conservators' opposition to an application from the Roebuck at Wych Cross for a full liquor licence in late 1929 that would entail 'a large number of cars and chars-a-banc'.[71] The Board minute is revealing: the chairman directed Fovargue:

> to oppose on behalf of the Board the granting of the licence on the grounds that (1) having regard to the very few inhabitants in the locality a licenced house is not required (2) that for the convenience of wayfarers there is ample provision within reasonable distance at Forest Row, Nutley and Chelwood Gate, and (3) that concentrating large bodies of beanfeasters at this spot will very greatly increase the risk of fires from cigarette ends thrown away in the undergrowth on this part of the Forest where forest fires have been especially frequent.

The application was also opposed by Douglas Freshfield and other residents. Initially the licence was refused, but in early 1931 it was granted with the conditions that only 'residents or those partaking of food should be served with intoxicating liquor'. Freshfield strongly felt that the Forest must be preserved against it becoming another Epping Forest or Hampstead Heath. It was not to be catering for 'a rather different class of people'.[72] In fact the Board's rather exclusionary stance caused even the Commons Preservation Society to ask in 1948 that the Conservators drop their resistance and recognise some of the Forest tracks as public rights of way.[73]

Another new development occurred in June 1938 involving the Citrone Ice Cream Company, a 'firm who were supplying ice cream in cartons from a horse drawn vehicle'.

---

70 ESBHRO, ACC 9997 Fred Marshall, Ashdown Forest oral history project 2005–6.

71 The strangely rendered plural shows the unfamiliarity of the term (at least to Fovargue) by the late 1920s.

72 Phelan, 'Economy to Amenity', 217–25.

73 Phelan, 'Economy to Amenity', 219.

It was admitted that no action could be taken under the byelaws and the only available option was to warn the company that a byelaw prohibited parking on the Forest. But in October 1939 Mr Keys of Lone Oak Hall, Chuck Hatch, refused to move his ice cream van off the Forest despite an order from the ranger. He was to be warned that proceedings would be taken against him if he persisted.

In 1929 Ivan Margary (1896–1976), the doyen of Sussex archaeology and Roman road specialist, had been living at Chartham Park, north of East Grinstead, since 1900, and now he sought permission, which was granted, to undertake non-intrusive archaeological investigations on Ashdown. His grandfather was Donald Larnach, the wealthy owner of Brambletye, and in 1927 Margary had inherited a vast fortune together with neighbouring Yew Lodge. He published the results of his investigations in 1930.[74]

Building on the Forest itself was restricted, although as we have seen many former cottages were being extensively renovated. But the interwar period was one of house building over much of the South East. This was the era of cheap farmland because of the agricultural depression, greater mobility through the use of cars, and easier mortgage facilities. There were certainly pent-up demands for housing. The East Grinstead Medical Officer of Health's reports in 1918 and 1919 refer to the poor housing conditions in much of the area and East Grinstead Rural District Council established a housing committee in 1919, allocating three sites in Hartfield for 14 houses; nine sites in Withyham for 40 houses; and, when suitable sites were forthcoming, there were to be 30 houses in Forest Row and 10 in West Hoathly . These were to be council houses for lower income residents, with three bedrooms, living room and scullery. Families began moving in during June 1921, although there were doubts that they could all afford the rents. From 1923 private developers also began to build in Ashdown's villages and in this way the local authorities hoped to satisfy the demand for better working-class housing.

But much was speculative building and not universally admired. Sheila Kaye-Smith wrote of the 'host of squalid little buildings ... around the stations of Uckfield, Horsted Keynes, and other sweet auburns, desecrated rather than deserted'.[75] The Board attempted to halt any visual intrusions but the pressures were now very insistent: more cars, telephones, water and gas supplies, metalled roads and the need for direction posts catering for visitors unfamiliar with the area. By 1930 it also became common for the Conservators to agree to noticeboards with multiple house names at the end of roads leading into the Forest, a feature still utilised today.

---

74 I.D. Margary, 'The Mystery Mounds on Camp Hill and Stone Hill, Ashdown Forest', *SNQ*, 3 (1930–31), 10–16. Margary was hugely influential (see Chapter 3). He purchased the stretch of Roman road at Holtye, which he presented to the Sussex Archaeological Society, where his unpublished papers can be consulted. The 40th anniversary of his death is commemorated in D. Rudling, 'Ivan Donald Margary: an Officer, Gentleman, Scholar and Philanthropist', *Sussex Past and Present*, 142 (August 2017), 4–5.

75 S. Kaye-Smith, 'Laughter in the South-East', in Clough Williams-Ellis (ed.), *Britain and the Beast* (London: J.M. Dent and Sons Ltd, 1937), 33.

The issue of household waste also became even more pressing than before the war. By the early 1920s the local authorities would not collect from households in more remote areas and much was being casually dumped on the Forest. In 1923 noticeboards were erected warning against rubbish tipping, but the problem persisted throughout the 1920s, with East Grinstead Rural District Council being slow to provide any kind of collection. On occasion the fires burning the vegetation exposed even more detritus. Rubbish was even brought from London to be fly-tipped. Nor had problems of sewage disposal been fully conquered: in 1943 'Lord Kindersley has had nothing done to prevent the sewage from Coldharbour emptying on to the Forest.' Furthermore, he had arranged for a small gate to be built onto the Forest near the sewage outfall, the boundary bank there had been flattened, and a path cut onto the main East Grinstead road.[76]

Problems with the uninitiated urban visitors were inevitable. Some visitors complained that footpaths might be blocked or, more abstractly, that their contribution to the national wealth was such that their outdoor recreation should be a reward and not impeded. Footpaths could indeed be controversial: one such path was the subject of a dispute in Pippingford Park in 1931, another in Buxted Park in 1934, and when one gate to a path was locked some residents complained that they were prevented from attending Withyham church in 1938.[77]

All the above is clearly indicative of what Phelan called the move 'from economy to amenity', and it was now well advanced.

THE IMPACT OF THE SECOND WORLD WAR

A minute from the Conservators' General Purpose Committee for 29 June 1938 reads:

> AIR RAID PRECAUTIONS
>
> A letter was read from the Organiser of the Uckfield Rural District Council suggesting that residents on Ashdown Forest might be allowed to prepare shelters now in anticipation of an emergency.
>
> RESOLVED that no action be taken.

That war was anticipated once more is well known, and this local refusal to act is strange. Nevertheless, the Conservators were almost immediately faced during late 1938 and early 1939 with demands once again for training space on Ashdown, this time for armoured vehicles, infantry and signal brigades and anti-aircraft crews, as the world moved inexorably towards a second conflagration. Once declared, the impact of war was quickly felt. In October 1939 the wealthy businessman Frederick Nettlefold

76 Robert Kindersley (1871–1954), first Baron Kindersley of West Hoathly from 1941, merchant banker, director of the Bank of England, who organised the National Savings movement. In 1936 he and his wife helped to raise £30,000 to resite the East Grinstead hospital, now the Victoria Hospital, the location from 1939 for Archibald McIndoe's pioneering plastic surgery techniques for treating allied aircrew during the Second World War.

77 Lockwood, 'The Changing Use of Land in the Weald', 192–6.

applied for permission to provide temporary accommodation for 50 evacuated children at his home of Chelwood Vachery. This was approved subject to a stipulation that the latrines be removed when the Board considered that the need no longer existed.[78] At the same time an application was approved from Water Farm, Colemans Hatch, for a temporary water supply from a spring to cater for 25 evacuees. And Wren's Warren was sought by the National Camps Corporation, with children from London's East End and the Medway Towns arriving by early 1940. Access had been approved by De la Warr and, again following discussion about drainage and disposal of effluent onto the Forest, the Board approved. The accommodation was in wooden huts, originally designed as a camp for underprivileged children, with the final layout for 'Holiday Camp No 17' devised by Serge Chermayeff in 1939, probably involved through his work on De la Warr's Bexhill Pavilion.[79]

At the October 1939 Board meeting it was unanimously resolved that Standing Orders be suspended; that the powers of the Conservators in all urgent matters be delegated to a committee of five members: the chairman, vice-chairman, the two Chairmen of the district committees and Mr Rowe; and that this committee should have full powers of action.

The meeting also heard that several sheep and one horse had already been killed by cars during the blackout on the Forest roads. By this time sheep were obviously ubiquitous on the Forest and the ancient prohibitions against them seemingly forgotten. But the blackout, combined with the military occupation, certainly produced problems. In July 1940 the Chairman reported a complaint:

> that on a recent evening ranger Nicholas whilst riding his motor cycle without a light was challenged by L.D.V. pickets. He failed to answer the challenge or to produce his identity card when asked and became abusive. He also failed to produce his pass as ranger. On receipt of the complaint the Chairman had asked the ranger for an explanation. The latter had at first denied the allegations but afterwards admitted them, and proffered his resignation. The Chairman reported that there had been previous complaints of the ranger's lack of manners. In other respects he had carried out his duties in a satisfactory manner and he had informed the ranger that the matter would have to be reported to the Board. It was RESOLVED that the head ranger be instructed to appear before the Board.[80]

Nicholas was put on probation for three months but in August he left for a better-paid job in Rye.

---

78 F.J. Nettlefold, a rhododendron enthusiast, had purchased the Vachery in 1925 from Sir Stuart Samuel. In the Second World War it was also used as a military hospital. A similar hospital was organised by Lady Dorothy Macmillan at Birch Grove.

79 ESBHRO, AMS 7005. Photographs and reminiscences of Wren's Warren Camp for Evacuees, Hartfield 1939–45.

80 The L.D.V. (the Local Defence Volunteers) was formed in May 1940 and known from July as the Home Guard. Ranger Nicholas was also a part-time telephonist at East Grinstead during part of the war.

And now the military use of Ashdown really began again. An application to use the southern part of the Forest for training for approximately 10,000 troops with mechanised transport, and with the right to dig trenches, was approved. The Board asked that an Area Commandant for the Forest be made available to liaise with the Conservators, and General Hildyard was duly appointed. An application from the Royal Engineers at Kidbrooke Park for digging trenches near the Isle of Thorns was also approved and the First World War camp at Maresfield was now expanded to accommodate extra troops. By late 1939 many, both British and Canadian, were billeted at various locations, such as the Isle of Thorns, Woodgate (Danehill) and Plaw Hatch, together with British troops at Sheffield Park by 1942.

And so the impact of the military was again substantial. As in 1914–18 there were occasional misunderstandings with troops, whose conduct was criticised by the Conservators. In December 1940, for example:

> representations had been made regarding the misuse of the Forest by certain colonial troops [probably Canadians] who were ignorant of local conditions, and as a result a Compensation Officer had now been appointed ... which it is hoped would avoid misunderstandings in the future.

The 5th Canadian Armoured Division were singled out for their tank manoeuvres, especially affecting the steep Kidd's Hill down from near Gills Lap to Newbridge.

Once again pines were being felled by private hauliers, with resulting damage to the Forest. In 1943 the Ministry of Supply informed the Conservators that following an inspection they would need to fell from land at High Beeches, north-east of Wren's Warren, between Hindleap and Press Ridge Warrens, and south-east of Wych Cross on the Nutley road. De la Warr had given permission for this felling.

The fact that Ashdown was a watershed between the Medway and the Ouse also entailed defensive work being undertaken against potential German invasion. Part of the General Headquarters (GHQ) Line defending London from the south ran close to the Forest and today the remains of the squat brick and concrete pillboxes, sited near road junctions, rivers and other strategic points, can still be seen. Guarding the Medway and the railway lines were more than 20 pillboxes in Withyham, with similar numbers in Crowborough and Rotherfield. One guarded the King's Standing radio station (see below), and a few were placed along the Upper Ouse in Fletching and Buxted. An anti-tank ditch ran through Rotherfield.[81]

From 1943 the military effectively controlled the Forest under the Defence Regulations. In fact, the military could have requisitioned the entire Forest, and such a proposal had indeed been discussed in October 1942, but instead they chose to use Defence Regulation 52, allowing them to use the area but with ongoing governance by the Conservators. Nevertheless, in 1943 Commoners were told that the military had ordered that common rights be suspended when firing was taking place. Even more

81 A.W. Foot, 'The Impact of the Military on the Agricultural Landscape of Britain in the Second World War' (Univ. of Sussex unpub. MPhil, 1998), 41.

threatening was a demand from the military that the Forest be enclosed, although this was withdrawn after local opposition and a public enquiry. The Conservators were effectively powerless and could really only monitor the activities through the reports of the rangers.

Undoubtedly the Forest was being damaged: gulley erosion was initiated or worsened; temporary fencing or even barbed wire was erected; private roads, culverts, footbridges and trees were damaged. Tanks chewed up the Forest between Chelwood Gate and the Five Hundred Acre Wood; bulldozers cleared roads stretching across the Forest. There was not only tank training but also mortar bomb practice and secret training areas set aside for gas attacks. As in 1918, by late 1945 live ammunition, in this case mortar bombs, was still being found on the open Forest. There was unease about plans to fire live ammunition across the open Forest and trenches had to be infilled, but by July 1946 only 300 acres of trenches had been dealt with. Additionally, much would require reseeding after the war.

Troops were supposed to avoid the fir clumps but there was inevitably some damage. So, by late 1944 Italian prisoners-of-war were replacing the fencing around Camp Hill and King's Standing clumps, using stakes and barbed wire. A prisoner-of-war camp, built at Fairwarp to house the Italians became a Displaced Persons Camp after the war. Sheffield Park housed German working prisoners.[82]

By March 1941 the Forest had been hit by 41 bombs and by 1944 anti-aircraft units had taken up positions and were now alerted to flying bombs, six of which were shot down onto the Forest by October that year. And during 1944 an emergency landing strip was cleared, running parallel with the Wych Cross–Chelwood Gate Road. All those interviewed in the oral history exercise in 2005 relate watching the Battle of Britain taking place overhead and the arrival of flying bombs, 'doodle-bugs', some crashing on Crowborough.

On 31 July 1941 a British Wellington bomber, part of a mission to bomb Cologne but which turned back because of poor weather, crashed on the Forest north of Spring Garden. The Conservators heard that the mothers of the six young airmen had erected a wooden cross at the spot, and by April 1945 a ring fence about 12 feet across had been erected with flowers planted inside. Written on the cross was:

> Sacred to the Memory of SG/P Victor Ronald Sutton, R.A.F. Age 24. Also his five Comrades who crashed on this spot on July 31st, 1941 after bombing Cologne. No greater Sacrifice. Mothers, Brothers, Sisters.

A notice on the fence warned: 'To the boys who would destroy around this spot is spilt the blood of Six Heroes, although therefore this spot is sacred. They died that you may

---

82 An excellent account of the war as it affected the Forest is M. Smith, 'Nutley at war 1939–1945', <http://www.nutleyhistorical.org.uk/index_htm_files/NUTLEY AT WAR 1939–1945 Revised.pdf> [accessed 28 September 2021]. The Fairwarp camp is not listed among the 1,000-plus POW camps in the UK (R. Thomas, *Prisoner of War Camps (1939–1948)* (Swindon: English Heritage, 2003). A tour of the overgrown remnants of the camp by Nigel Broach in 2018 is at <https://www.youtube.com/watch?v=asMx37Mcr-w> [accessed 14 June 2020].

Figure 11.7. The airman's grave, Misbourne Valley (source: Dave Brooker).

live.'[83] This is a memorial but it became known as the Airman's Grave (Figure 11.7). In 1954 a stone cross was erected and a stone wall built against animals.

In October 1941 a letter from the Government Communications Centre was read to the Conservators regarding what was to become a Forest landmark. This was an application for the erection of a radio station near the clump at King's Standing. This was approved in principle with any details to be referred to the Emergency Committee. And 'In view of the nature of the application those present gave a pledge of secrecy'. It was conceded that the presence of the station might increase the likelihood of enemy bombing, and the installation arrangements did not run entirely smoothly, since the military authorities and Conservators argued over substitute land for the 65 acres taken by the site, as well as the fate of the station after the war. Would it be retained

[83] On normal Remembrance Sundays upwards of 2,000 gather at the Airman's Grave to commemorate those killed in war. It now contains a small garden of remembrance, a white cross with the inscription to Sgt Sutton from his mother and a plaque to the whole crew with their details.

Figure 11.8. Radio station, King's Standing. The station prior to the removal of the masts in the 1980s. Crows Nest clump to the right (source: Bob Toynbee). See Figure 12.2 for an aerial view of the facility.

by the government or be returned to the Forest? Nevertheless, pylons were being erected during 1943 and concrete tank teeth and entrances built. This was for the top-secret 'Project Aspidistra', controlled by the Political Warfare Executive, a branch of the British Intelligence Service. There was an extremely powerful transmitter to intercept German domestic radio and to send black propaganda and confusing messages to help sabotage the Nazi war effort (Figure 11.8).[84]

As D-Day approached even more troops were assembled around the Forest as Ashdown became part of the offensive campaign gathering pace under Operation Bolero. The interviewees in 2005 also remembered the large number of gliders being towed out towards the Channel at night. One interviewee spoke of the Lysanders coming and going from the landing strip.[85]

Some of the Conservators were more directly involved in the war: H.R. Hardy became Lt Comdr. RN; together with Col. A.E. Barchard, Rear Admiral T.P.H. Beamish, Col. N.S. Clarke and Lt-Col. G.R. Miller, DSO, the latter a leading officer in the Home Guard for the area, who oversaw field exercises on the Forest.

FARMING AROUND ASHDOWN 1914–1945

Rarely can the impact of international politics and economic vicissitudes have been more keenly felt on British soils than during the short period 1914–45. By 1914 grassland was still being sown on the eastern flanks of Ashdown as a response to the prevailing

[84] The site was used by the BBC until September 1982 but in 1986 the associated bunker was commissioned by the War Office for an underground command post in the event of nuclear war. From 1988 it has been used by Sussex police for training purposes. In 2007 it was listed Grade II by English Heritage (R. Martin, 'King's Standing Transmitter Station, Crowborough', *Industrial Archaeology Review*, 24 [2002], 91–102).

[85] ESBHRO, ACC 9997, Roy Pipe interviewed in 2005, Ashdown Forest oral history project 2005–6.

poor arable prices. But during the following months, as the war began in earnest, it became clear that national food supplies were endangered. However, only belatedly did a campaign to increase home-grown food production begin, from December 1916. The first Cultivation of Lands order was given in January 1917, followed by the Corn Production Act guaranteeing minimum prices for wheat and oats. Production was now to be overseen and enforced by a new County Council body, the East Sussex War Agricultural Committee, one of the members being Maryon-Wilson.

The available land use data for individual parishes does not allow us to distinguish between land that was within the ancient Forest pale and land beyond. Now grassland was to be ploughed, wherever feasible, although the yields on land bordering Ashdown would never be high on the poor acidic soils. The War Agricultural Committee could over-ride stipulations in leases and compel cultivation, and farms around the Forest did manage to increase their wheat areas by about two-thirds. In the northern area the acreage of oats also increased substantially. But doubts were cast throughout the campaign over the wisdom of devoting large amounts of fertiliser and manpower into improving such marginal areas, rather than concentrating on more productive soils elsewhere.[86] Nevertheless, many farmers agreed to increase arable production: at Buckhurst the deer park, covered in bracken, brambles and gorse, was ploughed up. The results of this, and many more individual farming decisions, are shown for Hartfield and Withyham parishes (Table 11.4).

Table 11.4. Arable increases, 1917 and 1918 (acres).

	Hartfield		**Withyham**	
	1917	1918	1917	1918
Wheat	213	393	101	253
Oats	331	608	309	550

Source: TNA, MAF 68/2827, 2884.

On the Forest itself some Commoners responded to the threatened food shortages by wishing to fence and plough small areas. Sylvanus Ridley was certainly an enthusiast for such action from late 1916 onwards, but it was early 1918 before the Board was finally actively corresponding with the East Sussex War Agricultural Committee on this question. Eventually a small plot was allocated to Ridley near Whitehouse Farm, Horney Common.[87] But others were unimpressed, arguing that the loss of Commoners' grazing, together with fencing costs, would render the process uneconomic.

Nevertheless, by May 1918 the military were also wanting to cultivate 100 acres themselves. Under the Defence regulations they were allowed 50 acres for potatoes, parallel to the Wych Cross to Chelwood Gate road, and around Wych Cross and Hindleap

[86] A view expressed, for example, by Lord Goschen, Agricultural Commissioner for East and West Sussex, *Sussex Express*, 15 June 1917, 7.

[87] *Sussex Express*, 15 February 1918, 4.

Warren, on condition that there be no interference with the timber, that any fencing would be removed at the end of the war and the surface then be restored. This, and the Ridley plot, were the only examples of extended food production from the Forest itself, although Jesse Ridley wrote to Lloyd George asking him to allow cultivation of 1,000 acres on the Forest, presenting a petition signed by 115 Nutley residents. In reply he was informed that his letter 'will receive attention'.[88] Following the Armistice, Sylvanus Ridley proposed at a Commoners' meeting that wounded soldiers be permitted to continue working the plots that had been cultivated, to no avail.

The first post-war harvest in 1919 witnessed the peak of cereal production, but already there was more fallow and ley grassland on the exhausted soils. Nowhere had the plough-up resulted in cereal acreages to match those of 1870; indeed, around Forest Row the 1918 wheat area was less than half that of 1870, for example, and permanent pastures were soon reappearing. Milk production had been protected in wartime but shortages of feed and dairy workers had made even this a difficult proposition. Thus, at Alchorns Farm the dairy herd was soon replaced by stock rearing, and on the death of Lord Brassey in 1918 his pedigree Jersey herd was dispersed and decimated.[89]

The real change, however, came with the Agriculture Act 1920, 'the Great Betrayal', which brought lower guaranteed prices, and in 1921 all financial provision was repealed. The fall in the cereal acreage was as sharp as had been its wartime rise, as the marginal nature of the Ashdown area reasserted itself. The land 'tumbled down' to grass once again and wheat production never attained anything like pre-war levels.[90] Indeed, by the later 1930s over 80 per cent of the farmland around Ashdown was under permanent pasture, often of poor quality, suffering invasion from couch grass, bracken and briars, and sometimes barely distinct from the rough grasses of the Forest. At Maynards Gate Farm (Rotherfield) on the Abergavenny estate in 1927, 64 acres out of a total of 114 acres were deemed to be Forest.[91] There was even a drop of about two-thirds in the amount of litter cut from Ashdown between 1907 and 1925 since it had become uneconomic to spare labour for cutting and collecting, and cheaper to buy in material from elsewhere. So scrub encroached, and the quality of commons grazing deteriorated. Nevertheless, in 1928 nearly half of the 6,400 acres of common land was still covered by litter.[92]

The changes between 1920 and 1939 can be demonstrated with reference to Hartfield (Table 11.5). Longer-term structural changes are evident: the decline in the use of horses, for example, and the end of hop production by 1930. Other changes were shorter-term reactions to the interwar situation: declines in wheat and oats and the cropped area,

---

88 ESBHRO, ACC 6815/1/1.

89 Short, 'Agriculture in the High Weald', 221. For the 1870 acreages see Table 8.8, p. 238.

90 E.W.H. Briault, *The Land of Britain: Parts 83–84 Sussex (East and West)* (London: Geographical Publications, 1942), 555.

91 Short, 'Agriculture in the High Weald', 290. In East Sussex the arable area fell from 122,622 acres at its 1918 peak to 56,006 acres by 1939.

92 Phelan, 'Economy to Amenity', 217.

and increases in permanent grassland and dairy cattle. There was also the abandonment of the old heavy turnwrist plough by the mid-1930s, which, especially when used with oxen, had eased the cultivation of the area's steeper slopes.[93]

Table 11.5. Land use in Hartfield, 1905–39.

	1905	**1920**	**1930**	**1939**
Wheat	293	322	103	129
Barley	38	37	13	9
Oats	353	448	215	140
Potatoes	11	31	21	10
Hops	102	5		
Ley grasses	265	349	104	147
Permanent grass	3991	3477	4390	3909
Total crops area	5560	5222	5128	4537
Agricultural horses	210	154	121	81
Dairy cattle	420	484	584	644
Sheep	1663	1367	1979	1701
Pigs	386	258	129	281
Poultry (fowls)	no data	no data	12,621	14,921

All crops and grass in acres.
Source: TNA, MAF 68/2143, 2992, 3532, 3920.

Around Ashdown stock rearing and dairying did provide some income at this time. Elasticities of demand helped as dietary changes favoured more milk, butter, cheese, eggs and poultry. In 1923 national schemes introduced hygiene inspections for milk and provided Grade A licences (later Tuberculin Tested) to producers with decent accommodation who provided samples and veterinary reports and used sterilisation. The first Grade A licence issued by the county council in the High Weald went to Summerford Farm (Withyham) in early 1924, followed closely by Buckhurst, Old Lodge and Chartners Farms in 1926. From Ashdown the schemes promoting clean milk thereafter spread throughout the Sussex Weald.

Poultry and fruit also gained adherents on the smaller holdings, particularly around Maresfield village and in Horsted Keynes. At the latter, and as part of the Ashdown

93 Straker, 'Agricultural History in the Hundred of Hartfield', 172–7.

Figure 11.9. Diversification at Horsted Keynes, 1932 (source: the author, courtesy of Rory Clarke).

and General Land Company, John Clarke planted 170 acres of fruit trees on the 1400-acre Broadhurst Farm in the 1930s. This was otherwise a mixed farm with a herd of 300 dairy cattle and 400 acres of oak woodland, but just 50 acres of grain. Clarke also developed the Mid Sussex Canning and Preserving Company under the 'Blue Smock' brand, alluding to the traditional Wealden working garment, and also developed the Ashdown Seedling apple variety (Figure 11.9).[94]

Outside observers saw farming here as on a modest scale, somewhat unambitious but perhaps, through milk, fruit and poultry, possessing a stability that helped to counter successive crises in agriculture.[95] Nevertheless, those attempting to set up chicken coops on the Forest itself were promptly instructed to remove them. But there was nothing modest about the decision by the young Herbrand, ninth Earl De la Warr (1900–1976), and his trustees in 1920 to farm that part of the Buckhurst estate then in hand (Figure 11.10).[96] He had succeeded to the title as a minor on the death of the eighth earl in 1915. The park, then full of timber, bracken and gorse, had recently been ploughed for wheat in early wartime but by now much had reverted to bracken and scrub, and by 1925 about 25 per cent of the land was practically derelict. The new earl took over the 2,500-acre estate, appointed high-quality management and instituted heavy applications of chalk and lime, followed by rape, kale and clover. The Home Farm

94 Information from John Clarke's son Rory. See B. Short with P. May, G. Vines and A-M. Bur, *Apples and Orchards in Sussex* (Lewes: Action in Rural Sussex, 2012), 103–05.

95 An outside observer's account at this time is Lemaitre, *Le Weald*.

96 The first hereditary peer to join the Labour Party, he served under Ramsay McDonald in 1924. He was permanent secretary to the Ministry of Agriculture during the 1930s, serving in various capacities until his final departure in 1955. He was mayor of Bexhill 1932–34, laying the foundation stone for Mendelsohn and Chermayeff's modernist De la Warr Pavilion in 1935.

Figure 11.10. Herbrand, ninth Earl De la Warr (source: © National Portrait Gallery X30976).

itself then ran to 776 acres, but much cut up by woodland and streams. Silage-making (then an innovation) began in 1920, together with a demobilisation centre.[97] In 1933, advised by the grassland specialist George Stapledon, ley grasses were successfully introduced with high inputs of slag, potash and nitrogen. In that year experiments also proceeded with Clun ewes, rather than with the more traditional Romney Marsh or Southdowns, and a Jersey milking herd replacing shorthorn cattle.[98]

By contrast, many ex-servicemen had expected a better life after 1918 but were bitterly disappointed. To help men gain a foothold on the land the county councils initiated schemes to develop smallholdings and farms, as in East Sussex at Hole and Alchorns Farms (Maresfield), but generally there seemed to be no great enthusiasm. The farms, totalling 81 acres, were subdivided into three smallholdings but there was a

97 Short, 'Agriculture in the High Weald', 250.

98 The Jersey herd reached 140 head by the mid-1960s (unpub. booklet for visitors 'Buckhurst Farm' *c.* 1965).

rapid turnover of tenants. Additionally, there was criticism for this particular initiative from Maryon-Wilson, who objected to any development of this land under a Land Settlement scheme. He stated that any proposal would be unsuitable since the area already had many small tenanted farms – but particularly because the land adjoined his own property.[99]

By the late 1930s it was clear that another war loomed and that farmers would again be required to produce more home-grown food. The Agriculture Development Act 1939 offered £2 per acre for ploughing permanent grassland again and County War Agricultural Executive Committees (CWAECs) were revived, this time answerable to Whitehall and demanding suitable acreages to be ploughed on each farm. As war began, most farmers responded well to the demands, although after two or three years yields began to flatten out and it was clear that more land would have to be cultivated. Earl De la Warr, with his familiarity with Ashdown, met with Stapledon again to produce a report advocating expansion into more marginal environments for food production.[100] As chairman of the wartime Flax Board, De la Warr enthusiastically grew 50 acres himself, making some profit. He also grew cereals wherever possible at Buckhurst, where his agent, Pusey, was described by Stapledon as 'an artist that not only paints in green, but also in golden corn'. A self-propelled combine harvester was used for the 1944 harvest, a grass drier in 1946 and grain driers after 1950. Ley grassland remained prominent.[101]

Around Ashdown the picture mirrored national concerns. With the ploughing of grassland sheep numbers fell away: by 1945 there were only 450 sheep being commoned on Ashdown. One Commoner began a large-scale pig scheme on the Forest by 1942 using Forest resources with supplementary swill. But again there was only limited ploughing on Ashdown itself. In 1941 the Conservators did consider ploughing part of the Forest, being advised that, as in the First World War, sufficient soil depth existed at Chelwood Gate and Wych Cross. But the CWAEC did feel that the large amounts of lime and phosphate required were better used elsewhere, and that, once again, the fencing to protect crops from Commoners' sheep was a difficulty. However, following repeated local requests the Conservators eventually did allow an area at Green Hall (Forest Row) for potatoes from 1942 to 1946. As in the First World War, the Conservators remained hostile to any large-scale cultivation, despite the national emergency.

A National Farm Survey 1941–3, conducted by the CWAEC on every holding over five acres, offers much information about Ashdown's farms and farmers. To give just one example, Major T.M. White's Chartners Farm (Hartfield) was surveyed in July 1942,

99 Phelan, 'Economy to Amenity', 220; Lockwood, 'The Changing Use of Land in the Weald', 101–02.

100 B. Short, *The Battle of the Fields: Rural Community and Authority in Britain during the Second World War* (Woodbridge: Boydell, 2014), 230.

101 *Kent and Sussex Courier*, 19 September 1941, 6. And see E. King and G. Nightingale, 'Good estate management: the Buckhurst estate', *Agriculture*, 62 (1955), 230–35. The earl demanded the purchase and re-equipping of all derelict land by the state, although this was never implemented (TNA, MAF 38/80 Agricultural policy – memorandum by Earl De la Warr).

after the tenant had been there for just one year. There were 102 acres, including 11 acres of rough grazing and a surprising 52 acres of arable. In terms of the classification applied to all farms, this was a 'C' farm, the lowest category, because of a perceived lack of efficiency and organising ability. However, the surveyor continued:

> This farm was in a derelict state when taken over by the present owner. An improvement is shown, but the crops and fields are full of weeds. The fallows have not been effective, not enough cultivating and work put in. The milk yield is very low, many of the cows are autumn calvers. About £200 is spent on a drainage scheme, which is under the supervision of the Drainage Board.

There was wheat, barley and oats, all infested with rabbits and moles, rooks and wood pigeons, and with docks, thistles and charlock. There were 10 acres of bare fallow and one Fordson tractor. At least the farm avoided the suffering of Mrs Guggenheim's 52-acre Brick Kiln Farm, where 'Herds of deer frequent the cultivations'. But the story of former or present dereliction was common as the toll of interwar difficulties was laid bare. Tye Farm, Hartfield, 'was in derelict condition when taken over by present tenant' and at nearby Upper Parrock 'fields are in rough condition'.[102] This was a most unpromising environment for any form of more intensive farming and attempting to cultivate on the deep clays around Marlpits Farm, for example, invariably entailed wheels sinking to their axles.[103]

The CWAEC flexed its muscles. On 22 November 1939 they warned Pippingford that if cultivation was not undertaken as requested they would take over the farm. At Walesbeech Farm the committee was keen for a different tenant to be installed, although the Ministry warned that such a procedure could be undertaken only if the land was not being farmed 'according to the rules of good husbandry'. And Earl De la Warr himself was asked to reduce deer numbers in the Wych Cross area. By this time Countess De la Warr was appointed to the committee representing what was to become the Women's Land Army in East Sussex.[104]

But some actively disregarded the CWAECs. At Tye Farm, where an excellent dairy herd had been established on this formerly near-derelict holding, an order to plough up a six-acre field for kale, mangolds and peas was received. Three other fields had already been ploughed for wheat but the weeding and labour expense was not thought justifiable and the dairy herd would have been left short of grass. No action was taken and the farmer, summoned for failure to comply, was fined a nominal £1. The ploughing of land around Ashdown would always be a difficult decision.[105] Difficult too were the actions of some unscrupulous farmers. Both Dr Frederick Ridley, an ophthalmic

102 TNA, MAF 32/1008/24. Annotated Ordnance Survey maps are at TNA, MAF 73/41/16.

103 I am grateful to Kevin Skinner for this memory of farming by his father and grandfather at Marlpits.

104 TNA, MAF 80/2811 East Sussex War Agricultural Committee minutes 1939.

105 *Kent and Sussex Courier*, 21 July 1944, 2.

surgeon of Mudbrooks Farm, Forest Row, and Matthew Madge, of Brockhurst Home Farm, East Grinstead, were fined for watering down milk for sale.[106]

The war certainly brought changes to farming around Ashdown. By 1945 there was more ley grassland, more arable (although again among the small shaw-lined fields this would hardly survive the post-war ending of government financial support), more dairy herds but fewer sheep, poultry and pigs being raised or hops being grown, with redundant oast houses often converted to livestock housing or for residential purposes. Meanwhile, many 'backyarders' around Ashdown devoted small amounts of space to poultry and pigs during and after the war.

ASHDOWN UNDER PRESSURE STILL

In 1938 J.B. Priestley, recalling his recent journeys through Britain, wrote that unless something was done quickly 'beautiful England will soon be no more than a ghost haunting libraries and art galleries'.[107] Many amenity groups and local government officials did attempt to stem the tide of disfigurement, and Ashdown's Conservators were no exception. They were doubtless aware of a miasma of different opinions, including views in favour of the protection of England by the nationalisation of land; by town and country planning; through the good offices of estate owners; by means of education for townspeople; or by the reorganisation of agriculture.

We have seen that Ashdown escaped the excesses of rural idyllicism, which had lasted from the mid-nineteenth century until 1939, but the Conservators struggled to reconcile the conflicting demands of an affluent group of incomers seeking peace and rural surroundings but at the same time wanting better roads for their cars, better gas, electricity and telephones, a piped water supply, mains sewers, signposts and firmer access roads to their properties.

And all the while there were the ongoing problems of conservation: how to deal with the changes wrought by wartime, depression and more wartime; how to stop the fires; what tree planting to allow; how to regulate stone quarrying; and how to deal with the ever-encroaching cover of litter when few Commoners now seemed to want it. Their task was difficult and time-consuming and there were many who criticised their actions, believing the traditional Commoner, the 'small man', was now marginalised in his own environment.

---

106 *East Grinstead Observer*, 16 September 1944, <http://spartacus-educational.com/2WW rationing.htm> [accessed 27 September 2021].

107 J.B. Priestley, 'Messages on the Publication of this Book', in Clough Williams-Ellis (ed.), *Britain and the Beast* (London: J.M. Dent and Sons Ltd, 1937), viii.

CHAPTER 12

ASHDOWN'S HISTORIC PRESENT FROM 1945

> Landscapes can be deceptive. Sometimes a landscape seems to be less a setting for the life of its inhabitants than a curtain behind which their struggles, achievements and accidents take place.[1]

MUCH OF THE FOREGOING account of Ashdown is in large measure based around both overt and covert conflict. The Forest landscape, so admired today, has been forged over many centuries through power struggles. And in various forms these have continued over the past 75 years. The 2021 census estimated the population of the United Kingdom to be just over 68 million, and a strong background theme throughout this chapter results from the fact that by 2021 almost one third of this population lives in England's southeast, with nearly eight million in London alone.

Table 12.1. Population of the Ashdown area, 1951–2011.

	1951	**1961**	**1971**	**1981**	**1991**	**2001**	**2011**	**% change 1951–2011**
Buxted	2956	2503	3795	3030	3081	3095	3343	+13.1
E. Grinstead	10,942	15,448	18,632	22,394	23,865	23,942	25,332	+131.5
Fletching	1008	902	881	867	945	1027	1064	+5.6
Hartfield	2332	1953	2140	2179	1935	2120	2179	-6.6
H. Keynes	1148	1256	1315	1389	1403	1507	1586	+38.2
Maresfield	3172	3403	3286	1422	2993	3276	3636	+14.6
Rotherfield	2795	2852	2865	3027	2996	3099	3208	+14.8
W.Hoathly	1626	1665	1787	1651	2134	2121	2181	+34.1
Withyham	3041	3066	3781	4167	2429	2673	2854	-6.1
Crowborough	7126	8169	13,250	16,473	18,639	19,974	20,951	+194.0

1 J. Berger and J. Mohr, *A Fortunate Man* (London: Penguin, 1967), 11.

Danehill	1271	1276	1385	1476	1780	1935	1957	+54.0
Forest Row	3258	3523	4483	4270	4594	4867	4954	+52.1
Total	40,675	46,016	57,600	59,315	66,794	69,636	73,245	+80.1

Source: ONS, gov.uk/census/2011census

Note: in 2011 the population of Crowborough was given for five wards: Crowborough East (5,337), Jarvis Brook (2,645), Crowborough North (5,801), Crowborough St Johns (2,324) and Crowborough West (4,844). The ward areas have since been subject to boundary changes. The East Grinstead population was given for five wards: Ashplats (5,623), Baldwins (5,309), Herontye (4,906), Imberhorne (4,333) and East Grinstead Town (5,161). For 1981 and 1991 the figures are for the consistent 'population present' base, rather than the 'usual resident population' introduced in 1981. This allows consistency with census years up to 1971.

Overall, the population of the Ashdown parishes grew by 80 per cent between 1951 and 2011 (Table 12.1).[2] Comparisons over time are difficult because of changing boundaries but it is clear that 'counterurbanisation', by which non-metropolitan areas have grown in population relatively more quickly than cities, has been a prominent demographic feature of the post-war years (Table 12.2). When the patterns of population growth since 1801 are assessed in 50-year periods it shows quite starkly that while Ashdown's rate of population growth was far below that of England as a whole between 1801 and 1901, the twentieth century saw a dramatic reversal such that Ashdown's rate of population growth has far outstripped that of England as a whole as England's birth rate has slowed down.

Smaller country towns such as Crowborough, Uckfield or East Grinstead have been favoured and become foci for work and population increase, and East Grinstead's population is now estimated at over 30,000. The smaller settlements have tended to become dormitories for commuters or retirement villages. Illustrating this, a recent survey showed that in West Hoathly there were 72 daily trips into the village for work, but 498 trips leaving, mostly to Crawley or East Grinstead, although 17 per cent were to London.[3] West Hoathly, in common with many Ashdown settlements, has become more like a 'commuter village'.

Table 12.2. Rates of population growth around Ashdown Forest, 1801–2001.

	Percentage increase Ashdown Forest parishes	**Percentage increase England**
1801–51	57	97.2
1851–1901	35	96.7
1901–51	52	28.6
1951–2001	71	27.1

Source: <https://en.wikipedia.org/wiki/Demography_of_England>.

2 I am grateful to Professor Nigel Walford for help with the population data.

3 Land Use Consultants, *Sustainable Settlements in the High Weald: Draft Final Report* (Bristol, 2006), 56–9.

In East Sussex recent population growth has owed more to in-migration than to natural increase, which has frequently been negative. Wealden District's Core Strategy anticipates that population growth at Uckfield and, to a lesser extent, Crowborough will continue through to the 2030s. By contrast, the populations of some of Ashdown's more rural parishes have remained relatively stable, as at Fletching, Hartfield, Withyham or Horsted Keynes. The first set of county Structure Plans in 1953 sought to concentrate new housing and population growth into those centres with the potential for growth, and with greater service provision, as, for example, at Forest Row.

A DAMAGED FOREST

Already by mid-1945 the Conservators were again anxious about repairs to Ashdown, that were to be undertaken either directly by, or at the expense of, the military, with the former the preferred option. The forest was undoubtedly in a sorry state by this time.[4] So who was to pay? Who was to effect the repairs? How much compensation could be claimed from the government? And how would the replacement land agreed for the King's Standing radio station and St Johns camp areas be implemented? Above all, when would the Forest be handed back? The two Conservator MPs Beamish and Clarke pressurised civil servants and met the Minister for Town and Country Planning, while others sounded out likely supporters, such as the Commons Preservation Society.[5]

Altogether, during the immediate post-war years, the Conservators faced a bewildering succession of War Office proposals for the Forest linked with the army's retention of Pippingford for training exercises (Figure 12.1). Pippingford had been the base for Canadian and New Zealand troops and for tank training from early in the war.[6] Progress was slow; indeed, the military even implied that they might need to retain the training area for longer for an enlarged army, thereby reneging on assurances given at the beginning of the war that following the cessation of hostilities Ashdown would be released and its surface restored. The threat even remained that the military might purchase the Forest under powers contained in the Defence Regulations. By 1947 this threat was taken very seriously.

A concerned and influential group was formed to counter any such suggestion. A letter to *The Times* signed by Lord De la Warr, Viscount Cecil, Col. R.S. Clarke, Sir Lawrence Chubb and Sidney J. Marsh, the new chairman of the Conservators, set out their case and the loss, for all practical purposes, of this major open space in South East England. They were given valuable support by that influential paper in publishing

4 See, for example, I.D. Margary, 'War Damage to Antiquities on Ashdown Forest', *SNQ*, 11 (1946), 1–3.

5 The Ministry of Town and Country Planning was established in February 1943, and by late 1945 Beamish and Clarke were in private discussions with Lewis Silkin, the minister in Attlee's government.

6 Gilbert Sargent, working at wartime Pippingford as chauffeur, gamekeeper and water bailiff, recalled a prostitute setting up business on blankets in a layby just outside the gates of Pippingford, which was 'greatly appreciated by the three thousand Canadian troops billeted in the park at that time' (Arthur, *A Sussex Life*, xii).

Figure 12.1. Military camp at Pippingford Park 1947. Pippingford house is at top right corner (source: <sussex.ac.uk/geography/researchprojects/airphotos-historic/1940/>).

photographs of the Forest and also by the *News Chronicle* with a photograph and leader. On the same day that the letter was published *The Times* printed photographs of Ashdown tracks torn up by army vehicles and bren gun carriers moving along Forest roads, impacts that were highly visible from the air in early 1947 (Figure 12.2).[7] Support arrived from across the country. The East Sussex Town and Country Planning Committee also had the matter very much in hand. The issue had now been brought to national attention, and Lord Castle Stewart wrote to *The Times* on 17 December 1946 to highlight the perceived local injustices of ongoing military occupation, detailing examples of inconvenience, threats to safety, environmental destruction at Old Lodge, and irresponsible behaviour by the soldiers.[8]

In July 1947 an enquiry opened at Crowborough into proposals by the War Office to purchase Ashdown.[9] Local residents protested: the former wartime liaison officer

---

7 *The Times*, 4 October 1946, 6. *The Times* had previously given good coverage of Ashdown matters, including the fires and the changes rolled into the 1937 Ashdown Forest Act. See *The Times*, 1 November 1929, half-page photo taken by a staff photographer between Forest Row and Wych Cross, with caption pointing out that the Conservators strenuously defend the Forest against overhead cables. And see *The Times*, *British Background* (Focal Press, 1947).

8 Lord and Lady Castle Stewart had just lost two sons in wartime fighting. A headline in the *Daily Mirror*, 26 October 1946, reported a protest by Commoners the previous day, 'They tell army, don't take our forest'.

9 TNA, TS 46/85 Public enquiry concerning the use of Ashdown Forest by the War Department.

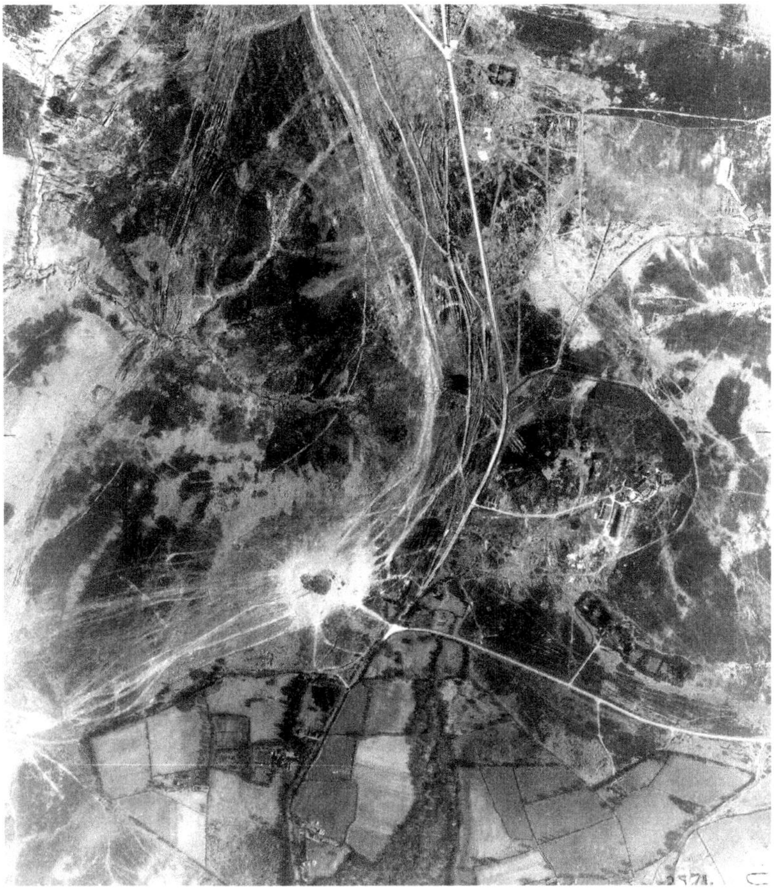

Figure 12.2. The relict landscape of the army presence on Ashdown Forest. Military vehicle tracks in sub-parallel lines north and south-west from Camp Hill, April 1947 (source: <sussex.ac.uk/geography/researchprojects/airphotos-historic/1940/>).

General Sir Reginald Hildyard of Hartfield referred to his move from London to find 'the poor man's Scotland', and the chairman of Uckfield District Council stated that it would be a shame to see the beautiful hotels of Crowborough turned into NAAFI canteens. The proposal was dropped but the army was still digging trenches and using the open Forest by March 1948, despite considerable local opposition. Yet another letter to *The Times*, signed now by Lords Cecil and Gage (chair of East Sussex Town Planning Committee) and others feared that Ashdown might become 'a miniature Aldershot'.[10]

[10] *The Times*, 31 March 1948.

Much of 1948 was taken up in negotiations relating to the ongoing military training, and later that year another Ashdown Forest Bill was prepared, receiving the Royal Assent in July 1949. It provided about 54 acres for permanent military training at St John's Camp, originally commandeered in 1939, with common rights extinguished and with finance for replacement land elsewhere. Access for military training was to be allowed over the Forest but with restrictions: no live ammunition, no fires, camping or trenches, and army vehicles to be limited to light carriers on four specified tracks. Compensation was to be paid for any animals injured. Some areas were specifically excluded from army use: the Royal Ashdown Golf Club; all dwelling houses; the recreation grounds; and seven fir clumps, for example at King's Standing.[11]

However, despite military assurances, Ashdown's landscape continued to be badly affected by army exercises. As late as December 1953 these were still causing damage, and a year later the Commoners were complaining of army vehicles not keeping to the agreed roads. In October 1957 a search around the Isle of Thorns perimeter found 15 more mortar bombs, and in the following April a bomb disposal unit was working near Camp Hill and west of Hindleap. In 1988 a rescue bid to unearth a Vickers tank near the Isle of Thorns ended with its disintegration.[12] Finally, to intense relief, in late 1958 a letter to the Conservators intimated that the MOD would be relinquishing Ashdown, except for access to the retained area of Pippingford. The Forest was returned to the Conservators in 1959.

ENVIRONMENTAL PROTECTION

The later 1940s witnessed significant legislation designed to protect the British countryside. One result has been to keep Ashdown's landscape within the medieval pale looking *superficially* unchanged, in contrast to marked changes in the social and economic make-up of the area. One far-reaching piece of legislation came in 1949, with the National Parks and Access to the Countryside Act, allowing Ashdown now to receive several layers of protection. As early as 1945 John Dower had advocated an enhanced landscape protection for the central Weald, agreed also by the Hobhouse Report (1947). The area pinpointed was in fact centred on Ashdown, and this was eventually to become the High Weald Area of Outstanding Beauty (AONB).

But designation was a tortuous process. Not until 1964 did Kent and East Sussex County Councils give serious consideration to creating a Forest Ridges AONB under the powers available in the 1949 Act. And not until 1969 did a proposal go to what then was the Countryside Commission, leading to a detailed analysis of the AONB boundaries beginning in the 1970s and with a public consultation in 1978. Many wanted additional land included, while the National Farmers Union felt the

---

[11] Ashdown Forest Act, 1949 [12 & 13 Geo. 6]. The Conservators, while not formally prohibiting games on Sundays on the recreation grounds, generally considered that no such games should take place until the afternoon.

[12] I am grateful to Jill Rolfe for her information on the tank and also on the King's Standing radio station.

designation to be unnecessary. Eventually, however, it was approved in late 1983, thereby becoming the 34th AONB. Fletching, originally excluded or on the boundary, was now included and urban Crowborough excluded.[13] The AONB designation now entailed a presumption against harmful development, the possibility of grant aid from the government and the requirement for its protection to be included in county structure and district local plans.[14]

Prior to that designation there had certainly been threats to Ashdown's landscape. In echoes of the early twentieth century, in 1954 the Anglo-Iranian Oil Company (which changed its name to British Petroleum [BP] that year) proposed to sink trial boreholes, and in the same year there were plans for gypsum mining at Pippingford. Both came to nothing, but the D'Arcy Exploration Company initially received permission to bore for natural gas, although the resulting damage led to the withdrawal of permission by the Conservators. In 1956 another survey, this time by the British Petroleum Exploration Company, was agreed on the basis that the sinking of wells would be resisted. The issue returned in December 1977 when Conoco requested permission to drill experimentally, which was actually agreed and duly undertaken. In fact, the greatest concentration of boreholes in Sussex has been in the Ashdown Forest area, particularly around Crowborough.

However, by the 1980s the general public had become better informed of the need for conservation, including the protection of lowland heathland, now seen as an important yet threatened habitat, especially in pressurised southern England. Recent years have therefore seen much effort focused on preventing natural plant succession to protect the heathland, a particular problem on Ashdown because of the spread of scrub and woodland. Aerial photographs from 1947 show 10 per cent of the Forest to be wooded, compared with 40 per cent in 2000. Very few Commoners now take advantage of their grazing rights and the cessation of grazing and cutting has allowed the heath to become old and woody, and bracken and scrubby birch to invade, threatening the many specialised and rare plants and animals that depend on the heathland (Figure 12.3). Between 1965 and 1995 alone nearly 1,000 acres of heathland were lost in this way. Furthermore, by late 1953 myxomatosis had reached South East England, decimating the rabbit population and removing another source of scrub and seedling control.[15] The Conservators have therefore been forced in recent years, prompted by the conservation designations, to intervene more proactively in combating the invasion of scrub and trees. By the early twenty-first century Ashdown still had 2.5 per cent of the UK heathland, but conservation has effectively taken on the former role of the Commoner in keeping the tree cover under control.

---

13 For a detailed account of the High Weald AONB designation process see Ray Woolmore, *Designation History Series: the High Weald AONB* (2013), <http://www.highweald.org/downloads/publications/2004-high-weald-aonb-designation-history-series-and-maps1/file.html> [accessed 12 October 2016].

14 Recreation, an important feature on Ashdown by this time, was not specifically an objective of AONB designation, which was 'to conserve and enhance natural beauty'.

15 It was estimated that the disease wiped out 99 per cent of the UK rabbit population, although numbers increased again soon afterwards.

Figure 12.3. Heathland succession 1947 to 2000 (source: University of Sussex air photo collection). The road top left runs between the Forest Centre (to the west out of the photograph) and Colemans Hatch. The buildings on the right are in Sandy Lane, with Stone Cottage, noted in Chapter 11, on the corner of the site. In 1947 (left) most of the landscape is treeless, except for the Sandy Lane properties and at bottom left, which is also privately owned. There is a band of woodland to the north, which by 2000 has become part of the general spread of woodland throughout.

There was one issue that all were still agreed upon, and this was minuted by the Board at a meeting in July 1966 following a visit made by the chairman, clerk and deputy clerk to Epping Forest: 'It is owned by the Corporation of London and caters for rather a different class of people, and as a result has to provide amusement and catering facilities.' Ashdown was to be different: still to be a place for quiet enjoyment and the contemplation of nature. But at the same time new car parks were needed, often with seating, and the Conservators seemed unable to deal with the plebeian ice cream vans that now parked around the Forest edges. Bowing to the inevitable, in 1967 it was finally decided that ice cream salespeople would henceforth be licensed, thereby also helping income in a difficult financial period. Mobile tea and coffee vans have now joined the applicants for licences. And, following the passing of the 1974 Ashdown Forest Act (see below) the obligation to assist visitors increased: by 1976 maps and literature were being made available and an information caravan proved a weekend success. Some thought the car parks too successful: at Easter 1989, for example, all 48 car parks were full and the Forest was deemed to have reached saturation point. At one time it was estimated that as many as 6,000 cars were parked here, many congregating around the access to Poohsticks Bridge. Furthermore, the Conservators were powerless to stop a music festival, 'Harmony Farm', during August

1971 at Whitehouse Farm, Duddleswell, since it was on private land. It was, in fact, later reported that there was no adverse effect on the Forest.[16]

So, how would the Conservators balance the increased public access and attendant pressures with their role in conserving a nationally (and indeed internationally) designated environment? In the face of ever-increasing numbers of visitors the Board initially saw the answer in strict conservation measures. In early 1977 these pressures were recognised as requiring a management plan to address issues such as visitor numbers and parking, as well as the optimum ratio between heathland and woodland on the open Forest. A discussion also began concerning the control of vegetation change through a grazing experiment, although hostility within the Board delayed any firm decision. But there was certainly great concern in January 1978 when the Royal Ashdown Forest golf club had reportedly sprayed DDT along the fairways to counteract leatherjackets.[17]

Ashdown's ecological importance as an area of lowland heath had in fact been recognised internationally by its early designation as a Site of Special Scientific Interest (SSSI) in 1953. It received European Union designations as a Special Protection Area (SPA) for birds in 1996 and a Special Area of Conservation (SAC) for its heathland habitats – wet and dry heathland and a site for populations of the great crested newt (*Triturus cristasus*). The SPA and SAC were also rolled into the European *Natura 2000* network (updated in 2015) as Ashdown hosts some of Europe's most threatened species and habitats, meriting strong legal protection and with responsibility to actively improve the Forest. Though not a statutory designation, Ashdown also forms part of the 'Western Ouse Streams and Ashdown Forest Biodiversity Opportunity Area', and thereby is an element in the Sussex Biodiversity Action Plan, intended to focus conservation bodies, local government and statutory agencies on work to conserve and enhance habitats and species in Sussex.[18]

The areas covered by the various statutory designations are generally larger than the area of Forest administered by the Conservators (Figure 12.4). The SSSI, for example, covers 7,741 acres, which includes the Ministry of Defence's Pippingford Park Dry Training Area (740 acres), Hindleap and Broadstone Warrens and the Old Lodge Local Nature Reserve. The latter was originally offered to East Sussex County Council by the Castle Stewart family, but it felt unable to take on the 250 acres, and management instead passed to the Sussex Trust for Nature Conservation (now the Sussex Wildlife Trust) in 1983. The SPA is slightly larger again at 7,921 acres and the SAC covers 6,740 acres.

---

16 <http://www.ukrockfestivals.com/duddleswellfreefest1971.html> [accessed 21 October 2018]. 'Harmony Farm on Ashdown Forest is a beautiful place and we want it to be yours and treat it as such. Please be litter conscious and mind the neighbours, then Her Majesty's constables will stay away' (excerpt from flyer advertising the 'sounds' and 'happenings').

17 The powerful insecticide DDT (Dichloro-diphenyl-trichloroethane) had been banned for agricultural use in the USA in 1972, although not in the UK until 1984.

18 The UK Biodiversity Action Plan (UK BAP) 1994 followed the 'Earth Summit', the UN Conference on Environment and Development in Rio de Janeiro in 1992. Lowland heath was one of 14 habitats singled out for priority conservation measures.

Figure 12.4. Ashdown: the complex of designated areas (source: author, based on <https://designatedsites.naturalengland.org.uk>).

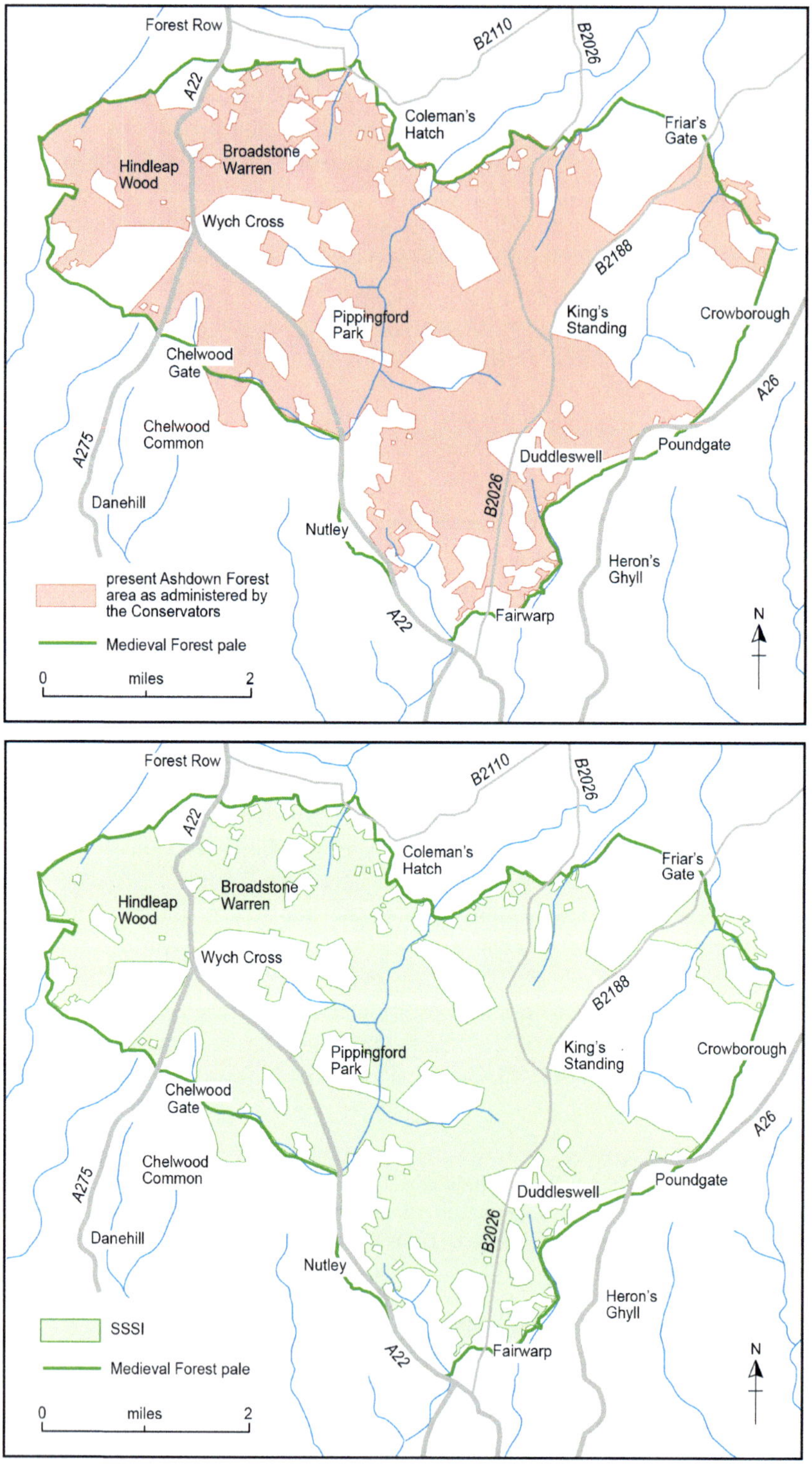

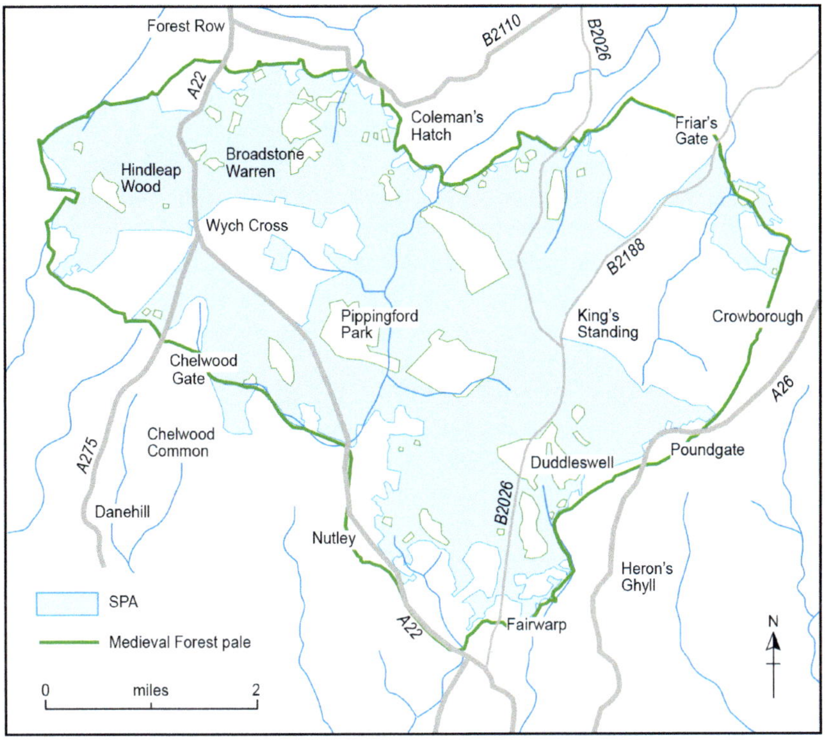
Forest Row
B2110
B2026
A22
Coleman's Hatch
Friar's Gate
Broadstone Warren
Hindleap Wood
Wych Cross
B2188
Pippingford Park
King's Standing
Crowborough
Chelwood Gate
A26
Chelwood Common
A275
Duddleswell
Poundgate
B2026
Danehill
Nutley
Heron's Ghyll
SPA
Medieval Forest pale
A22
Fairwarp
N
0
miles
2

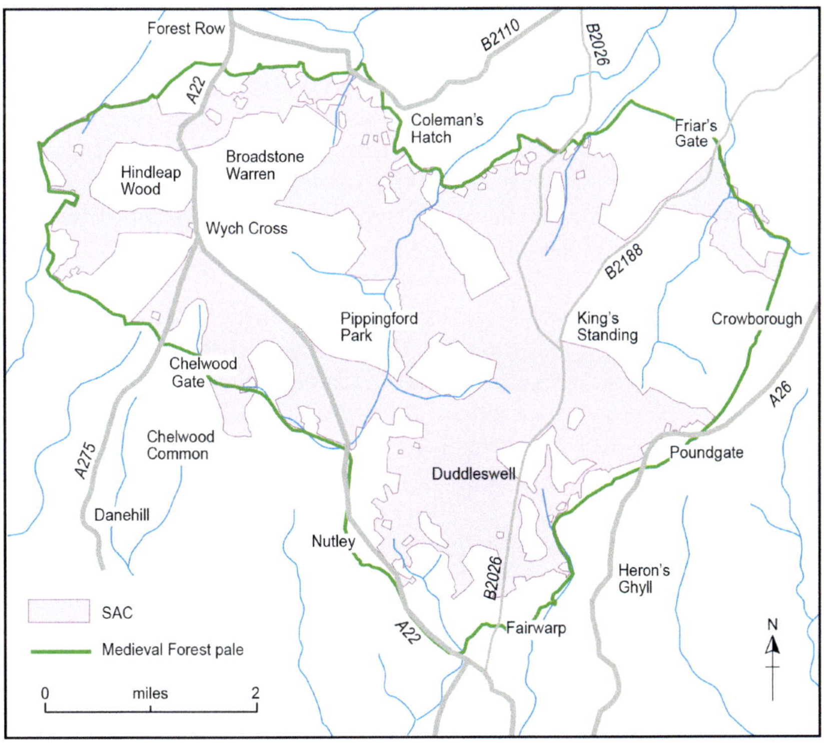
Forest Row
B2110
B2026
A22
Coleman's Hatch
Friar's Gate
Broadstone Warren
Hindleap Wood
Wych Cross
B2188
Pippingford Park
King's Standing
Crowborough
Chelwood Gate
A26
Chelwood Common
A275
Poundgate
Duddleswell
Danehill
Nutley
B2026
Heron's Ghyll
SAC
Medieval Forest pale
A22
Fairwarp
N
0
miles
2

The last 20 years have seen a range of grants resulting from these designations, most notably Higher Level Stewardship (HLS) from 2006 to 2016, to restore the Forest's internationally recognised heathland and to support conservation and restoration of this habitat, primarily by reversing scrub and successional woodland encroachment. A new emphasis also favours reducing soil fertility by more extensive grazing, burning, scraping and bracken spraying, and the removal of invasive non-native species such as rhododendron. Woodland, to be maintained at about 40 per cent of the land cover, is to be thinned and coppiced, and fallow deer numbers controlled using professional stalkers, mainly by reducing breeding females. As a result, the number of deer/vehicle collisions has fallen, plant communities are responding and the deer themselves have started to improve in condition. The conservation of key archaeological and landscape features was also a feature of the HLS grant.[19]

At the same time the Conservators aimed to improve public understanding of the Forest, to improve access for quiet recreation and to protect Ashdown against adverse development. Restoring the heathland to 'favourable condition' does involve a certain amount of tree felling and birch scrub removal, to the dismay of some residents who remain unfamiliar with the value of the heathland and the problems produced by unmanaged plant succession. In 2007, for example, a dispute arose between some residents and the Conservators, who were cutting down saplings, removing scrub and mowing the bracken. They complained that the results looked like a First World War battlefield. This is not a problem restricted to Ashdown, and similar debates have occurred at other heathland areas, such as the New Forest and in Surrey.[20]

A recent condition assessment found that only about 20 per cent of the Ashdown Forest SSSI was in 'Favourable Condition' – but for heathland, the priority habitat, only 10 per cent was in Favourable Condition. There is clearly work to be done in this respect, and it is notable that current best practice now revolves around a return to the traditional extraction of nutrients by grazing or collecting material to avoid the accumulation of soil nutrients that encourages woodland invasion onto heather. And, in a fascinating *volte-face*, rangers now use burning on a small scale (up to two or three acres at a time and 50 acres per annum) to rapidly remove nutrients, provide succulent early-bite for sheep and cattle and control accidental firings or arson.

Strangely, even grazing to maintain heathland has more recently been contested, despite its long historical tradition. By 1967 there were 12 grazier Commoners left on Ashdown, but the number was declining such that by the 1980s they had almost ceased turning out stock altogether. But a 1985 feasibility study concluded that a combination of grazing and mechanical measures would be very appropriate, and in 1989 a small area around Millbrook was enclosed to test the ecological and

---

19 Income from Countryside Stewardship has replaced the HLS income (Conservators of Ashdown Forest, *Achieving Favourable Condition: a Management Plan in support of Countryside Stewardship 2016–2020* [January 2016]). But in 2020 there were still 225 deer/vehicle collisions.

20 J. Brown, 'Oh bother! Nimbies do battle with council over Pooh's forest', The Independent, 'This Britain', 21 April 2007.

Figure 12.5. Hebridean sheep, Riggit Galloway cattle and Exmoor ponies on the Forest (source: courtesy of the Conservators of the Ashdown Forest).

community response, with Henry Osborne turning 97 sheep and 12 Welsh black cattle into the enclosure.[21] The fenced area was gradually extended such that by 1998 about one-third of the open heathland was fenced and grazed, and used by five Commoners. It was concluded by 2003 that the grazed area had an increased plant diversity: *Calluna* was more dominant and *molinia* more controlled. Trampling by the stock also helped in bracken control, and there had been increases in lapwing, the silver-studded blue butterfly and the marsh orchid.[22] From 2015 'invisible fences' were deployed using electronic tagging and a transmitter cable around the grazing area to keep cattle within bounds while not impinging on the landscape. But even with this initial success there were some who were opposed to any fencing of common land, and some who voiced criticisms of the grazing plans, especially the enclosures used in spring and summer to contain the small and hardy black Hebridean (formerly St Kilda) sheep, Exmoor ponies and the herd of the archaic Riggit Galloway cattle that now graze parts of the Forest, with the sheep controlled by a shepherd (Figure 12.5).

---

[21] Henry Osborne repeatedly complained to the Board about dogs worrying his sheep. In May 1979 he legally shot an Alsatian that had been worrying his sheep on his own land, an action which he (and his son) have had to repeat more recently. He also pressed for cattle grids to prevent sheep straying and was reportedly losing about 30 sheep a year through car accidents. In October 2017, at the age of 82, he was himself attacked and severely bitten on his face by one of two dogs worrying his sheep. East Sussex has the highest number of sheep-worrying incidents in the UK (*News from the High Weald AONB* [March 2020]).

[22] C. Marrable, *Ashdown Forest – a Review of Grazing*, English Nature Research Report 535 (Peterborough, 2003); C. Marrable, *Ashdown Forest Grazing Action Plan*, English Nature Research Report 602 (Peterborough, 2004).

These breeds can selectively graze tough acidic vegetation, allowing the heather to thrive.[23]

THE 1974 ASHDOWN FOREST ACT AND THE COMPLEXITIES OF FOREST GOVERNANCE

There was lively and sometimes acrimonious discussion in late 1972 following a presentation to the Commoners outlining a new Ashdown Forest Bill to better reflect the position of East Sussex County Council. The Commoners were uneasy at the proposal to hand both the main financial power and Conservation Board membership majority to East Sussex, with some mistrust of the county council being expressed. They also criticised the fact that sheep were now to be legally commonable, thereby formalising a situation that had in reality been ongoing for many years, as well as the power that the Conservators would now have to control where estovers might be cut, and that car parks and fire breaks were now deemed essential. The question of Commoners' power also remained a constant theme running through the meeting, with a one-man-one-vote system of Conservator elections particularly contentious.

But the democratisation of the landscape was now moving on apace, and a legal adviser warned of the dangers of trying to keep the Forest solely for the Commoners:

> If any of you think that any government of the future is going to allow 587 Commoners in a small area like this to keep control of Ashdown Forest, then you are crying to the moon. If you oppose it and this Act is not completed by 1974 you will lose the forest completely and utterly.

As so often in the past, relations between the Board and the Commoners descended into outright animosity. The Commoners' AGM in December 1972 witnessed 'a heated verbal assault on the Bill and on the Conservators' and was described as 'somewhat disorganised and unruly. For the first time in history the Chairman of the Board of Conservators had not been elected to the Chair [of the Commoners' meeting].' Although the Board's proposed wording for an Ashdown Forest Bill was finally voted through (still using the acreage method of voting), a group of Commoners continued their opposition by forming a new Commoners' committee, by organising a petition to the Department of the Environment, by statements to the media and even by accusations of fraud in the voting procedure. The Board was accused of abusing its power and of 'selling the Forest to the public'.

A meeting of residents and Commoners in December 1973 sought the Board's resignation and a petition signed by 350 Commoners demonstrated opposition to the Bill.[24] Board chairman Trevor Mountain was accused of misfeasance, abusing his power

[23] Hebridean sheep are effective grazers of young birch; Exmoor ponies will effectively control gorse, and cattle are best suited to areas dominated by *Molinia*.

[24] ESBHRO, AMS 6504 Papers of the Ashdown Forest Commoners' Committee. Opposition of the Commoners' Committee to the County Council's promotion of the Ashdown Forest Bill in the House of Lords 1973.

and wrongfully exercising his authority. The main issues of contention were now the fact that Commoners' representation on the Board was to be cut to five; the amount of the Forest rate; the voting procedures; the power now to be exerted by the county council; and the powers needed to exchange Forest land for new land.

The uncertainty was compounded since few could predict what changes might come with the reorganisation of local government in April 1974. Nevertheless, the Act was finally passed, with the last meeting of the Conservators under the old system taking place in January 1975.[25] Now East Sussex County Council and the newly formed Wealden District Council were to meet the Conservators' expenses in so far as they could not be recouped by income from the Forest rate and other sources, and to meet any shortfall in funding.[26]

The Conservators' duties were now re-established: setting the byelaws, managing the woodland and improving the grazing, with the ability to fence certain areas as required, but only up to 100 acres. They were to be advised by several committees: Conservation; Finance and General Purposes; and Roads, Planning and Amenities, which in turn could draw on outside expert opinion when needed.[27]

Under the Act, for the first time in its history, the public were given a legal right of access to the Forest, on foot and 'for quiet recreation and enjoyment'.[28] But, as the then chairman of the Conservators had previously remarked to the extraordinary general meeting of the Commoners at Forest Row in August 1972, 'No one, I think, would want an agricultural show, a point-to-point, or a hippy happening in the Forest.' The latter was a reference to 'Harmony Farm' the previous year. Nevertheless, the Commoners remained critical: in June 1975 at a Commoners' meeting they returned to the issue of Forest rates, which were declared to be 'manifestly absurd, morally obscene and indefensible'.

Under the 1974 Act 50 free car parks were now installed (now 47), some for summer use only, mostly chosen because they had previously been pull-ins but had damaged the Forest, especially in winter when they turned into muddy wastes. Now they were surfaced, supplied with picnic areas and summer extensions and surrounded by ditches and banks to curtail off-road driving. All have names, such as Stonehill or Gills Lap North (Figure 12.6). The western side of the Forest has fewer car parks, which is one way of maintaining this as a quieter zone.[29] Several additional byelaws have more recently been enacted to protect the Forest and to preserve its tranquillity, including prohibiting off-road driving, mountain-biking, horse-riding (except by permit), camping, the lighting of fires, digging and fly-tipping. Moving nearer to the present, drones are also prohibited.

---

25 Ashdown Forest Act 1974: Elizabeth II. Chapter xxi.

26 The paucity of funding for Ashdown was noted in E.M. Yates, 'The Management of Heathlands for amenity purposes in South-east England', *Geographia Polonica*, 24 (1972), 227–40.

27 The Conservation committee includes individuals from Natural England and DEFRA, as well as the county ecologist.

28 As such, the later Countryside and Rights of Way Act (CRoW) 2000, giving qualified statutory right of access on foot to all registered commons, did not introduce any new legislation relating to Ashdown.

29 Marrable and Muggeridge, *Information*, 53–4.

Figure 12.6. Gills Lap car park (source: author).

The years 1987–8 were exceptionally challenging. First BP announced once again that it wished to sink an exploratory borehole just off Kidd's Hill, although the proposal was rejected by the county council. On the death of Herbrand in 1976 his son, William Herbrand Sackville (1921–1988), had become tenth earl, and now in 1987 in his capacity of lord of the manor of Duddleswell he disclosed his intention to sell the Forest for £1.2 million, with first refusal to East Sussex County Council but with an underlying threat that the Forest might otherwise be sold off in lots.

Under the auspices of the Friends of Ashdown Forest (see below) an appeal was organised to raise the necessary money for the Forest to be put into local authority hands. There was a deadline of 24 November 1987. But then on 16 October came the Great Storm that struck the South East, ripping up ancient trees and devastating large areas of the Forest, with King's Standing the most affected clump. Travel was impossible and communications failed. Fallen timber was only gradually cleared and some was sold.

In fact, the money to purchase the Forest was raised by volunteers through a passionate public subscription, supported by Christopher Robin Milne and by hundreds of individuals as well as institutions. No fewer than 1,900 names were included. But before the contract with the county council could be signed came the news of the death of the tenth earl in February 1988.[30] It was reported that he had become anxious because

[30] A vivid account of the years 1987-88 by the renowned children's author is B. Willard, *The Forest: Ashdown in East Sussex* (Poundgate: Sweethaws Press, 1989), 164–78. Living at Nutley, she represented the Commoners on the Board 1975-85. Her *Mantlemass* novels contain much historical material relating to the Forest (A. K. Helbig, 'The Forest as Setting and Symbol

of the storm damage to his estate, and had thrown himself under an underground train. However, the sale, now for £1.4 million to take account of inheritance taxes associated with the earl's estate, proceeded by November 1988, with thanks sent to William Herbrand, the new eleventh earl, 'for fulfilling his father's wishes'.[31]

So now the Forest passed from the lordship of the De la Warrs into public hands. But the governance has become more complex and less stable. The lordship of the manor of Duddleswell now became invested in the Ashdown Forest Trust, a new charity acting for East Sussex County Council, with the council as the sole trustee.[32] The Conservators, 16 in number, remained responsible for the daily management of the Forest. There are currently nine appointed by the county council (one of whom still represents the lord of the manor, now the Ashdown Forest Trust) and two by Wealden District Council (from its inception in 1974), with the remaining five elected by the Commoners, of whom four must still themselves be Commoners.

The extent to which landscape conservation has become an inherently political process could not be more starkly illustrated than by the fact that 11 of the 16 Conservators are now political appointments. The balance of power within the Board had shifted further since 1938, when local authority representatives were first elected. At that time there were seven elected Conservators and eight local authority members. And it now followed that Ashdown's particular needs, requiring local authority inputs and public money, had to be weighed against other demands on the authority's resources.

There was one very positive development at this time. Having no formal role in the running of the Forest, but influential in promoting its interests, are the Friends of Ashdown Forest. In 1961 proposals were submitted for the management of Ashdown through the auspices of the Sussex Naturalists Trust.[33] Little came directly from this idea, but the issues raised certainly provoked action, and that year Ursula Ridley asked the Board, of which she was then a member, representing Cuckfield Rural District Council, whether they would support the idea of an association called the Friends of Ashdown Forest. This was welcomed, and the Friends, registered as a charity in 1962, have gone on to develop an important fundraising role, donating over £720,000 to date and recruiting many of the 100-plus conservation volunteers.

The 'Friends' now number almost 1,000 members. Their fundraising has helped towards the cutting of rides for fire breaks, purchase of capital equipment such as motor vehicles and enabling the purchase of parcels of land within the ancient pale for

in Barbara Willard's Mantlemass Novels', *Children's Literature Association Quarterly*, 7 (1982), 35-39): 'she has integrated the setting into the story to influence plot, shape the lives of the people who live within it, create mood, and embody theme'.

31 See T. Sackville, *Dumble: the Life and Death of Billy, 10th Earl De la Warr* (London: Zuleika, forthcoming 2022).

32 A similar purchase was by Surrey County Council in 1966 of several thousand acres of Chobham and adjoining commons near Woking from the sixth Earl Onslow, an area now managed by Surrey Wildlife Trust.

33 The management document was prepared by David Streeter (University of Sussex), the Trust's scientific officer. And see D.T. Streeter, 'Heathland conservation on Ashdown Forest', in *Society for the Promotion of Nature Reserves Handbook* (1964), 27–43.

Figure 12.7. The Ashdown Forest Centre soon after construction. The original local thatching has now been replaced. The Information Barn is in the centre with the Administration Barn to its right. The Education Barn was added later, to the far right of the buildings shown (source: author).

reincorporation into the Forest. They also contributed in 1964 towards the building of a new depot, now the Forest Centre at Wych Cross (Figure 12.7).[34] When the 'Friends' were formed their objective was 'the benefit of all who enjoy this place, and to ensure that others will enjoy it in the future'; as Garth Christian has noted, the word 'enjoy' was a rarity in earlier centuries, when the Forest was a working landscape.

Next chronologically came the Ashdown Forest Conservation Trust, established in October 1979, with Lord Abergavenny as president, and registered as a charity in 1980. Income from this fund has been used for the benefit of the Ashdown Forest Centre, albeit its objectives are wider than that.[35] The Countryside Commission helped to fund the information centre here, and the Forest Centre, initially thatched, was opened by the tenth Earl De la Warr in 1983.

So, rather confusingly, there were now three charities: the Ashdown Forest Trust, the Ashdown Forest Conservation Trust, and the 'Friends'. However, in 2016 the county council, as trustee of the Ashdown Forest Trust, considered an internal review of

34 Christian, *Ashdown Forest*, 3.

35 The philanthropist Alfred Wagg (1877–1969), from East Grinstead, was heavily involved with the Manor Charitable Trust, which donated the land for the Centre. The modern Information Barn probably originated from eighteenth-century Sussex; the Administration Barn was probably from late seventeenth-century Kent or the eastern Sussex Weald. The Education Barn, now much altered, was a free donation from the builders who reassembled the buildings, <www.ashdownforest.org/discover/docs/TheAFCentreBarns.pdf> [accessed 19 April 2021].

Forest governance arising from concerns expressed by the Charity Commission about overlap and duplication between the trusts. As a result, the council made a specific proposal to transfer ownership of the Forest to a new charitable company limited by guarantee, and to wind up the current charities. The Conservators agreed that a new charitable organisation was needed to own, run and manage the Forest independent of the local authorities and to develop a trading and fundraising platform.[36] This would not affect the 'Friends'. This was all the more urgent because East Sussex had also warned that because of

> reductions in Government funding and the Council having to make savings of between £70m-£90m over the current three-year financial period, it is highly likely that, irrespective of whether or not the governance arrangements change, funding to the Conservators from the Council (not the Trust) will cease at the end of 2018/19.

The fears of those who had been worried by the involvement of local authorities in the 1930s were now being realised. These potential funding cuts represent further significant challenges. As we have seen, finding adequate funding for the Forest has been a persistent issue. The income of the Conservators in 2009–10 was almost half funded from the Government's HLS scheme. Grants from the local authorities and the Ashdown Forest Trust accounted for another fifth. But, with East Sussex cutting its contribution, there was now a need to attract new sources of funding.

Instead, there is now one sole organisation, to comply with the Charity Commission and East Sussex County Council. This is The Ashdown Forest Foundation (TAFF), granted charity registration in June 2019. It is not involved in the day-to-day management of the Forest, but aims to support the Conservators in every way possible. Meanwhile, diversification of income sources has been a major theme. There is the sale of Christmas trees, making a virtue of clearing self-seeding Scots pine from the heathland, and the sale of bracken mulch or Halloween birch and hazel besom brooms. There are sales of Ashdown-reared Hebridean hoggett (one-year-old sheep) meat, three-year-old mature Galloway beef and venison packs resulting from the culling of fallow deer. Other sources of income include corporate sponsorship, a memorial programme, fungi trails, organised walks, a schools' outreach programme and many events at the Forest Centre. Sixty-two events were held there in 2016–17, underlining the importance of having a known administrative and educational base. Some activities do, however, raise issues of compatibility with conservation and reconciliation has not always been easy. Licences and permits also continue to bring in funds, with commercial filming now more important in this respect.

A Visitor Survey undertaken for Wealden District Council in late spring and summer 2016, further analysing visitor data from 2008, estimated that there were approximately 4,541 visits per day to Ashdown. Receiving something in the order of

[36] Conservators of Ashdown Forest, *Ashdown Forest Annual Review 2016–17* (2016), 9. The Conservators gained the Queen's Award for Voluntary Service in 2016, the highest possible award for voluntary service.

1.5 million visitors per year is a challenge, but also a fund-raising opportunity. During the Coronavirus pandemic from early 2020 visitor numbers to the Forest have soared as people seek quiet exercise, but the impact on wildlife such as deer and issues of dog control, rubbish dumping and random parking have been highlighted. Unfortunately, income has not increased alongside this, since the closure of the visitor centre has been necessary at various times.[37] A controversial scheme to charge for car parking in the Forest car parks has been agreed by the Conservators and will operate from summer 2022.

LEGISLATING THE COMMONS

A Royal Commission on Common Land met between 1955 and 1958 to recommend the changes deemed desirable in the law relating to commons 'to promote the benefit of those holding manorial and common rights, the enjoyment of the public ... etc.' Evidence was received from many organisations, including the Ashdown Conservators.[38] The Commission noted that East Sussex had 21 commons totalling 8,452 acres, of which Ashdown provided 6,400 acres, or 75.7 per cent of the county total, underlining its local importance.[39] There were also those surrounding commons excluded from the medieval park, such as Crowborough Common (190 acres), Holtye (100 acres), Chelwood Common (40 acres), Sheffield Green (15 acres) and Piltdown (150 acres).

In 1965 the subsequent Commons Registration Act came into force, requiring all reputed common land to be registered and forcing all Ashdown Commoners to reregister their rights by 1971. Approximately half did so, the bureaucracy deterring many who had anyway ceased exercising any rights. Consequently, they technically ceased to enjoy common rights, and once lost they could not be reinstated. Furthermore, land not registered under the 1965 Act cannot now be registered. The number of Commoners had grown to 1,160 prior to 1965, yielding a Forest rate of £1,531, but after registration this had fallen to 580 and a rate income of £700.[40] At one point in the 1960s financial difficulties led to some discussion about turning parts of the Forest over to the National Parks Authority – might it become a National Park, or a Country Park, to attract even more visitors?[41]

---

37 R. Clarke, J. Sharp and D. Lily, *Ashdown Forest Visitor Survey Data Analysis*, Natural England Commissioned Reports Number 048 (2010). The issues were highlighted in a report for BBC South East (18 January 2021).

38 TNA, MAF 96/151 Conservators' evidence to the Commission on Common Land 1956.

39 *Report of the Royal Commission on Common Land 1955–1958* (London: HMSO, 1958), 1, 202; W.G. Hoskins and L.D. Stamp, *The Common Lands of England and Wales* (London: Collins, 1963), 155.

40 Any increase in the current Forest rate (2021) is calculated with reference to the Consumer Price Index + 1 per cent, or 3–5 per cent, whichever is the greater.

41 Country Parks, designated in the 1970s, were local-authority controlled, primarily to provide outdoor recreation for urban centres. In East Sussex they include the Seven Sisters and Hastings Country Parks.

The registration of Ashdown itself became final on 7 October 1970.[42] The number of Commoners has now grown again to about 730, with grazing rights for 4,015 cattle and 20,140 sheep in numbers proportional to the holding size. Contrary to past practice, rights can now be relinquished by making a Statutory Declaration to the county council.

It remains the case that Commoners are entitled to vote for representatives on the Board of Conservators and must still pay a Forest rate. But the old voting system, favouring those with larger acreages, has been replaced by a one-person-one-vote system, as argued for in the early twentieth century by the Commoners' Protection Association. In recent years there have been two types of Commoner: those living in such places as Maresfield and Piltdown with little real interest in the Forest and those who continue to graze their animals on the Forest. However, there are also growing numbers of residents without common rights but with an interest in the wellbeing of the Forest and who might, for example, join the 'Friends' and who also might offer forceful views about the management of the Forest.

ASHDOWN UNDER PRESSURE: LOCAL AND NATIONAL FORCES

Over the last 70 years the national and local state has impinged to an increasingly greater extent upon the agency once exerted by the Conservators. Modernity has ushered in refinements to their methods: they now have management plans and strategic plans, and need to ensure compliance with an ever-increasing list of laws, guidance and charity regulations.[43] Committees, sub-committees and ad hoc committees offer advice. Care is taken to consult with local people, not just Commoners, through parish liaison meetings. Forest governance has been pulled into the orbit of local authority procedures, even though from 2019 the local authorities no longer control the Forest.[44]

National changes have certainly affected local farming, once the mainstay of employment and income. One important post-war theme has been the loss of jobs in farming, and the Ashdown area has been no exception. Using comparable data, and based on nine of the Ashdown parishes, numbers of farm workers fell from 1,344

---

[42] I am grateful to Natalie McLean, Legal Order Officer, Rights of Way and Countryside, East Sussex County Council. There are now some 90 registered East Sussex commons covering about 7,000 acres. The commons registers were lost in a fire in 1993 and a private act was passed in 1994 to reconstruct them, validated in 2004. In 2006 a new Commons Act was passed but with limited impact on Ashdown. As at December 2017 there was no 'commons council' for East Sussex, which is the registration authority.

[43] This section was written in January 2021 as the UK had just left the European Union, a situation complicated by the Coronovirus pandemic; there was, therefore, no certainty over future environmental legislation.

[44] The most recent statement of aims is to be seen in Conservators of Ashdown Forest, *Ashdown Forest Vision and Management Strategy 2021–2031* (October 2021), which emphasises the strategic aims to protect, enhance and broaden; towards collaboration and engagement; and to welcome and inspire.

in 1923 to 380 by 1988, by which time there were more farm holdings than hired workers, and the numbers will have fallen still further in the last 30 years.[45] Instead, more reliance has been placed on family labour or contractors. Thus, Imberhorne Farm (East Grinstead) had once employed 50 men but this was reduced to one full-time employee after the war, after whose death the land was leased out to various tenants.[46] And of course a related trend has been the loss of rural crafts and trades, such as blacksmiths, wheelwrights, thatchers and saddlers. Other key themes may be discerned: the number of holdings with no farmland rose from four in 1970 to 36 by 1988 as farms were converted into non-farmed residences. And of the remaining farms themselves, by 1988 over two-thirds were farmed part-time as the numbers of full-time farmers have fallen.

Such, then, has been the decline of the agricultural interest around the Forest. In line with historical land use, about 60 per cent of all farmland in the area remained under permanent or temporary grass by the late 1980s, much used for dairying, although that too declined severely, especially following the introduction of EU milk quotas in 1984. Instead, sheep are far more numerous but much currently depends on pricing policies, which at the time of writing were being formulated following the UK's exit from the EU. As with the Conservators, diversification has been one answer, and farmers have looked beyond traditional sources to attract income, including a greater concern for landscape conservation, and the redevelopment of redundant farm buildings for alternative commercial uses.[47]

As agriculture has declined, residential and other forms of development have multiplied and the post-war pressures on Ashdown have been unrelenting. This is in spite of the introduction of the Town and Country Planning Act (1947) and the resulting East Sussex Development Plan (1953), enhanced by a later generation of plans from the 1990s, including the East Sussex Structure Plan, now abandoned in favour of more local planning.

When peace was restored after the Second World War, London's eight designated New Towns were planned, one being at Crawley, about 12 miles from the Forest boundary. We do not need to describe the development of Crawley New Town here, but one issue that arose was the need for water. It was originally proposed that a storage reservoir be constructed on the Forest near Wych Cross. Objections were immediately raised by the Conservators and by Earl De la Warr, demanding that alternative sites be investigated. Eventually this entailed the construction of the Weir Wood Reservoir on the Upper Medway west of Forest Row, about one mile west of the Forest boundary, which involved the loss of valley bottom land, most notably from Walesbeech Farm (Figure 12.8).

---

45 B. Short, S. Swift and G. Swift, *The Ashdown Forest Area: Agricultural Change in the Western High Weald from 1970 to 1988* (Lewes: Sussex Rural Community Council Research Unit, 1991). The analysis omits East Grinstead, Crowborough and Rotherfield.

46 <http://www.felbridge.org.uk/index.php/exhibitions/imberhorne-farm-workers-past-present/> [accessed 15 October 2020].

47 Short *et al.*, *The Ashdown Forest Area*.

Figure 12.8. Weir Wood Reservoir (source: photograph courtesy of Trueplan (UK) Ltd). Looking west from above the dam at Weir Wood. The nature reserve is in the far distance.

Work began in 1951 and the reservoir eventually covered 279 acres by 1953, drowning the late sixteenth-century Stone furnace and forge at its west end. Interestingly, the loss of valuable farmland has been partly compensated more recently by the attraction of the reservoir for a rich community of breeding, migrating and wintering birds. The whole reservoir was declared an SSSI in 1966 and a Local Nature Reserve by 1994. The west end remains a bird sanctuary with restricted access.[48]

One largely unforeseen result of London's post-war Green Belt policy to curtail the outward expansion of the capital was a 'leapfrogging' effect as increasingly effective rail and car travel allowed commuters to move beyond its restricted area and into 'London's countryside'. This has brought more newcomers to Ashdown, especially to homes accessible from a railway station. Workers also commute to the mid-Sussex corridor area with its service industries at Crawley, Haywards Heath and Burgess Hill. And the growth of Gatwick as London's second airport from its opening in 1958, about 14 miles from Ashdown, has been very significant, leading to development pressures and commuting by pilots and other staff. By 1970 the Conservators were actively opposing any extension of Gatwick because of the anticipated associated population increases, and this issue resurfaces continually. Another development issue in the 1970s was, once again, the question of an East Grinstead bypass. In July 1972 the Board considered a proposed route that would take between 29 and 52 acres of Forest land

48 The reservoir is owned by Southern Water. East Sussex County Council manages the SSSI in consultation with Natural England and input from the Friends of Weir Wood.

near Wych Cross, depending on the final decision. In 1974 the issue arose again, but was 'shelved temporarily', as it was once more in the mid-1990s. But East Grinstead has been earmarked for some housing expansion and the issue of associated road improvements and a bypass continues to be discussed.

Many Ashdown villages now include post-war private estates of large houses with double garages and modern facilities, the occupants of which may not necessarily contribute much to village life. At Wren's Warren, for example, was 'your ultimate country home in the Ashdown Forest': seven houses with six to eight bedrooms and double or triple garages developed during the 1990s and valued by 2020 at an average £2.4 million.[49] Fully equipped interiors, sometimes with a 'rural idyll' theme, are now ubiquitous. Given the difficulty of finding new sites for building, house prices around Ashdown have become among the highest in the UK, effectively freezing out younger and local people from the housing market. A housing need survey conducted in 1988–91 for Wealden District showed that nearly two-thirds of the then-current needs for housing in the district came from the High Weald, especially from younger people, and from parishes such as Forest Row, Withyham and Hartfield.[50]

Since then, substantial housebuilding has occurred. To cite just one example from Maresfield, a housing association development was followed by larger detached housing. The mixture of larger and housing association properties caused tension in the 1990s, a common feature of village life at this time. More homes were subsequently approved in 2016, and still more are planned, including some expected to fetch £1 million. Some residents have objected to the ongoing developments on numerous grounds, including the impact on Ashdown Forest.[51] The full impact of the Ashdown Business Park at Maresfield is yet to be seen, but the mix of light industrial and office space is typical of the pressures upon the rural environment around the Forest.

A 2017 report from the CPRE claimed that the High Weald AONB had the second highest number of AONB housing schemes in the country following the Cotswolds, together with a large number of planning appeals.[52] Wealden District Council has to find 9,600 new dwellings up to 2030, with the anticipated development located within towns such as Crowborough and Uckfield, or in a key service centre such as Forest Row. The Mid Sussex District Plan considers that 13,600 homes will be required in the

---

49 *Daily Telegraph*, 20 July 2002. The development was on the 'brownfield' site, with 2020 house price estimate from Zoopla. By the 1970s the site had become a home for dogs and cats.

50 B. Short, *Rural Housing Need in South East England: Wealden District, East Sussex 1988–91* (Crowborough: Wealden District Council, 1991); B. Short, *Rural housing need in the 'affluent' South East*, Univ. of Sussex Research Papers in Geography (Brighton, 1992).

51 <http://uckfieldnews.com/twenty-five-new-homes-park-farm-2-maresfield/> [accessed 29 September 2021].

52 D. Dixon, N. Sinden and T. Crabtree, *An Independent Review of Housing in England's Areas of Outstanding Natural Beauty 2012–2017*, CPRE (2017), 8, <https://landscapesforlife.org.uk/application/files/5315/5552/0923/Housing-in-AONBs-Report.pdf> [accessed 1 February 2022].

plan period, mostly at East Grinstead, Haywards Heath and especially Burgess Hill – all relatively close to Ashdown.[53]

Arising from these anticipated pressures, new legislation seeks to protect Ashdown from increased nitrogen deposition from vehicles and from the increased recreational pressures arising from residential development. The Strategic Access Management and Monitoring Strategy (SAMMS) legislation is designed to counteract acid deposition and eutrophication, while the Suitable Alternative Natural Green Space (SANGS) measures are intended to provide sites acting as counter-magnets to Ashdown for recreation, much of which involves dog walking. Sites near East Grinstead and Crowborough have already been designated as SANGS, with funding supplied by a Community Infrastructure Levy on developers as part of a planning application. A design premium is also levied to ensure quality of design on new building near the Forest.

These strategies were brought forward because Ashdown, as we have seen, has numerous habitat designations and is protected by the European Habitats and Wild Birds Directives, transposed into UK law through the provisions of the Habitat Regulations.[54] These required Wealden District Council to avoid or mitigate any adverse impact that development may have on the conservation objectives of the Forest. Any local authority plan must take cognisance of a 7km (4.3-mile) Zone of Influence, which therefore stretches into Mid Sussex, Surrey and Kent.[55] In fact, Wealden District Council by 2018 was implementing an embargo on development within 25km (15.5 miles) of Ashdown Forest because of the unacceptable resulting traffic generation and increased nitrogen. And, controversially, they plan to protest against future housing development in any neighbouring districts which may affect the Forest.

NATURE, CULTURE, POLITICS, ECONOMICS

Ashdown's history is littered with disputes as notions of property, freedom, control and environmentalism jostle against one another. The post-war period has been no exception. Issues of access to the commons have been highlighted as more people seek rural lifestyles in an urbanising part of England. But politics and culture can clash. The Old Surrey, Burstow and West Kent Hunt is the only one allowed over Ashdown, but in 1994 East Sussex County Council attempted to ban hunting over Ashdown, in line with its ban on blood sports on land that it owned, but the Commoners – although not necessarily other residents – saw this as an attack on their traditional way of life.

53 *Mid Sussex District Plan 2014–2031: Submission Version Incorporating Focused Amendments and further proposed modifications* (2016), 15, 50–53, <https://www.midsussex.gov.uk/media/2989/bp1-submission-district-plan-august-2016.pdf> [accessed 1 February 2022].

54 EC Directive 92/43/EEC on the Conservation of Natural Habitats and of Wild Fauna and Flora; and EC Directive 79/409/EEC on the Conservation of Wild Birds. As at 2021 there remains uncertainty over much environmental legislation as the UK leaves the European Union.

55 I am grateful to Craig Steenhoff, Senior Planning Policy Officer, Wealden District Council, for alerting me to the actual Habitat Regulation Assessment (2018), and to Kelly Sharp, Planning Policy Manager, Wealden District Council, for information on the current SAMMS/SANGS policy around Ashdown as at February 2021.

As it happens, the matter is now settled by the ban on hunting with dogs introduced nationally in 2004, but many saw the Conservators as protecting the interests of the Commoners rather than the general public, even though the Forest was now owned by a public body. Commoners, residents and Conservators were thus brought into a fraught situation. What some people saw as repressive power others saw as an enlightened approach.

As ever, many issues are not purely local. Increasingly the Forest has been opened up to external influences: the modern state, media and culture have infiltrated Ashdown living. One early manifestation occurred on 29–30 June 1963, when the charismatic American president John F. Kennedy paid a fleeting visit to meet prime minister Harold Macmillan at his country house, Birch Grove. At the height of the Cold War the two men cemented the Anglo-American alliance. The next day Kennedy attended 8.30 Mass at Our Lady of the Forest, Forest Row (which closed in 2009). The Conservators decided that a new clump should be planted between Wych Cross and Chelwood Gate to commemorate the visit, which was financed by Lord De la Warr.[56]

So no longer can rural areas such as Ashdown be seen in isolation from wider environmental, social or political processes and events, but rather they are integral elements of regional, national or even international changes. And if ever Ashdown was in any sense a community that word, contested as it is, no longer resonates with the meanings it once held. Post-war Ashdown, as elsewhere, has undergone a social recomposition: more atomised households, connected loosely and occasionally through local groupings or events, but equally connected online to social relations in a wider world.[57] Not only do workers commute away from Ashdown, but now their links are far wider afield, frequently international, and, thanks to the internet, many can work from their converted forester's cottage, a trend heightened by the Coronavirus pandemic from early 2020.

There has rarely been a moment when Ashdown was apolitical, although not necessarily in relation to accepted party politics. The disputes, often bitter, over Ashdown's future that have been detailed here are often proxies for disputes of a wider nature between classes, between locals and non-locals, between residents and the impersonal forces of the State or between conflicting visions of rurality. Ashdown has been the stage but the actors are fighting over wider issues that must be seen in the broader context of social, economic and cultural influences that have increasingly came to dominate. Perhaps one event particularly symbolises this change: when Brigadier Scott became chairman of the Conservation Board in early 1977 it was the first time in its history that the chairman was not himself an Ashdown Commoner.

Thinking of Ashdown in the twenty-first century, it is perhaps appropriate to ask where power over the landscape now resides. Gone are the old oligarchic certainties of the lord and the manorial court; gone too is the power of the church and the parish

---

[56] A clump to commemorate Harold Macmillan was also located nearby, the site being agreed in late 1979. The house also saw other visitors to the Macmillans, such as de Gaulle, Eisenhower, Krushchev and Nehru. See P. Elgood, *Kennedy at Birch Grove* (privately published, 2012).

[57] See M. Castells, *The Rise of the Network Society* (Oxford: Blackwell, 1996).

vestry; and now diminished is the power of the farming lobby. So, does power over Ashdown actually even reside in and around Ashdown? This is doubtful. The real power to make decisions about the landscape comes instead via democratic representation and finance from central and local government, and, until recently, from Brussels. But things can change quickly. Austerity has forced new appraisals of the relationships between the local authorities and the Forest. East Sussex and Wealden have effectively pulled back from their financial commitments to Ashdown, and now maintain only their statutory duties of protection under the designations applied to the area.[58] Ashdown has to fund itself and state aid, either from local or central government, has to be applied for and cannot be guaranteed. The dramatic repercussions of the Coronavirus pandemic for the economy will make such income even more uncertain. Ashdown will need its 'Friends' and its visitors more than ever.

Furthermore, the relationship of Ashdown residents to their environment, once so intimate, is now far from strong. As one consultant noted for the High Weald AONB as a whole:

> The connections between the communities of the High Weald, and the protected landscape they live in, are often tenuous. The harmonious and productive relationship between communities and landscape ... is functionally weak for most of the lives of most of the people living in the High Weald. This is not to say that the High Weald's landscape is not important to its residents – it clearly is – but their presence in it offers little to its ongoing good heart ... the majority of the employment and service lives of the majority of the residents of the AONB are lived outside their home settlement, and frequently outside the AONB altogether. So their connections with local economies are only partial.[59]

If these findings for the High Weald as a whole are replicated for Ashdown, its 'ongoing good heart' will require sustainable good will from many different quarters.

---

58 For a case study of Devon in this context see N. Kirsop-Taylor, D. Russel and M. Winter, 'The Contours of State Retreat from Collaborative Environmental Governance under Austerity', *Sustainability*, 12 (2020), 2761, <https://www.mdpi.com/2071-1050/12/7/27610> [accessed 29 September 2021].

59 Land Use Consultants, *Sustainable Settlements*, 97.

CHAPTER 13

FOREST CONFLICTS: A CONCLUSION

> Landscapes are created by people – through their experience and the world around them.[1]

THE FOREGOING CHAPTERS HAVE revealed the 'unwitting autobiography' of Ashdown's people and their landscape.[2] Rich and poor have created the landscape, but people's lives were in turn influenced by their physical surroundings and attached meanings of the environment into which they were born or in which they made their living. Historically Ashdown's cottagers and small farmers were born into a hegemonic society in which the iconography of the wealthy was readily displayed: the plaques, memorials and interior or exterior furnishings of their parish church, for example, constantly reminding them of their own humble place within a highly stratified society. Work among the landscapes of farms, gardens and parks similarly reinforced their subservient relationship to the owners, and the manor house, schoolroom or almshouse would be unconsciously 'read' in the same way. They inherited attitudes to social place and obligation, but this is not to deny what Keith Snell has called 'a culture of deferential bitterness', a fierce undercurrent of protest and claims to ancient rights that we have charted and which were not so easily trampled.[3]

The idea of landscape in relation to its inhabitants has a long history. John Ruskin was struck by the different attitudes to 'natural beauty' between classical and modern writers. The former preferred landscapes of tamed and humanised or gently productive nature, whereas for Ruskin's contemporaries nature might also be sublime – wild and rugged. And if such a landscape, even a mimetic one, could be found and enjoyed within a train's ride from London, so much the better. And not too wild a heath, but one shaped by the inhabitants of the scattered picturesque but mean cottages, by a

1 B. Bender (ed.), *Landscape, Politics and Perspectives* (Oxford: Berg, 1993), 1.

2 See P. Lewis, 'Axioms for reading the Landscape: some Guides to the American Scene', in D. Meinig (ed.), *The Interpretation of Ordinary Landscapes* (Oxford: Oxford University Press, 1979), 11–32.

3 K.D.M. Snell, 'Deferential Bitterness: the Social Outlook of the Rural Proletariat in Eighteenth and Nineteenth Century England and Wales', in M. Bush (ed.), *Social Orders and Social Classes in Europe since 1500* (London: Longman, 1992), 158–84.

kind of 'primeval nobility'.[4] This was an age when 'cultural primitivism' as a fashion attracted the wealthy and leisured. The very word 'folklore' was coined in 1846. Failing this, the countryside was to be innocent, pastoral and reflective, part of a critical counter-current to the anxieties of urbanisation and especially life in London.

The writing of Thomas Hardy, as we noted in Chapter 1, makes use of the character of a heathland environment, Egdon Heath, most particularly in *The Return of the Native* (1878), and by the later nineteenth century heathland had taken its place in the list of escapee environments for wealthy city dwellers. True, it lacked the romanticism of the Alps and the allure of a really bucolic life. It was neither ruggedly wild nor presenting the gentle pastoral of a Constable painting. But, importantly, it was now accessible and available to those with some freedom from material want and with the educational attainment that ensured detachment from hard rural labour. Ashdown was now a 'landscape' to be admired and physically enjoyed – indeed, 'Scotland in Sussex'.

This book has sought to reveal something of the unfolding relationship between people and place through many centuries and, as mentioned in Chapter 1, owes something to Fernand Braudel's concept of *la Longue Durée*.[5] And throughout this narrative certain themes have arisen, upon two of which we might now reflect further.

THE COMMON OF ASHDOWN: A THEATRE OF CONFLICT?

Ashdown Forest was common land inasmuch as there were anciently claimed common rights over what became a private chase or park. But it was never technically public space before the twentieth century and only those with the right to do so could legally venture onto the Forest. Land, indeed all property within feudalism, was controlled through a relationship between the lord and his subordinates. The land might have rights to common enjoyed since time immemorable but it did have an owner, hence *Nulle terre sans seigneur.*

The latter point is instructive because we might now return to Garret Hardin's 'Tragedy of the commons', which we introduced at the beginning of this book.[6] We recall that in his view commons were inherently inefficient and their resources over-exploited in a robber-economy because those using them did not in turn repay commons with sufficient management to counter what was extracted. At first sight this might indeed seem useful in assessing the environmental damage being done to Ashdown at many periods in the past, as we have set out. However, such free-riding behaviour based on narrow self-interest, as decried by Hardin in 1968, has little place in the commons as described in this book. Instead, systems of coercion, management techniques and institutions were developed by groups of users to manage tensions and

4 A phrase from a discussion of historical and literary attitudes to Danish heaths in Jutland in K. Olwig, *Nature's Ideological Landscape* (London: Allen & Unwin, 1984), 31.

5 Braudel and the *Annales* School of French geographers recognised that historical changes proceed at different rates: namely, the short time-span of individuals and events; the longer time-span of economies and societies; and the slow fundamental structural changes of environments. And all also operate at different spatial scales.

6 Hardin, 'The Tragedy of the Commons'.

exclude non-Commoners. Thus, we have seen manorial control, although sometimes inconsistent or half-hearted, or Commoners' associations. Ashdown was by no means an area of free-for-all open access before the twentieth century, when it became subject to local government policy that did open it to the public – although even then there were byelaws to control public activities. This degree of self-government actually made commoning more optimal and self-sustaining. Management might, for example, involve stinting by various means: privilege by acreage size, by tenure, by other ancient rights, or by levancy and couchancy. This last became a widely accepted custom by the fourteenth century, giving way later to stinting in southern England as population pressure increased.[7]

Furthermore, the context will change (or, as Hardin caustically put it, 'morality is system-sensitive'), such that the commons' embeddedness within, for example, feudalism, bastard feudalism or capitalism made a difference, as indeed did location and habitat – the local context is always important. There were also always the opportunity costs of using the commons against other options. Thus, perceptions of the worth of the commons changed as modes of exploitation changed. For example, as agricultural output rose, perhaps by employing sown fallows, new fodder crops, mechanisation and so on, so the common might well seem less desirable, and become under-exploited. But again, context is important: as we have seen, social values change, such that the least productive land may become that which is most valued aesthetically or for recreation.

Of course, economic conditions may override such considerations and the drive to enclose the common for profit may become seen as a beneficial process – turning the unproductive heath to use for human food and seeing the resultant cultivated fields as aesthetically pleasing as well as economically beneficial. Various defensive strategies might then be deployed by those reliant on the common to a greater or lesser degree: occupation of the land in question, tearing down the enclosures either overtly or covertly or, in more recent centuries, appealing to any available legal protection and to those institutions that had arisen to protect open spaces, whether pressure groups or local authorities who concede the benefits of safeguarding open space.[8]

Hardin's views were embraced by many, following his various apocalyptic pronouncements about over-population, for example. However, after examining the history of the common uses of Ashdown Forest we can agree with Rackham and Grove that

> This pernicious notion was invented by an American with no experience of how commons actually work. Despite a lack of evidence, it has been repeatedly used by governments as an excuse to abolish or oppress commons, often with results very different from those intended.[9]

---

7 Shaw-Taylor, 'The Management of Common Land', 70.

8 On responses to enclosure see B. McDonagh and C. Griffin, 'Occupy! Historical Geographies of Property, Protest and the Commons 1500–1850', *Journal of Historical Geography*, 53 (2016), 1–19.

9 A.T. Grove and O. Rackham, *The Nature of Mediterranean Europe: an Ecological History* (New Haven, CT: Yale University Press, 2001), 88. And see the fuller discussion of Hardin's

Although Hardin later sought to clarify his position by insisting that he had been referring to *unmanaged* commons, his controversial views on this matter, and on many other things, can be largely dismissed.[10]

Today we are fortunate that Ashdown has survived thus far. Many other heaths have succumbed. Actually, Ashdown's modern landscape has resulted from what *did not* happen. Hans Christian Andersen wrote of the Jutland heaths: 'The heath, yes, it is hard to believe – but come yourself, look it over: the heather is a splendid carpet, flowers crowd for miles around. Hurry, come! In a few years the heath a grainfield will be.'[11] His words were prophetic, for the Danish heaths were transformed into family farms under an 'improving' logic. Heathland reserves were established but, like Ashdown, the problem of ecological succession has led to bush and scrub invasion necessitating artificial clearance – 'nature ceases to be natural'.[12]

The above discussion is directly relevant to an understanding of how the conflicts which have surrounded Ashdown over several hundred years have arisen. Individuals would attempt to maximise their own wellbeing through enclosing or extending small areas of land, grazing as many cattle as possible and taking as much timber, underwood, turf, game or litter as possible. It was an area for the 'poor and dispossessed' at the margins of settlement with limited resources: an area for the poor to seek some subsistence.[13] But, most importantly, as we have seen, and perhaps contrary to expectation, Ashdown was an area in which there was *organisation*. The Forest could be controlled, more or less effectively, because of several factors, not all of which operated at the same time or necessarily very well:

- it covered a large area but not so large that control was impossible;
- there was a legal, clear-cut boundary, the pale;
- evidence of over-use could be ascertained and monitored by accountable persons;
- there was some awareness, for better or worse, of inter-dependence within and between social groups;
- there were individuals prepared to initiate controls and fund controlling mechanisms;
- the number of Commoners, mostly living near the Forest edges, was not so large as to inhibit contact between them, allowing communication among a 'community of knowers' with 'a socially shared notion';[14]

---

ecological views in J. Radkau, *Nature and Power: a Global History of the Environment* (Cambridge: Cambridge University Press, 2008), 71–3.

10 G. Hardin, 'The Tragedy of the Unmanaged Commons', *Trends in Ecology and Evolution*, 9 (1994), 199.

11 Hans Christian Andersen, 'Jutland between two seas' (1859), cited in Olwig, *Nature's Ideological Landscape*, 52.

12 Olwig, *Nature's Ideological Landscape*, 103.

13 S. Birtles, 'Common Land, Poor Relief and Enclosure; the Use of Manorial Resources in Fulfilling Parish Obligations, 1601–1834', *Past and Present*, 165 (1999), 80.

14 The phrases are from Umberto Eco, in C. Birchall, 'Economic Interpretation', in C. Ross and R. Sibley (eds), *Illuminating Eco: on the Boundaries of Interpretation* (London–New York: Routledge, 2016), 81–3.

- there was a set of resources that was acknowledged as communal;
- there were institutions in place whose frameworks could be adapted to the specifics of the Forest – e.g., manorial courts – to help resolution of conflicts and with sanctions against transgressors;
- the use-rights were known to all users and could be transmitted to following generations, often through kinship, sets of conventions and routines;
- there was self-determination, with the degree of outside interference historically limited.[15]

While these elements can be shown to have been variously attributable to Ashdown over time, it remains unclear how they originally evolved, in part or whole, during its earliest days as an area used communally. Indeed, why communal at all? When did it appear that collective decision-making on Ashdown should replace individual grazing? At what stage did the costs and restrictions involved in communal decision-making outweigh individual benefits? Was there some incipient communal arrangement pre-dating the manors? Had some broad consensus developed by or during the Saxon years? The imposition of a manorial system did at least bring with it a communal decision-making arrangement, even if, at various times, it was top-down, inconsistant and somewhat ineffective.

Clearly, medieval Ashdown received added 'rules' as a result of the imposition of the pale and its function as a deer preserve, such that pannage times, for example, became more tightly controlled. These 'rules' changed through time: how did the 'rule' forbidding the grazing of sheep, for example, on the Forest break down? Indeed, one could posit transitions between ancient individualism, when anyone could try to establish a right to resources, through to a collective phase, and then to a more regulated and tightly administered phase in more recent decades, though tenaciously resisted by those who felt their ancient rights being threatened.

Regulation of Ashdown did not, of course, avoid environmental degradation, nor was it socially just. Exclusion was deemed necessary against many who sought the shelter and opportunities afforded by this more weakly manorialised and marginal environment. And we should also ask whether management of the Forest actually worked. By the late eighteenth and early nineteenth centuries there were claims that the commons generally were overstocked. In 1805 John Boys described the cattle on Kentish commons as 'half-starved', and it could well be that the decay of the old manorial courts had left a void in regulation, allowing real deterioration to occur. However, we should be aware that all too often these claims were made by enthusiasts for 'improvement' and enclosure.

---

15 These factors are partly inspired by E. Ostrom, 'The Rudiments of a Theory of the Origins, Survival, and Performance of Common-property Institutions', in D. Bromley (ed.), *Making the Commons Work: Theory, Practice, and Policy* (San Francisco, CA: ICS Press, 1992), 293–318; and E. Ostrom, *Governing the Commons: The Evolution of Institutions for Collective Action* (Cambridge: Cambridge University Press, 1990). The work of the Nobel prize winner Elinor Ostrom (1933–2012) has been fundamental in moving the commons debate on from that of Garret Hardin.

ASHDOWN'S COMMON: LIKE ALL OTHER COMMONS?

There is now a voluminous literature on the history of the English commons. Here we must be selective, relating Ashdown to its wider contexts, picking out that which has a direct bearing on the conflicts we have been discussing. So what follows is neither a literature review nor an all-encompassing assessment of work on commons, as that would remove us from the immediacy of the Forest itself.

One key theme in the Ashdown story is that of protests over attempted enclosure. And this, of course, has many echoes elsewhere. Disputes between lords and commoners over losses of rights are, for example, investigated for Shropshire's wood-pasture countryside in the sixteenth and seventeenth centuries by Bowen, who stresses the importance of the subtleties of the locality in understanding the complexities of such situations, and the fact that the protests were upheld by most sectors of society and not just the poorest.[16] The legal and illegal taking of wood was an ongoing theme elsewhere, as for example in the work of Reeves and also Cook in investigating the Commissions of Enquiry into waste and spoil in the New Forest in 1673 and 1677. The Commissions there also heard of the fall in red and fallow deer numbers during the Civil War and Commonwealth, again a familiar theme on Ashdown.[17] And allied to studies of protest is the large body of work on the dispossession of the poor from the commons, as in the work of Hindle for Rockingham Forest in Northamptonshire, Broad for Bernwood Forest, Buckinghamshire, or Neeson in a wide survey but also focusing on Northamptonshire. Winchester and Straughton have also focused on the management of and inherent conflict on the northern commons.[18]

Fundamentally, Ashdown escaped total enclosure. Exmoor, for example, was enclosed and reclaimed following the final 1819 enclosure award. This had been a royal forest of some 20,000 acres, leased out and agisted for sheep in summer and with cutting for fuel and litter. The new parish of Exmoor was created and the enclosed lands sold or allotted. Eventually John Knight came to own 16,000 acres, and others came to accumulate and consolidate the smaller allocations originally granted in lieu of various rights, such as turning out cattle. This was a fate familiar to many commons. A boundary wall was built, to the annoyance of the stag hunts, together with 22 miles

[16] J. Bowen, '"Before the breaking of the day, in a riotous manner and with great shouts and outcries": Disputes over Common Land in Shropshire in the Sixteenth and Seventeenth Centuries', *Rural History*, 26 (2015), 133–59.

[17] R.P. Reeves (ed.), *Use and Abuse of a Forest Resource: New Forest Documents 1632–1700* (Lyndhurst: New Forest Museum and Library, 2006). H. Cook, *New Forest: the Forging of a Landscape* (Oxford: Windgather Press, 2017).

[18] Hindle, 'Not by bread only?'; J. Broad, 'The Smallholder and Cottager after Disafforestation – a Legacy of Poverty?' in J. Broad and R. Hoyle (eds), *Bernwood: the Life and Afterlife of a Forest* (Preston: University of Central Lancashire, 1997), 90–107; A. Winchester, *The Harvest of the Hills: Rural Life in Northern England and the Scottish Borders, 1400–1700* (Edinburgh: Edinburgh University Press, 2000) and *Common Land in Britain: a History* (Woodbridge, Boydell, forthcoming); Neeson, *Commoners*; Straughton, *Common Grazing*.

of roads.[19] This was a far cry from Ashdown, where there was no enclosure award before the 1880s, and that only to seal the agreed status of the Forest; there was no wall built at this time by a landowner who owned the Forest as a freehold; there were few large tenant farms around its edge, only smaller all-grass or mixed farms; and, crucially, there was the bitterly contested continuation of common rights.

While other commons shared certain general structural features with Ashdown, the contingent and local features of each common were unique: what could be taken from the common, and at what times of the year; who were the lords and how did their policies, family histories and attempts at control affect the common; how did the location of the common impact upon its history? Contingency is certainly important: the death of a manorial lord could lead to a chain of later events through what has been called 'historical path dependency' as processes and landscape features followed the logic of the chain. But then another discontinuity could arise, setting the Forest on another path.

A TRANSGRESSIVE LANDSCAPE AT THE 'MARGIN'

In a sense this was a detached part of 'Marginal Europe', albeit not a landscape at the physical frontier but surrounded by more productive land nearer the coasts of South East England. Here core and periphery are inverted geographically and a process of internal colonisation could proceed inwards through numerous small-scale advances. Core and periphery might be seen as complementary and linked: the practice of transhumance along the drove roads is but one early example, the seasonal migration of labour from the Ashdown area to the coastal cereal farms and orchards another, and the investment of outside (London) capital into the Ashdown ironworks and residences yet another.[20] And the latter reminds us that Ashdown was and remains tied to global flows of people, information, capital and goods.

But over time the Forest became a place not for production but for consumption – of open space, for leisure time. But it was also a place of extraction, of nutrients, stone, vegetation, and over time this also has ended, not only through the intervention of the Conservators but also through the auspices of the national and local state. Instead, the Forest has become a 'sacred space', protected and even venerated, not just locally but internationally.

Above all, we should always remember that Ashdown is a heritage: made and remade by people, whose economic, cultural, political and ideological activity has 'sedimented' upon the area, and its extent and appearance cannot ever be regarded as either natural or neutral. Instead, the Forest landscape tells us about its people, if we can learn to understand what it is telling us. It was never enough to set out the material changes made to Ashdown; it was also essential to understand and interpret

19 C.S. Orwin and R.J. Sellick, *The Reclamation of Exmoor Forest*, 2nd edn (Newton Abbot: David & Charles, 1970).

20 S. Pollard, *Marginal Europe: the Contribution of Marginal Lands since the Middle Ages* (Oxford: Clarendon, 1997), 128–41.

the changes (and stabilities) resulting from the lives of its inhabitants and those with power and surveillance over its surface, as local ties came to be overlain by the national and global. It is to be hoped that in reading this book such an interpretation will be made all the more telling and enjoyable.

GLOSSARY

Advowson the right to present an incumbent to a living under ecclesiastical law

Agistment the pasturing of one person's animals on another's land on payment of a due

Alpine Orogeny occurred mainly between 65 and 2.5 million years ago, although it is still active today, with the collision of the African and Eurasian plates, giving rise in southern England to localised folds and faults, as in the Weald

Anthropogenic vegetation cover resulting from human activity which has modified any previous vegetation cover

Ashlar a block of hewn stone with straight edges for building purposes

Atlantic period a warmer and more moist climatic period yielding denser woodland cover, following on from the cooler Boreal/Atlantic transition of the seventh and sixth millennia BC, and giving way to the modern climatic period from 5,000 to 3,000 BC. It is generally associated with the Mesolithic cultures

Besoms brooms made from a bundle of twigs tied to a handle

***Blockfluren* fields** a network of blocks of fields with ratios generally of length to width of 1:1 or 1:1.25, as opposed to long strip fields

Bloomery ironmaking where a mass of directly reduced iron (the bloom) from smelting is produced in a small furnace, a precursor to the blast furnace

Brown earths more free-draining, darker and potentially more fertile soils, often associated with woodland cover

Bryophytes a group of plants that include mosses and liverworts, some uncommon in the South East, to be found in the deeper ghylls of Ashdown

Churchwardens' Presentments ecclesiastical courts became particularly active in enforcing personal morality, church attendance and other religious matters following the Reformation. Local churchwardens reported cases of misdemenours to a church court, here the Archdeaconry of Lewes or Chichester

Claustura brushwood for making fences, intended for an enclosure

Climax vegetation the dominant and stable vegetation combination achieved naturally under particular environmental conditions over a long period of time

Cotman land medieval term for a small area of land, possibly about eight acres, but somewhat variable (and see farthing land)

Custumal a survey of the rights and extent of customary tenants of a manor

Depocentre the part of a sedimentary basin, as in the Weald during a humid subtropical swampy environment, where Hastings Beds sediments accumulated to provide maximum thickness

Drift or driving of the Forest driving cattle on the Forest to establish ownership, with illegally grazed cattle being impounded

Drofmanni the drofman or drover herded swine to and from the Ashdown pastures for manorial lords elswhere

Escheator official taking note of escheats or failures of heirs within a manor, and thus reversion of property to the lord (including the crown)

Estovers wood from the manorial common land to be used for fuel, implements, cottage repairs, hedge making etc. (medieval French *estovoir*, to be necessary or Latin *est opus* there is a need)

***Extranei*/foreign tenants/strangers** those permitted to graze cattle for payment although living outside Duddleswell or Maresfield manors

Farthing land medieval term for a small area of land, perhaps a quarter-acre in extent, but variable (and see cotman land), a term used in manorial customals, as also yardland, a highly variable amount of land

Fence month the time of fawning in a deer forest, generally 6 June to 6 July, with tighter restrictions on tenants' activities

Ferme Ornée largely a fad of the eighteenth century but lasting into the nineteenth, for decorative farm buildings, with varying degrees of functionality

Forest eyres introduced in the late twelfth century involving royal justices travelling on circuits to county sessions, to be replaced by assizes in the later thirteenth century (Latin *errare* to wander)

General Headquarters (GHQ) Line v-shaped ditches and ramparts were dug to obstruct German military vehicles in the event of invasion in the Second World War. The General Headquarters Line ran from Scotland through eastern England and across the South East. There were also secondary stop-lines, together with concrete 'dragon's teeth', pillboxes, barbed wire etc.

Ghylls (also gills) steep-sided incised valleys resulting from headward erosion into the Hastings Beds by the smaller tributaries of the rivers Medway and Ouse

Guano dried droppings of South American seabirds imported for use as fertiliser. The most significant nitrogen Guano was imported from Peru, and in Europe the application of Guano emerged in the 1840s and lasted until the early twentieth century when it was replaced by industrial manufactured fertilisers

Heriot payment made by a tenant upon inheriting a property made vacant by the death of the previous occupier. Originally the best plough beast or cart horse but commuted to a money payment by the nineteenth century

Hockday or hocktide, festival celebrated on the second Monday and Tuesday after Easter. Attendant revelries were frowned upon by the church and the tradition died out

'Levant et [or 'and'] Couchant' a measure of the number of animals permitted on a tenement or the number allowed as commonable depending on the number that could be maintained on the holding during winter

Lidar 'light detection and ranging', a technique now more widely used based on airborne laser-focused imaging and the calculation of distances by measuring the time for a signal to return. It can accurately measure differences in terrain and thereby produce detailed images of topographic and archaeological importance, especially when using the infrared spectrum. This has the advantage of filtering out tree cover which would otherwise conceal detailed landscape features

Litter bracken, heather and gorse to thatch or to litter down cattle yards, and then as a manure when trodden down by cattle. Cut using a shorter and broader scythe than normal in the months from October to January. If the litter was mainly bracken this might be cut with a hook, knife or sickle from August onwards, often by women and children

Loch Lomond stadial a period of glacial readvance and colder conditions in northern Britain during the overall shrinking of the final ice sheet. By the end of the Loch Lomond Stadial all the ice fields and glaciers in Britain had entirely receded, and all the ice had gone by about 11,700 years ago

Loess wind-blown silty material that covered much of the South East at the end of the last glacial period, between 30,000 and 15,000 years ago. Often referred to as brickearth from its use in brick manufacture

Marl clay with varying degrees of calcium carbonate found on the Forest in the Wadhurst clay outcrops; it was dug from small pits and used to reduce acidity in soils and enhance their texture and, on lighter soils, moisture retentiveness

Microclimates highly localised atmospheric conditions affecting a small area, e.g. ghylls

Microliths small stone tools usually made of flint frequently associated with rock shelter sites resulting from the making of spear points and arrowheads from the Upper Palaeolithic through to the early Neolithic

'Northsealand' the area off the eastern coast of England, now submerged, but once land connecting Britain to the continent ('Doggerland'). It was lost with the rising sea levels associated with the warming following the last glaciation

Oligotrophic mires areas of poor drainage lacking in nutrients and plant life but rich in oxygen

Oyer and terminer Anglo-French 'to hear and determine'; justices of the royal court were occasionally commissioned to determine serious crimes within a particular county, often involving those of higher rank

Pallister a worker engaged in repairing and maintaining the pale around Ashdown, or the manorial pound, with the insertion of stakes yielding the word 'knockpyn'

Pannage payment for the right of agisting cattle, horses or swine on the common, mostly associated with the medieval practice of driving swine to Ashdown. Pannage times were not to compromise fence month

Periglacial landforms although escaping glaciation, the Ashdown area did experience permafrost throughout the Quaternary period and periglacial conditions (e.g. deep seasonal freezing) during slightly warmer, but still colder-than-present conditions. The landforms were affected by the growth and melt of ground ice, and by erosion and deposition of sediment. The transport of sediment acted to produce a more rounded topography

Pillow mound artificial mounds, sometimes also called 'berries', built within rabbit warrens into which rabbits could burrow and breed. Many more have come to light on Ashdown through lidar surveys

Plagioclimax a stable vegetation community that has been prevented from developing further. The ecosystem may have been stopped from reaching its full climatic climax because modified by human activities. Such intervention has affected, for example, most lowland heath communities of western Europe, including Ashdown, through forest clearance, grazing pressure and controlled burning

Podzolisation occurs where soils are more free draining, allowing the upper horizons to become bleached through the downward leaching of iron oxides, leading to an iron pan at lower levels and to degradation and depletion. Where Ashdown Forest heathland has developed on free-draining sands and silts, or where compact subsoils have formed, shallower stagnogley podzols develop. Gley soils arise where a clay layer results in impeded drainage, and can yield areas of peat

Purlieu land bordering a Forest, sometimes indicating land originally added to a Forest unlawfully

Purprestures encroachments into a Forest resulting in interference with the free movement and sustenance of deer. From *porprendre* to seize, occupy, enclose

Severalty land held by individual right, in separate and exclusive possession without a joint interest in any other person

Slag (iron) semi-liquid waste from the bloomery or blast furnaces process frequently found around ironworking sites where it had hardened and was discarded, often in heaps. That from blast furnaces was more glassy, with a silica content. Slag could be used for mending road surfaces or strengthening the bays (dams) at ironworking sites

Solifluction gradual processes in which overlying material, such as wind-blown silty loess, moves downslope, often with a frozen underlying base during periglacial periods. Both colluvium (the resulting deposit downslope of unconsolidated sediments) and alluvial action (carriage of unconsolidated sediments by water) are part of this overall process of resorting superficial material

Stagnogleys soils developed on heavy clays, such as the Wadhurst clays of Ashdown, and where drainage is poor, offering poor farming conditions

Stinting the controls imposed to regulate cattle numbers allowing tenants certain rights to specific numbers that could be grazed on the common. This more precise method gradually replaced the older 'levant et couchant' understanding

Subinfeudation the procedure by which a manorial lord granted manorial rights on a holding to a sub-tenant. This resulted in large numbers of manors, often quite small, by the later Middle Ages

Thermoluminescence a dating technique measuring the accumulated radiation dose of the time elapsed since material containing crystalline minerals was either heated (such as pottery) or exposed to sunlight (sediments). The amount of luminescence is proportional to the original dose of radiation received and this can be used to date the objects. The technique has become a common tool for dating pottery or other fired archeological materials

Tithing penny a small amount paid by adult members of a tithing (an area organised for the purpose of collecting tithes for religious maintenance) to a court official to defray expenses of arranging for a court to be held

Turbary the right to take turf (or in some cases peat) for fuel from common land. It was a right contested in the Ashdown Forest Case

Vaccary, beccaries a cattle-breeding area or enclosure and/or dairy farm, often associated with deer parks, as in Chelwood Vachery. The beccaries were equivalent sheep stations

Valley bulging and cambering the distortion of softer rock strata by discontinuities in the base of valleys. The processes leading to valley bulging can be complex, incorporating rebound due to erosion and periglacial freezing of the ground. Cambering involves large-scale stretching, tilting or rotation of rocks over softer strata such as clays. Cambered strata occur on valley crests and commonly the cambered blocks are tilted to the valley axis

Wandersiedlungen also referred to as the 'Middle Saxon shift', whereby settlements have migrated over time or shifted their constituent parts to different locations within the overall settlement. This has been held to describe apparent differences between the settlement patterns of early and late Saxon England. It now seems that most nucleated medieval villages are not necessarily the direct successors of earlier Saxon settlements, and that nucleation itself appears to be a remarkably late phenomenon

SELECT BIBLIOGRAPHY

Only those texts with some direct and sustained relevance to Ashdown are included here.

PRIMARY SOURCES: MANUSCRIPT AND ARCHIVAL COLLECTIONS

Arundel Castle archives

Box list 2, Part III. Post-medieval deeds, Buxted, D3127 (1880–92)

British Library

Add MS 11,571, A Traveller's reveries, or Journey through Surrey and Sussex, prefaced by a critical epistle in certain instruction of the Greek language, by John Burton, M.D. 1752

Add MS 32,732 Newcastle Papers XLVII 1753

Egerton MS 1967 Rental and customs of Barkham Manor, 1723; Dallingridge Manor; 1785–91; Gravetye, 1755; West Hoathly Rectory Manor, 1788

General Reference Collection 08226. bb. 53 London and Provincial Hop Growers and Pickers Defence League

Harleian MS, No. 1579, f.22 John Hatton survey of Ashdown Forest 1632

Maps 137 Sale catalogues, 137 b.10 (11) Sackville House estate, Buxted 1880; 137 b 9. Plan of … Kidbrooke Park, in the parish of East Grinstead, Sussex. For sale … 1874

MS Facsimile Suppl IV W Photographic copy of 1693 award map (see ESBHRO, DLW 465)

East Sussex Brighton and Hove Record Office

ACC 448/2/14 Sale of Ashdown House Estate 1910

ACC 1831/1–4 Ashdown Forest Case

ACC 1831/6 Copy of survey of the manor of Duddleswell, 1564 (made *c.* 1814)

ACC 2717/1 Robert Bingham at Maresfield to Lord Sheffield 1809

ACC 3402/8 Maresfield Park Estate

ACC 3715/1–5 Notebooks of William Augustus Raper

ACC 4669 A.J. Ridley, beekeeper etc of Horney Common 1883–1913

ACC 5411/120, Copies of decrees in the Duchy Court of Lancaster relative to Ashdown Forest, 1691, 1693

ACC 5640 CAF Conservators of Ashdown Forest (temporary reference as at 2021)

ACC 5786/55 Draft version Ashdown Forest enclosure award on 25-inch sheets 1887
ACC 6815 Minutes of the Ashdown Foresters Protection Association (included in Ridley and Lewis papers below)
ACC 6815 Ridley and Lewis families of Horney Common
ACC 9997 Ashdown Forest Oral History Project 2005–6
ACC 10360/2 Letters mainly to Mrs Elizabeth Davies (née Martin), the wife of Col. Francis Davies (notebook *c.* 1845, examples of Sussex dialect)
ADA 98–9 Rental and Custumal of the Manor of Duddleswell *c.* 1883–1915
ADA 100 Adams and Remers solicitors, rights of common on Ashdown Forest *c.* 1883–1915
ADA/ACC 3628, 3628–1 Minute Books Manor of Duddleswell 1745–1805
AMS 364 Manor of Ringmer 1666
AMS 683/1 counterpart lease land and furnace in Ashdown Forest 1717
AMS 715 Copy Decree of the Duchy of Lancaster Court relating to Ashdown Forest 8 July 1693
AMS 718 Letters of administration Alexander Staples 1709
AMS 3780–4104 Papers in De La Warr v Miles and Hale in chancery and the court of appeal (The Ashdown Forest Case)
AMS 4899 Proposal for the formation of [a Gilbert Union] based on the parish of Fletching to build a Workhouse at Piltown or Chelwood Common *c.*1810
AMS 5481 Marriage settlement Charles Sackville and Grace Boyle 1744 (list of CS possessions)
AMS 5762 Reminiscences of Harry John Peckham (1841–1922) Vicar of Nutley 1882–1913
AMS 5818, Sir Henry Fermor's endowment. Will dated 21 January 1732
AMS 6062/1/1 Notes and Recollections of Alterations and Changes in Hartfield since my Childhood: by George Jenner born at the Church Gates Cottage June 23rd 1817
AMS 6190/1–2 Mortgage assignment part of Ashdown Forest, Alexander Staples
AMS 6403 Narrative of the circumstances of the arrest of the Rev Robert Bingham, curate of Maresfield, 7 February 1811
AMS 6504 Papers of the Ashdown Forest Commoners' Committee. Opposition of the Commoners' Committee to the County Council's promotion of the Ashdown Forest Bill in the House of Lords 1973
AMS 6627/1 conveyance De la Warr estate 1920
AMS 7005 Photographs and reminiscences, Wren's Warren Camp for Evacuees, Hartfield 1939–45
AMS 7127 Court roll, manor of Duddleswell 1561
CLW 2/1–3 Baron Chelwood (formerly Sir Tufton Beamish MP)
DLW 283–93 Receipts and assignment of mortgage, Williams on Ashdown Forest 1680s
DLW 459–60 draft of an agreement to compromise in Hale and others v. De La Warr 9 December 1881
DLW 465 tracing of 1693 map of Ashdown Forest
DLW 466–8 Ashdown Forest. Earl de La Warr v. Miles and Hale. Book of Documents I and II (as also AMS 3959)

DW/A/15/286 Countess of Norbury, House at Forest Gate, 1912
DYK 784 conveyance Freshfield forge and hammer 1677
L/C/G/3/1–2 Proceedings under the Defence Acts 1801–06
PAR 286 Buxted parish records
PAR 360 Hartfield parish records
PAR 379 West Hoathly parish records (now transferred to WSRO)
PAR 420 Maresfield parish records
PBT Probate inventories, East Sussex parishes
QDD 6/E13 Enclosure of Chelwood Common 1864
QDD 6/E15 Ashdown Forest Inclosure Award 1887
RAF ACC654/138 plan of proposed light railway Nutley to Groombridge 1899
SAS AB/824 Manor of Duddleswell court roll 1782
SAS ACC7740 Fitzhugh Gates, solicitors: clients' papers including records of St Mary's Home, Brighton, Ovingdean and Buxted
SAS F1 Sir Thomas Williams and Joseph Fells of London, goldsmith, to Alexander Staples, lease
SAS G/ACC/903–4 Court book Manor of Maresfield 1822–41 and 1841–70
SAS G/6/50 Correspondence between Sackville and Gage re Maresfield etc. 1560–62
SAS PN/272–3, 733, Staples family mortgage details
SAS PS/334 Outlying portions of the Ashdown House Estate 1914
SAS RF/14/87 Correspondence Fuller and Earl of Dorset 1791–1822
SPK 1/88/9 Charles Abbot to Lord Sheffield re enclosure of Ashdown Forest, September 1808
SPK 1/98 Rev. Robert Bingham to Lord Sheffield 1809–11
SPK P/4, 6–12 Fletching poorhouse and poor 1812–19
SRL 2/16/18a and 18b Mivart to Maryon Wilson 1882
TDE 149 Tithe map and apportionment for Hartfield 1842
U of Sx Spec Colls, AFC I and II – University of Sussex Special Collections, 2 volumes prepared for the Ashdown Forest Case 1876–82

House of Lords Library

FCP/1/61a Forest and Commons Papers (formerly Box 25 Number 56)

Imperial War Museum

Documents 9714 Private papers of Major the Hon R.E.S. Barrington DSO

Kent History and Library Centre

U269/E352 Partition of the Sackville family estates between Lady De La Warr and Lady Plymouth as co-heiresses of the third duke of Dorset 1829–30
U269/E388/1 Legal opinions in connection with a dispute between the lord and commoners over cutting turf 1816
U269/L19/4–7 L92 Family disputes on Reginald's accession to the earldom

London Metropolitan Archives

MS 10274–97, A, Worshipful Company of Goldsmiths Calendar of the Records of the Hamsell and Birchden Estate, Sussex Compiled by J.R. Ede

The National Archives

Assi 35/58/8: East Grinstead Assizes 8 July 1616

C 7/2/2 Court of Chancery, estate of Alexander Staples 1710

C 133/8/7 Details from the IPM of Sir Jordan de Saukevyll alias de Sakevile 1274

C 145/16 Court of Chancery, Miscellaneous Inquisitions 1243

DL 29/441/7082 Sussex: Lordship of Pevensey and others: ministers' accounts

DL 42/112 Surveys of manors and hundreds belonging to the Duchy of Lancaster in the Rape of Pevensey by Edmund Twynyho ... 4 August 1564

E 101/136/19: Particular Forests (south of Trent): accounts relating to Ashdown forest, 24 Edw I (1295–6)

E 101/510/28, Account of Walter Prynkell, outlawed at the suit of John, duke of Lancaster, of the issues of his lands in Sussex. 7 Rich II (June 22 1383–June 21 1384)

E 134/33/34 Chas 2, Hil. 6 Exchequer Depositions

HO 67/7 Parish acreage returns 1801: Chichester Diocese

HO 107/1115 Hartfield census 1841; 1640 Maresfield census 1851 (and see RG below)

IR 18/10352 Hartfield Tithe File 1842

JUST 1/912a, m. 17d. Sussex Eyre 1262

JUST 1/947/5; 1/944 General oyer and terminer roll, Sussex, felonies and trespasses in Pevensey rape 1384

LRRO 67/1–3 Wilson v De la Warr. Hewlett Record Agency: Papers and evidences Ashdown Forest

MAF 1/377 Enclosure awards and regulation of commons under the Enclosure Acts 1845 to 1899: Ashdown Forest 1887

MAF 32/1008/24 National Farm Survey 1941–3, Hartfield

MAF 38/80 Agricultural policy – memorandum by Earl De la Warr 1938/9

MAF 68/2143 (1905), 2992 (1920), 3532 (1930), 3920 (1939) Parish summaries of agricultural returns, Sussex

MAF 73/41/16 National Farm Survey 1941–3 annotated maps, Sussex, East, number 16

MAF 80/2811 East Sussex War Agricultural Committee minutes 1939

MAF 96/151 Conservators' evidence to the Commission on Common Land 1956

MH 12/12854, 13138 Hawley to the Poor Law Commission 9 February and 15 August 1835

MPH 1/407 Plan of the camp in Ashdown Forest, showing the positions of the regiments (1793)

RG 10/1054 Buxted census 1871; 1055 Fletching 1871; 1056 Withyham 1871; 13/905, East Grinstead (Forest Row) 1901; 14/4977, Hartfield 1911; 4962, Maresfield 1911

SC 1/12/202 Edward I ... to John Lovel: venison from Ashdown forest 22 September 1302

SC 6/1027/20–22; 1028/1–2 Maresfield Manor 1290–1303

SC 6/1127/10 Ministers' and Receivers' accounts: Manor in Withyham called Monkenecourt 1 September 1431–31 August 1433
SC 8/36/1783 Petitioners: Reyner le Brom, Parson of the church of Hartfield etc. *c.* 1314
SC 12/15/46 Ashdown (Essedune) Forest: Extent 2 Edw 1 20 November 1273–19 November 1274
SP Henry 8, 1/231–2 letters and papers. Ashdown Forest enquiry 1509
STAC 2/23/69 Court of Star Chamber Henry VIII, Waleys v Sackville
TS 46/85 Public enquiry concerning the use of Ashdown Forest by the War Department
WO 100/25 Foot Guards 1854–5

West Sussex Record Office

Add MSS 30332 parochial totals 1801 census for Sussex
PAR 348/37/6/1 East Grinstead, Proposals for farming out the workhouse to Edward Head and William Purchase 1735; 348/37/8 Overseers of the Poor 1727; 348/37/7 'Return of all lunatics and dangerous idiots in the parish of East Grinstead to Edward Cranston, esq., with a report on their health by John Collins'
PAR 379/12/12 West Hoathly vestry minute books 1806–7; 379/7/8 parish register; 379/31/2/6 overseer of the poor ledger 1828–32; 379/34/4/1 bastardy order/bond; 379/12/26 Transcripts of extracts from parish and Quarter Sessions records, 1594–1807

Yale University Library

Gen Mss vol. 221 Repton's Red Book, *Hints, plans and sketches for the improvement of Stonelands*

PRIMARY SOURCES: OFFICIAL DOCUMENTS AND PUBLICATIONS

Ashdown Forest Act, 1937 [1 Edw 8 & 1 Geo. 6]
Ashdown Forest Act, 1949 [12 & 13 Geo. 6]
Ashdown Forest Act 1974: Elizabeth II. Chapter xxi
Commons Land Regulation (Ashdown Forest) Provisional Order Confirmation Act 1885 (48–9 Vict.c.56)
British Parliamentary Papers, *Report of the Commissioners concerning charities … on the Education of the Poor: Sussex*, 1820, 437–48
——, Report from HM Commissioners on the Administration and Practical Operation of the Poor Laws: Appendix (B1), Parts 1–4. Sussex. Answers to rural queries … 1834
——, 1851 census table: 6 (1851–3), 458–9
——, Turnpike Trusts County Report no. 3; Sussex (1852), 26–55
——, The Land Commission (England and Wales) – Ashdown Forest, 1 April 1886, 433–4
——, Report of B.H. Holland to the Enquiry on Wages and Conditions of Employment in Agriculture (1919)
——, *Census of England and Wales 1921*, County Report of Sussex, Table 17: occupations

Conservators of Ashdown Forest, *Strategic Forest Plan of the Board of Conservators of Ashdown Forest 2008–2016* (2008)

——, *Annual Report* (2015–16)

——, *Achieving favourable condition: a management plan in support of Countryside Stewardship 2016–2020* (2016)

——, *Ashdown Forest Annual Review 2016–17* (2016)

——, *Ashdown Forest Vision and Management Strategy 2021–2031* (October 2021)

Great Britain Poor Law Commissioners, *Extracts from the Information received by His Majesty's Commissioners as to the Administration and Operation of the Poor Laws* (1837) 'No. 1 Communication to the Poor Law Commissioners, from Ashurst Majendie Esq., Assistant Commissioner, on Kent, Sussex and Essex' (Report from the Disturbed Districts of Sussex)

Hansard, House of Commons Debates 14 June 1915 vol. 72 cc 481–3

High Weald AONB, *High Weald AONB Management Plan 2019–2024 Consultation Draft* (June 2018)

House of Lords L/PO/JO/10/1/329/138: Draft of an Act for the improvement of the forest ... (29 February 1668)

House of Lords, *Second Report of the Select Committee to examine the Turnpike Returns* (21 August 1833), evidence of J. Irish, 34–5

Poor Law Commissioners for England and Wales, *Second Annual Report*, June 1836, Report as to the operation and effects of the Poor Law amendment Act, in the County of Sussex, by William Henry Toovey Hawley Esq, Assistant Poor Law Commissioner, 214–40

Westaway, S., *Weald Ancient Woodland Survey A revision of the Ancient Woodland Inventory for Wealden District, East Sussex* (High Weald AONB, 2006)

Westaway, S., M. Grose and P. McKernon, *A Revision of the Ancient Woodland Inventory for Mid Sussex District, West Sussex* (High Weald AONB, 2006)

CONTEMPORARY BOOKS AND ARTICLES

Drayton, Michael, *Poly-olbion. Or A Chorographicall Description of Tracts, Rivers, Mountaines, Forests, and other Parts of this renowned Isle of Great Britaine* (printed by HL, 1613)

Marshall, William, *The Rural Economy of the Southern Counties*, vol. 1 (London: printed for G. Nicol; G.G. and J. Robinson; and J. Debrett, 1798)

Young, Arthur, 'A Tour in Sussex', *Annals of Agriculture,* 11 (1789), 170–304

Young, Rev. Arthur, *General View of the Agriculture of the County of Sussex. With Observations on the Means of its Improvement* (Board of Agriculture, 1793; 2nd edn 1813)

NEWSPAPERS AND PERIODICALS

Annals of Agriculture

Brookers Guide and Directory for Uckfield and District

Country Life

Daily Echo

Daily Mirror
Daily Telegraph
East Grinstead Observer
East Sussex Advertiser
The Echo
Illustrated London News
Independent
Kent and Sussex Courier
Law Journal Reports
The Law Times
Solicitor's Journal and Reporter
Spectator
Sussex Advertiser
Sussex Agricultural Express
Sussex Daily News
Sussex Express
Sussex Weekly Advertiser
The Times

SECONDARY SOURCES

Aalto, K., *The Natural World of Winnie the Pooh* (Portland, OR: Timber Press, 2015)

Abbs, B., 'A Maresfield entrepreneur: William Wood and Woodlands nurseries', *SAC*, 132 (1994), 161–72

Anon., 'Modern South Saxons: No.88 – Mr William Augustus Raper', *SCM*, 10.4 (April 1936), 224

Antram, N. and N. Pevsner, *The Buildings of England: Sussex: East* (New Haven, CT: Yale University Press, 2012)

Armitage-Smith, S. (ed.), *John of Gaunt's Register* II, Camden 3rd series XXI (London, 1911)

Arthur, D. (ed.), *A Sussex Life: the Memories of Gilbert Sargent Countryman* (Tunbridge Wells: Chapel Press, 1989)

Austen, B., 'East Sussex turnpikes of the High Weald', *Sussex Industrial History*, 43 (2013), 31–47

Awty, B., 'The Continental Origins of Wealden Ironworkers, 1451–1544', *Economic History Review*, 34 (1981), 524–39

——,'Aliens in the Ironworking Areas of the Weald: the Subsidy Rolls, 1524–1603', *Bulletin Wealden Iron Research Group*, 4 (1984), 13–78

——, 'A New Forgemaster, William Bassett and an old name, Grubsbars, for Crowborough Forge', *Bulletin Wealden Iron Research Group*, 9 (1989), 33–7

——, 'Henry VII's First Attempt to exploit iron in Ashdown Forest', *Bulletin Wealden Iron Research Group*, 11 (1991), 11–14

——, *Adventure in Iron*, 2 vols, Wealden Iron Research Group (Tonbridge, 2019)

—— and C. Whittick, 'The Lordship of Canterbury, Iron-founding at Buxted, and the Continental Antecedents of Cannon-founding in the Weald', *SAC*,140 (2002), 71–81

Barber, L. and J. Russell, 'Training for War: Plans of the three Great War Divisional Camps in Sussex', *SAC*, 153 (2015), 191–201

Barker, E.E., 'Sussex Anglo-Saxon Charters, I', *SAC*, 86 (1947), 42–101

Bateman, J., *The Great Landowners of Great Britain and Ireland. A List of all Owners of Three Thousand Acres and Upwards*, 4th edn (London: Harrison and Sons, 1883)

Berry, M., 'The Turnpike Roads of Ashdown Forest: a Forgotten History?' *Ashdown Forest News* (Autumn 2016)

Birch, R. and R. Cordiner, *Building Stones of West Sussex* (privately published, 2014)

Blaauw, W.H., 'Wakehurst, Slaugham and Gravetye', *SAC*, 10 (1858), 151–67

Blair, J., *Early Medieval Surrey: Landholding, Church and Settlement* (Gloucester: Sutton and Surrey Archaeological Society, 1991)

Blencowe, R.W., 'Extracts from the Parish Registers and other Parochial Documents of East Sussex', *SAC*, 4 (1851), 243–90

Boorman, W.J., 'The Sussex Moss Collectors', *SCM*, 11 (1937), 299–300

Bourne, George [George Sturt], *Change in the Village* (New York: G.H. Doran Company, 1912)

Boxall, J.P., 'The Sussex Breed of Cattle in the Nineteenth Century', *Agricultural History Review*, 20 (1972), 17–29.

Brandon, P., 'Medieval Clearances in the East Sussex Weald', *Transactions of the Institute of British Geographers*, 48 (1969), 135–52

——, *The Sussex Landscape* (London: Hodder, 1974)

——, 'Warrens and Fossil Fields on Broadstone Warren, Ashdown Forest', *SAC*, 112 (1974), 164

——, 'The South Saxon *Andredesweald*', in P. Brandon (ed.), *The South Saxons* (Chichester: Phillimore, 1978), 138–59

——, *The Kent and Sussex Weald* (Chichester: Phillimore, 2003)

——, *The Discovery of Sussex* (Andover: Phillimore, 2010)

——, *Sussex Writers in their Landscape: Self-fulfilment in the Age of the Machine*, ed. B. Short (Chichester, forthcoming 2022).

—— and B. Short, *The South East from AD 1000* (Abingdon: Longman, 1990)

Bristow, C. and R. Bazley, *Geology of the Country around Royal Tunbridge Wells* (London: HMSO, 1972)

Burchall, M. (ed.), *Sussex Hearth Tax assessments 1662: Pevesey Rape*, Sussex Genealogical Centre occasional papers 4 (Brighton, 1980)

Butler, C., V. Blandford and A. Locke, *An archaeological survey of the Old Lodge nature reserve, East Sussex* (Chris Butler Archaeological Services, 2010)

——, *Ashdown Forest Historic Environment Resource: a LiDAR-Enhanced Archaeological Survey* (Chris Butler Associates, May 2011)

Byford, E., 'Kidbrooke over 200 years', *Forest Row: Historical aspects and recollections*, 1 (1984), 5–17

Caffyn, J., *Sussex Schools in the Eighteenth Century*, SRS 81 (Lewes, 1998)

Campbell, B.M.S. and K. Bartley, *England on the Eve of the Black Death: an Atlas of Lay Lordship, Land and Wealth, 1300–49* (Manchester: Manchester University Press, 2006)

Charlesworth, A., B. Short and R. Wells, 'Riots and unrest', in K. Leslie and B. Short (eds), *An Historical Atlas of Sussex* (Chichester: Phillimore, 1999), 74–5

Chatwin, D. and M. Gardiner, 'Rethinking the Early Medieval Settlement of Woodlands: Evidence from the Western Sussex Weald', *Landscape History*, 27 (2005), 31–49

Cheal, H., 'A Buckhurst Diary', *SCM*, 12 (1938), 392–5 and 448–50

Chibnall, J. and B., *Great Holly Bush Farm, East Grinstead*, report for the National Trust (1988)

Christian, G., *Ashdown Forest* (Forest Row: Society of the Friends of Ashdown Forest, 1967)

Clarke, R., J. Sharp and D. Lily, *Ashdown Forest Visitor Survey Data Analysis*, Natural England Commissioned Reports Number 048 (2010)

Cleere, H. and D. Crossley, *The Iron Industry of the Weald* (Leicester: Leicester University Press, 1985; 2nd edn Cardiff: Merton Priory Press, 1995)

Coates, R., 'Corrections and Additions to Sussex Material in Later Volumes Published by the English Place-Name Society' (unpub. paper, Univ. of Sussex, 1986)

Cobbett, W., *Rural Rides in the Southern, Western and Eastern counties of England*, ed. G.D.H. and Margaret Cole (London: Peter Davies, 1930)

Cocks, R., 'The Great Ashdown Forest Case', in T.G. Watkin (ed.), *The Legal Record and Historical Reality* (London: Hambledon Press, 1989), 175–97

Combes, P., 'The Crowborough Warren estate, 1809–44', *SAC*, 123 (1985), 215–23

Coutin, K., 'Lord Sheffield's Model Farm', *Historical Farm Buildings Journal*, 4 (1990), 10–15

Crawfurd, R.P., *East Grinstead Parish Register, 1558–1661*, SRS 24 (Lewes, 1917)

Crossley, D., 'Cannon-Manufacture at Pippingford, Sussex: The Excavation of Two Iron Furnaces of *c.* 1717', *Post Medieval Archaeology*, 9 (1975), 1–37

Daniel Tyssen, J.R., 'The Parliamentary Surveys of Sussex', *SAC*, 23 (1871), 242–313; 24 (1872), 189–287 and 25 (1873), 23–61

Davey, R., *East Sussex Land Tax 1785*, SRS 77 (Lewes, 1991)

Dimmock, M., A. Hadfield and P. Quinn (eds), *Art, Literature and Religion in Early-modern Sussex: Culture and Conflict* (Farnham: Ashgate, 2014)

Dodgson, J.M., 'Place-names in Sussex: the material for a new look', in P. Brandon (ed.), *The South Saxons* (Chichester: Phillimore, 1978), 54–88

Drewett, P., D. Rudling and M. Gardiner, *The South East to AD 1000* (Harlow: Longman, 1988)

Dunkin, E.H.W. and E.W.D. Penfold, *Marriage Licences at Lewes, 1772–1837, A to L*, SRS 25 (Lewes, 1917); *M to Z*, SRS 26 (Lewes, 1919)

Everitt, A., *Continuity and Colonization: the Evolution of Kentish Settlement* (Leicester: Leicester University Press, 1986)

Eversley, Lord [G.J. Shaw-Lefevre], *Commons, Forests and Footpaths: the Story of the Battle during the last Forty-five years for Public rights over the Commons, Forests and Footpaths of England and Wales* (London: Cassell & Co., 1910)

Faith, R., *The English Peasantry and the Growth of Lordship* (London: Leicester University Press, 1997)

Farrant, J., '"Spirited and intelligent farmers": the Arthur Youngs and the Board of Agriculture's reports on Sussex, 1793 and 1808', *SAC*, 130 (1992), 200–12

——, 'Growth of communications 1720–1840', in K. Leslie and B. Short (eds), *Historical Atlas of Sussex* (Chichester: Phillimore, 1999), 78–9

Fielding, A.J., *Place, Work and Folk: a Study of the Development of Crowborough, Sussex*, Univ. of Sussex Research Papers in Geography 5 (Brighton, 1981)

Firmin, Boys, *Illustrated Guide to Crowborough* (London: Hansard Publishing Union, 1890)

Firth, J.B. and J.A. Royle, 'Romantic Ashdown Forest', *SCM*, 1 (1927), 360, 362

Fletcher, A., *A County Community in Peace and War: Sussex 1600–1660* (London: Longman, 1975)

Freeland, P., 'People who knew Ashdown Forest: (John) Richard Jefferies (1848–1887)', *Ashdown Forest News* (Spring/Summer 2017), 4–6

Fuller, G. Joan, 'The Development of Roads in the Surrey–Sussex Weald and Coastlands between 1700 and 1900', *Transactions Institute British Geographers* 19 (1953), 37–49

Gallois, R. and B. Worssam, *Geology of the Country around Horsham* (London: HMSO 1993)

Gardiner, M., 'Planned Medieval Land Division in Withyham, East Sussex', *SAC*, 123 (1985), 109–14

——, 'The archaeology of the Weald – a survey and review', *SAC*, 128 (1990), 33–53

——, 'The geography and peasant rural economy of the eastern Sussex High Weald, 1300–1420' *SAC*, 134 (1996), 125–39

——, 'Economy and landscape change in post-Roman and early medieval Sussex, 450–1175', in D. Rudling (ed.), *The Archaeology of Sussex to AD 2000* (Kings Lynn: Heritage Marketing & Publications, 2003), 151–60

Giuseppi, M.S., 'The Accounts of the Iron-works at Sheffield and Worth in Sussex, 1546–1549', *Archaeological Journal*, 69 (1912), 276–311

Greenaway, T., P. Roper and K. Ryland, *Wooded Heaths in the High Weald* (Henfield: Woods Mill Record Centre Survey Unit, 2004)

Griffin, C., *The Rural War: Captain Swing and the Politics of Protest* (Manchester: Manchester University Press, 2012)

Groves, J.A., M.P. Waller, M.J. Grant and J.E. Schofield, 'Long-term Development of a Cultural Landscape: the Origins and Dynamics of Lowland Heathland in Southern England', *Vegetation History and Archaeobotany*, 21 (2012), 453–70

Gulley, J.L.M. 'The great rebuilding in the Weald', *Gwerin*, III.3 (1961), 1–16

Hall, R., *I Never Played to many Posh Dances: Scan Tester, Sussex Musician, 1887–1972* (Rochford: Musical Traditions, 1990)

Hamerow, H., *Rural Settlements and Society in Anglo-Saxon England* (Oxford, 2012)

Hardin, G., 'The Tragedy of the Commons', *Science*, 162 (1968), 1243–8

Harding, A.F. and J. Ostoja-Zagorski, 'Excavations in Rocks Wood, Withyham, 1982', *SAC*, 125 (1987), 11–32

Harvey, S., 'Domesday England', in H.E. Hallam (ed.), *The Agrarian History of England and Wales* II *1042–1350* (Cambridge: Cambridge University Press, 1988), 127–31

Heath, R., *The English Peasant: Studies: Historical, Local, and Biographic* (London: T. Fisher Unwin, 1893; reprinted Wakefield: E.P. Publ., 1978)

Helbig, A.K., 'The Forest as Setting and Symbol in Barbara Willard's Mantlemass Novels', *Children's Literature Association Quarterly*, 7 (1982), 35–9

Hemingway, M.F., 'Preliminary explorations at the Rocks, Uckfield', *Sussex Archaeological Society Newsletter*, 31 (1980), 209–10

——, 'Further explorations at the Rocks Early Mesolithic site, Uckfield, East Sussex', *Sussex Archaeological Society Newsletter*, 34 (1981), 243

Hicks, A., 'Hunting, Status and Landscape in Fourteenth-Century England: A Dispute between John of Gaunt and Sir Edward Dallingridge', paper given at Australian and New Zealand Association for Medieval and Early Modern Studies third conference 5–8 July 2001, 'Metamorphoses – Peoples, Places, Times' (2001)

Hills, W.H., *The History of East Grinstead* (Lewes: Farncombe & Co., 1906)

Hindle, S., 'Not by Bread Only? Common Right, Parish Relief and Endowed Charity in a Forest Economy c.1600–1800', in S. King and A. Tomkins (eds), *The Poor in England 1700–1850: an Economy of Makeshifts* (Manchester: Manchester University Press, 2003)

Hobsbawm, E. and G. Rudé, *Captain Swing* (Harmondsworth: Penguin, 1973)

Hodgkinson, J., 'The Decline of the Ordnance Trade in the Weald: the Seven Years' War and its aftermath', *SAC*, 134 (1996), 155–67

——, *The Wealden Iron Industry* (Stroud: The History Press, 2008)

——, 'Field Notes: Ashdown Forest Iron Sites – updated locations', *Bulletin Wealden Iron Research Group*, 28 (2008), 3–4

Hooke, D., 'Old English *wald*, weald in Place-names', *Landscape History*, 34 (2013), 33–49

Hunnisett, R., *East Sussex Coroners' Records 1688–1838*, SRS 89 (Lewes, 2005)

Jacobi, R.M. and C.F. Tebbutt, 'A Late Mesolithic Rock Shelter site at High Hurstwood, Sussex', *SAC*, 119 (1981), 1–36

Jarvis, M., R. Allen, S. Fordham, J. Hazelden, A. Moffat and R. Sturdy, *Soils and their use in South East England* (Soil Survey of England and Wales: Bulletin 15, 1984)

Jones, G., 'A "common of hunting"? Forests, Lordship, and Community before and after the Conquest', in J. Langton and G. Jones (eds), *Forests and Chases of Medieval England and Wales c.1000–c.1500* (Oxford: St John's College Research Centre 2010), 36–67

Jurkowski, M. and N. Ramsay with S. Renton, *English monastic estates, 1066–1540: a list of manors, churches and chapels part 3*, List and Index Society Special Series 42 (London, 2007)

Kaye-Smith, S., *Isle of Thorns* (London: Cassell and Co., 1913; 'Sussex edition' 1924)

Kelly, C.A., 'Sussex Industries: Albert Turner and son Ltd, Timber Merchants', *SCM*, 2 (1928), 588–9

King, E. and G. Nightingale, 'Good estate management: the Buckhurst estate', *Agriculture*, 62 (1955), 230–35

King, S.H., 'Sussex', in H.C. Darby and E.M.J. Campbell (eds), *The Domesday Geography of South-East England* (Cambridge: Cambridge University Press, 1962), 407–82

Kirby, F., 'The Reeve's tale 1', *Ashdown Forest Newsletter*, 6 (1984), 14–16

Kirby, P., *Forest Camera: A Portrait of Ashdown* (Poundgate: Sweethaws Press, 1998)

Land Use Consultants, *Sustainable Settlements in the High Weald: Draft Final Report* (Bristol, 2006)

Langton, J., 'Medieval Forests and Chases: another Realm', in J. Langton and G. Jones (eds), *Forests and Chases of Medieval England and Wales c.1000–c.1500* (Oxford: St John's College Research Centre 2010), 22–4

Langton, J. and G. Jones (eds), *Forests and Chases of England and Wales c.1500 to c.1850* (Oxford: St John's College Research Centre, 2005)

Lea, D. and J. English, 'Analytical Survey and Landscape Context of Saxonbury Hillfort, Rotherfield, East Sussex', *SAC*, 153 (2015), 29–40

Lemaitre, G., *Le Weald des Comtés de Kent, Surrey, Sussex, Hampshire* (Paris: Les Presses Universitaires de France, 1931)

Leppard, M., *A History of East Grinstead* (Chichester: Phillimore, 2001)

——, 'Church origins in East Grinstead', *SAC*, 142 (2004), 148–9

——, 'Heretics and martyrs in Marian Sussex', *SAC*, 154 (2016), 209–26

——, 'Pigs and dens in East Grinstead and their parent manors', *East Grinstead Society Bulletin*, 118 (Spring 2017), 5–9

——, 'The unregarded Sussex of J.M. Neale', *SAC*, 155 (2017), 165–79

Lill, M., 'William Cooke, Steward of the Sheffield Estate, 1828–1832', *SAC*, 144 (2006), 177–90

Llewellyn, N., *East Sussex Church Monuments 1530–1830*, SRS 93 (Lewes, 2011)

Lodge, E.C. and R. Somerville (eds), *John of Gaunt's Register* I *1379–83*, Camden 3rd Series 56 (London, 1937)

Longenbach, J., *Stone Cottage: Pound, Yeats and Modernism* (Oxford: Oxford University Press, 1988)

Longstaff, T.G., 'Douglas Freshfield 1845–1934', *The Geographical Journal*, 83 (1934), 257–62

Lough, A.G. *John Mason Neale – Priest Extraordinary* (Newton Abbott, 1976)

Lowerson, J., 'Building on East Grinstead Common', *East Grinstead Society Bulletin*, 27 (1979), 5–8

——, 'The Aftermath of Swing: Anti-Poor Law Movements and Rural Trade Unions in the South East of England', in A. Charlesworth (ed.), *Rural Social Change and Conflicts since 1500* (Hull: Humberside College of Higher Education, 1982), 56

Lucas, E.V., *Highways and Byways in Sussex* (London: Macmillan, 1904)

Lucas, P.G. (ed.), *Church Lane, Danehill*, Danehill Parish Historical Society (2010)

McCrae, S.G and C.P. Burnham, 'The Soils of the Weald', *Proceedings of the Geologists Association* 86 (1975), 593–610

Madge, S.J., *The Domesday of Crown Lands* (London: Routledge, 1938)

Manning, R.B., *Religion and Society in Elizabethan Sussex* (Leicester: Leicester University Press, 1969)

Margary, I.D., 'A Celtic enclosure in Ashdown Forest', *SNQ*, 3 (1930), 71–2

——, 'King's Standing, Ashdown Forest', *SNQ*, 3 (1930), 72–6

——, 'The Mystery Mounds on Camp Hill and Stone Hill, Ashdown Forest', *SNQ*, 3 (1930–31), 101–6

——, 'Pack-horse Routes across Ashdown Forest', *SNQ*, 8 (1940), 34–40

——, 'The Parliamentary Surveys of Ashdown Forest – Topographical details', *SAC*, 81 (1940), 136–9
——, 'War Damage to Antiquities on Ashdown Forest', *SNQ*, 11 (1946), 1–3
——, *Roman Ways in the Weald* (London: Phoenix House, 1949)
——, 'Military Field Kitchens of the Eighteenth Century', *SAC*,103 (1965), 60–66
Margetts, A., *Wealdbæra: Excavations at Wickhurst Greeen, Broadbridge Heath and the Landscape of the West Central Weald* (Portslade: SpoilHeap Publications, 2018)
——, *The Wandering Herd: the Medieval Cattle Economy of South-East England c.450–1450* (Oxford: Oxbow Books, 2021).
Marrable, C., *Ashdown Forest – a Review of Grazing*, English Nature Research Report 535 (Peterborough, 2003)
——, *Ashdown Forest Grazing Action Plan*, English Nature Research Report 602 (Peterborough, 2004)
—— and N. Muggeridge, *Information on Ashdown Forest* (Wych Cross: Ashdown Forest Centre, 2001)
Martin, D., B. Short and P. Wilkinson, 'Population in 1676 and 1724', in K. Leslie and B. Short (eds), *Historical Atlas of Sussex* (Chichester: Phillimore, 1999), 66–7
Martin, R., 'King's Standing Transmitter Station, Crowborough', *Industrial Archaeology Review*, 24 (2002), 91–102
Mason, R.T., 'East Grinstead. Notes on its Architecture Part II. Medieval Farms', *SAC*, 81 (1940), 2–16
Mate, M., *Trade and Economic Developments 1450–1550: the Experience of Kent, Surrey and Sussex* (Woodbridge: Boydell, 2006)
Mawer, A. and F.M. Stenton, *The Place-names of Sussex*, 2 vols (Cambridge: Cambridge University Press, 1929–30)
Millett, M., 'Roman Kent', in J.H. Williams (ed.), *The Archaeology of Kent to AD 800* (Woodbridge: Boydell, 2007), 135–84
Milne, Christopher, *The Enchanted Places: A Memoir of the real Christopher Robin and Winnie-the-Pooh* (New York: E.P. Dutton, 1975)
Money, J.H., 'Medieval iron-workings in Minepit Wood, Rotherfield, Sussex', *Medieval Archaeology*, 15 (1971), 86–111
——, 'The Iron-Age Hill-fort and Romano-British Iron-working Settlement at Garden Hill, Sussex: Interim Report on Excavations, 1968–76', *Britannia*, 8 (1977), 339–50
Neeson, J., *Commoners: Common Right, Enclosure and Social Change in England, 1700–1820* (Cambridge: Cambridge University Press, 1993)
Olwig, K. *Nature's Ideological Landscape* (London: Allen & Unwin, 1984)
Parcell, M., 'The Case of the Missing Holly Sticks', *Hartfield Times*, 8 (n.d.), 1–4
Parsons, J.L., 'The Sussex ironworks', *SAC*, 32 (1882), 19–32
Patmore, Coventry, *How I managed and improved my estate* (London: George Bell and Sons, 1886)
Payne, M. and L. Batchelor, *Bygone Crowborough* (Chichester: Phillimore, 1987)
Pearce, R., *Facing Invasion: Proceedings Under the Defence Acts 1801–05*, SRS 99 (Lewes, 2019)
Pears, M., *Royal Foliage: a Brief History of Ashdown Forest* (privately printed 1959; reprinted 1983)

Pelham, R.A., 'Timber Exports from the Weald during the Fourteenth Century', *SAC*, 69 (1928), 170–82

Penn, R., *Portrait of Ashdown Forest* (London: Robert Hale, 1984)

Penninck, J., 'Royal Ashdown Forest golf clubs 1888–1953', *SCM*, 26 (1953), 366–72

Petley, Rev. J.L. Ward, *Maresfield Old and New* (Chichester: J.W. Moore, 1896)

Pilbeam, N. and I. Nelson, *Poor Law Records of Mid Sussex 1601–1835*, SRS 83 (Lewes, 2001)

Pinnegar, A.-N., D. and E., *Hammerwood Park East Grinstead, Sussex: 1792, First House of Benjamin Henry Latrobe* (East Grinstead: Hammerwood Park, 1992)

Prince, C.L., *Observations upon the Topography and Climate of Crowborough Hill, Sussex* (Lewes: H. Wolfe, 1885)

Pullein, C., *Rotherfield: the Story of some Wealden Manors* (Tunbridge Wells: Courier Printing and Publishing Co., 1928)

Rackham, O., *Ancient Woodland*, 2nd edn (Dalbeattie: Castlepoint, 2003)

Raper, W.A., 'Ashdown Forest', *SNQ*, 5 (1934–5), 146–7

Rawlings, D., 'Farming and land-owning in Victorian Danehill', *Danehill Historical Society Magazine*, 3 (1987), reprinted in P.G. Lucas, *Chelwood Common Project Study Notes* (2003), 29–31

Rawlings, H., 'Lord Sheffield's last Battle', *Danehill Historical Society*, 4 (1993), 10–28

Rich, T., P. Donovan, P. Harmes, M. McFarlane, C. Marrable, N. Muggeridge, R. Nicholson, M. and P. Reader, E. Rich and P. White, *Flora of Ashdown Forest* (privately printed, 1996)

Rider Haggard, H., *Rural England: Being an Account of Agricultural and Social Researches carried out in the Years 1901 and 1902*, 2nd edn (London: Longmans, 1906)

Robinson, D. and R. Williams, 'The Soils and Vegetation History of Sussex', in Geography Editorial Committee, *Sussex: Environment, Landscape and Society* (Gloucester: Alan Sutton Publishing 1983), 109–26

Rudling, D., 'The Excavation of a Roman Tilery on Great Cansiron Farm, Hartfield, East Sussex', *Britannia*, 17 (1986), 191–230

Russell-Goggs, M., 'The Ashdown Forest Pageant', *SCM*, 3 (1929), 410–11

Sackville, Reginald Windsor [seventh Earl De la Warr], *Historical Notices of the Parish of Withyham in the County of Sussex, With a Description of the Church and Sackville Chapel* (London: John Russell Smith, 1857)

Sackville, T., *Dumble: the Life and Death of Billy, 10th Earl De La Warr* (London: Zuleika, forthcoming 2022)

Sackville-West, R., *Inheritance: the story of Knole and the Sackvilles* (London: Bloomsbury, 2010)

Sackville-West, V., *Knole and the Sackvilles* (London: Heinemann, 1922)

Salzman, L.F., 'The Hundred Roll for Sussex Pt II Rape of Pevensey', *SAC*, 31 (1943), 35–54

——, *Record of Deputations of Gamekeepers*, SRS 51 (Lewes, 1951)

Sansum, P., *An Overview of the Character and Ecological Significance of Gill Woodland in the High Weald AONB, including Results of some Preliminary Baseline Surveys for Population Monitoring of Selected Rare Vascular Plants of Gill Woodland*, High Weald AONB Unit Commissioned Report (December 2014)

Saul, N., *Scenes from Provincial Life: Knightly Families in Sussex 1280–1400* (Oxford: Clarendon Press, 1986)

Saul, N., 'Sackville family (*per. c.*1066–*c.*1477)', *ODNB* (2004)

Scaife, R. and P. Burrin, 'Floodplain Development in and the Vegetational History of the Sussex High Weald and some Archaeological Implications', *SAC*, 121 (1983), 1–10

Shaw-Taylor, L., 'The Management of Common Land in the Lowlands of Southern England circa 1500 to circa 1850', in M. de Moor, L. Shaw-Taylor and P. Warde (eds), *The Management of Common Land in North-west Europe, c.1500–1850* (Turnhout: Brepols, 2002)

Short, B., *Rural Housing Need in South East England: Wealden District, East Sussex 1988–91* (Crowborough: Wealden District Council, 1991)

——, *The Geography of Local Migration and Marriage in Sussex 1500–1900*, Univ. of Sussex Research Papers in Geography 15 (Brighton, 1981)

——, *Rural Housing Need in the 'Affluent' South East*, Univ. of Sussex Research Papers in Geography (Brighton, 1992)

——, *The Ashdown Forest Dispute 1876–1882: Environmental Politics and Custom*, SRS 80 (Lewes, 1997)

——, 'Conservation, Class and Custom: Lifespace and Conflict in a Nineteenth-century Forest Environment', *Rural History*, 10 (1999), 127–54

——, 'Landownership in Victorian Sussex', in K. Leslie and B. Short (eds), *An Historical Atlas of Sussex* (Chichester: Phillimore, 1999), 98–9

——, 'Raper (William) Augustus (1845–1940)', *ODNB* (2004)

——, 'Environmental Politics, Custom and Personal Testimony: Memory and Lifespace on the late Victorian Ashdown Forest, Sussex', *Journal of Historical Geography*, 30 (2004), 470–95

——, S. Swift and G. Swift, *The Ashdown Forest Area: Agricultural Change in the Western High Weald from 1970 to 1988* (Lewes: Sussex Rural Community Council Research Unit, 1991)

Simpson, J., *The Folklore of Sussex* (London: Batsford, 1973)

Small, J.E., 'A Review of Ashdown Forest and the Common Rights Thereon', *SAC*, 126 (1988), 155–65

Somerville, R., *History of the Duchy of Lancaster* I *1265–1603* (London: The Chancellor and Council of the Duchy of Lancaster, 1953)

Stevenson, J.H., 'Alexander Nesbitt, a Sussex Antiquary, and the Oldlands Estate', *SAC*, 137 (1999), 161–73

Strachan, C.J.L., *Fair Ways in Ashdown Forest: the History of an Elite Golf Club playing on Common Land for 125 years* (St Andrews: St Andrews Golf Press, 2013)

Straker, E., *The Buckhurst Terrier*, SRS 39 (Lewes, 1933)

——,'Agricultural History in the Hundred of Hartfield', *SAC*, 76 (1935), 172–7

——, 'A Lost Tudor Iron Furnace found', *SNQ*, 6 (1936), 217–18

——, 'Ashdown Forest and its inclosures', *SAC*, 81 (1940), 121–35

——and B.H. Lucas, 'A Romano-British bloomery in East Sussex', *SAC*, 79 (1938), 224–32

Straughton, E., *Common Grazing in the Northern English Uplands, 1800–1965* (Lampeter: Edwin Mellen, 2008)

Streeter, D., 'Heathland Conservation on Ashdown Forest', in *Society for the Promotion of Nature Reserves Handbook* (1964), 27–43

——, 'Biogeography, Ecology and Conservation in Sussex', in Geography Editorial Committee, *Sussex: Environment, Landscape and Society* (Gloucester: Alan Sutton Publishing 1983), 127–47

Sutton, Rev. C.N., *A Short History of Withyham and Buckhurst* (Tunbridge Wells: A.K. Baldwin, 1893)

Sydney, W. Connor, 'Gossip in the Sussex Oberland', *The Gentleman's Magazine*, 296 (February 1904), 114–31

Tebbutt, C.F., 'Garden Hill Camp, Hartfield', *SAC*, 108 (1970), 39–49

——, 'Two Newly-discovered Medieval Sites', *SAC*, 110 (1972), 31–6

——, 'The Prehistoric Occupation of the Ashdown Forest area of the Weald', *SAC*, 112 (1975), 34–43

——, 'Iron Sites on Ashdown Forest', *Bulletin Wealden Iron Research Group*, 11 (1977), 9–11

——, 'The Excavation of Three Roman Bloomery Furnaces at Hartfield, Sussex', *SAC*, 117 (1979), 47–56

——, 'A Deserted Medieval Farm Settlement at Faulkner's Farm, Hartfield', *SAC*, 119 (1981), 107–16

——, 'Castle Field, Hartfield (TQ 481 361)', *SAC*, 118 (1981), 377–8

——, 'A Middle-Saxon iron smelting site at Millbrook, Ashdown Forest, Sussex', *SAC*, 120 (1982), 19–35

——, 'The One-night House and its Distribution', *Folklore*, 95 (1984), 122

—— and H. Cleere, 'A Romano-British Bloomery at Pippingford, Hartfield', *SAC*, 111 (1973), 27–40

Thomas-Stanford, C., *Sussex in the Great Civil War and the Interregnum 1642–1660* (London: Chiswick Press, 1910)

Thompson, E.P., *Customs in Common: Studies in Traditional Popular Culture* (New York: The New Press, 1993)

Turner, Betty, *Maresfield* (privately printed, 1991)

Turner, Rev. E., 'The Free Chapels of Maresfield and Dudeney', *SAC*, 9 (1857), 41–4

——, 'Ashdown Forest, or as it was sometimes called Lancaster Great Park', *SAC*, 14 (1862), 35–64

——, 'Maresfield', *SAC*, 14 (1862), 138–70

Vickers, J.A., *The Religious Census of Sussex 1851*, SRS 75 (Lewes, 1989)

Vincent, A., *Roman Roads of Sussex* (Midhurst: Middleton Press, 2000)

Walker, S., 'Lancaster v. Dallingridge: a Franchisal Dispute in Fourteenth-century Sussex', *SAC*, 121 (1983), 87–94

——, *The Lancastrian affinity 1361–1399* (Oxford: Clarendon Press, 1990)

Wallis, K., 'Village Vignettes of late Victorian Days', *SCM*, 13 (1939), 468–71

Walter, H., 'Some notable happenings through the years from 1804 to the present time on Ashdown Forest from old stories heard' (unpublished typescript, 1950)

Webb, N., *Heathlands* (London: Collins, 1986)

Weir-Wilson, M., 'The Making of St Leonard's Forest: a New Landscape History', *Southern History*, 38 (2016), 106–29

Welch, M.G., *Early Anglo-Saxon Sussex*, BAR British Series 112 (Oxford, 1983)

Wickenden, N., 'Excavations at Chelwood Gate: interim report', *Sussex Archaeological Society Newsletter*, 48 (1986), 467

____, 'Final interim report', *Sussex Archaeological Society Newsletter*, 54 (1988), 11

Wilde, Rev. S.D., 'Fletching, Parish and Church', *SAC*, 4 (1851), 232–42

Willard, Barbara, *The Forest: Ashdown in East Sussex* (Poundgate: Sweethaws Press, 1989)

Williams-Ellis, Clough (ed.), *Britain and the Beast* (London: J.M. Dent and Sons Ltd, 1937)

Wolff, H.W., *Sussex Industries: a Series of Articles on the Principal and Peculiar Industries practised in Sussex* (Lewes: *Sussex Advertiser*, 1882)

Wright, J., 'Coventry Patmore in Sussex', *SCM*, 21 (1947), 24–7

UNPUBLISHED THESES

Allen, S., 'Poaching and the Game Laws in East Sussex, 1830–1880' (Univ. of Sussex unpub. MA, 1982)

Brandon, P., 'The Common Lands and Wastes of Sussex'(Univ. of London unpub. PhD, 1963)

Brent, C., 'Employment, Land Tenure and Population in East Sussex 1540–1640' (Univ. of Sussex unpub. DPhil, 1974)

Douch, J., 'Reconstructing Change in Vegetation Distribution on Ashdown Forest since 1947' (Univ. of Sussex unpub. BSc project, 2002)

Gulley, J.L.M., 'The Wealden Landscape in the early Seventeenth Century and its Antecedents' (Univ. of London unpub. PhD, 1960)

Irons, J., 'Aspects of the Impact of Man on the Historical Ecology of Ashdown Forest, Sussex before 1885' (Univ. of Sussex unpub. DPhil, 1982)

Kaminski, J., 'The Environmental Implications of Romano-British Iron Production in the Weald' (Univ. of Reading unpub. PhD, 1995)

Lill, M., 'George Holroyd (1802–1876), Second Earl of Sheffield' (Univ. of Sussex unpub. MPhil, 2002)

Lockwood, C., 'The Changing Use of Land in the Weald Region of Kent, Surrey and Sussex, 1919–1939' (Univ. of Sussex unpub. DPhil, 1989)

McCorquodale, C., 'The Landscape of Enclosure on the Fringes of Ashdown Forest from 1780' (Univ. of Sussex unpub. BA dissertation, 2009)

Merricks, L., 'Forest and Waste in Seventeenth century England: the Enclosure of Ashdown Forest, 1600–1700' (Univ. of Sussex unpub. DPhil, 1989)

Parsons, R., 'Status, Occupations and Landholding. The Emergence of an Independent Peasantry in the High Weald of East Sussex in the Late Thirteenth century. The Evidence of Ashdown Forest and its Region' (Univ. of Leicester unpub. MA, 1996)

Phelan, L., 'Economy to amenity: the commons of the New Forest and Ashdown Forest, 1851–1939' (Univ. of Sussex unpub. DPhil, 2003)

Short, B., 'Agriculture in the High Weald of Kent and Sussex 1850 to 1953 (a case study in the application of multivariate techniques in the field of Historical Geography)' (Univ. of London unpub. PhD, 1973)

Thompson, J., 'Newbridge Alder Carr. Site survey' (Univ. of Sussex unpub. MSc, 2006)

WEB-BASED SOURCES

www.ashdownforest.org/discover/docs/TheAFCentreBarns.pdf

www.ashdownforest.org/enjoy/history/docs/Ashdown_Forest_Historical_Minute_Book_1887_1897.pdf

www.ashdownforest.org/living/past/decline.php

Ashdown Forest Research Group, *Men of Ashdown Forest who fell in the First World War and are commemorated at Forest Row, Hartfield and Coleman's Hatch: A Collection of Case Studies*, www.ashdownforest.org/enjoy/history/docs/WWICaseStudies-AshdownForest_000.pdf

http://www.ashdownpark.com/history-ashdown-park-hotel

Bannister, N., 'Field systems in the High Weald', http://www.highweald.org/high-wealdaonb-management-plan/evidence/420-home/research-reports/2298-field-systems-in-the-highweald.html

Blandford, V., *Tracing the Pale of the Ashdown Forest Deer Park* (Chris Butler archaeological services 2012), http://www.highweald.org/downloads/publications/1128-tracing-the-pale-of-the-ashdown-forest-deer-park/file.html or https://www.ashdownforest.org/enjoy/docs/PaleReport.pdf

Bodleian catalogue: https://archives.bodleian.ox.ac.uk/repositories/2/resources/3201

British History online, *Calendar of State Papers Domestic: James I, 1603–1610*, ed. Mary Anne Everett Green (London, 1857), *British History Online* http://www.british-history.ac.uk/cal-state-papers/domestic/jas1/1603-10

——, *Calendar of State Papers Domestic: Charles II, 1660–1*, ed. Mary Anne Everett Green (London, 1860), *British History Online* http://www.british-history.ac.uk/cal-state-papers/domestic/chas2/1660-1

——, *Journal of the House of Commons* 18 June 1657, 562–3. 'An Act for instructions, agreed upon in Parliament…Commissioners for Surveying the Forest of Sherwood… the Forest or Chase of Needwood etc', https://www.british-history.ac.uk/no-series/acts-ordinances-interregnum/pp1116-1122

——, *Journal of the House of Lords* 11: 9 May 1663, 520 Ashdown Forest, https://www.british-history.ac.uk/lords-jrnl/vol11/p520

——, *Journal of the House of Commons* 9: 15 April 1668, https://www.british-history.ac.uk/commons-jrnl/vol9/p81, Ashdown Forest

——, Report from Treasurer Clifford to the King on the petition of the four daughters of Col. Henry Washington (for a grant of certain lands, part of the Queen's jointure) House of Commons Entry Book: 28 January 1673, 16–31 (*Calendar of Treasury Books*, 4, 1672–1675), https://www.british-history.ac.uk/cal-treasury-books/vol4/pp37-55

——, State Papers Domestic Entry Book 26, f. 197, 15 August 1675. The King to Sir Robert Carr, Chancellor and to the Attorney of the Duchy of Lancaster, https://www.british-history.ac.uk/cal-state-papers/domestic/chas2/1675-6/pp243-281

'Buxted Park and the Medley Evelyn Connection', http://www.felbridge.org.uk/index.php/publications/buxted-park/

Coates, R., *Preparatory to a Dictionary of Sussex Place-names* (2017), https://core.ac.uk/reader/323894005

Cranfield University, *Soils Guide*, https://www.cranfield.ac.uk/themes/environment-and-agrifood/landis/soils-guide

Davey, R., 'East Sussex Window and House Tax Assessments 1747', http://www.sussexrecordsociety.org/olb/wtax/

https://diversionsinnaturalhistory.wordpress.com/2015/09/07/marsh-gentians-in-ashdown-forest/

East Sussex Historic Environment Record, https://www.thekeep.info/collections/getrecord/ESHER

http://www.felbridge.org.uk/index.php/exhibitions/imberhorne-farm-workers-past-present/

'Harmony Farm' rock festival August 1971, http://www.ukrockfestivals.com/duddleswell-free-fest-1971.html

Harris, R.B., *Crowborough: Historic Character Assessment Report* (Sussex Extensive Urban Survey, 2008), https://www.westsussex.gov.uk/media/2105/crowborough_eus_report_maps.pdf

High Weald AONB, http://www.highweald.org/images/droving_map_highweald_rc.jp

Historic England, 'Chelwood Gate Enclosure', https://historicengland.org.uk/listing/the-list/list-entry/1423862

——, 'Stoneland', historicengland.org.uk/listing/the-list/list-entry/1000230

——, 'Iron minepits in Tugmore Shaw', https://historicengland.org.uk/listing/the-list/list-entry/1002230

——, 'Pillow mounds on Ashdown Forest', https://historicengland.org.uk/listing/the-list/list-entry/1002234

——, 'Kidbrooke Park', https://historicengland.org.uk/listing/the-list/list-entry/1000305

——, 'Forest Row camp War Memorial', https://historicengland.org.uk/listing/the-list/list-entry/1441756

History of Parliament, http://www.historyofparliamentonline.org/volume/1386-1421/member/hungerford-sir-thomas1397

——, http://www.historyofparliamentonline.org/volume/1386-1421/member/sackville-sir-thomas-ii-1432

——, http://www.historyofparliamentonline.org/volume/1386-1421/member/broke-john-1417

——, https://www.historyofparliamentonline.org/volume/1386-1421/member/pelham-john-1429

——, http://www.historyofparliamentonline.org/volume/1754-1790/member/sackville-john-frederick-1745-99

——, 'Abbot, Charles (1757–1829), of Kidbrook, Suss.', historyofparliamentonline.org/volume/1790-1820/member/abbot-charles-1757-1829

——, http://www.historyofparliamentonline.org/volume/1820-1832/member/shelley-john-1808-1867

The Militia Lists 1803, https://theweald.org/Militia.asp?SourceId=Bu1803Dir; https://theweald.org/Militia.asp?SourceId=Ha1803Dir; https://theweald.org/Militia.asp?SourceId=Ro1803Dir; https://theweald.org/Militia.asp?SourceId=Wi1803Dir

Breakdown of the Militia Lists 1803, https://theweald.org/HCensus2.asp?HMatch=HMi031803Dir; https://theweald.org/HCensus2.asp?HMatch=Wi1803Dir

Poor Law records for East Grinstead and West Hoathly, http://www.sussexrecordsociety.org/dbs/pl/

Prosopography of Anglo-Saxon England, https://pase.ac.uk/jsp/pdb?dosp=VIEW_RECORDS&st=PERSON_NAME&value=1776&level=1&lbl=Sigeberht

Reader, M.E., 'Marsh Gentian on Ashdown Forest', *Sussex Botanical Recording Society Newsletter*, 51 (January 2001), https://www.sussexflora.org.uk/resources/articles/marsh-gentian-ashdown-forest/

St John the Evangelist, Crowborough, https://cee55446-cfb0-48da-bc19-83b209e46dba.filesusr.com/ugd/efe475_00cc66c99e444b6585478e98f1180d46.pdf

Smith, M., 'Nutley at war 1939–1945', http://www.nutleyhistorical.org.uk/index_htm_files/NUTLEY%20AT%20WAR%201939-1945%20Revised.pdf

——, 'The military on Ashdown Forest 1793-1925', http://www.nutleyhistorical.org.uk/index_htm_files/THE%20MILITARY%20ON%20ASHDOWN%20FOREST%20%201793-%201924.pdf

https://sussexparishchurches.org/church/withyham-st-michael-and-all-angels/

Sussex Wildlife Trust, *A Vision for Nature and Wellbeing in Sussex* (2017), 11–12, https://assets.sussexwildlifetrust.org.uk/Files/swtvssummary-reportfinalweb.pdf

The Trials of the Rev Robert Bingham (J.M. Richardson, 1811) https://archive.org/stream/trialsrevrobert01binggoog#page/n33/mode/2up

http://uckfieldnews.com/twenty-five-new-homes-park-farm-2-maresfield/

Wealden Iron Research Group, http://www.Wealdeniron.org.uk; http://www.wirgdata.org/searchpro2.cgi?personid=363

Woolmore, Ray, *Designation History Series: the High Weald AONB* (2013), http://www.highweald.org/downloads/publications/2004-high-weald-aonb-designation-history-series-and-maps1/file.html

http://www.workhouses.org.uk/EastGrinstead/EastGrinstead1881.shtml#Inmates; http://www.workhouses.org.uk/Uckfield/Uckfield1881.shtml#Inmates

INDEX

Place names are frequently rendered inconsistently across time. Wherever possible the modern names, as employed by the Ordnance Survey, have been used. Dates of birth and death are given here only where necessary for identification purposes. Bold type indicates presence in a figure or in the heading of a table.

Garden and Landscape History

Previously Published

1. *Designs Upon the Land: Elite Landscapes of the Middle Ages*
Oliver H. Creighton

2. *Richard Woods (1715–1793): Master of the Pleasure Garden*
Fiona Cowell

3. *Uvedale Price (1747–1829): Decoding the Picturesque*
Charles Watkins and Ben Cowell

4. *Common Land in English Painting, 1700–1850*
Ian Waites

5. Observations on Modern Gardening, *by Thomas Whately: An Eighteenth-Century Study of the English Landscape Garden*
Michael Symes

6. *The Landscape Studies of Hayman Rooke (1723–1806): Antiquarianism, Archaeology and Natural History in the Eighteenth Century*
Emily Sloan

7. *Transhumance and the Making of Ireland's Uplands, 1550–1900*
Eugene Costello

8. *The Foldcourse and East Anglian Agriculture and Landscape, 1100–1900*
John Belcher

9. *Erasmus Darwin's Gardens: Medicine, Agriculture and the Sciences in the Eighteenth Century*
Paul A. Elliott

10. *Rediscovering Lost Landscapes: Topographical Art in north-west Italy, 1800–1920*
Pietro Piana, Charles Watkins and Ross Balzaretti

11. *Cottage Gardens and Gardeners in the East of Scotland, 1750–1914*
Catherine Rice

12. *Territoriality and the Early Medieval Landscape: The Countryside of the East Saxon Kingdom*
Stephen Rippon